Landmark Expe
in Molecular Biology

Landmark Experiments in Molecular Biology

Michael Fry

AMSTERDAM • BOSTON • HEIDELBERG • LONDON
NEW YORK • OXFORD • PARIS • SAN DIEGO
SAN FRANCISCO • SINGAPORE • SYDNEY • TOKYO
Academic Press is an imprint of Elsevier

Academic Press is an imprint of Elsevier
125 London Wall, London EC2Y 5AS, UK
525 B Street, Suite 1800, San Diego, CA 92101-4495, USA
50 Hampshire Street, 5th Floor, Cambridge, MA 02139, USA
The Boulevard, Langford Lane, Kidlington, Oxford OX5 1GB, UK

British Library Cataloguing-in-Publication Data
A catalogue record for this book is available from the British Library.

Library of Congress Cataloging-in-Publication Data
A catalog record for this book is available from the Library of Congress.

ISBN: 978-0-12-802074-6

For Information on all Academic Press publications
visit our website at http://www.elsevier.com/

Working together
to grow libraries in
developing countries

www.elsevier.com • www.bookaid.org

Publisher: Sara Tenney
Acquisition Editor: Jill Leonard
Editorial Project Manager: Fenton Coulthurst
Production Project Manager: Melissa Read
Designer: Maria Ines Cruz

Typeset by MPS Limited, Chennai, India

Dedication

To my beloved Yonatan, Omri, Uri, and Nina

Contents

About the Author

Michael Fry is Professor Emeritus of Biochemistry at the Ruth and Bruce Rappaport Faculty of Medicine of the Technion—Israel Institute of Technology in Haifa, Israel. He received BSc, MSc, and PhD degrees in microbiology and molecular biology from the Hebrew University, Jerusalem, and did postdoctoral research at the Roche Institute of Molecular Biology in Nutley, New Jersey. While serving on the Technion faculty, he taught biochemistry to students of medicine and to graduate students and had trained in his laboratory a large number of graduate students. He also maintained for many years close working association with colleagues at the School of Medicine of the University of Washington, Seattle. His research work and his many publications center on eukaryotic DNA polymerases and helicases and on quadruplex DNA and its interacting proteins. Pursuing a long-standing interest in the history of biology, he focuses on the interplay between theory and experimentation in molecular biology. The present *Landmark Experiments in Molecular Biology* volume grew out of a graduate course that he taught at the Technion and at the Feinberg Graduate School of the Weizmann Institute of Science.

Preface

Only a few instances in the history of biological research, and indeed in the history of science, were as dramatic and revolutionary as the birth and the first decades of molecular biology. Among the landmark discoveries that were made between the early 1950s and late 1970s were the establishment of DNA as the carrier of genetic information, the discovery of its double helical structure, and the identification of its semiconservative mode of replication. Of equal importance were the deciphering of the genetic code, the discovery of bacterial messenger DNA, the elucidation of the regulation of bacterial gene expression, and the identification of the basic components of the protein synthesis machinery. Several years later the split nature of eukaryotic genes and RNA splicing were discovered and recombinant DNA technologies came into wide use. At the same time methods were developed to determine amino acid sequences in proteins and nucleotide sequences in DNA. Structural molecular biology had its parallel triumphs with the determination of the three-dimensional structure of the proteins myoglobin and hemoglobin. Not less importantly than its concrete achievements, molecular biology instituted the principles of reduction and of mechanism as a way to explain biological phenomena.

This book describes selected key experiments in molecular biology in their historical context. Its intended readers are graduate students in the Life Sciences and the History of Biology as well as practicing scholars in both fields. Although today's practitioners of molecular biology are technically skilled and knowledgeable about current developments in their specific areas of research, many are unfamiliar with the origins and the recent history of their own field. In 1953, at about the time of the birth of molecular biology, André Lwoff wrote (Ref. 1, p. 271):

> For many young scientists the future is more important than the past and the history of science begins tomorrow.

More than four decades later, well after molecular biology had already become a full-fledged forefront discipline of life sciences, Benno Müller-Hill commented (Ref. 2, p. 1):

> Molecular biology has no history for the young scientist. What happened ten years ago seems prehistoric and thus of no interest.

Remarking on Müller-Hill's stance, Sidney Brenner opined[3]:

I hold the somewhat weaker view that history does not exist for the young, but is divided into two epochs: the past two years, and everything that went before. That these have equal weight is a reflection of the exponential growth of the subject, and the urgent need to possess the future and acquire it more rapidly than anybody else does not make for empathy with the past.

This situation of technical proficiency but unfamiliarity with the history of molecular biology is inverted to an extent among historians of biology. Books and articles by professional historians of science on the history of molecular biology go deeply into various historical, sociological, economic, political, personal, and philosophical aspects of the subject. Yet, oftentimes the concrete theoretical and technical particulars of critical experiments are taking a secondary place to the mentioned other aspects of the history of molecular biology.

This *Landmark Experiments in Molecular Biology* book grew out of an eponymous graduate course that I have taught at the Faculties of Medicine and of Biology of my own institution and at the Weizmann Institute of Science. Similar to Lwoff, Müller-Hill, and Brenner, it was also my experience that graduate students in molecular biology and younger senior faculty were largely ignorant of the historical roots of their own field. However, contrary to the assertions of Lwoff, Müller-Hill, and Brenner, I discovered that many among those who attended the course were keenly interested in the history of molecular biology. Although I had no similar direct contact with historians of science, it is likely that they may find complementing interest in detailed analyses of historically important experiments.

This volume attempts to address the needs of both practitioners and historians of molecular biology by combining a history of important junctures in the field with technical analyses of concrete key experiments. The book focuses on critical points at the prehistory and dawn of molecular biology (late 19th century to the 1940s) and during its golden age (1950s–1970s). By combining appraisals of the history and of the science of selected milestone experiments, this book offers molecular biologists a primer on the history of their field and presents historians of biology with technical analyses of critical experiments.

One common deficiency of texts on molecular biology is that they offer incomplete description of the evolution of ideas and of empirical findings. This was best put by Müller-Hill (Ref. 2, p. 1):

In the textbooks, almost everything is solved and clear. Most claims are so self-evident that no proofs are given. Old, classical experiments disappear. They are self-evident. Old errors in interpretation not mentioned. Who cares?

In this book I tried to provide a more realistic description of the nonlinear progression of molecular biology. Thus, theories and hypotheses that had been proven incorrect are described alongside productive ones. Similarly, erroneous or lacking interpretations of experimental observations are examined side-by-side with ultimately fruitful interpretations.

Finally, even with its limited focus on landmark experiments, this book is definitively not comprehensive. The experiments that are described here were selected because of their fundamental contribution to our present understanding of key molecular elements of life. Yet, because of its limited space, this volume ignores many important theories and experiments. First, the book's main focus is on principal developments in the informational branch of molecular biology. Thus, except for some discussion in Chapter 5, "Discovery of the Structure of DNA" of the early phases of structural molecular biology, this volume does not cover advances in this equally important division of molecular biology. Second, due to its limited space, the book mentions only in passing or entirely omits cardinal discoveries in informational molecular biology. Among the many important developments that are not covered are the greatly influential operon model, mechanisms of regulation of gene expression and of enzyme action, the development and wide implications of the techniques of protein and DNA sequencing and of polymerase chain reaction, and the emergence of recombinant DNA and of cloning. The interested reader would certainly find material on these and additional facets of molecular biology in some of the books and articles that are listed in Chapter 1, "Introduction" of this volume.

Finally, it is a privilege to give thanks to individuals who made this book possible. I am indebted to my colleagues and friends, Drs Avram Hershko and Herman Wolosker, for patiently serving as a sounding board to my evolving ideas on different elements of this book. The able and dedicated help of members of the Elsevier/Academic Press editorial staff is gratefully acknowledged. Thanks are due to Jill Leonard, Senior Editorial Project Manager at Elsevier, for her early and sustained faith in this project. I am also grateful to the Editorial Project Managers who accompanied the book at its different stages of development: Elizabeth Gibson, Halima Williams, and particularly to Fenton Coulthurst and my Editor Melissa Read. Last but not least, in addition to a long list of my debts to her, I am grateful to Iris for standing by me throughout the often torturous writing of this volume.

REFERENCES

1. Lwoff A. Lysogeny. *Bact Rev* 1953;**17**(4):269–337.
2. Müller-Hill B. *The lac operon: a short history of a genetic paradigm*. Berlin: de Gruyter; 1996.
3. Brenner S. A night at the operon [review of "The *Lac* Operon: A Short History of a Genetic Paradigm" by Benno Müller-Hill]. *Nature* 1997;**386**(6622):235.

Chapter 1

Introduction

Origins, Brief History, and Present Status of Molecular Biology

Chapter Outline

Since its birth in the 1950s molecular biology has developed into a dominant and exceptionally productive branch of the life sciences. This introduction briefly traces the roots of the term "molecular biology"; it delineates its structural and informational branches and points to their principal tenets. Finally the reader is directed to selected scholarly articles on facets of molecular biology that are not covered in this book.

1.1 ORIGINS AND EARLY USES OF THE TERM "MOLECULAR BIOLOGY"

The term "Molecular Biology" was first coined by the mathematician and science administrator Warren Weaver (1894−1978). Between 1932 and 1955 he served as director of the Division of Natural Sciences at the Rockefeller

M. Fry: Landmark Experiments in Molecular Biology. DOI: http://dx.doi.org/10.1016/B978-0-12-802074-6.00001-1

Foundation, which was the dominant funding agency of scientific research in Europe and America between the two World Wars.[1-6] Weaver recognized early on that tools and techniques of chemistry and physics could be adopted to study life at the molecular level. In this context he introduced the term "Molecular Biology" in a 1938 annual report to the Rockefeller Foundation (cited in Ref. 7, p. 442):

> *And gradually there is coming into being a new branch of science—molecular biology—which is beginning to uncover many secrets concerning the ultimate units of the living cell [...] Among the studies to which the Foundation is giving support is a series in relatively new field, which may be called molecular biology, in which delicate modern techniques are being used to investigate ever more minute details of certain life processes.*

Notably, Weaver's proposal to employ the "delicate" modern techniques (implicitly of physics and chemistry) to study biology at "more minute" (molecular) details did not specify the domains of this new branch of science. When the term "Molecular Biology" began to take root in the 1950s, one of its earliest advocates, the British physicist and X-ray crystallographer William Astbury (1898−1961) (see Chapter 5: "Discovery of the Structure of DNA") professed in 1951[8] and reiterated in 1961[9] that the objective of molecular biology was the investigation of the *structure*, origins, and functions of biomolecules:

> [Molecular biology] *implies not so much a technique as an approach, an approach from the viewpoint of the so-called basic sciences with the leading idea of searching below the large-scale manifestations of classical biology for the corresponding molecular plan. It is concerned particularly with the forms of biological molecules, and with the evolution, exploitation and ramification of those forms in the ascent to higher and structural - which does not mean, however, that it is merely a refinement of morphology. It must at the same time inquire into genesis and function.*

While the structural view of molecular biology was taking shape a burgeoning *informational (genetic)* school also claimed proprietorship of molecular biology.[10-13] This school was hatched by the Phage Group, a loose association of researchers who used bacterial viruses (bacteriophages) as a model system (see Chapter 4: "Hershey and Chase Clinched the Role of DNA as the Genetic Material"). Adherents of this school viewed the goal of molecular biology to be the elucidation of the molecular mechanisms of heredity. Leading figures of the Phage Group; Max Delbrück (1906−81), Salvador Luria (1912−91) and Alfred (Al) Hershey (1908−97) exploited induced mutations and genetic recombination to define the nature of genes. Despite its early division into structural and informational branches, molecular biology was and still remains to this day an inexactly defined discipline.[14]

1.2   WHAT IS MOLECULAR BIOLOGY?

The vague definition of molecular biology was nicely illustrated by the explanation that was given by Francis Crick (1916−2004) of how he became a molecular biologist[15]:

> *I myself was forced to call myself a molecular biologist because when inquiring clergymen asked me what I did, I got tired of explaining that I was a mixture of crystallographer, biophysicist, biochemist, and geneticist − an explanation which in any case they found too hard to grasp.*

So what indeed is molecular biology? The historian Michel Morange defined it in its simplest form as: "[. . .] that part of biological research in which explanations are looked for at the level of molecules, by a description of their structure and interactions".[16] He was quick, however, to point out that the investigated "molecules" were actually *macromolecules*. More specifically, molecular biologists are interested in nucleic acids and proteins, whereas other macromolecules such as polysaccharides or long-chain lipids remain in the provinces of biochemistry and cell biology. Even after constraining molecular biology to nucleic acids and proteins and after it branched out into structural and genetic schools, its boundaries remained somewhat arbitrary. Thus, for instance, parts of photosynthesis that have all the attributes of molecular biology were never labeled as bona fide molecular biology.[17]

Evolved out of X-ray crystallography, structural molecular biology aimed at determining the atomic structure of macromolecules, their modes of action, and interaction with other molecules. Early spectacular achievements of this branch of molecular biology were the discovery in the 1950s of the double helix structure of DNA and the solution of the three-dimensional structure of the proteins myoglobin and hemoglobin (see Chapter 5: "Discovery of the Structure of DNA"). Employment in the next decades of spectroscopy, electron microscopy, nuclear magnetic resonance, and X-ray diffraction analysis solved the atomic structure of numerous DNA and RNA molecules and of more than 100,000 different proteins. The informational/genetic school of molecular biology used in its earliest phase bacteriophage and bacterial model systems. By combining genetic and mostly biochemical techniques, this school deciphered the genetic code,[18] identified major elements of the protein synthesis machinery, and outlined the mechanics of gene expression.[19] The development in the next decades of methods to sequence proteins and nucleic acid, the advent of recombinant DNA technologies, and the employment of diverse cellular and molecular biology approaches greatly expanded our understanding of the operation and regulation of the genetic machinery.

1.2.1   Fundamental Tenets of Molecular Biology

Because molecular biology is the product of historical circumstances and evolution, it eludes precise definition. Nonetheless, its concrete scientific achievements were not less than revolutionary and the molecular approach to biology brought to the fore several key principles.[16,20−24] Molecular biology is hallmarked by its adherence to *reduction*. By combining physics and chemistry with biochemistry and genetics, this discipline strives to burrow down to the lowest resolvable level of explanation. The conception of the reductionist approach is that all the biological phenomena would be ultimately explained by uncovering structures and mechanisms at the molecular level.[24] Molecular biology also introduced into biology the concept of *information*. Practitioners and philosophers of science offered different criteria for information in biology.[18] In its simplest form, however, it is the information that DNA stores and that is transcribed to RNA and then translated into proteins.[22] A third tenet of molecular biology is the concept of *mechanism* which provides *explanation* of biological phenomena.[22]

1.2.2 Prevalence and Continued Success of Molecular Biology

A frequently expressed view is that after its golden age of paradigm-changing discoveries, molecular biology entered into a phase which Kuhn labeled "normal science"[25] that foretold its decline.[16,19,26,27] It was mostly contended that the reductionist approach of molecular biology cannot explain life. Instead, integrative system strategies should be applied in the study of biology.

Doubts about the capability of reductionist molecular biology to explain life were raised by no other than Warren Weaver, the very same originator of the term "Molecular Biology." In a Jun. 16, 1964 letter to Crick he contrasted physics with biology and concluded that because of its complexity, life has to be investigated in integrative fashion at levels higher than the molecular (facsimile of the letter is in: http://profiles.nlm.nih.gov/ps/retrieve/ResourceMetadata/SCBBKF):

> *[...] in the realm of inanimate nature, one can successfully*
>
> a) *isolate a very small part of a great system, and then study that small part without taking any explicit account of the general system. Physics is the subject in which you can successfully separate variables.*
> b) *restrict that study to the consideration of a very small number of variables (often two and hardly ever more than four). Physics is essentially not a complicated subject.*
> c) *keep pushing the study to an ever smaller quantitative scale of length, time, and mass. The fun in physics occurs on a sub-microscopic level.*
>
> *[...] My net conclusion then is simply this—that in the effort to work out ways of moving from molecules to man, I think we ought to have around a person who is aware of larger-scale, more complex, perhaps even more subtle, relations than can be captured by the small number of variables that so magnificently handle the law of Newton, of Maxwell, of thermodynamics, of Einstein, of quantum theory ...*

As the argument for the need for holistic lines of attack became more pervasive in time, the more fashionable disciplines of synthetic and system biology have replaced molecular biology at center stage. Synthetic biology, however, is based to a large extent on molecular cell biology. As to system biology, Sydney Brenner strongly argued that it is incapable in principle to provide meaningful answers to biological questions.[28] Even if one does not adopt this position in full, the present record of system biology leaves in question whether it is capable of producing fundamental discoveries such as those that molecular biology had made. In any event, whatever its position is vis-à-vis synthetic or system biology, molecular biology is now established both as a powerful tool kit of experimental techniques and as a way to examine life through the lenses of reduction and mechanistic explanations. The pervasiveness of the molecular approach to life sciences is revealed by the inclusion of the word "molecular" in the titles of scores of biological journals. Thus, there are journals of molecular medicine, pathology, diagnostics, evolution, toxicology, endocrinology, biotechnology, pharmacology, neurobiology, metabolism, immunology, nutrition, psychiatry, urology, vision, and numerous other areas of biomedicine and biotechnology. It seems, therefore, that despite premonitions of its demise, molecular biology is still a vibrant key instrument for acquiring biological knowledge and insight.

1.3   FURTHER READING

The central themes of this book are the science and the historical background of selected milestone experiments in molecular biology. It does not cover, however, myriad scientific, historical, and philosophical facets of molecular biology. It is fitting, therefore, to conclude this introduction with a list of key publications on aspects of molecular biology that this book does not cover. Far from being an all-inclusive catalog, this list is just a primer to a much broader body of scholarship.

1.3.1   General History of Molecular Biology

Judson's *The Eighth Day of Creation*[29] tells the history of the "golden age" of molecular biology (principally the 1950s and 1960s). This book is mostly based on the author's extensive interviews with key proponents of the molecular revolution in biology. *History of Molecular Biology*[19] by Michel Mornage is a substantial introduction to the main developments in the field up to the end of the 20th century. Several chapters in Graeme Hunter's *Vital Forces: The Discovery of the Molecular Basis of Life*[30] review key crossroads in the history of molecular biology. Harrison Echols' *Operators and Promoters*[31] is a highly readable survey of the science and history of molecular genetics. *Discovering Molecular Genetics*[32] by Jeffrey Miller is an annotated and interpreted collection of original key papers on, mostly bacterial, molecular genetics. The history and science of the discovery of the double helix structure of DNA (see Chapter 5: "Discovery of the Structure of DNA") were comprehensively described in Robert Olby's classic book *The Path to the Double Helix.*[7] This fundamental discovery was also the subject of volumes by Portugal and Cohen[33] and by Lagerkvist.[34] Lili Kay authored books on the emergence of the molecular view of life[3] and on the history of the deciphering of the genetic code[18] (see Chapter 7: "Defining the Genetic Code" and Chapter 9: "The Discovery and Rediscovery of Prokaryotic messenger RNA"). Frederick Holmes analyzed in two books the discovery of the semiconservative mode of DNA replication[35] (see Chapter 6: "Meselson and Stahl Proved That DNA Is Replicated in a Semiconservative Fashion") and the mapping by Seymour Benzer of the rII region in phage genome[36] (see Chapter 7: "Defining the Genetic Code").

1.3.2 Monographs, Festschrifts, Autobiographies, and Biographies

1.3.2.1 The Discovery of the Structure of DNA

The largest corpus of books was dedicated to the discovery of the double helix structure of DNA. Key contributors to this landmark discovery described in autobiographical books their parts in it. The most famous is the highly personal *The Double Helix* by James Watson. Originally published in 1968, this book was reprinted with illuminating annotations in 2012.[37] Watson also authored two other autobiographic books.[38,39] Francis Crick wrote a more concise version of the DNA story in his 1988 book *What a Mad Pursuit.*[40] Other autobiographical accounts of this discovery were written by Wilkins[41] and by Chargaff.[42] Different aspects of the DNA story were also covered in biographies or festschrifts of Astbury,[43] Watson,[44−46] Crick,[44,47,48] Rosalind Franklin,[49,50] and Linus Pauling.[51−53]

1.3.2.2 Bacteriophages and the Birth of Molecular Genetics

Several books were dedicated to bacteriophages and to the central place of the Phage Group in the birth of molecular genetics. Biographies of Frederick Twort[54] and Felix d'Herelle[55] described their independent discovery of bacteriophage. A historical 1966 festschrift for Delbrück traced the origins and achievements of the Phage Group,[10] and a published symposium[56] celebrated Delbrück's centenary. Luria told his part of the phage story in an autobiography[57] and a festschrift was dedicated to Hershey.[58]

1.3.2.3 Molecular Biology at the Pasteur Institute

Leaders of the Pasteur school of molecular biology during the 1950s and 1960s; André Lwoff (1902−94), Jacque Monod (1910−76), and

Françoise Jacob (1920–2013) described their critical contributions to molecular genetics and discussed the broader philosophical implications of these discoveries in several books.[59-62] Personal aspects of work at the Pasteur were illuminated in Jacob's autobiography[63] and in a posthumous festschrift to Monod.[64]

1.3.2.4   Other Books

The substantial contributions to molecular biology by Sydney Brenner (1927–) were described in his autobiography[65] and in a more recent biography by Errol Friedberg.[66] Conversations with Fred Sanger (1918–2013), the two times Nobel Prize laureate who developed the first methods to sequence proteins and DNA, were the basis for his biography by George Brownlee.[67] Franklin Portugal wrote a biography of Marshall Nirenberg (1927–2010), the decipherer of the genetic code[68] and a book by Benno Müller-Hill (1933–), of the *lac* repressor described the science and history of this discovery.[69]

1.3.2.5   Structural Molecular Biology

Soraya de Chadarevian authored a history of structural molecular biology in England, particularly in the University of Cambridge, after World War II.[70] Biographies[43,71-73] and autobiographical assays[74,75] of leading X-ray crystallographers describe the early period of the structural analysis of proteins.

1.3.3   Articles on Different Facets of Molecular Biology

1.3.3.1   Origins of Molecular Biology

Several articles were dedicated to the origins of structural molecular biology.[8,9,51,76] Other studies examined the early use of bacteriophage as a model system and the emergence of the gene-centric version of molecular biology.[12,13,77,78] The demarcation of molecular biology apart from biochemistry was discussed by Pnina Abir-Am.[14]

1.3.3.2   Role of Physicists in the Birth of Molecular Biology

The contribution of physicists to the emergence of molecular biology was the subject of several scholarly articles.[1,79-81]

1.3.3.3   Techniques and Experimental Systems

Papers dedicated to the evolution of techniques that advanced molecular biology focused on spectroscopy,[2] nucleic acid hybridization,[82] electrophoresis,[83] and protein and nucleic acid sequencing.[84] Other specific experimental systems were reviewed for instance in Refs. 77, 85.

1.3.3.4   Patronage

A large body of scholarship was devoted to the impact on molecular biology of early funding by the Rockefeller Foundation,[1-6] the Atomic Energy Commission (AEC),[81,86] and the British Medical Research Council (MRC).[70,87]

1.3.3.5   National Programs of Molecular Biology

In addition to articles that surveyed molecular biology in Europe and the United States,[88,89] books and articles dealt specifically with molecular biology programs in France,[90-93] England,[70,87] Germany,[94,95] Japan,[96] and Spain.[97]

1.3.3.6   Institutional Programs

Articles were devoted to molecular biology programs in specific organizations and universities such as the European Molecular Biology Organization (EMBO) and the EMBO Laboratory,[98-100] the Royal Society,[101] Cambridge University and the MRC,[70] the California Institute of Technology (Caltech),[3] and the University of Manchester.[102]

1.3.3.7   Assessment of Present Day Molecular Biology

Decline of molecular biology was professed and its future directions were debated by some thinkers already at the time of its great triumphs in the 1950 and 1960s and also at later periods.[16,19,26-28,103]

1.3.3.8   Historiography

Several historians of science discussed the problematic issue of the historiography of the relatively recent developments in molecular biology.[104-108] Also, the philosophy and history of molecular biology were the subject of a compendium of articles.[109]

REFERENCES

1. Abir-Am P. The discourse of physical power and biological knowledge in the 1930s: a reappraisal of the Rockefeller Foundation's 'policy' in molecular biology. *Soc Stud Sci* 1982;**12**(3):341–82.
2. Zallen DT. The Rockefeller Foundation and spectroscopy research: the programs at Chicago and Utrecht. *J Hist Biol* 1992;**25**(1):67–89.
3. Kay LE. *The molecular vision of life: Caltech, the Rockefeller foundation, and the rise of the new biology*. New York and Oxford: Oxford University Press; 1993.
4. Abir-Am PG. The Rockefeller Foundation and the rise of molecular biology. *Nat Rev Mol Cell Biol* 2002;**3**(1):65–70.
5. Abir-Am PG. *Rockefeller Foundation: biomedical and life sciences offshoots. Encyclopedia of life sciences*. Hoboken, NJ: John Wiley & Sons, Ltd; 2010.
6. Abir-Am PG. The Rockefeller Foundation and the post-WW2 transnational ecology of science policy: from solitary splendor in the Inter-war era to a 'Me Too' agenda in the 1950s. *Centaurus* 2010;**52**(4):323–37.

7. Olby R. *The path to the double helix: the discovery of DNA.* New York, NY: Dover Publications; 1994.
8. Astbury WT. Adventures in molecular biology. *Harvey Lect.* 1950—1;3—44 Series 46
9. Astbury WT. Molecular biology or ultrastructural biology? *Nature* 1961;**190**(4781):1124.
10. Cairns J, Stent GR, Watson JD, editors. *Phage and the origins of molecular biology.* Cold Spring Harbor, NY: Cold Spring Harbor Laboratory; 1966.
11. Hayes W. Max Debrück and the birth of molecular biology. *J Genet* 1985;**64**(1):69—84.
12. Olby R. The origins of molecular genetics. *J Hist Biol* 1974;**7**(1):93—100.
13. Creager ANH. The paradox of the phage group: essay review. *J Hist Biol* 2010;**43**(1):183—93.
14. Abir-Am PG. The politics of macromolecules: molecular biologists, biochemists, and rhetoric. *Osiris* 1992;**7**:164—91.
15. Crick FHC. Recent research in molecular biology: introduction. *Br Med Bull* 1965;**21**(3):183—6.
16. Morange M. *History of molecular biology. Encyclopedia of life sciences.* Hoboken, NJ: John Wiley & Sons, Ltd; 2009.
17. Zallen DT. Redrawing the boundaries of molecular biology: the case of photosynthesis. *J Hist Biol* 1993;**26**(1):65—87.
18. Kay LE. *Who wrote the book of life? A history of the genetic code.* Stanford, CA: Stanford University Press; 2000.
19. Morange MA. *History of molecular biology.* Cambridge, MA: Harvard University Press; 1998.
20. Darden L. Flow of information in molecular biological mechanisms. *Biol Theory* 2006; **1**(3):280—7.
21. Tabery JG. Synthesizing activities and interactions in the concept of a mechanism. *Philos Sci* 2004;**71**(1):1—15.
22. Tabery J, Piotrowska M, Darden L. Molecular Biology. In: Zalta EN, editor. Stanford Encyclopedia of Philosophy (Summer 2015 edition). Stanford, CA; 2015 <http://plato.stanford.edu/entries/molecular-biology>.
23. Schaffner KF. Theory structure and knowledge representation in molecular biology. In: Sarkar S, editor. *The philosophy and history of molecular biology: new perspectives.* Dordrecht, Boston, London: Kluwer; 1996. pp. 27—65.
24. Rosenberg A. *Darwinian reductionism: or, how to stop worrying and love molecular biology.* Chicago, IL: University of Chicago Press; 2006.
25. Kuhn TS. *The structure of scientific revolutions.* 2nd ed. Chicago, IL: The University of Chicago Press; 1970.
26. Stent GS. That was the molecular biology that was. *Science* 1968;**160**(3826):390—5.
27. Morange M. The transformation of molecular biology on contact with higher organisms, 1960—1980: from a molecular description to a molecular explanation. *Hist Philos Life Sci* 1997;**19**(3):369—93.
28. de Chadarevian S. Interview with Sydney Brenner. *Stud Hist Philos Biol Biomed Sci* 2009;**40**(1):65—71.
29. Judson HF. *The eighth day of creation.* Expanded ed, New York, NY: Cold Spring Harbor Laboratory Press; 1996.
30. Hunter GK. *Vital forces: the discovery of the molecular basis of life.* San Diego, CA: Academic Press; 2000.
31. Echols H. *Operators and promoters: the story of molecular biology and its creators.* Berkeley and Los Angeles, CA: University of California Press; 2001.
32. Miller JH. *Discovering molecular genetics: a case study course with problems & scenarios.* Cold Spring Harbor, NY: Cold Spring Harbor Laboratory Press; 1996.
33. Portugal FH, Cohen JS. *A century of DNA.* Cambridge, MA: MIT Press; 1977.

34. Lagerkvist U. *DNA pioneers and their legacy*. New Haven, CT: Yale University Press; 1998.

35. Holmes FL. *Meselson, Stahl and the replication of DNA: a history of "the most beautiful experiment in biology"*. New Haven & London: Yale University Press; 2001.

36. Holmes FL. *Reconceiving the gene: Seymour Benzer's adventures in phage genetics*. New Haven, CT: Yale University Press; 2006.

37. Watson JD. *The annotated and illustrated double helix*. New York, NY: Simon & Schuster; 2012.

38. Watson JD. *Genes, girls, and Gamow: after the double helix*. New York, NY: Vintage Books; 2001.

39. Watson JD. *A passion for DNA: genes, genomes, and society*. Cold Spring Harbor, NY: Cold Spring Harbor Laboratory Press; 2001.

40. Crick FH. *What mad pursuit: a personal view of scientific discovery*. New York, NY: Basic Books; 1988.

41. Wilkins M. *The third man of the double helix: the autobiography of Maurice Wilkins*. Oxford: Oxford University Press; 2003.

42. Chargaff E. *Heraclitean fire: sketches from a life before nature*. New York, NY: Rockefeller University Press; 1978.

43. Hall KT. *The man in the monkeynut coat. William Astbury and the forgotten road to the double-helix*. Oxford: Oxford University Press; 2014.

44. Edelson E. *Francis Crick and James Watson: and the building blocks of life*. Oxford: Oxford University Press; 1998.

45. Inglis JR, Sambrook J, Witkowski JA, editors. *Inspiring science: Jim Watson and the age of DNA*. Cold Spring Harbor, NY: Cold Spring Harbor Laboratory Press; 2003.

46. McElheny VK. *Watson & DNA: making a scientific revolution*. New York, NY: Basic Books; 2004.

47. Ridley M. *Francis Crick: discoverer of the genetic code*. New York, NY: Harper Perennial; 2006.

48. Olby R. *Francis Crick: hunter of life's secrets*. New York, NY: Cold Spring Harbor Laboratory Press; 2009.

49. Sayre A. *Rosalind Franklin and DNA*. New York, NY: W. W. Norton & Co; 1975.

50. Maddox B. *Rosalind Franklin: the dark lady of DNA*. New York, NY: Harper Perennial; 2003.

51. Rich A, Davidson NR, editors. *Structural chemistry and molecular biology*. W. H. Freeman; 1968.

52. Serafini A. *Linus Pauling: a man and his science*. Paragon House Publishers; 1991.

53. Hager T. *Force of nature: the life of Linus Pauling*. New York, NY: Simon & Schuster; 1995.

54. Twort A. *In focus, out of step: a biography of Frederick William Twort, F.R.S. 1877–1950*. Dover, NH: Alan Sutton Publishing Ltd.; 1993.

55. Summers WC. *Felix dHerelle and the origins of molecular biology*. New Haven, CT: Yale University Press; 1999.

56. Shropshire WJ, editor. *Max Delbrück and the new perception of biology*. Bloomington, IN: AuthorHouse; 2007.

57. Luria SE. Alfred P. Sloan Foundation series. *A slot machine, a broken test tube: an autobiography*. New York, NY: HarperCollins; 1984

58. Stahl FW, editor. *We can sleep later: Alfred D. Hershey and the origins of molecular biology*. Cold Spring Harbor, New York, NY: Cold Spring Harbor Laboratory Press; 2000.

59. Lwoff A. *Biological order*. Cambridge, MA: MIT Press; 1962.

60. Monod J. *Chance and necessity*. New York, NY: Vintage Books; 1972.

61. Jacob F. *The possible and the actual*. New York, NY: Pantheon; 1982.
62. Jacob F. *The logic of life*. Princeton, NJ: Princeton University Press; 1993.
63. Jacob F. *The statue within: an autobiography*. New York, NY: Cold Spring Harbor Laboratory Press; 1995.
64. Ullmann A, editor. *Origins of molecular biology. A tribute to Jacques Monod*. Revised Edition ed. Washington, DC: ASM Press; 2003.
65. Brenner S. My life in science: Sydney Brenner, a life in science. *Biomed Central* 2001.
66. Friedberg E.C. Sydney Brenner: a biography. Cold Spring Harbor, New York: Cold Spring Harbor Laboratory Press; 2010.
67. Brownlee GG. *Fred Sanger—double Nobel laureate: a biography*. Cambridge, UK: Cambridge University Press; 2015.
68. Portugal FH. *The least likely man: Marshall Nirenberg and the discovery of the genetic code*. London: MIT Press; 2015.
69. Müller-Hill B. *The lac operon: a short history of a genetic paradigm*. Berlin: de Gruyter; 1996.
70. de Chadarevian S. *Designs for life: molecular biology after World War II*. Cambridge, UK and New York: Cambridge University Press; 2002.
71. Brown A. *J. D. Bernal: the sage of science*. Oxford: Oxford University Press; 2005.
72. Ferry G. *Dorothy Hodgkin a life*. Cold Spring Harbor, NY: Cold Spring Harbor Laboratory Press; 1998.
73. Ferry G. *Max Perutz and the secret of life*. Cold Spring Harbor, NY: Cold Spring Harbor Laboratory Press; 2007.
74. Perutz MF. *Science is not a quiet life: unravelling the atomic mechanism of haemoglobin*. Singapore: World Scientific Publishing; 1998.
75. Perutz MF. *I wish I'd made you angry earlier: essays on science, scientists, and humanity*. New York, NY: Cold Spring Harbor Laboratory Press; 2003.
76. Kendrew JC. How molecular biology started. *Sci Am* 1967;**216**(3):141−4.
77. Summers WC. How bacteriophage came to be used by the Phage group. *J Hist Biol* 1993;**26**(2):255−67.
78. Rheinberger H-J, Müller-Wille S, Meunier R. Gene. In: Zalta EN, ed. The Stanford Encyclopedia of Philosophy (Spring 2015 Edition); 2015 <http://plato.stanford.edu/entries/molecular-biology>.
79. Fleming D. Emigre physicists and the biological revolution. *Persp Am Hist* 1968;**2**:152−89.
80. Fox Keller E. Physics and the emergence of molecular biology: a history of cognitive and political synergy. *J Hist Biol* 1990;**23**(3):389−409.
81. Rasmussen N. The mid-century biophysics bubble: Hiroshima and the biological revolution in America, revisited. *Hist Sci* 1997;**35**(109):245−93.
82. Giacomoni D. The origin of DNA: RNA hybridization. *J Hist Biol* 1993;**26**(1):89−107.
83. Chiang HH-H. The laboratory technology of discrete molecular separation: the historical development of gel electrophoresis and the material epistemology of biomolecular science, 1945−1970. *J Hist Biol* 2009;**42**(3):495−527.
84. García-Sancho M. A new insight into Sanger's development of sequencing: from proteins to DNA, 1943−1977. *J Hist Biol* 2010;**43**(2):265−323.
85. Rheinberger H-J. Experiment and orientation: early systems of in vitro protein synthesis. *J Hist Biol* 1993;**26**(3):443−71.
86. Creager ANH. Nuclear energy in the service of biomedicine: the U.S. atomic energy commission's radioisotope program, 1946−1950. *J Hist Biol* 2006;**39**(4):649−84.
87. de Chadarevian S. Reconstructing life. Molecular biology in postwar Britain. *Stud Hist Philos Biol Biomed Sci* 2002;**33**(3):431−48.

88. Abir-Am PG. Molecular biology in the context of British, French, and American cultures. *Intl Soc Sci J* 2001;**53**(168):187−99.

89. Strasser BJ. Institutionalizing molecular biology in post-war Europe: a comparative study. *Stud Hist Philos Biol Biomed Sci* 2002;**33**(3):515−46.

90. Gaudillière JP. Molecular biology in the French tradition? Redefining local traditions and disciplinary patterns. *J Hist Biol* 1993;**26**(3):473−98.

91. Gaudillière J-P. Molecular biologists, biochemists, and messenger RNA: the birth of a scientific network. *J Hist Biol* 1996;**29**(3):417−45.

92. Gaudillière J-P. Paris−New York roundtrip: transatlantic crossings and the reconstruction of the biological sciences in post-war France. *Stud Hist Philos Biol Biomed Sci* 2002; **33**(3):389−417.

93. Gaudillière J-P. *Inventer la Biomédecine. La France, l'Amérique et la production des Savoirs du Vivant (1945−1965)*. Paris: Editions La Découverte; 2002.

94. Deichmann U. Emigration, isolation and the slow start of molecular biology in Germany. *Stud Hist Philos Biol Biomed Sci* 2002;**33**(3):449−71.

95. Lewis J. From virus research to molecular biology: Tobacco mosaic virus in Germany, 1936−1956. *J Hist Biol* 2004;**37**(2):259−301.

96. Uchida H. Building a science in Japan: the formative decades of molecular biology. *J Hist Biol* 1993;**26**(3):499−517.

97. Santesmases MJ. National politics and international trends: EMBO and the making of molecular biology in Spain (1960−1975). *Stud Hist Philos Biol Biomed Sci* 2002; **33**(3):473−87.

98. Kendrew JC. The European molecular biology laboratory. *Endeavour* 1980;**4**(4):166−70.

99. Krige J. The birth of EMBO and the difficult road to EMBL. *Stud Hist Philos Biol Biomed Sci* 2002;**33**(3):547−64.

100. Cassata F. "A cold spring harbor in Europe." EURATOM, UNESCO and the foundation of EMBO. *J Hist Biol* 2015;1−35.

101. Thomas JM. Peterhouse, the royal society and molecular biology. *Notes Rec R Soc Lond* 2000;**54**(3):369−85.

102. Wilson D, Lancelot G. Making way for molecular biology: institutionalizing and managing reform of biological science in a UK university during the 1980s and 1990s. *Stud Hist Philos Biol Biomed Sci* 2008;**39**(1):93−108.

103. Stent GS. Gunther stent. In: Larry RS, editor. *The history of neuroscience in autobiography*. Amsterdam, Boston, Heidelberg: Academic Press; 1999. p. 396−422.

104. Judson HF. Reflections on the historiography of molecular biology. *Minerva* 1980; **18**(3):369−421.

105. Halloran SM. The birth of molecular biology: an essay in the rhetorical criticism of scientific discourse. *Rhetoric Rev* 1984;**3**(1):70−83.

106. Abir-Am PG. Molecular biology and its recent historiography: a transnational quest for the 'Big Picture'. *Hist Sci* 2006;**44**(1):95−118.

107. Holmes FL. Writing about science in the near past. In: Söderqvist T, editor. *The historiography of contemporary science and technology*. Amsterdam: Harwood Academic; 1997. p. 165−77.

108. Rheinberger H-J. Recent science and its exploration: the case of molecular biology. *Stud Hist Philos Biol Biomed Sci* 2009;**40**(1):6−12.

109. Sarkar S, editor. *The philosophy and history of molecular biology: new perspectives*. Dordrecht, Boston, London: Kluwer; 1996.

Chapter 2

Prehistory of Molecular Biology: 1848−1944

The Discoveries of Chromosomes and of Nucleic Acids and Their Chemistry

Chapter Outline

M. Fry: Landmark Experiments in Molecular Biology. DOI: http://dx.doi.org/10.1016/B978-0-12-802074-6.00002-3
13

In tracing the earliest beginnings of molecular biology, one must contend with the question of what constituted nascent molecular biology before the field was even incepted. Should the focus be put on the early phases of biochemistry or of genetics, the two major disciplines that were eventually conjoined under the umbrella of molecular biology? Then again, should the early ideas on the structure and function of macromolecules be highlighted, as protomolecular biology? The inception of molecular biology is commonly dated to the 1944 discovery by Avery, MacLeod, and McCarty that DNA is the genetic material.[1] The cellular and chemical stage for the identification of DNA as the bearer of genetic information was set by the pre-1944 discoveries of chromosomes, chromatin, and nucleic acids. These landmark advances are thus considered here as the major prehistoric developments that heralded molecular biology.

2.1 DISCOVERY OF THE CELL NUCLEUS AND PROTOPLASM AND THE FORMULATION OF THE CELL THEORY

Although the Dutch microscopist Antonie van Leeuwenhoek (1632–1723) was the first to observe cell nuclei, he did not identify them as distinct subcellular organelles. The Austrian botanist Franz Andreas Bauer (1758–1840) made in 1802 accurate drawings of microscopically observed plant cell nuclei but also failed to classify them as discrete organelles. The first to name the new organelle "cell nucleus" was the Scottish botanist and microscopist Robert Brown (1773–1858) of the Brownian motion fame.[2] Reporting in 1831 to the Linnaean Society on the microscopy of orchid epidermal cells, he defined the cell nucleus. Although Brown mistakenly thought that the nucleus was not universal to all cell types and that it was mostly confined to monocotyledons, he perceptively suggested that this organelle plays important roles in the fertilization and embryonic development of plants (Ref. 3, pp. 108–125; Ref. 4, pp. 76–81). Brown's identification of the cell nucleus was complemented in 1839 by the description (and naming) of the cell protoplasm by the Czech anatomist Johannes Evangelist Purkinje (1787–1869). These independent observations were consolidated in the same year by the German physiologist Theodor Schwann (1810–82) who formulated the cell theory in his book *Microscopic Investigations on the Accordance in the Structure and Growth of Plants and Animals* (the 1847 English translation of the book is available at http://vlp.mpiwg-berlin.mpg.de/library/data/lit28715/index_html?pn=7). Declaring that all living things are composed of cells and cell products, Schwann showed that cells were the basic element of every tested plant and animal. He also demonstrated that the ovum was a single cell that could develop into a complete organism. The father of modern pathology, Rudolf Virchow (1821–1902) put in 1858 an authoritative seal of approval on these emerging perceptions by popularizing a maxim which was originally coined by the French naturalist François-Vincent Raspail (1794–1878): *Omnis cellula e cellula*; that is, every cell originates from another cell.

2.2 CHROMATIN, CHROMOSOMES, AND MITOSIS DISCOVERED

Friedrich Wilhelm Benedikt Hofmeister (Box 2.1), an entirely self-taught botanist, is virtually unknown today although his scientific achievements were enormous.

Some botanists even claim that Hofmeister's accomplishments are in the same league as those of Darwin's and Mendel's.[5,6] His most prominent achievement was the discovery of the alteration of generations in plant life, that is, the change between haploid and diploid states of the cell. Starting with mosses and ferns and advancing later to seed-bearing plants, Hofmeister demonstrated that regular alteration between sexual and nonsexual generations was common to the life histories of both lower and higher plants. It was noted that this discovery, which provided tangible evidence that higher plants were derived from lower plants, was published in 1851, 8 years before the publication of Darwin's *On the Origin of Species*.[5] Hofmeister also made fundamental contributions to the embryology of flowering plant. Despite his severe near-sightedness, he was a gifted microscopist who was able to discern the minutest possible details. In fact, his drawings of plant embryos are in perfect correspondence with what is now known about events that take place in developing ovules.[6]

The discovery of what would later be named chromosomes occupies a prominent place among Hofmeister's great achievements. It is commonly thought that the German botanist Carl Wilhelm von Nägeli (1817–91) was the first to report cell division in 1842. However, the process of cell division was understood only hazily and knowledge was limited to the recognition that nuclear multiplication involved disappearance of an existing nucleus and subsequent formation of two new nuclei.

In his first published paper of 1847, the 23-year-old Hofmeister touched on the problem of cell division in the fertilization in the *Onagraceae* (the evening primrose family of flowering plants). In his second paper of 1848 he dealt directly with pollen formation and cell division. Hofmeister's illustrations of the microscopic events during cell division (reproduced in Ref. 7) showed bodies that clearly represented chromosomes. Although an understanding of their significance could not be attained by Hofmeister and his contemporaries, based on this work he is nevertheless credited as the earliest discoverer of chromosomes.

Chromosomes were subsequently observed in 1872 by the Baltic-German botanist Edmund Russow (1841–97) who discerned during cell division microscopic rod-like bodies that he named *Stäbchen* (German for rods). The Belgian embryologist Edouard van Beneden (1846–1910) used in 1875

BOX 2.1 Pioneers of Chromosome Studies

Wilhelm Hofmeister was born in Leipzig to a father who was a book publisher and owner of a music shop. Hofmeister had no formal schooling beyond a vocational school from which he graduated at the age of 15. Working in his father's shop, he dedicated the early morning hours to microscopic studies of plant cells. In addition to becoming completely self-taught in all areas of botany, this true genius also achieved mastery of languages, history, poetry, art, and music. His trailblazing work on plant embryology, development, and growth made him a leading scientific figure of his time. Thus, despite his lack of formal education, he was appointed at the age of 39 Ordinarius Professor (the highest academic rank at a German university) and Director of the Botanical Gardens at Heidelberg University. Following the untimely death of his wife, Hofmeister moved in 1872 to the University of Tübingen. A series of family tragedies and ill health led to his early death at age 52.

Walther Flemming was trained in medicine and after serving as a military physician in the French-Prussian war, he taught at the University of Prague and later at the University of Kiel. Using basophilic aniline dyes, he identified thread-like chromatin in cell nuclei. Through his studies of the distribution of chromosomes between daughter cells, Flemming discovered mitosis. This led to his conclusion that cell nuclei do not form de novo but rather are derived from other nuclei. Flemming's discovery of mitosis is considered to be one of the most important landmarks in biology.

The German anatomist von Waldeyer-Hartz is most famous for his "neuron theory" that placed the neuron as the basic structural unit of the nervous system. In another line of research, he studied the basophilic stained chromatin that Flemming described. Based on Flemming's and his own studies, von Waldeyer-Hartz coined in 1888 the term "chromosomes" to describe the filaments that were observed during mitosis (Fig. 2.1).

Wilhelm Hofmeister
1824– 77

Walther Flemming
1843–1905

Heinrich Wilhelm Gottfried
von Waldeyer-Hartz
1836–1921

FIGURE 2.1   Pioneers of chromosome studies.

the term *bâtonnet* (short stick in French) to describe similar bodies. The French embryologist Edouard Balbiani (1825−99) wrote in 1876 that during cell division the nucleus dissolved into a collection of *bâtonnets étroits* ("narrow short sticks"). Those pioneering observations were greatly expanded and advanced by Walther Flemming (Box 2.1), a founding father of the field of cytogenetics. Using synthetic aniline basophilic dyes to stain the nucleus in dividing cells, he observed that the red dye was intensely absorbed by granular structures in the nucleus that he named *chromatin* (from the Greek *khrōma*—color). Flemming reported in 1884 that during cell division in salamander larvae, the stained chromatin coalesced into thread-like structures (his original German term was *Fäden*, ie, thread). Four years later the authoritative anatomist von Waldeyer-Hartz (Box 2.1) named these bodies *chromosomes* (ie, colored bodies). Flemming observed in stained dividing cells longitudinal splitting of the chromosomes to produce two identical halves and called this process *mitosis* (from the Greek *mitos*—thread) (Fig. 2.2). Flemming, Van Beneden, and the Polish-German botanist Eduard Strasburger (1844−1912) expounded the details of mitosis showing that it was characterized by equal distribution of chromosomes among the daughter cells.

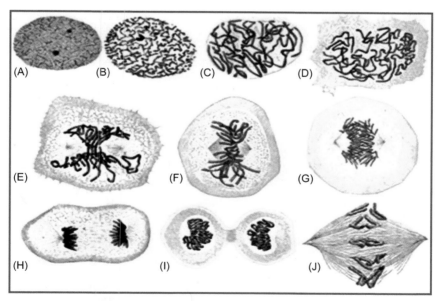

FIGURE 2.2   Flemming's hand-drawn sketches of the phases of mitosis. In this drawing from his 1882 book *Zellsubstanz, Kern und Zelltheilung* (*Cell Substance, Nucleus and Cell Division*), (Ref. 8 also available online at https://archive.org/details/zellsubstanzker02flemgoog), mitosis was followed from the condensation of chromosomes (prophase) to their separation into two "bundles of wool," and the emergence of separating daughter nuclei (A−J).

It should be noted, however, that despite his profound observations, Flemming did not comprehend the relationship between cell division and heredity. It fell onto others to link nuclei and chromosomes to heredity. The German physician, naturalist, and philosopher Ernst Haeckel (1834−1919) suggested that the nucleus contains factors that are responsible for the transmission of hereditary attributes (Ref. 9, pp. 287−288; also available at: https://ia600309.us.archive.org/31/items/generellemorphol01haec/generellemorphol01haec.pdf). Based on his independent studies of dividing fertilized sea-urchin eggs, the German biologist and leading cytologist of his time, Theodor Boveri (1862−1915) later proposed that the genetic material is contained in the chromosomes and that different chromosomes carry different pieces of the hereditary information.[10−12] Independently, the chromosome theory of heredity was reinforced by the American Walter S. Sutton (1877−1916) who was the first to show that chromosomes obey Mendel's rules.[13]

2.3 THE HOPPE-SEYLER SCHOOL OF PHYSIOLOGICAL CHEMISTRY: INCUBATOR TO THE DISCOVERY OF NUCLEIC ACIDS

Ernst Felix Immanuel Hoppe-Seyler (1825−95) (Fig. 2.3) and his competitor and adversary Willie Kühne (1837−1900) were among the most influential German and European physiological chemists (or biochemists in today's terms) of the second half of the 19th century. Hoppe-Seyler's life and scientific contributions were reviewed by Kohler (Ref. 14, pp. 22−24) and by Fruton, who also described the achievements of his scientific progeny and his disputes with Kühne (Ref. 15, pp. 76−79 and 82−101). Hoppe-Seyler's prominence was due to his personal scientific achievements and equally

| Felix Hoppe-Seyler 1825–95 | Friedrich Miescher 1844–95 | Miescher's lab in Tübingen which was a converted kitchen of a former castle, was considered state of the art at the time |

FIGURE 2.3   Felix Hoppe-Seyler, Friedrich Miescher, and the laboratory that Miescher occupied in Hoppe-Seyler's Tübingen department of applied chemistry.

to his nurturing of students who eventually reached eminence on their own. Among his most distinguished trainees were Friedrich Miecher, the discoverer of nuclein (DNA in today's term), and Albrecht Kossel who, among his other achievements, determined the chemical structure of purines and pyrimidines.

Born in Freiburg, Germany, Felix Hoppe who was orphaned at the age of 10 had added "Seyler" to his last name after having been adopted by his brother-in-law, the clergyman Dr Seyler. Although he received a doctor of medicine degree in 1850, Hoppe-Seyler was not attracted to the practice of medicine and chose instead to dedicate his career to laboratory work. In 1856 he was first appointed as an assistant to Rudolf Virchow and then as the head of the chemical laboratory in the Institute of Pathology that Virchow founded in Berlin. In 1864 Hoppe-Seyler became full Professor of applied chemistry in the University of Tübingen. There he was a member of the first Department of Natural Sciences in Germany which was home to some of the foremost natural scientists of the day.[16] In 1872 he was recruited with some other leading scientists to the new German University of Strasbourg at which he remained until his death in 1895.

In his capacity as a leader of 19th century German biochemistry, Hoppe-Seyler advocated the establishment of separate research-oriented university departments. He had also founded in 1877 a leading biochemical journal, *Zeitschrift für Physiologische Chemie*, and remained its editor until his death. The journal was renamed in 1895 in his honor *Hoppe-Seyler's Zeitschrift für Physiologische Chemie* and was edited, until his death in 1927, by Hoppe-Seyler's former student and the 1910 Nobel laureate Albrecht Kossel (1853–1927). The journal was renamed once again in 1985; *Biological Chemistry Hoppe-Seyler* and since 1996 its title became *Biological Chemistry*.

Hoppe-Seyler's scientific work was dedicated in large part to the study of proteins, which he named at the time "proteids." The focus of his investigations was the study of the chemistry of hemoglobin to which he also gave its name. Showing that hemoglobin was the constituent of erythrocytes that bound oxygen, he also demonstrated its tighter association with carbon monoxide (CO). He also discovered that hemoglobin contained iron and conducted spectroscopic measurements that revealed relationships between hemoglobin and its related molecules hemochromogen, hemin, and hematoporphyrin. In addition, Hoppe-Seyler investigated the chemistry of milk and bile and was responsible to the development of many new preparative and analytical biochemical methods. Arguably, however, he impacted biochemistry to an equal or greater extent by having been teacher and guide to students and associates. Two scientists stand out among his trainees: Friedrich Miescher and Albrecht Kossel who laid the foundations to the field of nucleic acids research.

2.3.1 1871: Isolation of Nuclein by Friedrich Miescher

Although the term DNA is commonly associated with the names of Watson and Crick, it was the Swiss biochemist Friedrich Miescher (Fig. 2.3) who was the first to isolate DNA 82 years before Watson and Crick presented in 1953 their model of the DNA double helix. Miescher's work consisted of the isolation and characterization of the basic properties of a nuclear substance that he named "nuclein" that was actually crude DNA.

Several excellent reviews described the state of biochemistry in Miescher's time[17] and his own life and work[16,18–20] (Ref. 18 is also available at: http://www.bizgraphic.ch/miescheriana/html/the_man_who_dicovered_dna.html). Of particular interest are articles by Dahm that uncovered a treasure-trove of Miescher's correspondence with his parents and his uncle, the anatomist and physiologist Wilhelm His (1831–1904). These letters shed light on much of Miescher's life and on his scientific progress and thinking.[16,19,20]

Miescher was born to a respected academic family. His father was Professor of Pathologic Anatomy at Basel University and his maternal uncle Wilhelm His was Professor of Anatomy and Physiology in the same institution. Although he was hard-of-hearing and myopic, Miescher excelled in his studies, graduating in 1868 from medical school with an outstanding thesis. Miescher realized that because of his physical handicaps he would not be able to practice his preferred medical specialty of otology. Having been passionate about science he chose, therefore, to dedicate his career to research work and soon after completing his medical studies he moved to Tübingen. There he initially spent one semester in the laboratory of the leading chemist Adolph Strecker (1822–71) and then joined Hoppe-Seyler's laboratory in the autumn of 1868. Working in a laboratory that was housed in a castle (Fig. 2.3), Miescher started by studying proteins of leukocytes that were collected from pus that he washed off soiled bandages of infected wounds. In these investigations he noticed that acidification of the solution precipitated a substance that could then be solubilized in alkali. Assuming that the precipitated material was of nuclear origin, Miescher developed a procedure to isolate leukocyte nuclei. Extraction of the isolated nuclei with alkaline solution yielded a yellow solution that upon acidification generated insoluble white flocculent precipitate (for Miescher's detailed experimental protocol, see Ref. 19). Realizing that the precipitated material differed from known proteins, Miescher named it *nuclein*. In fact, nuclein was later recognized to be mostly DNA with traces of associated proteins.

To characterize nuclein, Miescher had to modify the isolation procedure such that larger amounts of material of greater purity could be obtained. The modified nuclein isolation protocol included steps of lysis of the nuclei with warm alcohol and removal of residual cytoplasmic proteins by digestion with pepsin. Lipids were also separated from the sedimented nuclein by extraction with ether.[16] Using the better-purified nuclein, Miescher determined its elementary composition. Analyses detected elements expected in an organic molecule: carbon, hydrogen, oxygen, nitrogen, and also sulfur, which in retrospect was derived from contaminating proteins. However, one identified element that was unexpected was phosphorus whose ratio to other elements in the nuclein was much higher than in any organic molecule that was known at the time.[21] Since incinerated nuclein did not contain phosphoric acid, Miescher concluded that the phosphorous evaporated during combustion. This indicated that it was not present as an inorganic element but rather that phosphorous was organically bonded to the nuclein.[21] Becoming convinced that nuclein represented a new substance that differed from all the known proteins, Miescher stated (Ref. 21; translation into English in Ref. 16):

> *I believe that the given analyses, as incomplete as they may be, allow the conclusion that we are not dealing with some random mixture, but, apart from at most low level of contamination, with a chemical individual or a mixture of very closely related entities [...] We rather have here entities sui generis not comparable to any hitherto known group.*

Finding that nuclein was also present in liver, kidney, testes, and nucleated erythrocytes, Miescher also predicted that[16,21]:

> *[...] an entire family of such phosphorous-containing substances, which differ slightly from one another, will reveal itself, and that this family of nuclein bodies will prove tantamount in importance to proteins.*

Miescher's letters to his uncle Wilhelm His reveal that he understood that the chemical composition of nucleus and cytoplasm were different and that nuclein could, therefore, be used to define the nucleus as a distinct subcellular compartment.[20] In contrast to these perceptive insights, Miescher was completely off-the-mark in his conjectures on the metabolism and possible function of nuclein (DNA). Having been impressed by recent studies of Strecker and Hoppe-Seyler on the phosphorous-containing lecithin that they have isolated, he speculated that nuclein was the precursor of lecithin. He also proposed that the probable cellular function of nuclein was the storage of phosphorous.

2.3.2 Publication of the Discovery of Nuclein

After concluding his initial work on nuclein, Miescher felt that before embarking on a career of independent investigator he should seek broader training. He thus moved to the Leipzig University laboratory of the eminent physician and physiologist and 1884 laureate of the Copley medal, Carl Ludwig (1816−95). There he worked for less than 2 years on projects unrelated to nuclein. Still, while working in Leipzig he also drafted his first scientific manuscript on the isolation and properties of nuclein. In late 1869 Miescher sent the completed paper to Hoppe-Seyler for critique and approval. Hoppe-Seyler suspected, however, that nuclein might have been an artifact that was generated by the combination of phosphorous with products of the digestion of protein by pepsin.[17] Because of this misgiving, Hoppe-Seyler suspended the publication of Miescher's manuscript until he could repeat the experiments with his own hands. Only after he successfully replicated Miescher's results was his paper published 1871—almost 2 years after it was written.[21] Two confirmatory papers were published back-to-back with this paper.[22,23] In one, Pál Plósz (1844—1902), a Hungarian trainee of Hoppe-Seyler, proved that the source of nuclein was the nucleus. Essentially, he showed that nuclein was present in nucleated avian and reptilian erythrocytes but not in bovine erythrocytes that lacked nuclei.[22] In the other article[23] Hoppe-Seyler himself corroborated Miescher's data, asserting that: "[...] I have therefore repeated parts of his experiments, mainly the ones concerning the nuclear substance, which he has termed nuclein; I can only emphasize that I have to fully confirm all of Miescher's statements that I have verified."[16]

2.3.3 Expanding His Studies of Nuclein, Miescher Came Up With a New Theory on Its Function

Ending his sojourn in Leipzig, Miescher was offered professorship in the University of Basel. Backed by the enthusiastic sponsorship of two pillars of German science, Hoppe-Seyler and Carl Ludwig, Miescher became Professor in Basel at the exceptionally young age of 28. Inheriting his father's Chair of Physiology in 1872, Miescher's appointment was not without problems. First, there was talk at the University of Nepotism as both Miescher's father and uncle were Professors in the same institution. Second, the working conditions in his new independent laboratory were inferior in comparison to the ones that he enjoyed in Hoppe-Seyler's laboratory. In Basel he was assigned a converted corridor for a laboratory and his help consisted of a single part-time technician. Miescher was also laden with teaching duties and although he invested considerable effort in the preparation of lectures, teaching was not one of his strengths. Being hard-of-hearing and shy, he was not a good public speaker and he also was inept at personally engaging with students. Nevertheless, he resumed in Basel his studies on nuclein, focusing mostly on its isolation from salmon sperm. Fortunately, salmon fish were abundant in the Rhine River where they came to spawn. Using sperm that was extracted from freshly caught fish in the nearby river, Miescher refined the protocol for the isolation of sperm nuclein. Obtaining highly purified material, he found that its content of phosphorus pentoxide (P_2O_5) was 22.5% of the total mass of nuclein—a value closely similar to its actual content of 22.9% in DNA.[16] Miescher also identified the acidic nature of nuclein and, based on its inability to diffuse through parchment paper, he concluded that it had high molecular weight. However, his measurements yielded grossly off-the-mark molecular weight of only 500−600 Da.[16]

In light of his newer findings on salmon sperm DNA and based on similar results that he obtained with sperm of frog, chicken, and bull, Miescher changed his thinking on the possible function of nuclein. The profusion of nuclein in spermatozoa of diverse species led him to abandon the hypothesis that it served merely as a store for cellular phosphorous. Rather, he now speculated that nuclein was involved in fertilization, stating that (Ref. 24; translation into English in Ref. 18):

> If one were to assume that the specific cause of fertilization depends on a single substance, for instance through an enzyme or in some other way, perhaps through a chemical stimulus, then one would have to think without a doubt principally of nuclein.

Clearly, however, this new conjecture did not mean that Miescher associated nuclein with heredity. In mentioning fertilization, he adhered to an idea that was widely held at time (and that was also shared by his influential uncle Wilhelm His) that "fertilization is a physical process of motion" and that contact of the sperm with the egg transmitted the needed "motion stimulus." Hence, as the above citation suggests, Miescher proffered that nuclein was the required "chemical stimulus" of motion.[16,18] As to heredity, he considered the possibility that the wide variety of genetic traits could be encoded by the enormous number of stereoisomers that can be generated by asymmetric carbon atoms in the amino acids residues in proteins. He calculated that a protein molecule with 40 asymmetric carbons would have up to a trillion isomers.[18] As we shall see in subsequent chapters, variations on the idea that protein is the genetic material persisted until the early 1950s.

Not only was Miescher unmindful of the possibility that nuclein was the bearer of genetic information, he was also oblivious its connection to chromatin and chromosomes. Although Edmund Russow reported in 1872 that similarly to nuclein, chromosomes were also dissolved at a basic pH, Miescher had not linked nuclein to chromosomes. It was left to Walther Flemming to suggest in 1882 that nuclein and chromatin were the same.[20] Last, although Miescher noticed that the nuclein content increased in dividing cells and that this increase was a prerequisite to cell proliferation, he never published these findings and did not associate them with a possible function of nuclein as the carrier of hereditary traits.[20]

Nuclein was finally linked to chromosomes and to heredity by Theodor Bovery[10,12] and by his friend, the American embryologist and geneticist and the discoverer of sex chromosomes, Edmund Beecher Wilson (1856−1939). In his book *The Cell in Development and Inheritance*, Wilson wrote[25]:

> Now chromatin is known to be closely similar to, if not identical with, a substance known as nuclein which analysis shows to be a tolerably definite

chemical compound of nucleic acid and albumin. And thus we reach the remarkable conclusion that inheritance may, perhaps, be affected by the physical transmission of a particular compound from parent to offspring.

2.3.4   After an Initial Controversy, Nuclein Was Verified as a Distinct Chemical Entity

Despite Miescher's meticulous experimental work and its careful verification by Pál Plósz and Hoppe-Seyler, many researchers held a negative view of the claim that nuclein was an entity separate and distinct from protein. Indeed, even the better-purified preparations of nuclein were often positively stained by protein-specific dyes.[17] Adding to the uncertainty was the fact that Miescher detected purine bases in protamine that he had separated from sperm nuclein. Later Miescher put his student and a future Professor of chemistry at Basel, Jules Piccard (1840−1933) to look again into this question. Although Piccard also found that purines were present in the acid extracted protamine, he also showed that their probable source was nuclein.[26]

The uncertainty whether nuclein was identical to or distinct from proteins was ultimately removed by the University of Leipzig pathologist Richard Altmann (1852−1900) who succeeded in isolating protein-free nuclein (DNA). Moreover, like Miescher before him,[21,24] Altmann documented the acidic nature of nuclein and thus renamed it nucleic acid (*Nucleïnsäure* in German).[27] As told by Dahm,[16] Miescher was displeased by the growing acceptance of Altmann's new nomenclature, feeling divested of his cherished nuclein.

2.3.5   Miescher's Later Work in Basel

Miescher's humble initial working conditions in Basel improved with time. With an expanded research group his investigations encompassed subjects other than nuclein. As indicated by his correspondence,[19] despite his continued interest in nuclein he did not make much progress in elucidating its properties. Turning to other areas, his investigations on the development of sperm and oocytes of spawning salmon led to the finding that as the fish were traveling without feeding up the Rhine, they produced gonads while their trunk muscles degenerated. Miescher showed that muscle tissue degradation products were utilized to build up their ovaries and sperm, a process that he named "liquidation." In these studies which were published during his lifetime[28] and posthumously,[29] (also available online at http://www.bookprep.com/read/uc1.b4106983), Miescher provided detailed statistical data on changes in the weights of the salmon body, trunk muscles, ovaries, and spleen at different stages of the migration and spawning of the fish. In another line of inquiry he investigated respiration physiology and looked into changes in blood composition as a function of altitude.[30,31] Miescher's expanding duties also included research projects that were not of his own choice. Thus, to his displeasure, he was assigned to conduct an extensive study on the nutrition needs of Swiss prisoners. He was also burdened in the early 1880s with the heavy responsibility of erecting Basel's new institute of anatomy and physiology. The institute was inaugurated in 1885 and Miescher served as its head until his death.

Being obsessive and a perfectionist in his work, Miescher bowed under the weight of his many responsibilities. He contracted tuberculosis and had to leave his laboratory for a clinic in the Alps. All this led to his untimely death in Aug. 1895 at the age of 51. His legacy as the discoverer of nucleic acids is well-remembered and celebrated as is exemplified for instance by the establishment in 1970 of the Basel Friedrich Miescher Institute (http://www.fmi.ch/).

2.3.6   Early Identification and Isolation of Purines

Long before it became recognized that nucleic acids are comprised of purine and pyrimidine nucleobases, scientists isolated some members of the purine family. The first purine (derived from the Latin *purus*—pure) was isolated in 1776 by the Swedish pharmacist Carl Wilhelm Scheele (1742−86). This eminent pharmaceutical chemist was the first to identify oxygen, molybdenum, tungsten, barium, hydrogen, and chlorine and to isolate organic acids such as tartaric, oxalic, lactic, and citric acids. Among the latter he also identified the purine uric acid that he isolated from bladder calculi.[32] The composition of uric acid was determined in the 1830s by Justus von Liebig (1803−73) and Friedrich Wöhler (1800−82) who also instituted nomenclature for other members of the purine group. The formula of uric acid was suggested in 1875 by the Würzburg University chemist Ludwig Medicus (1847−1915). Additional purines, caffeine, xanthine, hypoxanthine, and theobromine, had been isolated in the first half of the 19th century. The German chemist and physicist Heinrich Gustav Magnus (1802−70) and his student Unger were first to isolate guanine from deposited feces of the Guano sea bird in the North Chincha Island of Peru.[33]

2.3.7   The Chemistry of Purines and Pyrimidines Defined: The Work of Albrecht Kossel

The isolation and chemical definition of purines and pyrimidines, the basic building blocks of nucleic acids, was largely the fruit of the work of Hoppe-Seyler's former student, the German biochemist Ludwig Karl Martin Leonhard Albrecht Kossel (Fig. 2.4; for a comprehensive review of Kossel's life and work, see Ref. 34).

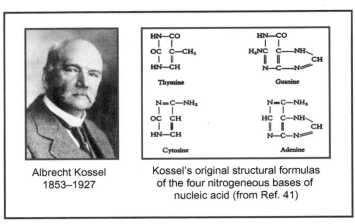

Albrecht Kossel 1853–1927	Kossel's original structural formulas of the four nitrogeneous bases of nucleic acid (from Ref. 41)

FIGURE 2.4 Albrecht Kossel and the structural formulas that he determined for the four nucleobases of DNA.

Between 1885 and 1901, Kossel characterized the nucleobases adenine, cytosine, guanine, and thymine (some of his primary papers on this subject are, for instance, Refs. 35–39). Summaries of Kossel's investigations on purines and pyrimidines can be found in his 1881 book[40] and in his Nobel[41] and Harvey[42] lectures. Uracil, the fifth nucleobase, was discovered by Kossel's Italian student, the later serologist and biochemist, Alberto Ascoli (1877–1957).[43]

Albrecht Kossel attended the University of Strasburg where he worked under Hoppe-Seyler at his Department of Biochemistry. He concluded his studies of medicine at the old university of his birth town, the Baltic city of Rostock. Rather than starting a practice of medicine, he chose to return in 1877 to Strasburg to serve as an assistant to Hoppe-Seyler. There he first studied the diffusion and dialysis of salts and peptones. After publishing his findings in the *Zeitschrift für Physiologische Chemie*, Kossel undertook in 1878 the task of characterizing nuclein that had been isolated some years before by Miescher in Hoppe-Seyler's Tübingen laboratory.[21–23] In a first 1878 paper on yeast nuclein[44] and in a second 1879 paper[45] Kossel described the preparation and analysis of nuclein and the isolation of hypoxanthine and xanthine from products of its hydrolysis. Essentially, these results confirmed Piccard's previous findings on salmon sperm nuclein and protamine.[26] In 1881 Kossel published his first book in which he argued that an observed higher content of nucleic acid in liver or spleen relative to muscle suggested that rather than serving as reserve material, nucleic acids were formed during tissue regeneration.[40]

In 1884 Kossel left Strasburg to become the Director of the Division of Chemistry at the Berlin Physiological Institute and in 1887 he was promoted to Professor Extraordinarius in physiology at that institution. Continuing his studies on the constituents of nucleic acid in Berlin, he discovered adenine that he isolated from nuclein of pancreas and yeast cells.[36–38] After the 1889 announcement by Altmann of the isolation of protein-free nucleic acid,[27] Kossel reported the isolation of phosphoric acid, guanine and adenine from this pure material. Together with his student Neumann he next isolated from nuclein the nucleobases thymine and cytosine.[39,46] Kossel predicted that xanthine and hypoxanthine were breakdown products of nucleic acids and that hypoxanthine was a secondary product of adenine and not a primary constituent of nucleic acids. He also showed that hydrolysis of Altmann's DNA yielded substance with the properties of carbohydrate. More specifically, he suggested that the carbohydrate moiety of yeast nucleic acid was a pentose—an early hint for the existence of deoxyribose in DNA.

With the four nucleobases of DNA thus isolated and characterized, and uracil later defined by Ascoli,[43] Kossel, who was appointed in 1895 to be Professor of physiology at the University of Marburg, moved on to the study of proteins, and in particular of protamine (which he named histone), and their constituent amino acids. After moving once again in 1901 to Heidelberg because his dislike of the domineering Prussian ways in Marburg,[34] he continued his work on products of hydrolysis of proteins and on the structure of the liberated amino acids. He identified among hydrolysis products of goose protamine the amino acids leucine and tyrosine whereas sperm protamine that consisted by weight of 87% arginine, was found to also contain lysine and a new amino acid, histidine that Kossel was the first to identify.

In recognition of his contributions to the understanding of the chemistry of the basic elements of nucleic acids and proteins, Kossel was awarded in 1910 the Nobel Prize in Physiology and Medicine.

2.3.8   Kossel's "Bausteine" Concept of the Building Blocks of Nucleic Acids and of Proteins

Based on his work on nucleobases and on amino acids, Kossel crystalized his idea of *Bausteine* (building bricks (blocks) in German), as the basic constituents of infinitely variable larger structures. He realized already in the early 1900s that although hydrolysis of different proteins yielded a limited number of different amino acids, the proportions of individual amino acids differed in different proteins. He thus concluded that proteins were distinguished from one another by the composition of their amino acids. Kossel distilled his idea of Bausteine by writing in 1912 about the amino acid building blocks of protein[47]:

> *The number of Bausteine which may take part in the formation of protein is as large as the number of letters in the alphabet. When we consider that through the combination of letters and infinitely large number of thoughts can be*

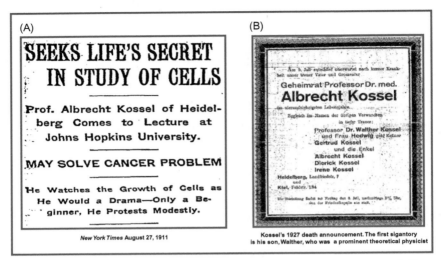

FIGURE 2.5   (A) Headlines to an interview that Kossel gave to an unsigned *New York Times* reporter during his 1911 visit to the United States (http://query.nytimes.com/mem/archive-free/pdf?res=F20F14FF3D5813738DDDAE0A94D0405B818DF1D3). (B) Kossel's 1927 death announcement.

expressed, we can understand how vast a number of properties of the organism may be recorded in the small space which is occupied by the protein molecule.

The idea of amino acids as the building blocks of proteins was also valid for nucleobases as primary elements of nucleic acids. This view was articulated by Kossel himself in an interview that he gave to the *New York Times* during his 1911 visit to the United States. This interview indicated that the unnamed reporter felt that despite his being admired as one of the greatest scientists of his era, Kossel was truly modest (Fig. 2.5A).

Opening with a description of Kossel and his eminent standing in the world of science, the reporter went on to present to the layman reader the essence of his work:

He is one of the greatest living scientists, and is the foremost authority in the world on cellular life, and has been covered with honors which he bears with a notable modesty. [...] He is seeking not the secret of life but the secrets of cellular composition, and this knowledge is of vital importance to every branch of science.

Attempting to describe in a way that would be understood by nonspecialists his concept of the primary elements that build up complex cellular structures, Kossel was then cited:

The processes of life are like a drama [...] and I am studying the actors, not the plot. There are many actors, and it is their characters which make the drama. I seek to understand their habits, their peculiarities.

2.3.9   Kossel's Later Years

Since 1901 Kossel was the incumbent of the Chair of Physiology and the Director of the Physiological Institute at Heidelberg University. This was one of the most prominent positions in German Science since Heidelberg was considered at the time to be the preeminent center of physiology in Germany. From 1907 until his death he held the title of *Geheimrat* (the highest advising official at the Imperial court; the equivalent of the English *Privy Councilor*). In this capacity he became a life-member of the ruling board of the University of Heidelberg and in 1908 and 1909 he was also its pro-rector.

Despite his growing burden of administrative duties and although he was increasingly showered with awards and accolades, Kossel continued his active research work. Having been followed to Heidelberg by his Marburg students, his research group consisted by 1905 of about 30 students from different countries.[48] His work was mostly dedicated to the study of proteins although he maintained an interest in nucleic acids. Among his later achievements were the discoveries of catabolic enzymes of DNA and of amino acids. Kossel's student de la Blanchardière identified in tissue extracts nuclease(s) that decreased the viscosity of nucleic acid in solution.[49] In the course of their studies on amino acids Kossel and his student Dakin discovered arginase in the liver.[50] In his investigations on proteins Kossel and his associates isolated protamine peptides and studied the optical rotations of proteins. They also showed that treatment with alkali caused racemization of amino acid within intact proteins and together with Dakin, Kossel studied crystalline forms of basic amino acids.

Many of Kossel's students became leading scientists on their own. His student at Marburg, the Jewish-Russian-American Phoebus Levene (1869–1940), made important contributions (and a critical mistake) to the chemistry of nucleic acids (*vide infra*). The English-American Henry Drysdale Dakin (1880–1952) and the Swede-American Otto Knut Folin (1867–1934) contributed to the chemistry of amino acids and of proteins. Many of Kossel's other students such as Hermann Steudel (1871–1967), Walter Jones (1865–1935), Albert Mathews (1871–1957), Edwin Hart (1874–1953), Lawrence Henderson (1878–1942), and Alexander Cameron (1882–1947) also made remarkable careers as investigators and academic teachers.

Kossel was weighed down first by the death of his wife in 1913, and then by Germany's role in World War I. He became Professor emeritus in 1924 but continued to teach physiological chemistry and to conduct research work at the institute that he founded. The last months of his life were dedicated to writing a monograph on protamines and histones that was published shortly after his death on Jul. 5, 1927 at the age of 74 (Fig. 2.5B). Kossel was commemorated by the creation at the

University of Rostock, his town of birth, the *Albrecht Kossel Institute for Neuroregeneration* (http://www.albrecht-kossel-institut.de/).

2.4 UNRAVELING THE CHEMISTRY OF NUCLEIC ACIDS—THE WORK OF PHOEBUS LEVENE

Although Kossel and his associates identified and determined the structure of the DNA four nucleobases and of uracil, their work left open the question of the chemical structure of DNA and RNA. Although many scientists contributed to the growing knowledge on nucleic acids, the scientist who more than anyone is remembered for his pioneering investigations on the structure of nucleotides and of their linkage to one another in DNA and RNA was the American biochemist Phoebus Aaron Theodor Levene (1869–1940).[51] An excellent biography and concise summary of Levene's scientific work was published after his death by the National Academy of Science of the United States.[52] For a thorough analysis of his erroneous but highly influential hypothesis of the tetranucleotide structure of nucleic acids (*vide infra*), see Ref. 53.

Phoebus Levene (his birth name was Fishel Aaronovich Levin) (Fig. 2.6A) was born to a Jewish family in Sagor, Tsarist Russia.

Levene's family moved to St Petersburg when he was 2 years old. There he graduated from the Classical Gymnasium and then attended the Imperial Military Medical Academy of that city. Levene had illustrious university teachers such as his Professor of Chemistry, (and more famously the composer), Alexander Borodin (1833–87) and the physiologist and 1904 Nobel laureate Ivan Pavlov (1847–1936) who was then still a Privatdozent (a university teacher without a professorial chair). While Levene was still a student of medicine he had his first taste of research work in the laboratory of Borodin's son in law, the organic chemist Alexander Dianin (1851–1918). Because of mounting anti-Semitism in Russia, Levene's family immigrated to the United States in 1891 and a year later, after having been granted a medical license, he followed them to America. Levene started by practicing medicine in the lower east side of New York but at the same time he also did research work in the Department of Physiology of Columbia University College of Physicians and Surgeons and took classes to round his chemical education. In 1894 he left medical practice and was appointed to a position of investigator in the New York Pathological Institute of the New York State Hospitals. There he developed an interest in nucleic acids which eventually became a major piece of his scientific oeuvre. To expand his scientific horizons Levene traveled to conduct research in the Marburg laboratory of Kossel, then in an electrochemical laboratory in Munich and last in the Berlin laboratory of Emil Fischer. Having been impressed by the professional achievements of the 35-year-old Levene, Simon Flexner (1863–1946), the founding director of the then

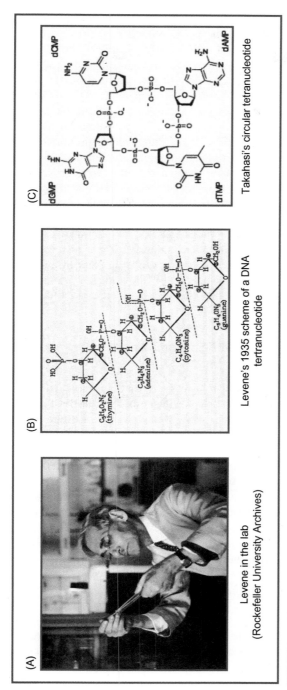

FIGURE 2.6 Phoebus Levene and schemes of tetranucleotide structures of nucleic acids. (A) Levene was in the habit of running experiments with his own hands.[53] (B) Levene's late scheme of a DNA tetranucleotide. Note the three phosphodiester links between the four nucleotides and the primary phosphoryl group. (C) Takahashi's later model of a circular tetranucleotide. There is no terminal primary phosphoryl group and all four phosphoryl groups are engaged in internucleotide phosphodiester linkages. *Panel (B) from Levene PA, Tipson RS. The ring structure of thymidine. J Biol Chem 1935;109(2):623–630.*

recently created Rockefeller Institute, recruited him in 1905 to the scientific staff of the new Institute that was to become his lifetime professional home. Leading a large research group, Levene was a prolific author of scientific papers—publishing about 700 papers that were mostly original reports of experimental results (a full bibliography of his contributions from 1894 to 1941 can be found in Ref. 52). Levene was regarded as one of the leading chemists of his time. It is ironic that although he did not have a chemistry degree, he was awarded in 1931 the Willard Gibbs Medal of the American Chemical Society with the citation: "[the] outstanding American worker in the application of organic chemistry to biological problems."

Levene was a true Renaissance man; a polyglot, lover and scholar of European art, bibliophile, and warm and friendly person. He was a gifted teacher and a dedicated researcher, not just supervising and advising numerous students, associates, and colleagues but also doing bench work with his own hands.[52] Levene's investigations ranged over a wide array of biochemical subjects. He contributed studies on enzymes, proteins and amino acids, glycoproteins and nucleoproteins, and on the stereochemistry of hexosamines. Arguably, however, the part of his work that had the greatest impact was the deciphering of the chemistry of nucleotides and of their linkage in nucleic acids.

2.4.1   Levene Identified the Base, Pentose, and Phosphorous Components of Nucleotides in RNA and in DNA

After establishing the empirical formula of yeast nucleic acid (RNA) as $C_{38}H_{50}O_{29}N_{15}P_4$, Levene and his student Walter Abraham Jacobs (1883–1967) determined that its carbohydrate moiety was the pentose sugar ribose.[55,56] Twenty years later Levene and London identified 2-deoxyribose as the carbohydrate in thymus DNA.[57] In 1919 Levene reported that hydrolysis of yeast RNA with ammonia yielded AMP, CMP, GMP, and UMP but no TMP (these compounds were termed in the paper guanosinphosphoric acid, adenosinphosphoric acid, etc.).[58] From these results Levene defined correctly the structure of the basic component of RNA,[58] and later of DNA,[59] as a phosphate−sugar−base unit. Levene and Jacobs also coined the term nucleoside to denote the purine−carbohydrate compounds and the term nucleotide to define the phosphate ester of a nucleoside.

2.4.2   The Mode of Linkage Between Nucleotides in RNA and DNA Was Debated

In a series of papers that were published over almost two decades, Levene addressed the question of the nature of the linkage between individual nucleotides in RNA and DNA. Essentially, his final conclusion was that RNA or DNA were, respectively, ~ 1300 Da units of four linked ribonucleotides or deoxyribonucleotides—two purine and two pyrimidines all in equimolar proportions.[60] This erroneous hypothesis, that Levene named the tetranucleotide theory, is fully discussed in the next section. Based on different pieces of experimental data,[58,59,61–63] some of which misinterpreted, Levene suggested that individual nucleotides were linked by three phosphodiester bonds with one primary phosphoryl group forming a phosphomonoester at the end of the tetranucleotide (Fig. 2.6B). Later on, when Levene and Tipson identified the ring structure of deoxyribose and ribose, they correctly identified the phosphodiester linkage to be $3'-5'$ in DNA but they mistakenly concluded that it was a $2'-3'$ diester bond in RNA.[54] Interestingly, prior reports by Gerhard Schmidt (1901–81), a student of Gustav Embden (1874–1933) who escaped Germany to later work with Levene in New York, indicated that ribonucleotides in RNA were linked by $3'-5'$ phosphodiester bonds and not by $2'-3'$ bonds as was proposed by Levene and Tipson.[64,65]

The question of the structure of the hypothesized tetranucleotide remained unsettled for some time. Whereas Levene claimed that a linear structure included three phosphodiester bonds and a single monoester primary phosphoryl group (Fig. 2.6B), newer published evidence in the 1930s argued for a circular structure in which all the phosphoryl groups were doubly linked (Fig. 2.6C). Takahashi in Japan followed the liberation of phosphoric acid from yeast RNA that he attacked with each one of the enzymes; phosphomonoesterases, phosphodiesterase, or pyrophosphatases. His results indicated that the RNA contained neither a phosphomonoester nor pyrophosphoric ester groups and that all the phosphoryl groups were present as diesters of two nucleotides.[66] Using combinations of different enzymes, Gulland and Jackson later reached the same conclusion.[67] While these authors remained within the convention of the tetranucleotide structure of nucleic acids, they interpreted their results as indication that rather than being a linear molecule with one monoesterified phosphoryl group, the tetranucleotide was a circular structure that contained four phosphodiester bonds and no phosphomonoester (Fig. 2.6C).

2.4.3   Levene's Tetranucleotide Hypothesis of the Structure of Nucleic Acid Persisted for Decades Despite Accumulated Contradicting Evidence

Levene was not the first to propose that nucleic acids are literal tetranucleotides. He was preceded by the Americans Thomas Osborne (1859–1929) and Isaac Harris[68,69] and by Hermann Steudel (1871–1967)[70] in Germany who concluded on the basis of their coarse measurements that nucleic acids were composed of equal molar amounts of the different bases. In fact, although Levene is thought now to be the instigator and proponent of the tetranucleotide hypothesis of the structure of DNA and RNA, his initial belief was that nucleic were pentanucleotides.[71,72] Only later did he arrive at the conclusion that they are "literal tetranucleotides" composed of equimolar amounts of the four bases (for a nuanced review of Levene's changing views on the constitution of nucleic acids, see Ref. 53). The tetranucleotide hypothesis maintained that nucleic acids were tetrads of the four bases—two purines and two pyrimidines in equimolar proportions. Thus, nucleic acid molecules were thought to consist of equal amounts of guanine, cytosine, adenine, and thymine or uracil in DNA or RNA, respectively. Historically, the proposed monotonous tetranucleotide structure invalidated for many years the possibility that nucleic acids could serve as the genetic material. By default it strengthened, therefore, the misleading notion that the genetic material was protein. As a matter of historical fact, the tetranucleotide model of the structure of nucleic acids remained the accepted convention until 1950 when it was experimentally disproved by Erwin Chargaff (Ref. 73, see Chapter 5).

The most striking peculiarity about the relatively long survival of the tetranucleotide hypothesis is that evidence was accumulating between 1909 and 1944 that contradicted this model. One conflicting body of results showed that in many cases the ratio of bases in DNA and RNA deviated significantly from the expected 1:1:1:1 ratios of the four nucleotides in a tetranucleotide (Fig. 2.7A).

How, then, did the tetranucleotide model endure in the face of these conflicting data? As pointed out by Olby[74] and by Hunter,[53] the task of hydrolyzing nucleic acids was technically unwieldy and the determination of the molar amounts of the liberated nucleotides was grossly inaccurate. It is tempting to think that having been aware of these difficulties investigators in the first half of the 20th century were indulgent about the highly diverse base ratios that were obtained in different laboratories. In any case it is a historic fact that the biochemical community had not felt that the apparent lack of nucleotide equimolarity in DNA or RNA was a good enough reason to abandon the tetranucleotide model.

A second line of evidence that was in conflict with the tetranucleotide model was the growing awareness that DNA was a macromolecule much larger than a tetranucleotide. Interestingly, although Miescher thought that nuclein was a nondialyzable, multibasic, phosphorous-containing acid of high molecular weight, this conception was abandoned at later times and was replaced by the tetranucleotide model of nucleic acids.[75] Yet, with time the development of less drastic methods for the isolation of DNA and of more accurate measurements of its size,[76] led to a growing awareness that DNA had much higher molecular size than that of a tetranucleotide. Data assembled by Olby show that the measured molecular weight of DNA increased dramatically from 1909 through the 1930s and onward (Fig. 2.7B). Thus, for instance, in 1938 the University of Berne chemist Rudolph Signer (1903−90) used flow birefringence to measure the molecular weight of DNA that was prepared by the Karolinska Institute investigators Torbjörn Caspersson (1910−97) and Einar Hammarsten (1889−1968). The molecular

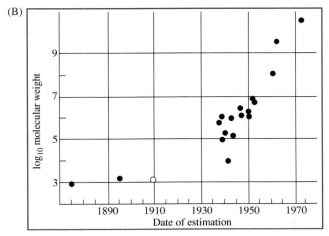

(A)

TABLE 1 Base analyses of nucleic acids from 1902 to 1950						
	Molar ratios					Nucleic
Data	A	G	C	T/U	Authors	Acid
1902	1.0(?)	0.98	—	1.17	Osborne and Harris	RNA
1905	0.47	0.23	1.35	1.0	Levene	DNA
1906	1.30	0.98	0.63	1.09	Steudel	DNA
1908	0.98	0.98	1.02	1.02	Levene and Mandel	
1909	1.0	1.3	1.7	?	Levene	RNA
1947	0.8–0.9	1.0	1.1–1.2	1.0	Gulland	DNA
1948	0.8	2.0	1.0	0.25	Chargaff	RNA
1949	1.6	1.3	1.0	1.5	Chargaff	DNA
1950	0.29	0.18	0.18	0.31	Chargaff	DNA

(B)

FIGURE 2.7   (A) Ratios between the four nucleic acid bases as determined between 1902 and 1950. (B) The experimentally measured molecular size of DNA increased over time, exceeding by orders of magnitude the expected ~1300 Da size of a tetranucleotide. *From Olby R. DNA before Watson–Crick. Nature 1974;248(5451):782–785.*

size that he determined for that DNA was $0.5-1.0 \times 10^6$ Da.[77] William Astbury (1898–1961) at the University of Leeds and his graduate student Florence Bell employed in the same year X-ray fiber diffraction to find a similar molecular size for DNA.[78] Most interestingly, Levene himself, together with Schmidt, followed in 1938 the migration of native DNA in the ultracentrifuge to conclude that it behaved as molecule with a size of $0.2-1.0 \times 10^6$ Da.[79] Therefore, still during Levene's period of active research it became evident to him and to others that the molecular weight of DNA exceeded considerably the predicted ~1300 Da size of a tetranucleotide. How then had the tetranucleotide hypothesis persisted in the face of such discrepant results?

At the time, colloidal chemistry attempted to explain the presence of giant molecules such as starch, cellulose, proteins, and nucleic acids. Essentially, the consensus opinion was that large chemical entities were aggregates of smaller units. This idea was based in part on the 1902 chemical bonds theory of the Swiss chemist Alfred Werner (1866–1919). This theory distinguished *Hauptvalenzen* from *Nebenvalenzen* (respectively, primary (main) and secondary (next) valences). Not considering valence as unit force, Werner allowed for a secondary force that he conceived to be residual attraction after a primary valence was filled. Under this theoretical framework the high molecular weight form of DNA was thought to consist of multiple units of primary-bonded tetranucleotide that were held together by weaker secondary bonds. Misinterpretation of experimental results that were gathered in the 1930s gave credence to this view of the structure of DNA. Thus, in 1935 the German chemist Robert Feulgen (1884–1955) digested high molecular size thymus DNA, which he named thymonucleic acid, with crude pancreatic DNase that he termed nucleogelase. He reported that the enzyme converted the parent gelatinous DNA that he dubbed "a-DNA" and which was in hindsight large-size native DNA, into a non-gel, acid-precipitable form that he termed "b-DNA" and that in retrospect consisted of shorter fragments of DNA.[80] Complete hydrolysis of both the a- and b-forms of the DNA liberated from both the same proportions of the same products: guanine, adenine, cytosine, and thymine. Feulgen interpreted the results of the partial digestion of the a-DNA as indications that the enzyme severed the weaker links between tetranucleotide units that were the product b-DNA.[80] Schmidt and Levene replicated Feulgen's results and showed by ultracentrifugation that pancreatic DNase, (which they named depolymerase), converted native DNA (Feulgen's "a-DNA") of $0.2–1.0 \times 10^6$ Da into a DNA form that did not sediment and which they assumed to be Feulgen's b-DNA. Although they could not determine size of the nonsedimentable digestion products, Schmidt and Levene presumed that they represented the core tetranucleotide.[79]

The prolonged survival of the tetranucleotide hypothesis and its resistance to mounting contradictory evidence stands as an instructive general lesson on the power of preconceived ideas and the difficulty to replace a deep-rooted paradigm. Examined in the context of the history of genetics and biochemistry, and eventually of molecular biology, the tetranucleotide concept was responsible to an extent for the deferred realization that nucleic acids are the carriers of hereditary information. Being a "dull" molecule of just the four nucleotides or a polymer thereof, the tetranucleotide could not be imagined as the genetic material. Instead, the highly variable proteins in chromatin were deemed to be the expected bearers of genetic information. This view changed, albeit not instantaneously, only after Avery, MacLeod, and

McCarty discovered in 1944 that a purified DNA fraction could alter some genetic properties of bacterial cells.[1] This revolutionary development is the subject matter of Chapter 3, "Avery, MacLeod, and McCarty Identified DNA as the Genetic Material."

REFERENCES

1. Avery OT, Macleod CM, McCarty M. Studies on the chemical nature of the substance inducing transformation of pneumococcal types: induction of transformation by a deoxyribonucleic acid fraction isolated from pneumococcus type III. *J Exp Med* 1944;**79**(2):137–58.
2. Ford BJ. Browninan movement in clarkia pollen: a reprise of the first observations. *Microscope* 1992;**40**(4):235–41.
3. Oliver FW. *Makers of British botany*. Cambridge, MA: Cambridge University Press; 1913.
4. Harris H. *The birth of the cell*. New Haven, CT: Yale University Press; 2000.
5. Campbell DH. The centenary of Wilhelm Hofmeister. *Science* 1925;**62**(1597):127–8.
6. Kaplan DR, Cooke TJ. The genius of Wilhelm Hofmeister: the origin of causal-analytical research in plant development. *Am J Bot* 1996;**83**(12):1647–60.
7. Darlington CD. *The facts of life*. London: Allen & Unwin; 1953.
8. Flemming W. *Zellsubstanz, Kern, und Zelltheilung*. Leipzig: P. C. W. Vogel; 1882.
9. Haeckel E. *Generelle Morphologie der Organismen*. Berlin: G. Reimer; 1866.
10. Boveri T. Zellenstudien II. Die Befruchtung und Teilung des Eies von Ascaris megalocephala. *Jena Zeit Naturwiss* 1888;**22**:685–882.
11. Boveri T. Über mehrpolige mitosen als mittel zur analyse des zellkerns. *Verh Phys Med Ges* 1902;**35**:67–90.
12. Boveri T. Zellenstudien VI: Die Entwicklung dispermer Seeigelier. Ein Beitrag zur Befruchtungslehre und zur Theorie des Kernes. *Jena Zeit Naturwiss* 1907;**43**:1–292.
13. Sutton WS. The chromosomes in heredity. *Biol Bull* 1903;**4**:231–51.
14. Kohler RE. *From medical chemistry to biochemistry: the making of a biomedical discipline*. Cambridge, MA: Cambridge University Press; 1982.
15. Fruton JS. *Contrasts in scientific style: research groups in the chemical and biochemical sciences*. Philadelphia, PA: *Memoirs Am Philosoph Soc*; 1990.
16. Dahm R. Discovering DNA: Friedrich Miescher and the early years of nucleic acid research. *Hum Genet* 2008;**122**(6):565–81.
17. Olby R. Cell chemistry in Miescher's day. *Med Hist* 1969;**13**(04):377–82.
18. Wolf G. Friedrich Miescher, the man who discovered DNA. *Chem Herit* 2003; **21**(10-11):37–41.
19. Dahm R. Friedrich Miescher and the discovery of DNA. *Dev Biol* 2005;**278**(2):274–88.
20. Dahm R. From discovering to understanding. Friedrich Miescher's attempts to uncover the function of DNA. *EMBO Rep* 2010;**11**(3):153–60.
21. Miescher F. Ueber die chemische Zusammensetzung der Eiterzellen. *Med Chem Untersuchungen* 1871;**4**:441–60.
22. Plósz P. Ueber das chemische verhalten der kerne der vogel- und schlangenblutkörperchen. *Med Chem Untersuchungen* 1871;**4**:461–2.
23. Hoppe-Seyler F. Ueber die chemische zusammensetzung des eiters. *Med Chem Untersuchungen* 1871;**4**:486–501.
24. Miescher F. Die Spermatozoen einiger Wirbeltiere. Ein Beitrag zur Histochemie. *Ver Nat Ges Basel* 1874;**6**:138–208.

25. Wilson EB. *The cell in development and inheritance.* New York, NY: Macmillan; 1896.

26. Piccard J. Ueber protamin, guanin und sarkin als bestandtheile des larchssperma. *Ber Dtsch Chem Gesellsch* 1874;**7**:1714–19.

27. Altmann R. Ueber nucleinsäuren. *Arch Anat Physiol Physiol Abt* 1889;524–36.

28. Miescher F. Ueber das Leben des Rheinlachses im Süsswasser. *Arch Anat Physiol Anat Abt* 1881;193–218.

29. Miescher F. *Statistische und biologische Beiträge zur Kenntniss vom Leben des Rheinlachses im Süsswasser. Die Histochemischen und Physiologischen Arbeiten von Friedrich Miescher.* Leipzig: F.C.W. Vogel; 1897. p. 116–91.

30. Miescher F. Bemerkungen zur Lehre von den Athembewegungen. *Arch Anat Physiol Physiol Abt* 1885;355–80.

31. Miescher F. *Bemerkungen zur Physiologie des Höhenklimas. Die Histochemischen und Physiologischen Arbeiten von Friedrich Miescher.* Leipzig: F.C.W. Vogel; 1897. p. 502–28.

32. Scheele VQ. Examen chemicum calculi urinari. *Opuscula* 1776;**2**:73.

33. Unger B. Das guanin und seine verbindungen. *Justus Liebigs Ann Chem* 1846;**59**:58–68.

34. Jones ME. Albrecht Kossel, a biographical sketch. *Yale J Biol Med* 1953;**26**(1):80–97.

35. Kossel A. Ueber guanin. *Hoppe-Seyler's Z Physiol Chem* 1883-1884;**8**:404–10.

36. Kossel A. Ueber das adenin. *Ber Dtsch Chem Ges* 1885;**18**:1928–30.

37. Kossel A. Ueber das adenin. *Ber Dtsch Chem Ges* 1887;**20**:3356–8.

38. Kossel A. Ueber das adenin. *Hoppe-Seyler's Z Physiol Chem* 1888;**12**:241–53.

39. Kossel A, Neumann A. Darstellung und spaltungsproducte der nucleinsäure (adenylsäure). *Ber Dtsch Chem Ges* 1894;**27**:2215–22.

40. Kossel A. *Untersuchungen über die Nucleine und ihre Spaltungsprodukte.* Strassburg: K.J. Trübner; 1881.

41. Kossel A. *Albrecht Kossel—nobel lecture: the chemical composition of the cell nucleus.* Nobel Media AB 2013. Nobelprize.org. <http://wwwnobelprizeorg/nobel_prizes/medicine/laureates/1910/kossel-lecturehtml>; 1910.

42. Kossel A. *Harvey Lectures, The chemical composition of the cell.* Philadelphia and London: J.B. Lippincott; 1911–2. p. 33–51.

43. Ascoli A. Ueber ein neues spaltungsprodukt des hefenukleïns. *Z Physiol Chem* 1900;**81**:161–4.

44. Kossel A. Ueber das Nuclein der Hefe, I. *Zschr Physiol Chem* 1879;**3**:284–91.

45. Kossel A. Ueber das Nuclein der Hefe, II. *Zschr Physiol Chem* 1880;**4**:290–5.

46. Kossel A. Ueber das thymin, ein spaltungsproduct der nucleinsäure. *Ber Dtsch Chem Ges* 1893;**26**:2753–6.

47. Kossel A. The proteins. *Bull Johns Hopkins Hosp* 1912;**23**:65–76.

48. Kennaway E. Some recollections of Albrecht Kossel, Professor of Physiology in Heidelberg, 1901–1924. *Ann Sci* 1952;**8**:393–7.

49. de la Blanchardière P. Über die wirkung der nuclease. *Z Physiol Chem* 1913;**87**:291–309.

50. Kossel A, Dakin HD. Über salmin und clipein. *Z Physiol Chem* 1904;**41**:407–15.

51. Hargittai I. The tetranucleotide hypothesis: a centennial. *Struct Chem* 2009;**20**(5):753–6.

52. Van Slyke DD, Jacobs WA. Phoebus Aaron Theodor Levene 1869–1943. *Biogr Mem Natl Acad Sci* 1943;**23**:73–126.

53. Hunter GK. Phoebus Levene and the tetranucleotide structure of nucleic acids. *Ambix* 1999;**46**(2):73–103.

54. Levene PA, Tipson RS. The ring structure of thymidine. *J Biol Chem* 1935;**109**(2):623–30.

55. Levene PA, Jacobs WA. Über die pentose in den nucleinsäuren [I]. *Ber Chem Ges* 1909;**42**(2):2102–6.

56. Levene PA, Jacobs WA. Über die pentose in den nucleinsäuren [II]. *Ber Chem Ges* 1909;**42**(3):3247–51.

57. Levene PA, London ES, Chem JB. Guaninedesoxypentoside from thymus nucleic acid. *J Biol Chem* 1929;**81**(3):711–12.

58. Levene PA. The structure of yeast nucleic acid: IV. Ammonia hydrolysis. *J Biol Chem* 1919;**40**(2):415–24.

59. Levene PA, London ES. The structure of thymonucleic acid. *J Biol Chem* 1929;**83**(3):793–802.

60. Levene PA. Über die hefennucleinsäuren. *Biochem Z* 1909;**17**:120–31.

61. Levene PA. On the structure of thymus nucleic acid and on its possible bearing on the structure of plant nucleic acid. *J Biol Chem* 1921;**48**(1):119–25.

62. Levene PA, Simms HS. Nucleic acid structure as determined by electrometric titration data. *J Biol Chem* 1926;**70**(2):327–41.

63. Levene PA, Bass LW. *Nucleic acids*. New York, NY: Chem. Catalog. Co; 1931.

64. Schmidt G. Über fermentative desaminierung im muskel. *Z Physiol Chem* 1928;**179**:243–69.

65. Embden G, Schmidt G. Über muskeladenylsaure und hefeadenylsiure. *Z Physiol Chem* 1929;**181**:130–9.

66. Takahashi H. Über fermentative dephosphorierung der nucleinsäure. *J Biochem (Japan)* 1932;**16**(2):463–82.

67. Gulland JM, Jackson EM. The constitution of yeast nucleic acid. *J Chem Soc* 1938;1492–8.

68. Osborne TB, Harris IF. Die nucleinsäure des weizenembryos. *Z Physiol Chem* 1902;**36**:85–133.

69. Osborne TB, Harris IF. The nucleic acid of the embryo of wheat. *Rept Conn Agr Expt Sta* 1902;**25**:365–430.

70. Steudel H. Die zusammensetzung der nucleinsäuren aus thymus und aus heringsmilch. *Z Physiol Chem* 1906;**49**(4–6):406–9.

71. Levene PA, Mandel JA. Über die darstellung und analyse einiger nucleinsäuren. XIII. Über ein verfahren zur gewinnung der purinbasen. *Biochem Z* 1908;**10**:215–20.

72. Levene PA, Mandel JA. Über die konstitution der thymo-nucleinsäure. *Ber Dtsch Chem Ges* 1908;**41**(2):1905–9.

73. Chargaff E. Chemical specificity of nucleic acids and mechanism of their enzymatic degradation. *Experientia* 1950;**6**(6):201–9.

74. Olby R. DNA before Watson–Crick. *Nature* 1974;**248**(5451):782–5.

75. Gulland JM, Barker GR, Jordan DO. The chemistry of nucleic acids and nucleoproteins. *Ann Rev Biochem* 1945;**14**:175–206.

76. Cohen JS, Portugal FH. The search for the chemical structure of DNA. *Conn Med* 1974;**38**(10):551–2 4–7.

77. Signer R, Caspersson T, Hammarsten E. Molecular shape and size of thymonucleic acid. *Nature* 1938;**141**(3559):122.

78. Astbury WT, Bell FO. X-ray study of thymonucleic acid. *Nature* 1938;**141**(3573):747–8.

79. Schmidt G, Levene PA. The effect of nucleophosphatase on "native" and depolymerized thymonucleic acid. *Science* 1938;**88**(2277):172–3.

80. Feulgen R. Über a- und b-thymonucleinsäure und das die a-form in die b-form überfahrende ferment (nucleogelase). *Z Physiol Chem* 1935;**237**(5–6):261–7.

Chapter 3

Avery, MacLeod, and McCarty Identified DNA as the Genetic Material

A Celebrated Case of a Clinical Observation That Led to a Fundamental Basic Discovery

Chapter Outline

M. Fry: Landmark Experiments in Molecular Biology. DOI: http://dx.doi.org/10.1016/B978-0-12-802074-6.00003-5
41

The Rockefeller Institute bacteriologist Oswald Theodore Avery and his two associates, Colin MacLeod and Maclyn McCarty, published in Feb. 1944 a groundbreaking paper that showed for the first time that DNA acted as carrier of genetic information. Specifically, Avery, MacLeod, and McCarty reported that purified DNA dictated the synthesis of a particular antigenic type of capsular polysaccharide of the pathogen *Streptococcus pneumoniae* (*pneumococcus*).[1] Considered today to be the harbinger of molecular biology, the finding that genes are comprised of DNA defied the common belief at the time that proteins are the genetic material. Oddly, this landmark discovery was somewhat unintended since Avery's initial objective had not been to identify the chemical basis of inheritance. Rather, he was a bacteriologist whose life work was dedicated to the immunochemistry of *pneumococcus* with the ultimate goal of developing an effective vaccine against this pathogen. Yet, this clinically oriented research effort took an unexpected direction that culminated in the unpredicted finding that DNA was responsible for heritable traits of *pneumococcus*. Because of the prominence of Avery's discovery and its unusual evolution, its scientific, historical, philosophical, and personal facets were explored in a sizeable body of studies.[2–15] An excellent resource of relevant original correspondence, manuscripts, images and other documents is to be found online: The Oswald T. Avery Collection—Profiles in Science, National Library of Medicine National Library (http://profiles.nlm.nih.gov/ps/retrieve/Narrative/CC/p-nid/38).

To fully appreciate the groundbreaking discovery of Avery, MacLeod, and McCarty, this chapter opens with the description of works that preceded this key achievement and it then traces the lengthy road that ultimately led to the recognition that DNA is the genetic material.

3.1 PNEUMONIA—"THE CAPTAIN OF THE MEN OF DEATH"

Although physicians recognized pneumonia since antiquity, its clinical basis remained obscure until the respective introduction in the early and second half of 19th century of the recording of body temperature and auscultation. With no existing vaccination to prevent the spread of the disease and no antibiotics to treat patients, the mortality rate was 25%. In fact, bacterial pneumonia surpassed tuberculosis as a leading cause of death in the late 19th and early 20th centuries. This was well-exemplified by William Osler's naming of pneumonia as "The Captain of the Men of Death"; a paraphrase on John Bunyan's 17th-century poetic description of tuberculosis as the "Captain Consumption, with all his men of death."

3.1.1 Epidemiology of Pneumonia

Bacterial pneumonia became the leading cause of death with the onset of the Spanish influenza pandemic (Box 3.1).

BOX 3.1 The Spanish Influenza Pandemic

Lasting from 1917 to 1920, the Spanish influenza pandemic spread throughout the world from the Arctic Circle down to remote Pacific islands. The number of Spanish influenza fatalities in India alone was about 23 million (5% of the total population) and the 675,000 American fatalities exceeded the combined number of Americans killed in combat in World War I, World War II, Korea, and Vietnam (423,000). All in all, the estimated number of fatalities in the world was between 50 and 100 million, making it one of the deadliest natural disasters in human history. At the lower estimate of 50 million dead, 3% of the world's population (1.8 billion at the time) were killed by the disease and some 500 million (28%) were infected (Fig. 3.1).

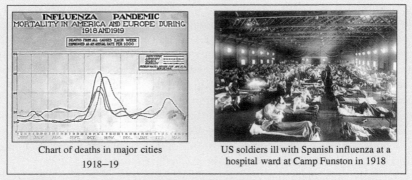

| Chart of deaths in major cities 1918—19 | US soldiers ill with Spanish influenza at a hospital ward at Camp Funston in 1918 |

FIGURE 3.1   Chart of Spanish influenza fatalities in major cities 1918—19 and a 1918 photograph of influenza patients in an American military hospital.

Recent analyses indicated that bacterial pneumonia was the major direct cause of death in the 1918 influenza pandemic. In 2008 the NIH-NIAID researchers Morens, Fauci, and Taubenberger reviewed results of 8398 individual autopsies that were performed between 1918 and 1929 on Spanish influenza victims from 15 countries. These analyses were complemented by reexamination of paraffin block-preserved lung tissue samples from 58 soldiers who died of influenza in 1918 and 1919 at various US military bases.[16] In most cases, changes characteristic of primary viral pneumonia were found to be also associated with severe, acute, secondary bacterial pneumonia. It was concluded that most of the influenza fatalities involved secondary bacterial pneumonia caused by common upper respiratory flora. Thus, while the virus landed a first strike, pneumonia-causing bacteria delivered the deadly blow.[16]

Pneumonia fatalities declined precipitously in the 1940s, first because of the introduction of sulfonamides and then of penicillin. However, when drug-resistant strains of *pneumococcus* emerged, the disease became again a major health problem.[17] A resurgence of fatalities in the last quarter of the 20th century accelerated the development and application of an effective multivalent vaccine targeted against multiple antigenic types of the *pneumococcus* capsule.[18] Yet, pneumococcal infections are still a considerable health problem, being responsible in the United States for 0.5×10^6 annual cases of pneumonia, $0.5-1.25 \times 10^5$ cases of bacteremia, 7×10^6 cases of otitis media, and 3×10^3 cases of meningitis (cited in Ref. 17).

3.1.2 The Causative Agent—*Streptococcus pneumoniae*

The causative agent of the majority of cases of bacterial pneumonia, *Streptococcus pneumoniae* (*pneumococcus*), was independently isolated in the very same year, 1881, by the American military surgeon George Sternberg[19] and by Louis Pasteur (1882—95) in France.[20]

Sternberg (Box 3.2) reported that following the inoculation of rabbits with his own saliva, micrococci proliferated to high densities in their blood

---

BOX 3.2 George Sternberg

Having authored in 1892 the *Manual of Bacteriology*, George Sternberg is considered the first US bacteriologist. After attending a seminary, Sternberg was medically trained; first in Buffalo and later in the College of Physicians and Surgeons of New York. After obtaining an MD degree in 1860, he joined the US army, where he had a lifetime career as a military physician. Sternberg took part in various battles of the Indian and American Civil Wars reaching the station of the Army Surgeon General at the rank of Brigadier General. In his capacity as bacteriologist, Sternberg identified *pneumococcus* in 1881 and was the first in the United States to demonstrate that *Plasmodium* is the cause of malaria (1885) and to confirm the causative roles of bacteria in tuberculosis and typhoid fever (1886) (Fig. 3.2).

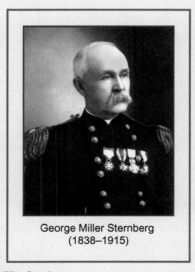

George Miller Sternberg
(1838–1915)

FIGURE 3.2   George Miller Sternberg.

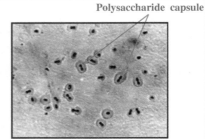

Polysaccharide capsule

| *Streptococcus pneumoniae:* A mucoid strain on blood agar showing alpha hemolysis (green zone surrounding the colonies). | The existence of pneumococcal capsule is demonstrated by antibody-induced capsule-swelling reaction (Quellung reaction). Note that pneumococci usually appear in pairs (diplococci). |

FIGURE 3.3   Blood agar culture of a mucoid strain of *pneumococcus* and visualization of the pneumococcal capsule. Note that the encapsulated bacteria are paired (diplococci).

usually resulting in death of the animals within 48 hours.[19] Pasteur reported that rodents which were inoculated with the saliva of a child killed by rabies developed a new disease (*"une maladie nouvelle"*) that was associated with the appearance of paired cells ("diplococci") that had a characteristic capsule around each cell[20] (Fig. 3.3). Soon thereafter the German bacteriologist Carl Friedländer (1847−87) identified pneumococci in stained exudates of pneumonia patients.[21,22] Although he later classified *Klebsiella pneumoniae* as another causative agent of pneumonia,[23] concurrent survey of 129 cases of pneumonia found *pneumococcus* to be the major cause of bacterial pneumonia. At about the same time, Albert Fränkel (1864−1938) produced the disease in rabbits that were inoculated with pneumococci that he isolated from a fatal case of pneumonia. Using this animal model of human pneumonia, he satisfied Koch's postulates and definitely established *pneumococcus* as an etiological agent of pneumonia.[24,25] Late 20th-century comparative analysis of nucleotide sequences of 16S ribosomal RNA showed *pneumococcus* to be a member of the *Streptococcus* genus.[26] Streptococci are distinguished by their proclivity to form chains when grown in culture (from Greek: *streptos*—warped; *kokkos*—grain). However, *S. pneumoniae* usually appear as pairs of cells (thus, *Diplococcus pneumoniae*).

3.1.3   The *Streptococcus pneumoniae* Capsule

Fred Neufeld (Box 3.3) was the first to report strain-specific agglutination of pneumococci by immune antiserum. He isolated a mucoid (encapsulated) strain of *pneumococcus* from a patient and solubilized the cells in bile. Horse and rabbit antisera against the bacterial lysate agglutinated the original pathogen as well as pneumococci that were isolated from other individuals.[27]

BOX 3.3 Three Pioneers of *Pneumococcus* Research

A notable German bacteriologist and *pneumococcus* researcher, Neufeld started as an assistant to Robert Koch and later became professor and director of the Koch Institute (1917–33). Although a Protestant, he was expelled from the Institute in 1933 by the Nazis. Neufeld perished of starvation in 1945 in the besieged Berlin.

Griffith was a career civil servant in the British Ministry of Health. As a consequence of his interest in the epidemiology of *pneumococcus*, he was the first to discover in 1928 bacterial transformation. An extremely modest person, Griffith did not win in his lifetime the recognition that he deserved. He was killed in 1941 in an air raid during the blitz on London.

Except for a short period in the Hoagland Laboratory in Brooklyn, Avery spent his entire career at the Rockefeller Institute in New York. He was the leading American bacteriologist/immunologist student of *pneumococcus*. His interest in the antigenicity of the *pneumococcus* capsule led unexpectedly to the identification of DNA as the genetic material (Fig. 3.4).

| Fred Neufeld (1869–1945) | Frederick Griffith (1879–1941) | Oswald Avery (1877–1955) |

FIGURE 3.4   Three leaders of *pneumococcus* research.

In subsequent work Neufeld found that other isolates of pathogenic pneumococci were of different antigenicity. Thus each bacterial strain was agglutinated only by homologous antiserum without cross-immunity.[28] In this work Neufeld identified the first two serotypes of *pneumococcus* that he labeled types I and II. Valuably, he also showed that exposure of the bacteria to homologous antiserum led to swelling and to altered refractive properties of the capsule. This so-called Quellung reaction ("*Quellung*" is German for swelling) afforded clear visualization of the capsule (Fig. 3.3). Subsequent work in the laboratory of Alphonse R. Dochez (1882–1964) at the Rockefeller Institute Hospital expanded the catalogue of antigenic types of *pneumococcus* to types I, II, and III and to an additional group IV that consisted of a mixture of several serotypes.[29] Continued research identified many more strains of the bacterium such that more than 90 distinct serotypes

of *pneumococcus* that differ in their antigenicity, virulence, prevalence, and degrees of drug resistance are known today.

Chemical and immunological analyses of constituents of the pneumococcal capsule were accomplished in the laboratory of Oswald Theodore Avery (Box 3.4) at the Rockefeller Institute Hospital in New York. As early as 1917 Dochez and Avery reported that blood and urine of infected rabbits contained a soluble component of *pneumococcus* that was agglutinated by antiserum raised against the intact infecting bacterium.[30] Avery's later work with Renè Dubos (1901−82) showed immunologically that the soluble material was derived from the bacterial capsule.[31] Importantly, they also demonstrated that removal of the capsule abolished the ability of strain III bacteria to cause systemic infection, thus proving that the capsule was essential for pneumococcal virulence.[32]

Realizing the importance of the capsule, Avery set out to determine its chemical composition. To this end, he collaborated with the chemist, and later eminent immunologist, Michael Heidelberger (1888−1991) who was then working in the Rockefeller Institute laboratory of the great immunologist and 1930 Nobel laureate Karl Landsteiner (1868−1943). Famously, Avery approached Heidelberger holding a vial of dark gray suspension of *pneumococcus* cells saying: "the whole secret of bacterial specificity is in this little vial. When are you going to work on it?" Responding to the challenge, Heidelberger first precipitated the soluble substance of *pneumococcus* Type II with specific antiserum and after removing the protein residue from the precipitate he showed that the remaining antigenic substance was nitrogen-free polysaccharide.[33-35] He then demonstrated that the repeating sequences of monosaccharides in capsular polysaccharides differed in different types of *pneumococcus*, indicating that the chemical composition of the capsule polysaccharide was the source of antigenic diversity (and virulence) among different strains of the bacteria.

3.2 FROM A CLINICAL OBSERVATION TO THE DISCOVERY OF BACTERIAL TRANSFORMATION: THE WORK OF FREDERICK GRIFFITH

3.2.1 Griffith's Initial Clinical Observation

In his service as an officer of the British Ministry of Health, Frederick Griffith (Box 3.3) was engaged in epidemiological studies of pneumonia and in the identification of serological types of *pneumococcus* that were responsible for the disease. Remarkably, these studies led to a series of elegant experiments that constituted the first demonstration of transformation in bacteria. Griffith had noticed already in 1922 that in a number of instances sputum samples from pneumonia patients contained pneumococci of several serological types.[36] Later he serotyped pneumococci in samples of sputum that were collected from a single pneumonia patient every 2 or 3 days for the duration of the disease.[37] At the early stages (days 4 and 6 after the onset of pneumonia), he isolated only Type I pneumococci from the patient. However, at later times (days 15 and 17) the pneumococci that were detected in the patient's sputum were of Types IV and Pn. 160 (a subtype of IV) (Fig. 3.6A). Similar successive appearances of multiple antigenic types of pneumococci in the course of pneumonia were recorded in five additional patients.[37]

3.2.2 Demonstration of the Conversion of *Pneumococcus* Serotype in Mice

To lend additional support to the observed changing serotypes, Griffith serotyped bacteria after their passage through mice. Type I pneumococci that he isolated at days 4 or 6 from a patient (Fig. 3.6A) were either injected directly into mice or inoculated only after they were first mixed with anti-Type I antiserum that neutralized any Type I bacteria. In both cases the mice died within 24 hours and bacteria were recovered from their peritoneal and serotyped.

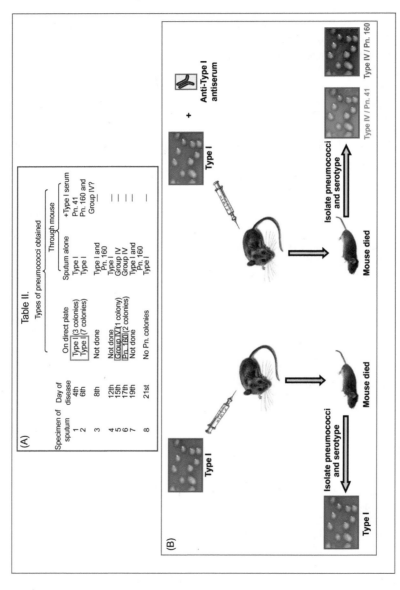

FIGURE 3.6 (A) Different serotypes of *pneumococcus* were detected by direct plating of sputum samples that were collected from a single patient during the course of pneumonia (Table II from Griffith F. The significance of pneumococcal types. *J Hyg (Lond)* 1928;**27**(2):113–59. The different serotypes that were detected in the patient at different times are boxed. Also listed are the types of *pneumococcus* that were recovered from mice which were inoculated with Type I bacteria without or with anti-Type I antiserum as illustrated in panel (B). (B) Scheme of the procedure and results of inoculation of mice with Type I pneumococci without or with anti-Type I antiserum (see text).

As listed in Fig. 3.6A and illustrated in Fig. 3.6B, mice that were inoculated with Type I pneumococci alone yielded Type I bacteria, whereas mice that were injected with Type I cells that were pretreated by anti-Type I antiserum yielded identified subtypes: Type IV; Pn. 41 and Pn. 160, and additional unidentified subtypes of group IV.[37] It appeared, therefore, that selective elimination of Type I pneumococci by specific antiserum exposed coexistent bacteria of other types. Griffith raised two alternative explanations for the observed presence of several *pneumococcus* variants in a single patient. One possibility was that different types of pneumococcus independently infected and cohabited in patients. Alternatively, the patient could have hosted only a single initial variant that was transformed in time into other subtypes.

Considering the low probability of infection with multiple strains of bacteria Griffith wrote[37]:

> On a balance of probabilities interchangeability of type seems a no more unlikely hypothesis than multiple infections with four or five different and unalterable serological varieties of pneumococci.

3.2.3 Isolation of Capsule-Free Rough Strains of *Pneumococcus*

The Lister Institute bacteriologist Joseph Arthur Arkwright (1864−1944) isolated in 1921 virulent and attenuated strains of several bacterial species. Colonies of the virulent strains were regular, domed, and smooth, whereas the attenuated bacteria formed colonies that were irregular, flat, and granular.[38] Arkwright named the virulent strains Smooth ("S" in short) and the avirulent bacteria Rough ("R" in short). Shortly after this initial description, additional bacterial species, including *pneumococcus*, were also found to form S and R strains.

To decide experimentally between the competing hypotheses of infection of a single patient by multiple strains or transformation of types in the course of the disease, Griffith put to use the morphologically distinct R and S strains of *pneumococcus*. Having reported already in 1923 that S strain pneumococci could revert in culture or in inoculated mice into R bacteria,[39] Griffith was familiar with the isolation and identification of virulent S and attenuated Type R *pneumococcus*. However, to reliably use R bacteria, he had to prepare attenuated bacterial strains that after losing their capsule, they formed stable nonvirulent R colonies that did not revert back to S-form cells. Starting with encapsulated S strain of defined serotype, cells were grown in culture either on chocolate blood agar medium or in the presence of selective specific antiserum. During early generations, cultures were mixtures of both S and R bacteria but after several passages in the selective environment, the cultures consisted of pure R strains that were unable to cause pneumonia and death in mice.[37] In one example (Fig. 3.7), a stable R strain was obtained as follows: Virulent Type II cells were

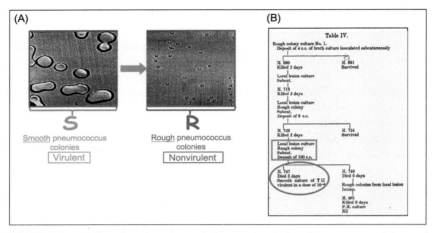

FIGURE 3.7   (A) Transformation of smooth into rough colonies of *pneumococcus* (rearranged photograph from Avery OT, Macleod CM, McCarty M. Studies on the chemical nature of the substance inducing transformation of pneumococcal types: induction of transformation by a desoxyribonucleic acid fraction isolated from pneumococcus type III. *J Exp* Med 1944;**79**(2):137−58. (B) A Type II rough strain of *pneumococcus* maintained its morphology and remained nonvirulent when deposits of bacteria were serially passaged in mice (Table IV from Griffith F. The significance of pneumococcal types. *J Hyg (Lond)* 1928;**27**(2):113−59.

mixed with anti-Type II antiserum and were repeatedly passaged for six generations in the presence of the immune serum.

Rough colonies that were obtained (Fig. 3.7A) were found to be nonvirulent upon inoculation in mice. Bacteria that were isolated from killed asymptomatic mice maintained their rough morphology and remained non virulent
when further inoculated in mice (Fig. 3.7B). Smooth virulent bacteria emerged
only in mice that were injected with massive, >12-fold larger, inoculum of
bacteria (*red box* in Fig. 3.7B). Thus, starting with Type II encapsulated and
virulent strain of bacteria, Griffith established a stable line of rough bacteria
that maintained their nonvirulence throughout serial passages in mice.[37]

3.2.4   Griffith's Decisive Experiment: The Discovery of Bacterial Transformation

Griffith noticed that while some serotypes of stable R variants of *pneumococcus* did not revert in the mouse into encapsulated S forms, revertants
were detected when the R cells were coinoculated with heat-killed S bacteria. Expanding on these preliminary observations, he defined experimental
conditions that maximized the capacity of heat-killed S cells to affect the
transformation of R cells into virulent encapsulated S pneumococci. In these
preparatory experiments Griffith identified combination of serotypes of the
heat-killed S cells and living R bacteria that yielded significant R to S

transformation in the mouse. Thus, for instance, R cells that were derived from Type SII cells were successfully transformed when they were coinjected with heat-killed SI bacteria. Griffith also defined optimal temperature, usually 60°C that effectively killed the S bacteria while allowing for significant transformation. He also found that transformation occurred only when the inoculum size of the heat-killed S cells was much larger than that of the living R cells. Importantly, using heat-killed S cells and alive R cells of two different serotypes, Griffith noticed that under proper conditions the R to S transformed cells acquired the serotype of the heat-killed S cells rather than maintaining the serotype of the original R cells.[37] Next, Griffith conducted fully controlled

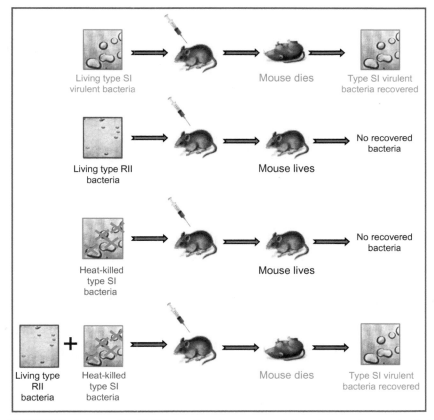

FIGURE 3.8   Scheme of the Griffith transformation experiment. In control experiments mice that were infected with encapsulated virulent Type S pneumococci (serotype I) died and SI bacteria were recovered from the dead animals. By contrast, injected heat-killed SI bacteria did not cause death and no bacteria were detected in sacrificed animals. Similarly, capsule-less R bacteria (serotype II) did not cause death and no bacteria were detected in sacrificed mice. By contrast, mice that were co-inoculated with mixture of RII pneumococci and of large excess of heat-killed SI cells did die and the bacteria that were recovered from the dead animals were serotype I of the heat-killed S cells.

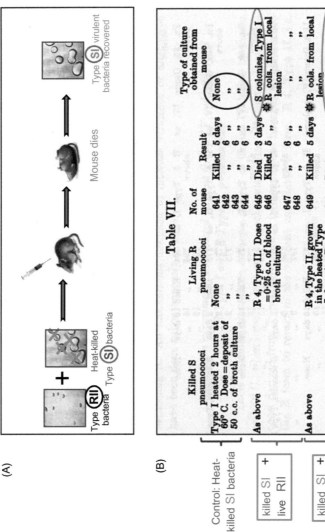

FIGURE 3.9 Results of typical transformation experiments. (A) Scheme of successful transformation by mixture of living RII pneumococci and heat-killed SI bacteria. (B) Summary of the actual results (Table VII from Griffith F. The significance of pneumococcal types. *J Hyg (Lond)* 1928;**27**(2):113–59. Control mice that were injected with only heat-killed SI bacteria and remained unaffected are encircled in red. Successfully transformed colonies are encircled in black. *Red asterisks* mark untransformed R cells that were recovered from sacrificed unaffected mice in cases of ineffective transformation.

experiments that are schematically depicted in Fig. 3.8. Multiple experiments (see Fig. 3.9B for representative original results) showed that neither R cells nor heat-killed S pneumococci by themselves could produce disease and death in mice. By contrast, injected R cells that were mixed with a large excess of heat-killed S bacteria caused death in the mice. The serotype of bacteria that were recovered from the dead animals was that of the heat-killed S cells and not the different original serotype of the R cells. Typical features of the transformation of *pneumococcus* in Griffith's hands (and in the hands of researchers who later replicated his results) are shown in Fig. 3.9B.

First, whereas heat-killed S cells alone did not produce disease and death in mice, injection of their mixture with living R cells resulted in some, but not all, cases in death of the animals.

Importantly, the newly encapsulated transformed bacteria that were recovered from the hearts of infected dead mice were of serotype I of the heat-killed S cells and not of Type II from which the R cells were originally derived. Notably, Griffith also executed the reverse experiment, that is, mice that were inoculated with a mixture of heat-killed SI cells and living RII pneumococci died and produced living SI cells.

Two features of the results shown in Fig. 3.9B are noteworthy. First, transformation did not always occur. Rather, whereas it was observed in some cases (mice 645 and 650 in Fig. 3.9B), in other instances the mice remained asymptomatic and the sacrificed animals either had no detectable bacteria (mouse 651) or they bore some untransformed R pneumococci (mice 646−648 and 652). Second, to affect even this inconsistent transformation, the heat-killed S cells had to be added at an excess of about 200-fold over living R bacteria.

Although the source for the incomplete efficiency of transformation and for the need for a large excess of heat-killed S cells was unclear at the time, these features are easily understood in hindsight. As will be discussed later, transformation of pneumococci by DNA was inefficient, especially under the less than optimal conditions that were employed by Griffith and his immediate successors. It was impossible, therefore, to obtain uniformly successful transformation even when heat-killed S cells were added at a great excess over R cells. Significantly, Griffith also tried to achieve transformation in vitro by mixing in the test tube heat-killed S cells with living R bacteria. However, neither R to S conversion nor transformation of serotype was attained. This failure to achieve transformation in vitro was most likely due to experimental conditions that were suboptimal for efficient transformation and to degradation of the transforming material by enzymes that were released from autolyzed R cells.[40]

3.2.5   Griffith's Interpretation of Rough to Smooth Transformation

Considering the state of biology at the time and taking into account that Griffith's was not a geneticist, it is not surprising that he did not raise the possibility that transformation reflected transmission of genetic material from the heat-killed S cells to recipient R cells. How then did he view the generation of R cells and their transformation into S-type cells of different antigenicity? First, he explained that R cells emerged from S bacteria as a way to escape agglutination and phagocytosis under the selective pressure of anti-S cell antiserum. Yet, in his mind the evolved R cells did retain residual S antigenicity[37]:

> By assuming the R form of the pneumococcus has admitted defeat, but has made such efforts as are possible to retain the potentiality to develop afresh into virulent organism. The immune substances do not apparently continue to act on the pneumococcus after it has reached the R stage, and it is thus able to preserve remnants of its important S antigens and with them the capacity to revert to the virulent form.

Griffith next proposed that the excess of heat-killed S cells provided the recipient R strain with "pabulum" (ie, food or nutrient), which enables restoration of S phenotype of the same antigenicity as that of the heat-killed S pneumococci[37]:

> When the R form of either type is furnished under suitable experimental conditions with a mass of the S form of the other type, it appears to use that antigen as a pabulum from which to build up similar antigen and thus to develop into an S strain of that type.

Because R cells could be transformed by S bacteria that were heated to 60°C but transformation was abolished when the S pneumococci were killed at 100°C, Griffith assumed that the material that the S cells supplied was protein[37]:

> By S substance I mean that specific protein structure of the virulent pneumococcus which enables it to manufacture a specific soluble carbohydrate. This protein seems to be necessary as material which enables the R form to build up the specific protein structure of the S form. But it appears that this material may be modified by heat in such a way that the R form cannot utilise it for the reconstruction of its own internal structure.

Griffith also addressed the question of the acquisition of the serotype of the heat-killed S bacteria by the transformed R cells instead of preservation of the antigenicity of their strain of origin[37]:

> In order to reconcile the experimental data referred to above with the hypothesis that the R pneumococcus which reverts in the mouse to the S form has synthesised its S antigen from similar material in the heated virulent culture injected at the same time, it is necessary to assume that a virulent Type I pneumococcus contains some S antigen of Type II. An alternative hypothesis would be that the R form of Type II is able to reconstruct its virulent S form from either the S substance of Type I or that of Type II. One is then faced with the difficulty of accounting for the failure of an R form of Type I to build up its S form from Type II S substance.

As will be seen hence, Griffith was not alone in his attempt to explain transformation as transmission of material that was necessary for the synthesis of the capsular polysaccharides. Variations on this line of thought persisted in the *pneumococcus* research community practically until the transforming material was identified as DNA.

Despite its technical imperfections and incorrect interpretation, Griffith's experiment stands out in the annals of molecular biology as an exceptionally elegant one. Most importantly, it served as a springboard for a series of increasingly focused experiments that culminated in the identification of DNA as the genetic material just 16 years after the 1928 publication of Griffith's landmark paper.

3.2.6   Epilogue: Griffith's Personality, Scientific Style, and Work Environment

Although Griffith's work is recognized today as an essential prelude to the discovery of DNA as the genetic material; his career-long interest was the epidemiology of *pneumococcus* rather than genetics. That his principal concern was the implications of the transformation phenomenon to the epidemiology of bacterial pneumonia is well-reflected by the concluding comment of his transformation article[37]:

> *The results of the experiments on enhancement of virulence and on transformation of type are discussed and their significance in regard to questions of epidemiology is indicated.*

Indeed, after the publication of his transformation results, Griffith resumed his work on the epidemiology of pneumonia and never returned to the study of transformation.

Griffith was a dedicated and extremely modest scientist in the service of the public. He tragically perished in 1941 alongside his friend and director, William McDonald Scott (1884–1941), when a bomb hit their laboratory during the German blitz on London. Griffith's contemporaries remembered this life-long bachelor as a refined and reticent gentleman (for lively descriptions of this "most English of Englishmen," see Ref. 6, p. 132 and Ref. 41). Griffith's personality and commitment to work were succinctly described after his demise in a 1941 *Lancet* obituary signed under the acronym H.D.W.[42]:

> *[...] he was a recluse, known to few. To these, however, his quiet, kindly manner and his devotion to his life job, made him a lovable personality. Outside his work he found his pleasure in his winter skiing holiday in the Alps, in walks with his dog on the Sussex downs, and in the cottage he had built there. [...] For over thirty years of his working life, writes L. C., Fred Griffith followed a single star. He believed that progress in the epidemiology of infectious diseases would come—and only come—with more precise knowledge about the microorganisms responsible for those diseases.*

Griffith's scientific achievement is magnified by the subsequent portrayal of his and Scott's humble working environment and severe dearth of physical means[42]:

> *A foreigner, visiting for the first time the laboratory of the Ministry of Health in London, must have been little short of appalled to see how meanly this fundamental activity of a great department of a wealthy country was housed, and must have wondered how, thus cabined and confined, the two world-famed workers managed to exist, let alone function, in such a chaotic environment. But he would not be long there before he knew that they did so because they were Scott and Griffith who, like C. J. Martin, could do more with a kerosene tin and primus stove than most men could do with a palace. [...] one found there a compendium of knowledge, a wealth of experience and, above all, a willingness to help, at whatever cost of time and trouble, the like of which could rarely be found anywhere, here or abroad.*

3.3   TRANSFORMATION WAS PROMPTLY CONFIRMED

Although European and American members of the *pneumococcus* research community held Griffith in high esteem, many bacteriologists found it difficult to accept the idea that one type of bacteria can convert to another. Griffith raised already in 1922 the idea that the transformation of R to S pneumococci reflected change of mutable properties *within* the species.[36] However, many experts of *pneumococcus* regarded the conversion of serotype as contradiction of the commonly held doctrine of bacterial monomorphism.[6] Fortunately, despite this dogmatic opposition, doubts about Griffith's report were dispersed when three different laboratories promptly confirmed his findings on pneumococcal transformation. Fred Neufeld learned of the transformation results ahead of their formal publication during his visit to Griffith's London laboratory. Returning to Berlin, he initiated with Walter Levinthal (1886–1963) a series of experiments that quickly confirmed Griffith's report. Due to this early head start, Neufeld and Levinthal were able to announce their assenting results just several months after the publication of Griffith's pioneering paper.[43] About a year later, Hobart Ansteth Reimann (1897–1986) from the Peking Union Medical College in China published a similarly affirmative report.[44]

Experiments that were conducted in the Rockefeller Institute Hospital laboratory of Oswald Avery, who was then the leading American expert on *pneumococcus*, also validated Griffith's findings. Renè Dubos, then a fellow in Avery's laboratory and later his colleague, proclaimed that although Avery was well aware of the phenomenon of spontaneous reversion of R to S cells of the same serotype, he had difficulty in accepting that pneumococci could alter their antigenicity. This position was rooted in Avery's adherence to a dogma of fixated immunological type of bacteria.[6] His explanation of

Griffith's findings was, therefore, that despite the evidence that all the S-form pneumococci were completely killed at 60°C, a small number of live bacteria still persisted in the mass of heat-killed S cells. If so, propagation in the mouse of these surviving bacteria produced what appeared to be R to S transformed cells. However, a young Canadian fellow in Avery's laboratory, Martin Henry Dawson (1896−1945) who was working on R to S reversion since 1926 (Ref. 6, p. 137) immediately believed Griffith's published findings.[6] According a 1972 letter that Dubos wrote to the historian Robert Olby (Ref. 45, p. 179), Dawson took advantage of a 6-month long medical leave of the skeptical Avery to independently repeat, verify, and expand on Griffith's experimental results. Similar to Griffith's, Dawson's motivation to study transformation was mainly epidemiological[46]:

The conversion of relatively avirulent pneumococci into highly virulent organisms is obviously a matter of considerable biological and epidemiological significance.

The essential finding of his work was that transformation occurred when the R pneumococci were grown in a medium that contained homologous anti-*pneumococcus* antiserum. In summarizing his Jan. 1930 confirmatory report Dawson declared[46]:

The majority if not all of avirulent R cells have the ability, under the proper conditions, to revert to virulent, type-specific capsulated organisms.

Dawson's success in replicating Griffith's results and his follow-up work (*vide infra*) was the point at which the Avery laboratory took over the *pneumococcus* transformation system and advanced it step-by-step to a point at which it served to identify DNA as the genetic material.

3.4 ESTABLISHMENT OF AN IN VITRO TRANSFORMATION SYSTEM OFFERED A FIRST GLIMPSE OF THE TRANSFORMING PRINCIPLE

3.4.1 Dawson and Sia Transformed Pneumococci In Vitro

After he verified Griffith's transformation results, Dawson next collaborated with Richard Ho Ping Sia (1895−1970) to demonstrate that transformation could be accomplished in vitro. First, Dawson and Sia tested the idea that in order for transformation to occur in vitro, the heat-killed S-type cells should be preincubated in living tissue prior to their addition to recipient R cells. To examine this possibility, heat-killed SIII pneumococci were inoculated for 2−6 hours in mice and bacteria were collected from the peritoneal fluid of the killed animals. A large excess of the recovered heat-killed cells was then mixed with RII pneumococci and anti RII antiserum and the mixtures were seeded on solid growth medium (Fig. 3.10).

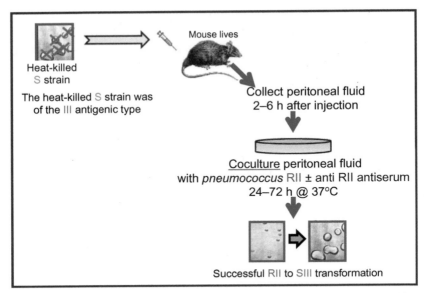

FIGURE 3.10 Scheme of in vitro transformation experiment that included passage of heat-killed SIII cells through mice prior to their seeding in culture together with RII cells.[47]

Results showed that when a suitable excess of heat-killed SIII cells was cocultured with the RII bacteria, the Type II rough cells were transformed to produce colonies of smooth Type III bacteria.[47]

In the second step of their study, Dawson and Sia proved that transformation in vitro could also be achieved without the need to first pass the heat-killed SIII cells through mice. In this experiment (Fig. 3.11A), aliquots of a small inoculum of RII pneumococci were mixed with different excessive amounts of heat-killed SIII bacteria.

Anti RII antiserum was added to the cell mixtures that were then seeded on solid growth medium and incubated at 37°C for 48 hours. Control cultures of heat-killed SIII cells and anti RII antiserum without living RII bacteria did not produce viable bacterial colonies (Fig. 3.11B, lower section of the Table). By contrast, cocultures of live RII cells and heat-killed SIII cells in mixture with anti RII antiserum, generated colonies of transformed SIII pneumococci. Markedly, the yield of transformed colonies was directly proportional to the excess of heat-killed SIII over live RII bacteria such that the number of transformed colonies diminished with decreasing ratio of heat-killed S to R cells (Fig. 3.11B, upper section of the Table). Last, Dawson and Sia found that whereas SIII cells that were heated at 80°C sustained their transforming activity, heating the bacteria at 100°C abolished their transforming capacity. In retrospect the requirement for high excess of killed donor S cells over recipient R bacteria is explainable by the low efficiency of *pneumococcus* transformation under the conditions that Dawson and Sia used. Also it is likely that autolysis of the

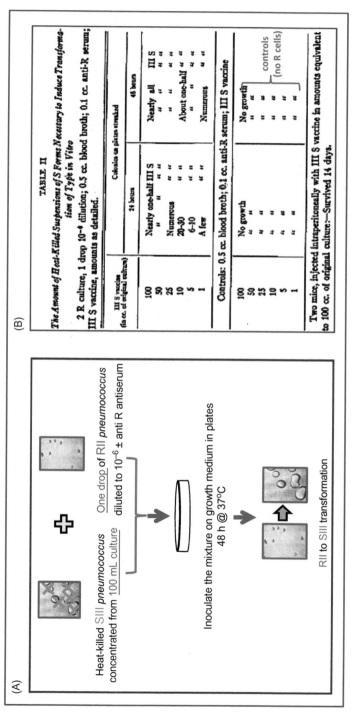

FIGURE 3.11 (A) Scheme of in vitro transformation experiment without passage of heat-killed SIII cells through the mouse. (B) Representative results of the experiment. Suspensions of heat-killed SIII cells are listed as "IIIS vaccine" and relative numbers of transformed cells were scored after 24 and 48 hours of incubation (Table from Dawson MH, Sia RH. In vitro transformation of pneumococcal types: I. A technique for inducing transformation of pneumococcal types in vitro. *J Exp Med* 1931;**54**(5):681−99.

R cells released DNase that destroyed the transforming principle and that DNA digestion was minimized by the use of low inoculum of R cells. Last, loss of transforming activity in cells that were heated at 100°C was retrospectively understood to be the result of DNA denaturation.

3.4.2 Lionel Alloway Isolated a Soluble Cell–Free Transforming Fraction

After Dawson left the Rockefeller Institute for the College of Physicians and Surgeons of New York, the work on transformation was taken over by Thomas J. Francis (1900—69) who joined the Avery laboratory in 1928. His attempts to build on Dawson's work were unsuccessful[4] and it next fell on another fellow; James Lionel Alloway (1900—54) to move the work forward. In two remarkable research articles that he published in 1932 and 1933, Alloway made a considerable leap forward by preparing cell-free extracts of S bacteria that effectively transformed R pneumococci. Citing Dawson and Sia, Alloway outlined the objective of his first paper[48]:

> In the studies just described, transformation of pneumococci of one specific type into another type was accomplished only by adding to living R cultures the whole heat-killed bodies of S pneumococci. In attempting to analyze the nature of this phenomenon it seemed desirable to determine whether the active principle responsible for the transformation could be extracted in soluble form from the S cells.

To attain this goal, Alloway developed a cell extraction procedure as outlined in Fig. 3.12A. Following the disruption of S-form pneumococci by freezing and thawing, the resulting bacterial extract was heated to 60°C and passed through a porous porcelain Berkefeld filter that separated a soluble fraction from any remaining intact cells and cell debris (Fig. 3.12A).

The concentrated clear soluble fraction was then shown to be capable of transforming Type II R cells into Type III encapsulated S bacteria.[48] Results such as those tabulated in Fig. 3.12B indicated that the cell-free extracts converted some of the RII pneumococci into SIII cells but that transformation was not uniformly successful and some of the R cells retained their rough phenotype. This experiment provided the first demonstration that intact S cells were not essential for transformation and that a soluble cell-free fraction from S bacteria could transform R cells. It was presumed that the cell-free fraction contained a transforming substance that was initially called "transforming factor"[49] and was renamed soon thereafter "transforming principle."

In their respective memoirs,[6,40] Renè Dubos, who was a member of the Avery laboratory since 1927, and Maclyn McCarty, who together with Avery and MacLeod identified the transforming principle as DNA, wrote that prior to Alloway's successful experiment Dawson attempted without success to obtain a soluble transforming fraction from S cells. They both indicated that Avery was closely associated with the work of both Dawson

(A)

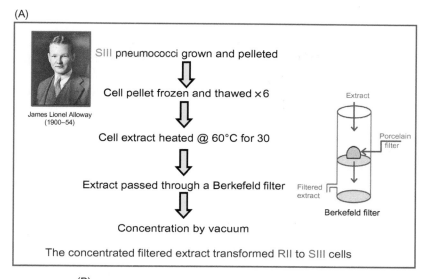

SIII pneumococci grown and pelleted

⇩

Cell pellet frozen and thawed ×6

⇩

Cell extract heated @ 60°C for 30

⇩

Extract passed through a Berkefeld filter

⇩

Concentration by vacuum

The concentrated filtered extract transformed RII to SIII cells

James Lionel Alloway
(1900–54)

Extract

Porcelain
filter

Filtered
extract

Berkefeld filter

(B)

TABLE I

Conversion of Strain of R Pneumococcus Derived from Type II S into Type III S by Means of a Filtered Extract of Type III S Pneumococci

Tube	Amount of broth	Strain of R pneumococcus	Filtered extract of Type III S pneumococci	Anti-R serum (normal hog)	Type of colonies*	Specific agglutinability of S colonies
	cc.		cc.	cc.		
1	1.5	D 39 R**	1.0	0.3	R and S	Type III
2	1.5	D 39 R	1.0	0.3	R and S	Type III
3	1.5	D 39 R	1.0	0.3	R and S	Type III
4	1.5	D 39 R	1.0	0.3	R and S	Type III
5	1.5	D 39 R	—	0.3	R only	—
6	1.5	—	1.0	0.3	Sterile	—

FIGURE 3.12   (A) Scheme of Alloway's procedure for the preparation of cell-free transforming material. (B) Quadruplicate results of transformation by filtered extract of SIII pneumococci. Note that some cells were transformed, whereas other retained the R phenotype. However, those cells that were transformed switched from the source Type II of the R cells to Type III of the extracted S cells. Controls devoid of either filtered S cell extracts or of recipient R cells, respectively, produced untransformed R colonies or remained sterile (Table I from Alloway JL. The transformation in vitro of R pneumococci into S forms of different specific types by the use of fltered pneumococcus extracts. *J Exp Med* 1932;**55**(1):91−9.

and Alloway and that his good advice was critical to Alloway's success. As with Dawson before him, Alloway's initial attempts to prepare a transforming soluble fraction have failed. Saving the day, Avery advised Alloway to raise the pH of the cell extract in order to minimize adsorption of material to the porcelain filter. As a result, the yield of transforming principle increased and in vitro transformation took place.[40] The reader must have noticed, however, that Avery was not listed as a coauthor of either the Dawson or

Alloway papers. This was due to his remarkable practice to be named author only of papers to which he personally contributed experimental work.

Following his successful demonstration of the capacity of a cell-free soluble fraction from heat-killed S pneumococci to transform R into S bacteria of different antigenicity, Alloway proceeded to partially purify the transforming principle from S cell extracts.[50] His purification procedure (Fig. 3.13) involved complete lysis of SIII pneumococci by the detergent sodium deoxycholate. The lysate was next heated at 60°C to kill any remaining intact bacteria and alcohol was added to precipitate "a thick stringy" fraction. The precipitate was centrifuged, further washed with alcohol, resuspended in 0.85% NaCl solution, and reheated at 60°C.

The absence of remaining live S cells in the lysates was ascertained by the absence of bacterial growth when the solubilized alcohol precipitates were seeded in culture or inoculated in mice. Thus, emergence of S cells following exposure of R cells to the alcohol-precipitated lysates could only be due to transformation and not to the presence of living S cells in the lysates. Indeed, Alloway found that recipient RII cells were transformed at high efficiency into SIII pneumococci when they were cocultured with the alcohol-precipitated cell lysates[50]:

Extracts made from S pneumococci by the action of sodium desoxycholate proved much more effective in inducing transformations in type than did the extracts which were formerly prepared from frozen and thawed organisms. They were effective in high dilution, and even failed to induce a change if present in too great concentration. Conversion of R pneumococci derived from Type II S organisms into Type III pneumococci could be accomplished in almost all instances by means of an extract of Type III organisms.

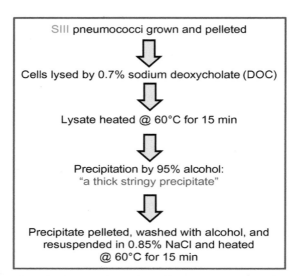

FIGURE 3.13   Alloway's procedure for partial purification of the transforming principle.[50]

A similar effective transforming fraction was obtained when the S cell lysates were precipitated with acetone instead of alcohol. However, the transforming activity was completely abolished after heating the partially purified fraction at 100°C.[50]

Precipitation of the active transforming material by alcohol or acetone, the "thick stringy" appearance of the precipitate and inactivation of the transformation principle at 100°C immediately raises the image of DNA in today's reader mind. However, for numerous reasons, this possibility was not raised at all in 1933. Since the focus of attention at that time was the pneumococcal polysaccharide capsule and its antigenicity, it was only natural for Alloway to link the transforming activity directly to the capsular polysaccharides[50]:

> *The exact nature of the active material in these extracts still remains to be determined. That it acts as a specific stimulus to the R cells which have potentially the capacity of elaborating the capsular polysaccharides of any one of the several types of pneumococci seems clear.*

Indeed, the transforming principle was perceived in the early 1930s to be directly involved in the synthesis of capsular polysaccharides. The initial conversion of S into R cells was rightly interpreted as an outcome of selection by environmental pressure such as growth in chocolate agar medium or negative selection by antiserum against the original capsular polysaccharide of the R cells (Fig. 3.14, upper panel). It was also assumed that although the

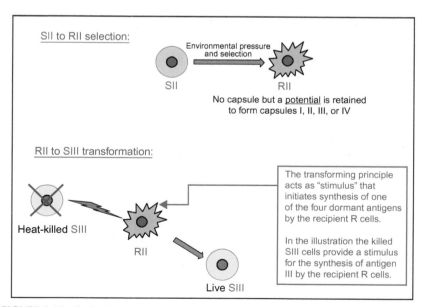

FIGURE 3.14 Early ideas on the mechanism of conversion of S to R pneumococci and on the possible mode of action of the S to R transforming principle. Notably, it was thought that R cell retained a potential to synthesize any of the capsular antigens. Upon transformation the heat-killed S cells furnished the R cells with a *stimulus* to make an antigen whose type was that of the S cells.

R cells lost their capsule, they did retain a *potential* to reform capsule of any of the known antigenic types. Further, it was thought that material ("transforming principle") that was released from the heat-killed S pneumococci enabled R cells to realize their potential to synthesize specific capsular polysaccharide antigen (Fig. 3.14, lower panel).

What ideas were raised about the chemical nature of this transforming material? The Rockefeller Institute chemist and biologist Rollin Hotchkiss (Box 3.6) who personally witnessed the evolution of ideas and experimental work in the Avery laboratory contended that whereas Dawson and Sia speculated that the transforming material was the capsular material itself, Alloway tended to think that it was a protein—polysaccharide complex[6] or protein.[2] Independent experiments by several members of the Avery laboratory, Rogers in 1933, MacLeod in 1935, and McCarty in 1941, showed that the transforming activity was unaffected by digestion of the SIII capsular polysaccharide by a polysaccharidase that was named then "Dubos enzyme SIII." Thus the possibility that the transforming principle was a polysaccharide became less likely.[40] Indeed, Hotchkiss wrote that already in 1936, 8 years before the transforming material was formally demonstrated to be DNA, Avery opined that its properties did not match carbohydrate or protein and that it might be nucleic acid.[2]

3.5 THE LATENT PERIOD: NO PUBLISHED RESULTS BETWEEN ALLOWAY'S 1933 REPORT AND THE 1944 ANNOUNCEMENT THAT THE TRANSFORMING PRINCIPLE WAS DNA

Hotchkiss described in a 1972 letter to Robert Olby the thinking at the Avery lab in the 1930s on the chemical nature of Alloway's "thick stringy alcohol precipitate." This material was thought to be a mucous linear polymer which did not contain DNA and that was named then "renosin."[45] Nevertheless, while the possibility that DNA was the transforming material had not been broached yet, once Alloway's procedure was published, the Avery laboratory appeared to be poised for rapid purification of the transforming principle and for the discovery of its chemical nature. Yet, despite this seemingly advanced starting point, it took Avery, MacLeod, and McCarty additional 11 years to develop the purification procedure and the in vitro transformation assay to a degree that allowed the identification of the transforming principle as DNA. What were the reasons for this long hiatus? As described later, the slow progress was due to personal and organizational circumstances as well as to technical difficulties inherent to the *pneumococcus* transformation system.

3.5.1 Avery's Personal Difficulties

Health problems curtailed Avery's ability to actively participate in the early stages of work on transformation by cell-free extracts. Having developed symptoms of Graves' disease in late 1931, Avery became mentally depressed and was physically handicapped by hyperthyroidism tremor that limited his ability to conduct bacteriological work.[40] After being operated on, probably in 1934, he went through a protracted period of slow convalescence. It so happened therefore, that Alloway and later Edward Rogers and Colin MacLeod investigated aspects of transformation for some years without a significant contribution by their mentor Avery.

3.5.2 Identification of the Chemical Nature of the Transforming Principle Was Not a Priority for Avery

Throughout his career, Avery's primary interest has been the clinical problem of pneumonia. This question continued to be the major trajectory of research in his laboratory during the 1930s and in the early 1940s. He thus regarded the question of transformation to be part of a larger effort to understand the pathogenesis of *pneumococcus*. In his 1935 Report of the Director of the Hospital to the Corporation of the Rockefeller Institute for Medical Research (cited henceforth as: "Report of the Director"), Avery stressed the relevance of the transformation work to the development of modalities of suppression of the pathogenicity of *pneumococcus* (Ref. 40, p. 92). Hotchkiss recalled that even later, in 1938, when he sought to work on transformation, Avery who was more interested in studying blood proteins in acute infection, asked Hotchkiss to wait, saying "we will get to this later."[2]

Avery's dedication to the task of eradicating pneumonia led to abandonment of the study of the transformation between the late 1930s and 1940. During that period, the laboratory focused on the examination of the possibility of replacing immunotherapy of pneumonia by chemotherapy with the then recently introduced sulfonamides.[40] This neglect of the transformation question is evinced by complete absence by its mention in the 1938 and 1940 Reports to the Director by the Avery laboratory. Interestingly, in putting aside for years their work on transformation, MacLeod and Avery appeared not to have been troubled by the prospect of competition by other laboratories. Indeed, although the advances that Dawson and later Alloway have made were placed in the public domain in the early 1930s, no other laboratory took on the challenge of attempting to define the chemistry of the transforming material. One reason for this lack of incentive might have been the considerable conceptual and technical difficulties that were inherent to pneumococcal transformation (*vide infra*).

3.5.3   The In Vitro Transformation System Needed Improvements

Side-by-side with the personal and organizational causes for the slow progress toward the identification of the transforming principle, this undertaking was also hindered by technical difficulties inherent to the transformation system. Alloway's procedure of partial purification and assay of the transforming principle was fraught with problems and was unreliable. Although it was not disclosed in his published papers,[48,50] oftentimes preparations of the soluble fraction were devoid of transforming activity.[6,40] Thus, in order to identify the chemical nature of the transforming material, the assay system had to be made more dependable. Indeed, although it remained unpublished except for mentions in some internal Reports to the Director, much work was invested since 1934 to improve the transformation assay. After Edward Rogers, a new fellow in the laboratory had failed to increase the reproducibility of the transformation assay system,[40] some progress began to be made when yet another fellow, Colin MacLeod (Box 3.5), entered the group in Aug. 1934. Joined at the bench in the fall of 1935 by Avery himself (Ref. 45, p. 186), MacLeod approached the problem methodically. He and Avery first used Dawson's in vitro transformation assay system to test several R strains for their extent of spontaneous reversion and transformability by S cell extracts. These efforts led to the identification of a specific RII cell strain, marked "strain R36," that did not revert to SII bacteria and that was efficiently transformed by extracts of heat-killed S pneumococci into S cells of the corresponding antigenicity.[40] Laboratory notes and Reports to the Director reveal that this strain of RII cells and extracts of SIII bacteria were established as standards. The reliability of the transformation assay was improved by the introduction of this and other modifications such as the stabilization of the transforming material by serum factors that were later replaced by pure albumin. However, despite these advances, the assay was still not entirely dependable and was not ready yet to reveal the nature of the transforming material.

3.5.4   1938–40: Little or No Work on Transformation

Although MacLeod drafted in mid-1936 a paper on his interim advances in making the in vitro transformation assay more reproducible, he never submitted the manuscript for publication.[40] Further, MacLeod who regarded at that point the study of transformation as unproductive, put it aside to pursue more fruitful lines of research on sulfonamides[51] and on the acute infection induced appearance of C-reactive protein.[52,53]

3.5.5   1940–41: MacLeod Returned to Transformation

Despite the slow and nonlinear progression of the search for the nature of the transforming principle and the absence of published reports on this

BOX 3.5 Colin MacLeod and Maclyn McCarty

Born in Nova Scotia, Canada, Colin MacLeod was a gifted student who skipped three grades to enter McGill University at the age of 15 years and to complete his medical studies by age 23. In 1934 he joined Avery's group at which he laid the groundwork for the identification of DNA as the transforming principle of *pneumococcus.* He was nominated in 1941 to be the chairman of the Department of Microbiology at New York University. During and after World War II, MacLeod served as consultant on microbial diseases to the US military. In the 1960s he served as scientific advisor to both Presidents Kennedy and Johnson.

Born in Indiana, United States, McCarty graduated from Stanford University at which he had already conducted some research in biochemistry. During his medical studies at Johns Hopkins University, McCarty became interested in infectious diseases and after doing research work at New York University he joined Avery's laboratory in 1941. There he performed with Avery the critical experiments that identified DNA to be the transforming principle. McCarty went on to spend his entire career at the Rockefeller Institute where he established himself as a world authority on *streptococci*, their antigens and enzymes. In his later years, he assumed various responsibilities as statesman of the biomedical sciences (Fig. 3.15).

Colin MacLeod
(1909–72)

Maclyn McCarty
(1911–2005)

FIGURE 3.15 Colin MacLeod and Maclyn McCarty.

problem until 1944, crucial concrete advances marked the period between 1933 and 1941. Laboratory records that McCarty reviewed[40] indicated that MacLeod and Avery resumed in late Oct. of 1940 their active pursuit of the question of the nature of the transforming material. By being able to handle only relatively small batches cells (3–5 L of liquid culture), MacLeod was constrained in the 1930s by the available limited amounts of SIII cell lysates. Experiments in the late 1940 were greatly enhanced when he began using a steam-driven centrifuge to collect bacteria from much larger volumes ($\sim$40 L) of liquid cultures. The large batches of cells that he amassed were then divided into smaller lots that were subjected in parallel to different procedures of purification of the transforming material. This and additional technical modifications improved the reliability of the system although transformation was still not uniformly successful for every SIII cell lysate.

MacLeod and Avery concentrated the transforming activity by precipitating the SIII lysates with calcium chloride or in most cases, by alcohol. Large amounts of RNA were detected in the precipitate by the RNA-specific Bial reaction. However, Avery and MacLeod found in early 1941 that the transforming activity was likely to be associated with DNA and not RNA. Hydrolysis of 75% of the RNA in the precipitates by a crystallized form of pancreatic ribonuclease (RNase) did not diminish the transformation activity. Also, RNase-treated or untreated extracts developed equally positive colorimetric reaction with diphenylamine that specifically detects deoxyribose. In addition to RNA, the alcohol precipitates also contained Type III polysaccharides. This capsule-derived component was largely removed in latter stage of the work (see below).

Parallel to the described piecemeal advances that Avery and MacLeod have made, they also confronted the problem of the inconsistent in vitro transformation by SIII cell lysates. This difficulty was overcome to some extent when they abandoned the Alloway procedure of first lysing the cells and then heating the lysates. Rather, they found that extracts displayed higher transforming activity when the bacterial pellets were first heated at 65°C for 30 minutes and only then lysed with high concentration of sodium deoxycholate. In retrospect, it is likely that heat inactivation of deoxyribonuclease before the cells were lysed allowed for greater yield of active transforming principle (DNA).[6,40]

As pointed out by McDermott,[54] MacLeod's efforts rendered an unreliable phenomenon of transformation by a partially purified cell-free fraction into a dependable and more robust assay system. Thus, when MacLeod departed the Rockefeller laboratory in 1941 to become chairman of the Department of Microbiology at New York University, he left in the hands of McCarty and Avery a more mature system that they profitably exploited in the next 2 years to ultimately identify the pneumococcal transforming principle as DNA.

3.6   ENTERS MCCARTY: CRITICAL EXPERIMENTS IDENTIFIED THE TRANSFORMING PRINCIPLE AS DNA

After MacLeod's departure Maclyn McCarty (Box 3.5) who joined the Avery group in Sep. 1941 took over the transformation work. He started his part of the work by preparing and stockpiling large batches of heat-killed

SIII pneumococci that were then lysed by deoxycholate and precipitated by alcohol.

The alcohol precipitates were redissolved in salt solution to remove protein and polysaccharides, precipitated again, pelleted, air-dried, and stored until used. At a later stage of his work, McCarty introduced improvements to this procedure such as reduction of the amounts of contaminating proteins and polysaccharides in the final transforming fraction by winding the stringy alcohol precipitate around a stirring rod (see below). Yet, despite all the improvements, the transforming activity of different batches was unpredictable and inconsistent such that some preparations had little or no transforming activity. On such unhappy occasions, Avery used to repeat his saying "Disappointment is my daily bread."[40]

3.6.1   Accumulating Evidence Indicated That the Transforming Substance Was DNA

In the course of his work, McCarty found that he could reduce the amounts of capsular polysaccharides by growing the SIII cells in media that contained lowered concentration of glucose. Also, washing the cells with salt solution prior to their lysis by deoxycholate further decreased the amount of residual capsular material. This diminution of polysaccharides was found to increase the transforming activity of the alcohol precipitated extracts.[40] Despite this observation, the possibility that the alcohol-precipitated fibrous material was DNA had not been raised by early 1942, and McCarty and Avery were still focused on the alternative possibilities that the transforming material was polysaccharide, protein, or RNA. According to McCarty, the realization that the transforming principle was DNA came about only gradually. Indeed, it was not until this prospect became a serious option that it was put to experimental testing.[40]

McCarty got an early intimation that the alcohol-precipitated stringy transforming substance might be DNA after he communicated with Alfred Ezra Mirsky (1900—74), an expert on the cell nucleus and nuclear proteins, whose Rockefeller Institute laboratory was located two floors above Avery's quarters. Together with the Columbia University cytologist Arthur W. Pollister (1903—93), Mirsky isolated nucleoprotein complexes from mammalian tissues. These complexes that represented today's chromatin were initially called "plasmosin" and later "chromosin."[55,56] Mirsky and Pollister found that the protein and nucleic acid components of chromosin could be dissociated from one another in high-salt solutions and that a stringy material that formed when the solutions were diluted could be wound around a glass rod. By communicating with Mirsky, McCarty was alerted to a similarity between his alcohol-precipitated material and the nucleic acid component of chromosin. In late Mar. 1942 he found that the alcohol precipitated transforming principle could also be wound around a stirring rod. When redissolved, the rod-wound substance was found to retain comparable proportions of the transforming activity and of the DNA in the original solution.[40] With this initial hint that the transforming principle might be DNA, McCarty assessed the molecular size of the transforming substance. Using an early model of the analytical ultracentrifuge that was built in-house at the Rockefeller Institute, he found that the material had very high molecular size. Based on this finding, he then applied ultracentrifugation to collect the transforming material as a concentrated pellet. Gaining increased confidence that the transforming principle might be DNA, Avery and McCarty next conducted experiments to determine which crude enzyme preparations destroyed the transforming activity. Mammalian DNA that was provided by Mirsky was used as control substrate to test which enzymes were able to degrade DNA. As is described henceforth, the results of these experiments were consistent with the idea that the transforming substance was DNA. However, even at that early stage the emergent idea that the transforming principle was DNA was met with skepticism. Although pneumococcal DNA that Mirsky purified by salt extraction also possessed robust transforming activity, he contested the inference that the transforming principle itself was DNA. Based on Phoebus Levene's tetranucleotide model of DNA (see Chapter 2: "Prehistory of Molecular Biology: 1848–1944"), he argued that because of the supposedly monotonous uniformity of DNA it was unlikely that it could serve as the transforming agent.[40]

In a last stretch toward the conclusion of their work, Avery and McCarty determined the chemical composition of the transforming substance. In essence, they aimed to assess the relative amounts of DNA, RNA, and carbohydrates in their preparations of the transforming principle. To this end, McCarty isolated DNA from batches of 200 L each of SIII *pneumococcus* cultures and employed a modified purification procedure that reduced as much as possible the amounts of protein and polysaccharides in the final DNA preparation. Analysis of the purified material revealed that the nitrogen and phosphorous content of samples that possessed transforming activity were close to respective theoretical values calculated for DNA (see later).

3.6.2   Avery Stated in a 1943 Internal Report That the Transforming Material Was DNA

In Apr. 17, 1943 Avery submitted a Report to the Director of the Rockefeller Institute Hospital on the progress of various research projects that were conducted at the time in his laboratory.[57] The first section of this report was titled *Study on the Chemical Nature of the Substance Inducing Transformation of Specific Types of Pneumococcus*. In it Avery described the progress of his work with McCarty and its opening statement clearly expressed his view that transformation reflected genetic change of phenotype[57]:

> Biologists, especially the geneticists, have long attempted by chemical means to induce in higher organisms predictable and specific changes which thereafter could be transmitted in series as hereditary characters. Among microorganisms the most striking and perhaps the only known example of inheritable and specific alterations in cell structure and function that can be experimentally induced and that are reproducible under well defined and adequately controlled conditions is the transformation of specific types of Pneumococcus.

In an interpretative summary of this part of the report Avery wrote:

> If the present studies are confirmed and the biologically active substance isolated in highly purified form as the sodium salt of desoxyribonucleic acid actually proves to be the transforming principle, as the available evidence now suggests, then nucleic acids of this type must be regarded not merely as structurally important but as functionally active in determining the biochemical activities and specific characteristics of pneumococcal cells.

It is obvious that in this platform of an internal report Avery allowed himself to explicitly assert the genetic implications of the chemical identification of the transforming principle. Interestingly, Avery stated his case more cautiously in the formal publication that shortly followed this report.[1] Indeed, explicit interpretation of experimental results was quite out of stride for the extremely cautious Avery who, according to McCarty,[40] had held to the belief that it was enough to present facts and leave their interpretations to others.

3.6.3 Avery Summarized His Findings and Ideas in a Letter to His Brother Roy

Shortly after submitting his Report of the Director, Avery delineated more expansively his ideas about the transforming principle as DNA in a personal letter that he wrote in May 1943. The letter was addressed to his brother Roy who was then a Professor of bacteriology at Vanderbilt University in Nashville, Tennessee (for facsimile of the letter, see http://profiles.nlm.nih. gov/ps/access/CCBDBF.pdf). The circumstances surrounding the writing of this letter call for some clarification. Avery, who reached the age of 65 in Oct. 1942, was scheduled to retire from the Rockefeller Institute on Jun. 30, 1943. His intended retirement plan was to settle in Nashville where he could be close to his brother's family. However, considering the exciting developments in the identification of the chemical constitution of the transforming principle, he decided to delay his retirement and continue his research work for a while. His request to stay on at the Rockefeller was granted by the Directors of the Institute Drs Gasser and Rivers who also provided him with budget and technical assistance to continue his studies. To explain this change of plans to his brother, Avery presented in the letter the background to the transformation work, the recent discovery, and its possible implications. With the letter having been an informal and personal communication, Avery allowed himself greater freedom in presenting the problem that he, MacLeod and McCarty were struggling with their emerging ideas, speculations, and lingering doubts. After detailing Griffith's discovery of transformation and the developments that ensued, he described the problem that he, MacLeod, and McCarty were facing:

For the past 2 years, first with MacLeod and now with Dr McCarty—I have been trying to find out what is the chemical nature of the substance in the bacterial extract which induces this specific change. The crude extract (Type III) is full of capsular polysaccharide, C (somatic) carbohydrate, nucleoproteins, free nucleic acids, of both the yeast and thymus type, lipids and other cell constituents. Try to find in that complex mixture the active principle!! Try to isolate and chemically identify the particular substance that will by itself when brought into contact with the R cell derived from Type II causes it to elaborate Type III capsular polysaccharide, and to acquire all the aristocratic distinctions of the same specific type of cells as that from which the extract was prepared! Some job—full of heartache and heart breaks. But at last perhaps we have it . . ."

After next describing the accumulated enzymatic and chemical evidence that the transforming principle was DNA, Avery, with his characteristic restraint, went on to implicate DNA as the vehicle of inheritance:

If we prove to be right—and of course that a big if—then it means that both the chemical nature of the inducing stimulus is known and the chemical structure of the substance produced is also known—the former being thymus nucleic acid—the latter Type III polysaccharide and both are thereafter reduplicated in the daughter cells—and after innumerable transfers without further addition of the inducing agent, the same action and specific transforming substance can be recovered far in excess of the amount originally used to induce the reaction—Sounds like a virus—may be a gene. But with mechanisms I am not now concerned—one step at a time and the first step is, what is the chemical nature of the transforming principle?—Some one else can work out the rest.

Of course the problem bristles with implications. It touches the biochemistry of thymus type nucleic acids which are known to constitute the major part of chromosomes, but have been thought to be alike regardless of origin and species. It touches genetics, enzyme chemistry, cell metabolism and carbohydrate synthesis—etc.—But lately it takes a lot of well documented evidence to convince anyone that the sodium salt of deoxyribose nucleic acid, protein free, could possibly be endowed with such biologically active and specific properties, and that evidence we are now trying to get. It's lots of fun to blow bubbles—but it's wiser to prick them yourself before someone else tries to.

Contrary to the claim by some that Avery, MacLeod, and McCarty were not fully conscious of the genetic implications of their discovery, the cited excerpts from the letter and from Avery's Report of the Director show unequivocally that they were fully aware of the significance of the finding that DNA was the bearer of genetic information. Rollin Hotchkiss, who personally witnessed the unfolding of the transformation story, also confirmed that Avery was completely mindful of the genetic implications of transformation by DNA.[2]

3.6.4 Purification and Characterization of Potent Transforming Material

While Avery was writing to his brother, McCarty purified from Type SIII pneumococci an especially potent batch of transforming principle. Performance of most of the purification steps in the cold, rather than at room temperature, resulted in significantly higher transforming activity of the purified material. As Avery's exceptionally neat laboratory notebook record shows (Fig. 3.16A), just 0.003 µg of this particular purified material sufficed to induce transformation. With an assumed molecular weight of the DNA of 1 million, Avery and McCarty calculated that their assay contained about 10^9 DNA molecules. This was a reasonable number considering the presumed low efficiency of penetration of the DNA into the recipient cells (Ref. 40, p. 173).

The purified material, whose chemical constitution was consistent with DNA, migrated in both analytic ultracentrifuge and electrophoresis as a single boundary that coincided with the transforming activity. The observed overlap between the seemingly homogeneous material and the measured biological activity suggested that the transforming principle was DNA itself and not an accompanying marginal contaminant.

3.6.5 Active Transforming Principle Could Be Isolated From Transformed Cells

Using the potent transforming principle to convert RII to SIII cells, McCarty showed then that a highly active transforming substance could be purified from the transformed cells. The persistence of the transforming principle through many generations indicated that it became fixed in the bacteria and was inherited (*vide infra*).

3.6.6 Despite It All, Avery Was Undecided Whether or Not the Transforming Material Was Truly DNA

Although their experimental data pointed to DNA as the transforming substance, Avery and McCarty acknowledged the possibility that transformation

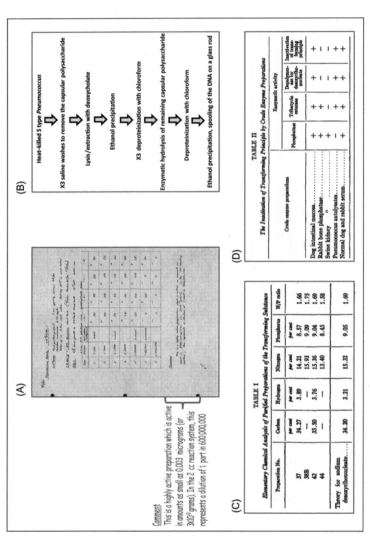

FIGURE 3.16 Avery's original record of measurements of the biological activity of purified transforming principle and major published results.[1] (A) A page from Avery's laboratory notebook dated Jun. 7, 1943, with results of titration of the biological activity of a highly potent preparation of transforming principle (http://profiles.nlm.nih.gov/ps/access/CCAADE.jpg). The handwritten comment at the bottom was magnified on the left for ease of reading. (B) Scheme of the transforming principle purification procedure. (C) Results of elemental analysis of the transforming principle are consistent with DNA (Table I from Avery OT, Macleod CM, McCarty M. Studies on the chemical nature of the substance inducing transformation of pneumococcal types: induction of transformation by a desoxyribonucleic acid fraction isolated from pneumococcus type III. *J Exp Med* 1944;**79**(2):137–58. (D) The transforming principle was exclusively inactivated by deoxyribonuclease-containing crude enzyme preparations (Table II from Avery OT, Macleod CM, McCarty M. Studies on the chemical nature of the substance inducing transformation of pneumococcal types: induction of transformation by a desoxyribonucleic acid fraction isolated from pneumococcus type III. *J Exp Med* 1944;**79**(2):137–58.

could be mediated by minor non-DNA component that was not detected by ultracentrifugation or electrophoresis. Because of this caveat, and although MacLeod and McCarty were eager to write up and publish their findings, Avery was still not ready to go public and kept vacillating between the two possibilities of the transforming principle being DNA or a chemically different minor element. Before committing to one particular conclusion, he consulted with Wendell Stanley (1904−71) and with John Northrop (1891−1987), the co-laureates of the 1946 Nobel Prize in chemistry for, respectively, crystallizing the tobacco mosaic virus and pepsin. These prominent scientists also encountered difficulties in convincing others that the crystalline forms of the virus or pepsin, and not minor contaminants, were responsible for their biological activities. However, except for words of encouragement neither scientist provided practical advice on ways to resolve the problem. Similar consultations with the Rockefeller Institute biochemist Donald Van Slyke (1883−1971) and the protein chemist Max Bergmann (1886−1944) also did not yield concrete guidance. In the end, although Avery did not find a satisfactory way to put to rest his lingering doubts, he nevertheless decided that it was time to write up a paper with the evidence that the transforming principle was DNA.[40] However, as will be described later in this chapter, the alternative possibility that the transforming principle was not DNA but rather minute amount of DNA-associated protein haunted Avery, MacLeod, and McCarty for years after they formally announced their discovery.

3.7   1944: DNA WAS FORMALLY PROCLAIMED TO BE THE TRANSFORMING MATERIAL

In Jul. 1943 Avery and McCarty started to draft a manuscript that documented the identification of the transforming principle as DNA. While Avery undertook to write the introduction and discussion sections during his annual vacation in Maine, McCarty drafted the methods chapter. The task of writing became more serious when Avery returned to New York. In Avery's habitual careful manner, he and McCarty proceeded with the writing very slowly, weighting every phrase. Just before completing the manuscript, McCarty added to it a photograph of rough and smooth colonies of *pneumococcus* (shown in a rearranged form in Fig. 3.7A) that underscored the distinctly different phenotypes of the two types of bacteria.

On Nov. 1, 1943 Avery delivered the manuscript by hand to the editor of the *Journal of Experimental Medicine*, the future (1966) Nobel laureate, Peyton Rous (1879−1970) who was Avery's colleague and friend at the Rockefeller Institute. Assuming the roles of both editor and reviewer, Rous offered multiple, mostly minor, suggestions which Avery and McCarty readily answered. By Dec. 1, 1943, the accepted manuscript was sent to the printer and it appeared in print in the Feb. 1, 1944, issue of the Journal.[1]

3.8   THE AVERY, MacLEOD, AND McCARTY LANDMARK 1944 PAPER

The paper that Avery and McCarty wrote, and that included MacLeod's essential input, was exemplary in its great care for the experimental details, the lucid presentation of the results, and their careful analysis. More than 6 pages of a text of less than 19 pages were dedicated to the experimental methods. This perhaps reflected the authors' recognition of the precariousness of the *pneumococcus* transformation system and their sense of obligation to describe every procedure and step in the most precise manner. Another remarkable aspect of the paper was the low-key style of statement of the work's objectives and the modest to a fault pronouncement of the conclusions. This manner of writing reflected Avery's reserved personality and cautious approach to science.

3.8.1   Stated Objective

Avery, MacLeod, and McCarty opened the paper with a concise description of the discovery of the transformation phenomenon by Griffith,[37] its subsequent confirmation[43,44,46] and the stepwise development of in vitro systems of transformation.[47,48,50] They then stated in a remarkably modest manner the objective of their work:

> *The present paper is concerned with a more detailed analysis of the phenomenon of transformation of specific types of Pneumococcus. The major interest has centered in attempts to isolate the active principle from crude bacterial extracts and identify if possible its chemical nature or at least to characterize it sufficiently to place it in a general group of known chemical substances.*

3.8.2   Methodology

Cognizant of the complexity of the transformation system and its inconsistent performance, the authors were careful to present highly detailed accounts of each of its essential components.

3.8.2.1   Growth Media and Antisera

Transformation occurred only when pleural or ascetic fluid was added to the medium. The requirement for a serum helper factor that was distinct from the need for selective anti R-form antiserum was considered a mystery for long time until Hotchkiss and Harriet Ephrussi-Taylor (1918−68) showed in 1951 that purified albumin could replace the pleural or ascetic fluids.[58] Avery, MacLeod, and McCarty also indicated that batch-to-batch variability of *pneumococcus* growth was eliminated by using charcoal to adsorb sulfonamide inhibitors that were present in the nutrient broth.[59]

Also, the sera and fluids were heated at 60°C to inactivate an enzyme (later recognized as deoxyribonuclease; DNase) that destroyed the transforming activity.

3.8.2.2 Selection of a Stable Recipient R Cell Strain

Criteria for true transformation of nonencapsulated RII to capsulated SIII pneumococci and the choice of a consistently stable R strain were described in detail. First, the criteria for bona fide transformation were laid out. Only conversion of the antigenic type of the transformed cells counted as true transformation, whereas mere acquisition by R cells of capsule of the same serotype signified reversion and not transformation. Avery, MacLeod, and McCarty then stipulated the stable RII strain (R36A) that they used in the study. This strain, which was the product of many repeated passages in selective medium, never reverted spontaneously to Type SII, even under transforming conditions.

3.8.2.3 Purification of the Transforming Substance

The methods section also included a detailed description of the procedure that was employed to purify highly active transforming principle from heat-killed SIII pneumococci (Fig. 3.16A).

As outlined in Fig. 3.16B, steps of mechanical and enzymatic removal of much of the capsular polysaccharides were followed by deproteinization by chloroform of the cell-free extract and ethanol precipitation of the transforming substance. DNA fibers in the final ethanol precipitate were spooled on glass rod and dissolved in saline or water.

3.8.3 Experimental Results

3.8.3.1 Properties of the Purified Transforming Material

Activity of the transforming substance was better preserved in salt-containing solutions than in water. Although the temperature resistance of the material was not systematically studied, it was inactivated at higher temperatures and activity was also lost in solutions at pH lower than 5.0. These observations were consistent with what is well-recognized today as denaturation of DNA by heat or alkali.

3.8.3.2 Protein, RNA, and Lipid Could Not Be Detected in the Transforming Material by the Then Available Methods

Chromogenic tests for DNA (diphenylamine), protein (Biuret and Millon), and RNA (Orcinol) suggested that protein and RNA were absent from the purified transforming material. Resistance of the transforming activity to repeated extractions with alcohol and ether was taken as evidence that it also did not contain lipids. However, it is clear in hindsight that because of the relatively low sensitivity of these assays the possibility that the transforming substance did contain trace amounts of protein, RNA, or lipid could not be definitely excluded.

3.8.3.3   Elemental Content of the Transforming Substance Was Consistent With DNA

More convincing evidence that the transforming material was DNA was gained by analysis of its elemental content. Four different batches of the material were analyzed for their relative amounts of carbon, hydrogen, nitrogen, and phosphorous. Results showed that the relative proportion of each element and the ratios of nitrogen to phosphorous in the four batches of transforming material were in very good agreement with corresponding theoretical values calculated for DNA (Fig. 3.16C).

3.8.3.4   The Transforming Principle Was Selectively Inactivated by DNase

The effect of hydrolyzing enzymes of different specificities on the biological activity of purified transforming material was consistent with its being DNA. Prolonged incubation with crystalline trypsin, chymotrypsin, and ribonuclease did not decrease the biological activity of the pneumococcal transforming substance, suggesting that it was not protein or RNA. At the time, crystalline forms of DNase, phosphatase, or esterase were unavailable, and Avery and McCarty were compelled to use extracts of different tissues and cells as sources for enzyme activities. Standard substrates were used to measure in the different tissue extracts activities of phosphatase, esterase, and DNase (named by the authors "depolymerase for desoxyribonucleate"). DNase activity was detected in dog-intestinal mucosa, in pneumococcal autolysates, and in dog or rabbit sera, whereas it was absent from preparation of bone phosphatase and from swine kidney extract. Significantly, the transforming activity was destroyed only by those extracts that contained DNase, whereas it remained undiminished after incubation with extracts that were devoid of DNase activity. By contrast, there was no correlation between the presence or absence of phosphatase or esterase in extracts and their ability to inactivate the transforming principle (Fig. 3.16D).

To further establish that the inactivation of the transforming principle was indeed due to the action of DNase, experiments were performed with untreated and heat-inactivated DNase. Whereas incubation of the transforming material with untreated dog serum completely abolished RII to SIII transformation, serum that was heated at 60°C or 65°C for 30 minutes and thus lost its DNase activity had no effect on the transforming activity. Rabbit serum differed from dog serum in that its DNase activity resisted heating at 60°C for 30 minutes and was destroyed only by incubation at 65°C for a similar period of time. In this case the transforming principle was inactivated by untreated rabbit serum or by serum that was heated at 60°C. By contrast, serum that was preheated at 65°C did not diminish the transforming activity.[1] In a similar vein, sodium fluoride that had been shown to protect the transforming principle against inactivation by different cell extracts was also shown to inhibit the activity of DNase in those extracts. In sum, Avery, MacLeod, and McCarty felt that the case for DNA as the transforming substance was strengthened by the parallelism between the presence or lack of DNase activity and the corresponding abolishment or maintenance of transforming activity.[1]

3.8.3.5   The Transforming Substance Was Devoid of Detectable Immunologically Reactive Protein

The reactivity of transforming principle from SIII pneumococci with anti SIII antiserum diminished progressively with its increasing purity without corresponding loss of biological activity. In fact, anti SIII antiserum that could detect SIII protein at a dilution of 1:50,000 did not produce immune reaction with the highly purified transforming material.[1] The contrast between the high biological activity of the purified transforming material and its lack of significant reactivity with the anti SIII antiserum was in line with the notion that the transforming material was not protein.

3.8.3.6   The High Molecular Size of the Transforming Substance and Its UV Absorbance Were Characteristic of Nucleic Acid

The transforming material migrated in both the ultracentrifuge and electrophoresis as a single biologically active high molecular weight boundary had a maximum absorbance at 2600 Å and a minimum at 2350 Å. Both the high molecular size of the material and its UV absorbance were consistent with its nucleic acid nature.

3.8.4   Interpretation of the Experimental Results

In stating the crux of their results, Avery, MacLeod, and McCarty stressed that nucleic acids dictated the nature of capsular polysaccharide that are of different chemical category[1]:

> [...] highly purified and protein-free material consisting largely, if not exclusively, of desoxyribonucleic acid is capable of stimulating unencapsulated R variants of Pneumococcus Type II to produce a capsular polysaccharide identical in type specificity with that of the cells from which the inducing substance was isolated. Equally striking is the fact that the substance evoking the reaction and the capsular substance produced in response to it are chemically distinct, each belonging to a wholly different class of chemical compounds.

And:

Thus, it is evident that the inducing substance and the substance produced in turn are chemically distinct and biologically specific in their action and that both are requisite in determining the type specificity of the cell of which they form a part.

Avery, MacLeod, and McCarty were mindful of the question of the chemical basis for the capacity of nucleic acids to dictate biological specificity. Being an immunochemist, Avery considered the idea of applying serological methods to solve this problem but and then rejected it because of the weak antigenicity of nucleic acids. Conceding that the mechanism of action of the transforming principle remained unknown, Avery, MacLeod, and McCarty nevertheless noted that the changes that the transforming material induced in the recipient R cells were inheritable. Also, since an active transforming principle was isolated from the transformed cells, it appeared to have been propagated in daughter cells for many generation[1]:

The biochemical events underlying the phenomenon suggest that the transforming principle interacts with the R cell giving rise to a coordinated series of enzymatic reactions that culminate in the synthesis of the Type III capsular antigen. The experimental findings have clearly demonstrated that the induced alterations are not random changes but are predictable, always corresponding in type specificity to that of the encapsulated cells from which the transforming substance was isolated. Once transformation has occurred, the newly acquired characteristics are thereafter transmitted in series through innumerable transfers in artificial media without any further addition of the transforming agent. Moreover, from the transformed cells themselves, a substance of identical activity can again be recovered in amounts far in excess of that originally added to induce the change. It is evident, therefore, that not only is the capsular material reproduced in successive generations but that the primary factor, which controls the occurrence and specificity of capsular development, is also reduplicated in the daughter cells. The induced changes are not temporary modifications but are permanent alterations which persist provided the cultural conditions are favorable for the maintenance of capsule formation.

In concluding their discussion, Avery, MacLeod, and McCarty were careful to caution that it was still possible that the transforming principle was not DNA but some non-DNA minor component. Yet, if proven in the end to be DNA, then it should be regarded not as structural component of the cell but rather as determinant of functions and attributes of pneumococci:

> *It is, of course, possible that the biological activity of the substance described is not an inherent property of the nucleic acid but is due to minute amounts of some other substance adsorbed to it or so intimately associated with it as to escape detection. If, however, the biologically active substance isolated in highly purified form as the sodium salt of desoxyribonucleic acid actually proves to be the transforming principle, as the available evidence strongly suggests, then nucleic acids of this type must be regarded not merely as structurally important but as functionally active in determining the biochemical activities and specific characteristics of pneumococcal cells.*

Taking into account that the consensual wisdom at the time was that DNA was "filling material," the proposal that it acted to dictate cellular functions and properties was not less than revolutionary. Avery, MacLeod, and McCarty remarkably went on to suggest that if the identification of the transforming principle as DNA would be confirmed, the next step should be the elucidation of the "chemical basis" for its biological action:

> *If the results of the present study on the chemical nature of the transforming principle are confirmed, then nucleic acids must be regarded as possessing biological specificity the chemical basis of which is as yet undetermined.*

This far-sighted recommendation remained unheeded in subsequent years. It was only after the 1953 elucidation of the structure of DNA by Watson and Crick that a quest had started to unravel the genetic code and expound the modes of transmission of genetic information.

3.9 RECEPTION OF THE AVERY, MacLEOD, AND McCARTY PAPER

As noted, Avery, MacLeod, and McCarty were careful to forewarn that their identification of DNA as the transforming material needed confirmation. In addition, their observations did not even hint at a possible mechanism by which DNA mediated transformation. The molecular biologist Gunther Stent (1924−2008) later promoted the idea that the Avery, MacLeod, and McCarty discovery was premature and that it was ignored by the scientific community for a considerable length of time.[60−62] However, examination of publications of the period reveal that Avery's contemporaries were familiar with and had keen interest in his discovery and that they held diverse opinions on its credibility and significance.[63] Yet, it remains a historic fact that no other group attempted to verify the identification of DNA as the material that dictated phenotypic change in pneumococci and more generally, to explore experimentally the contention that DNA is the genetic substance. The following section describes different views that scholars had on the

Avery, MacLeod, and McCarty discovery and assesses possible causes for the surprising lack of recognition of its broader significance.

When evaluating retrospectively the contention that DNA dictated the synthesis and antigenicity of capsular polysaccharides, today's reader should take into account the prevailing notions of the time on inheritance and on nucleic acids. The consensus was that genes were comprised of proteins and that being a monotonous polymer of tetranucleotide units, DNA lacked the required specificity of genetic material. Considering these deep-rooted ideas, it is not surprising that the research community was not in one mind about the Avery, MacLeod, and McCarty discovery. There were those who questioned the claim that the transforming material was DNA suggesting instead that protein that remained tightly associated with the DNA was a more likely carrier of genetic information. Other researchers offered different ideas about possible mechanisms by which DNA alone or a nucleoprotein complex acted indirectly to affect transformation. Some other scientists accepted DNA as the directly acting transforming agent but they regarded *pneumococcus* transformation to be a private case that did not necessarily implicate DNA as the universal carrier of genetic information. Finally, few scholars did comprehend the profound significance of the Avery, MacLeod, and McCarty results and accepted DNA as the material that genes are made of. It took almost a decade until the cumulative impact of findings by Chargaff,[64] Hershey and Chase,[65] and Watson and Crick[66,67] swayed the scientific community to accept DNA as the bearer of genetic information.

3.9.1 Mirsky Questioned the Claim That the Transforming Material Was DNA

Avery's colleague at the Rockefeller Institute Alfred Ezra Mirsky (Box 3.6) dedicated the early part of his scientific career to the study of the folding of native proteins and of their unfolding upon denaturation.[68]

In 1936 he coauthored with Linus Pauling an influential paper proposing that coiling of native proteins is stabilized by hydrogen and disulfide bonds.[69] After isolating in 1938 a nucleoprotein complex of nuclear origin, his interest turned to the structure and function of the cell nucleus and to the protein and DNA constituents of chromatin. Because of his expertise in the purification of DNA, Mirsky became familiar with the progress of the transformation work in the neighboring laboratory of Avery. McCarty used mammalian DNA that Mirsky contributed as reference for the work on pneumococcal DNA (Ref. 40, p. 137). Moreover, Mirsky extracted DNA from pneumococci, using NaCl rather than deoxycholate that was McCarty's reagent of choice. McCarty found that this DNA, which contained a higher proportion of residual protein, also transformed R cells into S-form pneumococci (Ref. 40, p. 146). After this initial collaboration and with the passing of time, the connection between the two laboratories was severed and

BOX 3.6 Alfred Mirsky and Rollin Hotchkiss

Born in New York and trained at Harvard and Cambridge Universities, Alfred Mirsky was for most of his career a member of the Rockefeller Institute. Beginning with his PhD thesis on hemoglobin, Mirsky initial focus was on protein folding and denaturation. In 1938 he embarked on structural and functional studies of the cell nucleus, becoming a leading expert in this area. Determining that the content of DNA in cells was constant, he deduced that DNA was part of the gene substance. However, he insistently opposed the view that DNA alone was the genetic material, and firmly held to the belief that the protein component in nucleoprotein complexes was the genetic material.

After concluding in 1935 his PhD thesis in chemistry at Yale, Rollin Hotchkiss joined the Rockefeller Institute at which he remained until his retirement in 1982. He initially worked in Avery's laboratory and collaborated with Renè Dubos in isolating and characterizing gramicidin that became the first commercially available antibiotic. His chemical analysis of the purity of the *pneumococcus* transforming DNA served to counter Mirsky's argument that it contained significant amount of tightly associated protein. In subsequent work Hotchkiss demonstrated that DNA could transfer penicillin resistance from one bacterial strain to another. Consequently, he became an early proponent of bacterial genetics showing that many attributes of classical genetics were applicable in bacteria (Fig. 3.17).

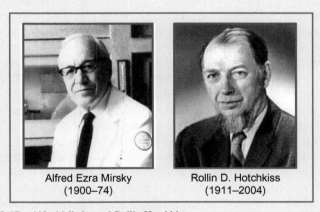

Alfred Ezra Mirsky
(1900–74)

Rollin D. Hotchkiss
(1911–2004)

FIGURE 3.17   Alfred Mirsky and Rollin Hotchkiss.

Mirsky became a vocal opponent of the claim that DNA was the transforming substance.[40] Instead, he asserted that rather than DNA, the transforming agent was more likely to be residual protein that remained tightly bound to the impure DNA. Apart for possible personal motivations, Mirsky's position was shaped by the prevailing conceptions of his time on expected properties of genetic material. The complexity and known biological specificities of proteins made them more palatable candidate carriers of genetic information.

By contrast, because of its seeming chemical monotony and lack of known specificity, DNA was deemed inadequate for this role. On top of this theoretical framework, Mirsky raised several technical points arguing that Avery's DNA preparations were impure and were likely to contain protein. First, he argued that the sensitivity of the Millon chromogenic test was not sufficiently high and low levels of protein could have escaped detection[56]:

> One of the most sensitive direct tests for protein is the Millon reaction, but in our experience a nucleic acid preparation containing as much as 5 per cent of protein would give a negative Millon test.

Considering the elemental analysis of the transforming substance (Fig. 3.16C), Mirsky proposed that the reported elementary makeup of the transforming material would not be significantly different if 2% of protein were present in the seemingly pure DNA[56]:

> At present the best criterion for the purity of a nucleic acid preparation is its elementary composition and especially the nitrogen:phosphorous ratio. Presence of 2 per cent of protein would increase this ratio, but only by an amount that is well within the range of variation found for the purest nucleic acid preparations.

As to the demonstrated resistance of the transforming material to trypsin and chymotrypsin, Mirsky argued that some proteins were resistant to digestion by these enzymes and especially that these proteases were capable of hydrolyzing only partially or completely denatured proteins. Closer examination of this point reveals Mirsky's arguments to have been self-contradictory. Mirsky regarded his procedure of DNA extraction by NaCl, which kept proteins in their native state to be superior to McCarty's method of DNA extraction by deoxycholate that denatured proteins. Yet, he maintained at the same time that trypsin and chymotrypsin were unable to digest McCarty's DNA-associated proteins because they were *not* denatured.

Although the case of DNA as the transforming material was later strengthened by additional evidence that was gathered in the Avery laboratory and by new findings of the French bacteriologist Andrè Boivin (1895–1949), Mirsky held on to his position that small amounts of proteins were the likely transforming agent. Thus, when Boivin announced in the Cold Spring Harbor Symposium of 1947 his success in transforming *Escherichia coli* with purified DNA, Mirsky commented[70]:

> In the present state of knowledge it would be going beyond the experimental facts to assert that the specific agent in transforming bacterial types is a desoxyribonucleic acid.

Mirsky's insistent position was quite influential. As reconstructed by Robert Olby, the eminent geneticist and Nobel laureate Herman Muller (1890–1967) adopted Mirsky's point of view arguing that it remained an open question whether nucleic acids determined biological specificity

(Ref. 45, p. 193). Despite the inconsistencies in Mirsky's arguments and although he was eventually proven wrong, there was, in principle, substance to his position. It is a challenge to achieve complete homogeneity of any biochemical even with today's advanced techniques and the presence of minute amounts of impurities is hard to be categorically excluded. Thus, linking biological activity to a purified biochemical is often questioned because of the possibility that the active component might be a minor impurity rather than the main purified entity. As will be described later, experiments that were performed by Hotchkiss in the late 1940s demonstrated that the transforming DNA was practically devoid of associated protein. Historically, however, there is little doubt that Mirsky's authoritative antagonism contributed at the time to the skeptical reception of the claim that the transforming agent was DNA.

3.9.2 The Evidence Convinced Other Scholars That the Transforming Agent Was DNA

Just 5 months after the publication of the Avery, MacLeod, and McCarty paper, Sir Alexander Haddow (1907−76) of the Chester Beatty Research Institute accepted DNA to be the transforming agent and pointed out that it was distinguished from the dictated product polysaccharides.[71] The nucleic acids chemists Johan Mason Gulland (1898−1947) and Dennis Oswald Jordan (1914−82) from University College Nottingham wrote in a 1945 review article that Avery, MacLeod, and McCarty provided strong evidence of that the transforming agent was DNA.[72] Both the Danish biochemist Herman Kalckar (1908−91)[73] and the Oxford University microbiologist Donald Woods (1912−65)[74] were also of the opinion that the transforming agent had been convincingly shown to be DNA.

3.9.3 Some Scientists Did Grasp the Broader Significance of the Avery, MacLeod, and McCarty Discovery

Few scientists not only accepted the conclusion that the transforming agent was DNA but were also alerted to the boarder significance of DNA as genetic material. In fact, these scholars were so indelibly impressed by the Avery, Macleod, and McCarty discovery that it affected their future course of research.

Reading the Avery, Macleod, and McCarty paper in Jan. 1945, the 19-years-old student and future (1959) Nobel laureate Joshua Lederberg (1925−2008) made the following note in his diary (see facsimile in http://profiles.nlm.nih.gov/ps/access/CCAAAB.pdf):

I had the evening all to myself, and particularly the excruciating pleasure of reading Avery '43 on the deoxyribose nucleic acid responsible for type transformation in Pneumococcus. Terrific and unlimited in its implications.

Lederberg later assessed the impact that the Avery, MacLeod, and McCarty discovery had on his future course of research (http://profiles.nlm.nih.gov/ps/access/CCAAAB.pdf):

> [...] the work had set me on the path of looking for DNA transformation in Neurospora, and eventually to my studies of genetic recombination in E. coli.

Indeed, working at the time in the laboratory of Francis J. Ryan (1916−63) at Columbia University, Lederberg attempted to transform *Neurospora* with DNA. Although these experiments were not successful, the Avery discovery inspired his foray into the genetics of bacteria. His celebrated studies in this area were culminated by the discovery that some *E. coli* strains can go through a sexual stage, that they mate and exchange genes.

Another eminent scientist who was inspired by the discovery that the transforming agent was DNA was the Columbia University biochemist Erwin Chargaff (1905−2002). Up to 1944 his area of research was mostly the biochemistry of *Mycobacterium tuberculosis* lipids. However, after his exposure to Avery's 1944 paper, he changed fields and focused on the biochemistry of nucleic acids.[75] Describing in 1971 this decision, he wrote[76]:

> This discovery, almost abruptly, appeared to foreshadow a chemistry of heredity and, moreover, made probable the nucleic acid character of the gene [...] Avery gave us the first text of a new language, or rather he showed us where to look for it. I resolved to search for this text.

Chargaff's insight and his switch to the study of nucleic acids led to his experimental refutation of the tetranucleotide hypothesis of the structure of nucleic acids[64,77] and to the discovery of the "Chargaff rules" of the equal amount in double-stranded DNA of adenine and thymine and of guanine and cytosine.[64] These rules were critical to the Watson and Crick insight of the pairing of purines and pyrimidines in the DNA double helix.[66]

Last, in explaining his early attraction to DNA, James Watson (1928−) mentioned that his doctoral thesis advisor and the future (1969) Nobel laureate Salvador Luria (1912−91) felt that "Avery's experiment made [DNA] smell like the essential genetic material" (Ref. 78, p. 17).

3.9.4 Scholars Raised Different Ideas on the Possible Mode of Action of the Transforming DNA

Mirsky's refusal to accept the identification of DNA as the transforming agent notwithstanding, other scientists who did agree that it was DNA that mediated transformation had different ideas on the mechanism of transformation. Whereas some considered DNA to be the directly active agent, others saw it as an element in more involved processes.

3.9.4.1 DNA as "Pure Gene"

The prominent Australian virologist and immunologist and future (1960) Nobel laureate, Macfarlane Burnet (1899–1985), was perhaps the most astute early interpreter of the Avery, MacLeod, and McCarty discovery. After paying a visit to Avery's laboratory in 1943, he wrote back to his wife in Australia to declare that in his opinion, DNA constituted "pure gene":

> *Avery had just made an extremely exciting discovery which, put crudely, is nothing less than the isolation of a pure gene in the form of desoxyribonucleic acid*

In his 1968 autobiography (Ref. 79, p. 81) Burnet later put the importance of his initial impression in historical perspective:

> *[...] Nothing since has diminished the significance or importance of Avery's work. Neither he not I knew at the time but in retrospect the discovery that DNA could transfer information from one pneumococcus to another almost spelt the end of one field of scholarly investigation, medical bacteriology, and heralded the opening of the field of molecular biology which has dominated scholarly thought in biology ever since.*

A similar idea that likened DNA to "gene in solution" was expressed in 1946 by the 1936 Nobel laureate Sir Henry Dale (1875–1968).[80]

3.9.4.2 DNA as Mutagen

An indirect mode of action of the transforming DNA was offered early on by the Columbia University geneticist Theodosius Grygorovych Dobzhansky (1900–75). As he told Olby in a 1968 interview (Ref. 45, p. 189) after paying a visit to Avery's laboratory in the early 1940s, Dobzhansky tried to convince him that transformation by DNA represented mutagenesis of the pneumococci. By regarding DNA as a mutagen, analogous to a chemical mutagen, it was not perceived as the genetic material itself but rather as an agent acting on some other (most likely protein) carrier of genetic information.

This idea appeared attractive to geneticists as is illustrated in the following 1948 statement by the geneticist and future (1958) Nobel laureate George Wells Beadle (1903–89)[81]:

> *[...] Pneumococcus type transformation, which appears to be guided in specific ways by highly polymerized nucleic acids, may well represent the first success in transmuting genes in predetermined ways.*

It should be pointed out in fairness that Beadle's position was modified with time. Already in 1948 he wrote that DNA was *either* a transmuting agent *or* a part of the genetic system.[82] By 1952 he already stated that type-specific DNA was a component of the gene[83] and in 1956 he recognized it as being the primary genetic material.[84]

3.9.4.3   Chromosomal Crossing Over

Herman Muller drew an analogy between pneumococcal transformation and phage recombination. He proposed that both phenomena could be explained as manifestations of events of crossing over between chromosomes of donor and recipient strains (Ref. 45, p. 190). It must be noted, however, that Muller did not view DNA as the principal carrier of information but rather, he like others,[85] saw it as part of a chromosomal nucleoprotein complex. Thus, when alluding to crossing over he most likely thought of whole chromosomes recombining. This notion was much in line with Mirsky's view of the transforming principle as protein that was tightly associated with DNA in chromosomes (*vide infra*).

3.9.4.4   The Transforming Principle as a Virus

The Rockefeller Institute pathologist and cancer researcher James B. Murphy (1884–1950) was the first to propose that viruses that cause tumors in fowl were analogous to the pneumococcus transforming principle.[86] At a later time the eminent virologist Wendell Stanley, who was in the 1940s a Rockefeller Institute investigator at Princeton, expanded this idea. Similar to Murphy, he was familiar with Avery's transformation work and because of his own focus on viruses he suggested that the transforming principle was a virus. A section of a chapter that he wrote in 1938 on the biochemistry and biophysics of viruses was dedicated to *pneumococcus* transformation about which he wrote (Ref. 87, p. 491):

> *This phenomenon is virus-like, and it is because of this and the fact that it may become important from the standpoint of the chemistry of viruses that a discussion is included here.*

After describing the purification procedure of the transforming material (which at that stage had not been identified yet as DNA) Stanley continued:

> *No chemical tests were made on these purified preparations, hence nothing is known about the nature of the active agent. It is to be hoped that the study of this agent will be continued because of its virus-like nature.*

The Berkley virologist Claude Arthur Knight (1914–83) also compared the transforming agent to virus.[88] Based on this analogy, he abandoned his studies of proteins of plant viruses and investigated instead their nucleic acid component.

With time Stanley withdrew his idea of the transforming agent as virus. In 1965, McCarty invited him to speak at a Rockefeller Institute ceremony in honor of Avery. In his letter of acceptance he confessed to McCarty how insistent he was at the time about the analogy between the transforming material and viruses and how he resisted the advice of Thomas Rivers (1888–1962), the director of the Rockefeller Institute Hospital, to drop this idea (for a facsimile of the letter, see http://profiles.nlm.nih.gov/ps/retrieve/ResourceMetadata/CCAAHB#transcript):

> *Tom Rivers gave me a pretty bad time, for I had asked him to look over my chapter and he felt that the section on "The transformation agent of the pneumococcus" should be deleted for, as he put it, it had absolutely nothing to do with viruses. Much argument could not convince him, but I felt so strong about it that the section was left in. Needless to say, because of this argument, I followed the work in which you and MacLeod participated with great interest. It is nice to know that Fess realized the significance of the work from the beginning.*

In point of fact, in his address at the Jun. 1965 dedication ceremony of the Avery Memorial Gateway at the Rockefeller Institute in New York, Stanley publicly apologized for his failure to recognize in time the significance of the Avery, MacLeod, and McCarty discovery.[4]

3.10   MULTIPLE FACTORS CONTRIBUTED TO THE BELATED RECOGNITION OF THE SIGNIFICANCE OF THE DISCOVERY OF AVERY, MacLEOD, AND McCARTY

Although a considerable number of scholars accepted Avery's 1944 claim that seemingly pure DNA transformed pneumococci, this discovery failed to prompt the scientific community to examine experimentally the proposition that DNA might be the genetic material. In fact, DNA was recognized as carrier of genetic information only after the 1952 demonstration by Hershey and Chase that it was responsible for phage reproduction[65] and the subsequent 1953 elucidation by Watson and Crick of its structure.[66,67] Why was the significance of Avery's discovery recognized so belatedly? Historians of biology identified several contributing factors to the delayed acceptance of his results and conclusion.

3.10.1   Disconnection Between the Disciplines of Genetics and Biochemistry

Many historians of 20th-century biology pointed to the gulf that existed between of the disciplines of genetics and biochemistry. This gap was not bridged until the birth of molecular biology in the 1950s. In many cases there was even institutional separation between geneticists and biochemists. Thus, at the time that Avery stumbled upon DNA as the transforming material, there was not even a single geneticist to consult with on the staff of the Rockefeller Institute. Beyond the lack of communication between practitioners of genetics and biochemistry, there was absence of active interest in both camps in the chemical nature of the genetic material. Geneticists of the first half of the century were mostly engaged in the mapping of genes and mutations with no concern to the material constitution of genes. Alfred Hershey, who together with Martha Chase, demonstrated in 1952 that the genetic material of bacteriophage was DNA (see Chapter 4: "Hershey and Chase Clinched the Role of DNA as the Genetic Material"), enumerated factors that contributed to the late acceptance of the Avery discovery. One reason that he mentioned was the general atmosphere of indifference toward the question of the chemical nature of genes (cited in Ref. 89, p. 105):

[...] as long as you're thinking about inheritance, who gives a damn what the substance is—it's irrelevant.

On the other camp, biochemists were focused on the study of metabolism and of smaller molecules and had less interest in macromolecules. Without much cross-talk between the two disciplines, there was no chance for a concerted effort to combine genetic and biochemical approaches to search for the chemical nature of genes.

3.10.2 A Misleading Preconception: Protein, and Not DNA, Was Thought to Be the Carrier of Genetic Information

As already mentioned, the claim that the transforming material was DNA collided with the contrasting consensual belief that genetic information is stored in protein molecules. Here two factors must be taken into consideration. First, the complexity of organisms and their wide phenotypic variability led to the thought that genes must also be structurally complex. Made of 20 amino acids that combined into infinitely diverse molecules, proteins appeared to fulfill this requirement. By contrast, nucleic acids, which were thought to comprise of repeating tetranucleotide units, were viewed as too monotonous to function as genes. Second, whereas proteins were recognized for their diverse biological specificities, functioning as enzymes, antibodies, and structural components of the cell, no known biological role was assigned to nucleic acids. This state of affairs is illustrated by a comment that Pollister, Swift, and Alfert made in a paper on nucleic acids metabolism which they presented at an Apr. 1950 Oak Ridge meeting[90]:

Nucleic acids bulk so large in the composition of cells that it seems certain that these substances are involved in some important biological functions. Nevertheless, with one exception [bacterial transformation], there has been no proof of a specific biological function mediated by one of these polynucleotides; no enzyme, hormone, vitamin, or even vague 'growth substance' has been found to be a nucleic acid.

It must be noted, however, that although the overwhelming consensus was that genes are comprised of proteins, some did not exclude the possibility that nucleic acids might function as genetic material. After he found that the transforming principle was DNA, McCarty was encouraged to discover in the 1929 textbook *Outlines of Biochemistry* by Ross Aiken Gortner (1885–1942), some support to the idea that nucleic acids could be potential gene material. Writing that chromosomes determine "our bodily characteristics down to the color of our eyelashes," Gortner noted that nucleic acids comprise 40% of the chromosomes. He then ventured that "[...] it becomes a question whether the virtues of nucleic acids may not rival those of amino acids" (cited in Ref. 40, p. 145). Also, prior to the publication of their 1944 paper, Avery and McCarty consulted with the Rockefeller Institute protein chemist Max Bergman. As cited by McCarty, Bergman insightfully asserted that, contrary to the commonly held view, nucleic acids are infinitely complex (Ref. 37, p. 163):

> In the light of present knowledge, the statement that all nucleic acids are the same regardless of the source from which they are derived is nonsense. If they are large polymeric compounds, there is an endless number of combinations all of which would possess the same elementary composition but would differ in chemical structure none the less.

Finally, it is interesting to note that at a later time, and before he became convinced that transformation was mediated by DNA, in discussing Boivin's 1947 report on transformation in *E. coli*, Mirsky himself opined that nucleic acids may not be after all uniform[70]:

> The work on transformation of bacterial types has been a healthy stimulus to the chemical investigation of nucleic acids. Discussions at this symposium have clearly shown that these wonderful bacteriological discoveries have caused chemists to consider critically the evidence for uniformity among nucleic acids, and the generally accepted conclusion is that the available chemical evidence does not permit us to suppose that nucleic acids do not vary.

3.10.3 Mirsky's Criticism

As told, Alfred Mirsky conducted for several years a campaign within and outside the Rockefeller Institute proclaiming that some proteins resisted Avery's purification process and that minute amounts of presumably information bearing protein remained associated with the DNA. His antagonism to the proposal that the transforming material was DNA and not protein had considerable impact. Although Avery was well-respected for his bacteriological and immunochemical contributions, he was not an expert biochemist or geneticist. Mirsky, by contrast, was an authority on the cell nucleus, on proteins, and on nucleoproteins. His judgment was thus valued and many were swayed by his opinion. As will be described later, McCarty and Avery addressed Mirsky's criticism in 1946 by improving their procedure of extracting transforming DNA and by providing more convincing enzymological evidence for its identity. McCarty presented parts of this work in the 1946 Cold Spring Harbor symposium.[91] The biologist H. Bentley Glass (1906–2005) who attended the presentation recalled later that McCarty's demonstration that the transforming principle was DNA was very strong. He added however that "the purification is not yet so complete that everyone is convinced that some protein does not remain in the preparation."[5] Although Glass was quick to point out that Mirsky was not present at the symposium and thus was not directly responsible for this opinion, it is unlikely that his well-publicized criticism did not influence the attitude of McCarty's audience.

3.10.4 Pneumococcal Transformation Was a Difficult Experimental System

The described excruciatingly slow development of the pneumococcal transformation system marked it as an unusually difficult experimental system. The complexity of the transformation assay and the batch-to-batch variability of the biological activity of the transforming material made results hard to reproduce. A revealing point is that no other laboratory attempted to replicate the results of Avery, MacLeod, and McCarty even years after they were published. This lack of follow-up was likely due in part to the myriad technical problems that the transformation system posed. Alfred Hershey identified retrospectively the technical complexity of the *pneumococcus* transformation system as one of the factors that contributed to the delayed acceptance of the Avery discovery (cited in Ref. 89, p. 105):

> *This was an extremely awkward system to work with, the pneumococcus, and a close homolog of course was the tobacco mosaic virus which was the first vehicle for demonstrating the role of nucleic acid in viral infection. These are the last systems you would have chosen, if you'd been looking for material to study from this point of view.*

3.10.5 Some Viewed Pneumococcal Transformation to Be a Private Case of Limited Significance

The Avery, MacLeod, and McCarty discovery remained for several years the only experimentally documented evidence for the function of nucleic acid as carrier of genetic information. It is not surprising therefore that instead of regarding DNA as the universal genetic material some scholars considered pneumococcal transformation to be a private case that was restricted to a limited number of bacterial species. This position is well-reflected in a summary of a 1950 Oak Ridge Conference on Nucleic Acids. The review that was authored by Arthur Pollister of Columbia University, Hewson Swift

(1920−2002) of the University of Chicago, and Max Alfert (1921−) of the University of California, Berkeley, declared[90]:

> [...] with one exception [bacterial transformation], there has been no proof of a specific biological function mediated by one of these polynucleotides; no enzyme, hormone, vitamin, or even vague 'growth substance' has been found to be a nucleic acid.

3.10.6   Inopportune Timing

The long search for the chemical nature of the transforming principle came to fruition in 1940−43 and was brought to the attention of the scientific community in 1944. During that period Europe, and later the United States were engulfed in war. Under these historical circumstances, many laboratories put on hold pure scientific research and scientific communication was severed within Europe and between Europe and the United States. Hence, the potential audience for Avery's discovery was distracted and decimated.

It should be mentioned here that the war also directly affected McCarty's work and that he personally experienced the effect of the breakdown of communication between Germany and the United States. In 1945 he invested much work in the development of purification procedure for pancreatic DNase that was needed for the demonstration that the transforming factor was DNA. Only after dedicating a great deal of effort to this venture did he discover that a purification procedure for DNase had already been developed and published in Germany in 1941.[92] Even then, in 1945 it was still impossible to get hold of the German journal in New York and McCarty had to turn to the office of the Alien Property Custodian in Washington to obtain a copy of the paper (Ref. 40, p. 185). In fact, at that time McCarty had already been in uniform for about 3 years. In May 1942, while he was engaged in the purification and testing of the transforming material, he was commissioned as a lieutenant J.G. in the US Naval Reserve and was assigned to the Naval Hospital in Annapolis (Ref. 40, p. 140). Fortunately, the Director of the Rockefeller Institute Hospital, Thomas Rivers, succeeded in changing the orders such that McCarty could stay on at the Rockefeller. Indeed, McCarty's military rank (Lieutenant Commander) was indicated in papers that he published in 1945 and 1946 and his affiliation was identified there as "The United States Navy Research Unit at the Hospital of the Rockefeller Institute for Medical Research."[93−96] Having been in uniform until 1946, McCarty later wrote that he had been "mentally conflicted" about doing basic research at a time of war. To be more useful to the war effort, in parallel to his work on transformation he also undertook a research project on viral pneumonia and was on duty doing clinical rotations.

3.10.7   Avery's Reticent Personality

Much had been written about Avery's personality and how it impacted his way of conducting science. His associates and trainees were of one voice in pointing to his low-key and extremely careful approach to scientific research. Many commented on the effect that Avery's personality and style of doing science had on the reception of the transformation work. In addition to objective technical difficulties that slowed and sometimes blocked the progression of the transformation work, it was also Avery's extreme caution that delayed its completion. Understandably, the careful researcher that he was, Avery hesitated for a considerable period of time before going public with the claim that DNA was responsible for the synthesis and antigenicity of pneumococcal capsular polysaccharides. This assertion went radically against the grain of the commonly held belief that genes were made of protein. However, there might have been another dimension to Avery's caginess. In a 1916 paper Dochez and Avery claimed that resistance to pneumococci, defined as antiblastic immunity, resulted from inhibition of pneumococcal enzymes by the serum of infected humans or animals.[97] A year later they reported that the active agent could be isolated from urine by precipitation with alcohol and acetone.[30] Soon thereafter, however, the Rockefeller Institute Hospital researcher Francis Blake (1887−1952) showed that although sera agglutinated the bacteria they did not inhibit their enzymes.[98] Antiblastic immunity was also refuted by experiments that another Rockefeller Institute Hospital bacteriologist, Marshall Albert Barber (1868−1953), conducted.[99,100] Although Dochez and Avery never retracted their papers, it probably caused the normally careful Avery to be ever more cautious in executing experiments and interpreting their outcome. This might have led to Avery's slow and hesitant acceptance of DNA as the transforming agent. It was also argued that the antiblastic immunity episode could have affected Avery's scientific credibility such that the 1944 DNA transformation paper was met with greater skepticism.[15]

Another facet of Avery's personality that probably contributed to the indeterminate reception of the transformation work was his aversion to travel and evasion of scientific meetings. He rarely attended scientific conferences and even did not travel abroad to receive the German Paul Ehrlich Gold Medal (1933) or ventured to England to accept an honorary doctorate from Cambridge University (1944), to London to be awarded the Copley Medal (1945), or to Sweden to receive the Pasteur Gold Medal (1950).[15] This uncommon behavior restricted Avery's capacity to engage with peers and defend and publicize his work. Alfred Hershey retrospectively cited Avery's refusal to advertise as a contributing factor to the muted reception of his discovery (cited in Ref. 89, p. 105):

> *Another reason for the work of Avery and his colleagues not attracting quite as much attention as they might was that they were just too modest. They refused to advertise.*

3.11   THE SCIENTIFIC ESTABLISHMENT WAS AMBIVALENT ABOUT THE IMPORTANCE OF AVERY'S DISCOVERY

Whereas some learned societies and institutions promptly recognized and acknowledged the importance of the discovery of DNA as genetic material, others ignored it. This mixed reaction is well illustrated by the citations of three important awards that were conferred on Avery in the 1940s. In Oct. 1944 just 7 months after the publication of the transforming DNA paper,[1] Avery was presented with the Gold Medal of the New York Academy of Medicine. The citation of this award read in part[101]:

> [...] During this present year with Mac Leod and MacCarty, you have studied the transformation of one type of pneumococcus into another and isolated the "transforming principle" as thymonucleic acid. This discovery has very far-reaching implications for the general science of biology [...]

Contrasting this almost instantaneous recognition of the significance of the transformation work, the Royal Society committee that bestowed on Avery the Copley Medal chose to ignore the transformation work and to focus on his contributions to immunochemistry. After being unsuccessfully proposed twice for the Medal, Avery finally won it in Nov. 8, 1945, almost 2 years after the publication of his paper with MacLeod and McCarty.[11] In bestowing this distinguished Medal, which had been given in the past to luminaries such as Benjamin Franklin, Davy, Darwin, Mendeleev, Einstein, and Bohr, the Royal Academy omitted any mention of transformation, citing instead (see facsimile in: http://profiles.nlm.nih.gov/ps/access/CCBDBM.pdf):

> In recognition of his success in introducing chemical methods in the study of immunity against infective diseases.

This oversight was somewhat rectified in an address that the president of the Royal Society Sir Henry Dale delivered in Nov. 30, 1945, at an anniversary meeting of the Society (for a facsimile of the Dale's speech, see http://profiles.nlm.nih.gov/ps/access/CCGMGY.pdf):

> [...] only last year, Avery, with Macleod and McCarty, has been able to isolate and to characterize a chemical principle acting in minute dosage as the specific stimulus to such a transformation. An unencapsulated, avirulent, typeless pneumococcus derived from a specific strain of type II, responds to this stimulus by acquiring and retaining the capsule and specific polysaccharide, with the virulence and the cultural characters, of a fully specific strain of type III. Here surely is a change to which, if we were dealing with higher organisms, we should accord the status of a genetic variation; and the substance inducing it-the gene in solution, one is tempted to call it—appears to be a nucleic acid of the desoxyribose type.

The committee that elected Avery to be the laureate of the 1947 Albert Lasker Basic Medical Research Award totally ignored the transformation work. The award citation focused on his contributions to the immunochemistry of *pneumococcus* without even a mention of the transformation work. Under the header "Oswald Avery—For distinguished service through studies on the chemical constitution of bacteria," the citation reads (Ref. 102, also see a facsimile of the citation in: http://www.laskerfoundation.org/awards/1947_b_description.htm):

> *[...] he undertook to elucidate in logical sequence the biological activities, the immunological characteristics and the pathogenic properties of any pneumococcus. [...] he discovered and identified the capsular polysaccharides and demonstrated their role in determining this specificity. He furthermore succeeded in throwing light on the immunological relationships of the intracellular constituents of this organism. Through these discoveries he laid the foundation for his brilliant analysis in chemical terms of the antigenic constitution of the whole pneumococcus. He thus established a perfect pattern for the antigenic analysis of other microorganisms both by himself and by others who have followed in his footsteps.*
>
> *Among Dr. Avery's many brilliant contributions to scientific knowledge, none has been more outstanding than his studies on the antigenic constitution of bacteria. Through them, he is one of the founders of the science of immunochemistry [...].*

In not choosing to award Avery a Nobel Prize in physiology and medicine, the Karolinska Institute Nobel Assembly displayed similar lack of appreciation for the discovery of DNA as the transforming agent (for historical records and commentary on the denial of a Nobel Prize to Avery, see Refs. 13–15). Ironically, according to Judson,[103] the Swedish 1948 Nobel laureate Arne Tiselius (1902–71), who was in 1960–64 the Chairman of the Board for the Nobel Foundation, commented that Avery was "the most deserving scientist not to be awarded a Nobel Prize for his work."

3.12   AFTER 1944: ADDITIONAL EXPERIMENTS STRENGTHENED THE CASE FOR DNA AS THE TRANSFORMING AGENT

A major obstacle to the acceptance of DNA as the transforming material was the possibility that the biologically active component was a tiny amount of protein that remained associated with the pneumococcal DNA. It was realized that even if protein comprised only 0.1% of the mass of the purified transforming material, there would still be millions of protein molecules present in even the lowest transforming dose (0.003 μg) of the most potent DNA preparation that McCarty and Avery isolated (Fig. 3.16A). There was, therefore, a need to substantiate the claim that the transforming agent was

DNA itself and not contaminating protein. In the few years that followed the 1944 publication of the Avery, MacLeod, and McCarty paper,[1] several researchers provided experimental evidence in support of the role of DNA as the transforming material.

3.12.1   McCarty and Avery's Added Experiments

After his decision to postpone his retirement and to stay on for a while at the Rockefeller Institute, Avery continued to work with McCarty on the transforming DNA. To strengthen the case for DNA as the transforming agent, they supplied better enzymological evidence and improved the DNA purification procedure.

One indication in 1944 that the transforming material was DNA was its selective inactivation by crude extracts that contained DNase (Fig. 3.16D). However, because of the impurity of the crude enzyme preparations, a possibility remained that hydrolase other than DNase destroyed the transforming agent. To address this problem, McCarty undertook to prepare and employ pure DNase. Consulting with the Rockefeller Institute master enzyme crystallizer Moses Kunitz (1887–1978), McCarty spent much of 1945 developing a procedure for the purification of DNase from beef pancreas. Although his hope to isolate the enzyme in crystal form was deflected, the DNase that he isolated was of high purity. This enzyme degraded DNA only in the presence of magnesium or manganese ions and their binding by the chelator citrate completely inhibited the nucleolytic activity.[96] In a 1946 paper that McCarty and Avery marked as the second installment of their studies on the chemical nature of the transforming substance,[94] they reported that minute amounts of the purified DNase sufficed to completely inactivate the transforming principle. Furthermore, the transforming activity remained intact when DNase activity was inhibited by citrate. Importantly, traces of proteolytic activity in the purified DNase could be detected only in amounts of the enzyme that were 100,000-fold greater than those needed to completely inactivate the transforming material.

The inhibition of DNase by citrate was also exploited by McCarty to develop a more efficient purification procedure of the transforming material. Addition of this chelator during the preparation of extracts of heat-killed pneumococci prevented breakdown of the bacterial DNA and increased its yield. In addition, the presence of calcium ions during the precipitation of the DNA by alcohol improved the separation of the DNA from residual polysaccharides and increased its purity. The pure DNA, which was obtained at a high yield, potently transformed pneumococci. The improved purification protocol of transforming DNA was published in a second 1946 paper that was marked as the third installment of the studies on the chemical nature of the transforming substance.[95]

In discussing their newer results, McCarty and Avery confronted directly the view that nucleic acids could not serve as the transforming agent[94]:

> *The objection can be raised that the nucleic acid may merely serve as "carrier" for some hypothetical substance, presumably protein, which possesses the specific transforming activity. [...] There is no evidence in favor of such a hypothesis, and it is supported chiefly by the traditional view that nucleic acids are devoid of biological specificity. On the contrary, there are indications that even minor disruptions in the long-chain nucleic acid molecule have a profound effect on biological activity. Thus treatment of the transforming substance with concentrations of deosoxyribonuclease so small that only a slight fall in viscosity occurs causes a marked loss of biological activity. It is suggested that the initial stages of enzymatic depolymerization which are reflected only by minimal changes in the physical properties of the nucleate are sufficient to bring about destruction of specific activity.*

Moreover, they also ventured to suggest that the transformation of pneumococcal antigenicity was just one of countless genetic traits that the transforming material may dictate[94]:

> *It is possible that the nucleic acid of the R pneumococcus is concerned with innumerable other functions of the bacterial cell, in a way similar to that which capsular development is controlled by the transforming substance. The deoxyribonucleic acid from Type III pneumococci would then necessarily comprise not only molecules endowed with transforming activity, but in addition a variety of others which determine the structure and metabolic activities possessed in common by the encapsulated (S) and unencapsulated (R) forms.*

They ended this assertion with the insightful statement that if DNA would prove to dictate multiple bacterial functions, then the next task would be to determine the chemical basis for its biological specificity.

3.12.2   Hotchkiss' Experiment: The Amount of Protein in the Transforming DNA Was Negligible

Rollin Hotchkiss (Box 3.6) who worked in the Avery laboratory in the 1930s rejoined it in 1946 after the end of World War II. There he joined the work on transformation by firstly uncovering the identity the mysterious factor of pleural or ascetic fluids that was obligatory for transformation. Collaborating with Avery's postdoctoral associate Harriett Ephrussi-Taylor (1918–68), he showed that chest fluid from infected patients could be replaced with crystalline serum albumin.[58]

After Avery's retirement in 1948, Hotchkiss continued on his own studies on transformation. By that time Kunitz had already crystallized pancreatic DNase which Hotchkiss used to show that the crystal form of the enzyme inactivated the pneumococcal transforming DNA.[104] Next, he confronted directly the haunting question of the possible presence of protein impurity in the transforming DNA. Purified transforming DNA, adenine, or reference proteins were hydrolyzed in parallel by 6 M HCl at 120°C and release of α-amino acid was measured by the Van Slyke ninhydrin method. Results showed that the rate of liberation of the amino acid from the transforming DNA was much slower than its rate of release from reference proteins or from adenine.[104] At a later time Hotchkiss showed that the product of decomposition of adenine by acid hydrolysis was glycine and that 100% of the amino acid nitrogen that was released from the transforming DNA came from glycine. Thus, nitrogen that was suspected to originate from protein contaminant in the DNA was actually the product of disintegration of the adenine base in DNA into glycine. Hotchkiss conservatively estimated that the maximum possible proportion of protein was at most 0.02% of the total mass of the transforming DNA. Unfortunately, because Hotchkiss published these results only in 1952,[105] Mirsky's reservations about the purity of the transforming DNA remained uncontested for several additional years.

It is instructive to mention two reactions to the Hotchkiss analysis. Although his results were not published until 1952, the rumor of their existence spread through the grapevine. Al Hershey, who was in 1949 at Washington University, was assigned to participate in a roundtable on nucleic acids. Preparing for this symposium, he wrote to Hotchkiss in April of that year to ask about his alleged evidence that "Avery's stuff is really DNA." In response, Hotchkiss sent to Hershey his yet unpublished data showing that possible protein contaminant comprised no more than 0.02% of the transforming DNA. On May 3, 1949 Hershey wrote back that "My own feeling is that you have cleared up most of the doubts. Some people may cling to the virus theory a little longer perhaps" (for facsimiles of the letters and commentary, see Hotchkiss' retrospective assay[106]).

Much later, Mirsky himself acknowledged that the Hotchkiss results were critical to his retraction from the position that the transforming DNA contained protein. Jack S. Cohen of the NIH asked Mirsky in 1973 for comments on a draft of a paper that he was writing with Franklin Portugal on the search for the chemical structure of DNA.[107] Responding in early 1974, Mirsky asked that the paper be modified to better reflect his past claim that the transforming substance could be protein that remained associated with the pneumococcal DNA. He argued that his view at the time was that until the DNA was proven to be free of protein, the possibility that transformation was mediated by protein could not be ruled out. He claimed, however, that once Hotchkiss showed that the amount of protein in the transforming DNA was negligible, he fully accepted that the transforming material was DNA and not protein (for facsimiles of the correspondence between Mirsky and Cohen, see http://profiles.nlm.nih.gov/ps/retrieve/ResourceMetadata/CCAAHR):

> From the beginning, I considered DNA as an essential part of the transforming principle and after it was proven by Hotchkiss that there was practically no protein present, (which was my original question) I accepted the conclusion without reservations.

3.12.3   Boivin's Claimed Transformation of *E. coli* by Purified DNA

During the war years the Pasteur Institute group of André Boivin (1895–1949) claimed to have developed a system by which nucleoprotein fraction from toluene-lysed encapsulated S-form *E. coli* cells transformed nonencapsulated R type cells to S-form cells and changed their antigenicity.[108] Boivin and his associate Roger Vendrely (later at the University of Strasburg) were exposed to Avery's 1944 paper and to their identification of DNA as the transforming agent of pneumococci only after the restoration of scientific communication in the late stages of World War II. Guided by this finding, they removed protein from the *E. coli* nucleoprotein fraction and showed that the purified DNA mediated R to S transformation. Publishing these results in a series of papers, Boivin presented them in full at the 1947 Cold Spring Harbor Symposium.[70] Although the title of that paper read in part "Directed *mutation* in colon bacilli, by an inducing principle of deoxyribonucleic nature..." Boivin referred to DNA in the body of the paper as the genetic material and not as a mutagen. His view of DNA as the hereditary substance was reflected by the title of this[70] and a previous paper on the subject.[109] There he explicitly used the respective phrases "...Meaning for the general biochemistry of heredity" and "...Significance for the biochemistry of heredity." More specifically, in discussing his own and Avery's results and also incorporating the works of Beadle and Tatum and of several cytochemists,[110] he made an astonishingly accurate prediction for that time of the primary role of DNA in the flow of genetic information[70]:

> We may, at the most, catch a glimpse of a series of catalytic actions, which set out from primary directing centers (the desoxyribonucleic genes), proceed through secondary directing centers (the ribonucleic microsomes-plasma genes) and thence through tertiary directing centers (the enzymes), to determine finally the nature of the metabolic chains involved, and to condition by this very means, all the characters of the cell in consideration.

It is obvious that Boivin's interpretation of his results was more daring than Avery's reserved and cautious statements. Boivin was also an imaginative scientist and an exciting speaker and his presentation at the Cold Spring

Harbor Symposium raised a great deal of interest. Unfortunately, Boivin's own work ended with his untimely death in 1949. Also, his transformation assay was hard to reproduce and to add insult to injury, his original transformation-competent *E. coli* strains were lost. Using different strains, Lederberg and Tatum could not replicate Boivin's results, and for many years afterward, *E. coli* was regarded to be refractory to transformation. It was only in the early 1970s that penetration of bacteriophage and plasmid DNA into *E. coli* cells was achieved by the addition of calcium chloride to the cell suspension.[111,112]

3.13   EPILOGUE: AVERY'S LEGACY

After extending his work at the Rockefeller Institute beyond the mandatory retirement age of 65, Avery moved in 1948 to Nashville to be closer to the family of his brother, Roy, who served on the faculty of the Vanderbilt School of Medicine as Professor of Bacteriology. Renting a home down the street from his brother, Avery became a well-liked figure in the community. He died of cancer of the liver in Nashville in 1955 at the age of 77 and is buried there in the Mt Olivet Cemetery. It is fitting to conclude this chapter with some observations on Avery's legacy and on his rare personal and scientific qualities.

3.13.1   Avery's Distinctive Scientific Style and Personality

The indelible impression that Avery left as scientist, mentor, colleague, and individual is evident in memoirs that several of his associates wrote. Perhaps, the most evocative is a chapter titled *As I remember Him* in Renè Dubos' book *The Professor, The Institute and DNA* (Ref. 6, pp. 161–179). Dubos, who knew Avery for many years and worked closely with him, presented a masterful portrait of Avery's way of doing science, his critical mind, modesty, interpersonal gifts, the paradoxes of his personality, and his guarded private life. The picture that emerges from this portrayal, from descriptions by other contemporaries of Avery,[40,113,114] and from multiple primary documents in the Oswald Avery Collection (http://profiles.nlm.nih.gov/ps/retrieve/Narrative/CC/p-nid/223) is that of a model scientist and a person marked by generosity of spirit. His longtime collaborator, colleague, and personal friend Alphonse Dochez concisely portrayed Avery's stature as scientist and individual[114]:

> *Dr. Avery was a true scientist with insatiable curiosity and a powerful and unremitting urge to discover the innermost mechanisms of biological facts that came under his observation. His approach to the solution of a problem was characterized by a logical simplicity of thought and a perfection of technical*

procedure, combined with a complete objectivity that guaranteed the soundness of his deductions. Most of Avery's associations were with his colleagues, by whom he was greatly revered and who have preserved for him a timeless affection. He was truly lovable person, humble, disinterested, and generous. The inspiring and friendly quality of his leadership is conspicuously represented by his many former associates who now occupy distinguished positions in medicine in the Unites States and elsewhere, so that his light still burns in places of learning and research.

3.13.2   The Scientific Community Held Avery in High Esteem

The considerable legacy that Avery had left at the time of his passing grew with time. His way of thinking about science and performing experiments, his professional achievements, personality, and the imprint that he had made on coworkers and peers were described and analyzed in books and in many assays.[2–15,107,110] Although his discovery together with MacLeod and McCarty that the transforming principle was DNA had gained due appreciation only after his death, Avery had been a venerated scientist already during his lifetime. Despite his diffident personality, and although he rarely traveled and refused to advertise his work and himself, his contemporaries did value and acknowledge his fundamental contributions to the immunochemistry of *pneumococcus*. He was elected to the National Academy of Sciences in 1933 and later to several foreign societies and he served as President of the Society of American Bacteriologists in 1941. Among the many honors that he received were several honorary degrees: the Joseph Mather Smith Prize, Columbia University (1930); John Philips Memorial Award from the American College of Physicians (1932); the Paul Ehrlich Gold Medal from University of Frankfurt (1933); the Copley Medal from the Royal Society of London (1945); the Kober Medal from the Association of American Physicians (1946); the Albert Lasker Award (1947); the Passano Award (1949); and the Pasteur Gold Medal from the Swedish Medical Society (1950). Avery was also recognized posthumously by naming a small lunar crater near the eastern limb of the Moon, the Avery Crater.

Although the significance of Avery's discovery of DNA as the determinant of phenotypic properties of *pneumococcus* became fully appreciated only posthumously, it is considered today as his crowning scientific achievement and the harbinger of molecular biology.

3.13.3   Avery Was Celebrated by Rockefeller University

The high esteem with which Avery was held is best illustrated by the special recognition that his institution, the Rockefeller Institute (later, Rockefeller

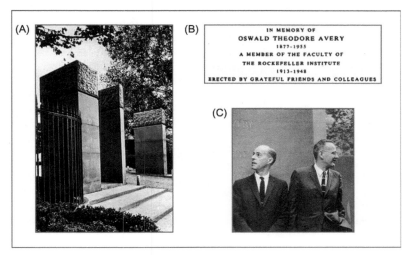

FIGURE 3.18   (A) The Avery Memorial Gateway to the Rockefeller University campus. (B) The inscription on the gateway. (C) Colin MacLeod (left) and Maclyn McCarty (right) at the Sep. 1965 dedication ceremony of the Avery gateway.

University), bestowed on him. Although Avery was not among the 24 graduates and faculty of the Rockefeller that won a Nobel Prize, his special place in the history of the Institute, and indeed in science, was epitomized by the decision of the Rockefeller University to erect in his honor the Avery Memorial Gateway on the northwest corner of its Manhattan campus (Fig. 3.18).

Speakers at the Sep. 29. 1965 ceremony included the President of Rockefeller University Detlev Bronk (1897−1975); Maclyn McCarty; Colin MacLeod; Wendell Stanley; Theodosius Dobzhansky; and the future (1968) Nobel laureate Robert Holley (1922−93) (for their addresses at the ceremony, see http://profiles.nlm.nih.gov/ps/retrieve/ResourceMetadata/ CCAADT).

3.13.4   Avery Nurtured an Exceptionally Distinguished Scientific Progeny

Alongside Avery's own contributions to bacteriology, immunochemistry, and the nascent field of molecular genetics, the unusually large number of prominent scientists that he trained or collaborated with was also a part of his legacy. As noted by McCarty, 10 of 46 members of the Section of Pathology of the National Academy of Science were in 1963 past associates of Avery. Also, 4 of the 18 full members of the Rockefeller Institute in 1953 were past trainees of the Avery laboratory (Ref. 40, p. 123).

REFERENCES

1. Avery OT, Macleod CM, McCarty M. Studies on the chemical nature of the substance inducing transformation of pneumococcal types: induction of transformation by a desoxyribonucleic acid fraction isolated from pneumococcus type III. *J Exp Med* 1944;**79**(2):137−58.
2. Hotchkiss RD, Oswald T. Avery: 1877−1955. *Genetics* 1965;**51**(1):1−10.
3. Coburn AF. Oswald Theodore Avery and DNA. *Perspect Biol Med* 1969;**12**(4):623−30.
4. Stanley WM. The "Undiscovered" discovery. *Arch Environ Health* 1970;**21**(3):256−62.
5. Glass B. The long neglect of genetic discoveries and the criterion of prematurity. *J Hist Biol* 1974;**7**(1):101−10.
6. Dubos RJ. *The professor, the institute, and DNA*. New York, NY: The Rockefeller University Press; 1976.
7. Diamond Jr AM. Avery's "neurotic reluctance". *Perspect Biol Med* 1982;**26**(1):132−6.
8. Russell N. Oswald Avery and the origin of molecular biology. *Br J Hist Sci* 1988;**21** (4):393−400.
9. Amsterdamska O. From pneumonia to DNA: the research career of Oswald T. Avery. *Hist Stud Phys Biol Sci* 1993;**24**(Pt 1):1−40.
10. Lederberg J. The transformation of genetics by DNA: an anniversary celebration of Avery, MacLeod and McCarty (1944). *Genetics* 1994;**136**(2):423−6.
11. Bearn AG, Oswald T. Avery and the Copley Medal of the Royal Society. *Perspect Biol Med* 1996;**39**(4):550−4.
12. Austrian R, Oswald T. Avery: the wizard of York Avenue. *JAMA* 1999;**107**(1):7−11.
13. Reichard P, Osvald T. Avery and the Nobel Prize in medicine. *J Biol Chem* 2002;**277** (16):13355−62.
14. Ghose T. Oswald Avery: the professor, DNA, and the Nobel Prize that eluded him. *Can Bull Med Hist* 2004;**21**(1):135−44.
15. Portugal F, Oswald T. Avery: Nobel laureate or Noble luminary? *Perspect Biol Med* 2010;**53**(4):558−70.
16. Morens DM, Taubenberger JK, Fauci AS. Predominant role of bacterial pneumonia as a cause of death in pandemic influenza: implications for pandemic influenza preparedness. *J Infect Dis* 2008;**198**(7):962−70.
17. Austrian R. The pneumococcus at the millennium: not down, not out. *J Infect Dis* 1999;**179**(Suppl. 2):S338−41.
18. Grabenstein JD, Klugman KP. A century of pneumococcal vaccination research in humans. *Clin Microbiol Infect* 2012;**18**(Suppl. 5):15−24.
19. Sternberg GM. A fatal form of septicemia in the rabbit produced by subcutaneous injection of human saliva. *Natl Board Health Bull* 1881;**2**:781−3.
20. Pasteur L. Sur une maladie nouvelle, provoquée par la salive d'un enfant mort de la rage. *Compt Rend Acad Sci (Paris)* 1881;**92**(3):1259−60.
21. Friedländer C. Über die scizomyceten bei der acuten fibrosen pneumonie. *Arch Pathol Anat Physiol Klin Med* 1882;**87**(3):319−24.
22. Friedländer C. Die mikrokokken der pneumonie. *Fortschr Med* 1883;**1**(22):715−33.
23. Friedländer C. Weitere arbeiten über die schizomyceten der pneumonie und der meningitis. *Fortschr Med* 1886;**4**(21):702.
24. Fränkel A. Über die genuine Pneumonie. *Verh Congr Innere Med Dritter Congr* 1884;**3**:17−31.
25. Fränkel A. Weitere beiträge zur Lehre von den mikrococcen der genuinen fibrinosen pneumonie. *Z Klin Med* 1886;**11**:437.

26. Kawamura Y, Hou XG, Sultana F, Miura H, Ezaki T. Determination of 16S rRNA sequences of *Streptococcus mitis* and *Streptococcus gordonii* and phylogenetic relationships among members of the genus *Streptococcus*. *Int J Syst Bacteriol* 1995;**45**(2):406−8.

27. Neufeld F. Über die agglutination der pneumokokken und über die theorien der agglutination. *Z Hyg Infektionskr* 1902;**40**:54−72.

28. Neufeld F, Händel L. Weitere untersuchungen uber pneumokokken heilsera. III. Mitteilung. uber vorkommenen und bedeutung atypischer varietäten des pneumokokkus. *Arb a d Kaiserl Gesund* 1910;**34**:293−304.

29. Dochez AR, Gillespie LJ. A biologic classification of pneumococci by means of immunity reactions. *Jama* 1913;**61**(10):727−32.

30. Dochez AR, Avery OT. The elaboration of specific soluble substance by *pneumococcus* during growth. *J Exp Med* 1917;**26**(4):477−93.

31. Dubos R, Avery OT. Decomposition of the capsular polysaccharide of *pneumococcus* type III by a bacterial enzyme. *J Exp Med* 1931;**54**(1):51−71.

32. Avery OT, Dubos R. The protective action of a specific enzyme against type III *pneumococcus* infection in mice. *J Exp Med* 1931;**54**(1):73−89.

33. Heidelberger M, Avery OT. The soluble specific substance of pneumococcus. *J Exp Med* 1923;**38**(1):73−9.

34. Heidelberger M, Avery OT. The soluble specific substance of pneumococcus: second paper. *J Exp Med* 1924;**40**(3):301−17.

35. Heidelberger M, Goebel WF, Avery OT. The soluble specific substance of a strain of Friedländer's bacillus. Paper I. *J Exp Med* 1925;**42**(5):727−45.

36. Griffith F. Types of pneumococci obtained from cases of lobar pneumonia. *Rep Public Health Med Subj* 1922;**13**:20−45.

37. Griffith F. The significance of pneumococcal types. *J Hyg (Lond)* 1928;**27**(2):113−59.

38. Arkwright JA. Variation in bacteria in relation to agglutination both by salts and by specific serum. *J Path Bact* 1921;**24**(1):36−60.

39. Griffith F. The influence of immune serum on the biological properties of pneumococci. *Rep Public Health Med Subj* 1923;**13**:20−45.

40. McCarty M. *The transforming principle: discovering that genes are made of DNA.* New York and London: W. W. Norton & Company; 1986.

41. Pollock MR. The discovery of DNA: an ironic tale of chance, prejudice and insight. *J Gen Microbiol* 1970;**63**(1):1−20.

42. H.D.W. Obituary. *The Lancet* 1941;**237**(6140): 588−589.

43. Neufeld F, Levinthal W. Beiträge zur variabilität der pneumokokken. *Z Immun Forschg* 1928;**55**:324−40.

44. Reimann HA. The reversion of R to S pneumococci. *J Exp Med* 1929;**49**(2):237−49.

45. Olby R. *The path to the double helix: the discovery of DNA.* New York, NY: Dover Publications; 1992.

46. Dawson MH. The transformation of pneumococcal types : I. The conversion of R forms of pneumococcus into S forms of the homologous type. *J Exp Med* 1930;**51**(1):99−122.

47. Dawson MH, Sia RH. In vitro transformation of pneumococcal types : I. A technique for inducing transformation of pneumococcal types in vitro. *J Exp Med* 1931;**54**(5):681−99.

48. Alloway JL. The transformation in vitro of R pneumococci into S forms of different specific types by the use of filtered pneumococcus extracts. *J Exp Med* 1932;**55**(1):91−9.

49. Avery OT. *Report of the Director of the Hospital to the Corporation of the Rockefeller Institute for Medical Research.* http://profilesnlmnihgov/ps/access/CCAAMVpdf; 1933.

50. Alloway JL. Further observations on the use of pneumococcus extracts in effecting transformation of type in vitro. *J Exp Med* 1933;**57**(2):265−78.

51. Macleod CM. The inhibition of the bacteriostatic action of sulfonamide drugs by substances of animal and bacterial origin. *J Exp Med* 1940;**72**(3):217−32.

52. Macleod CM, Avery OT. The occurrence during acute infections of a protein not normally present in the blood: II. Isolation and properties of the reactive protein. *J Exp Med* 1941;**73**(2):183−90.

53. Macleod CM, Avery OT. The occurrence during acute infections of a protein not normally present in the blood: III. Immunological properties of the C-reactive protein and its differentiation from normal blood proteins. *J Exp Med* 1941;**73**(2):191−200.

54. McDermott W. Colin Munro MacLeod 1909−1972. *Biogr Mem Natl Acad Sci* 1983;**54**:181−219.

55. Mirsky AE, Pollister AW. Nucleoproteins of cell nuclei. *Proc Natl Acad Sci USA* 1942;**28**(9):344−52.

56. Mirsky AE, Pollister AW. Chromosin, a desoxyribose nucleoprotein complex of the cell nucleus. *J Gen Physiol* 1946;**30**(2):117−48.

57. Avery OT. *Report of the Director of the Hospital to the Corporation of the Rockefeller Institute for Medical Research.* http://profilesnlmnihgov/ps/access/CCAADSpdf; 1943.

58. Hotchkiss RD, Ephrussi-Taylor H. Use of serum albumin as source of serum factor in pneumococcal transformation. *Fed Proc* 1951;**10**:200.

59. Macleod CM, Mirick GS. Quantitative determination of the bacteriostatic effect of the sulfonamide drugs on pneumococci. *J Bact* 1942;**44**(3):277−87.

60. Stent GS. *The coming of the golden age: a view of the end of progress.* New York, NY: The Natural History Press; 1969.

61. Stent GS. Prematurity and uniqueness in scientific discovery. *Sci Am* 1972;**227**(6):84−93.

62. Stent GS. Prematurity and uniqueness in scientific discovery. *Adv Biosci* 1972;**8**:433−49.

63. Hotchkiss RD. The identification of nucleic acids as genetic determinants. *Ann N Y Acad Sci* 1979;**325**(1):321−44.

64. Chargaff E. Chemical specificity of nucleic acids and mechanism of their enzymatic degradation. *Experientia* 1950;**6**(6):201−9.

65. Hershey AD, Chase M. Independent functions of viral protein and nucleic acid in growth of bacteriophage. *J Gen Physiol* 1952;**36**(1):39−56.

66. Watson JD, Crick FH. Molecular structure of nucleic acids; a structure for deoxyribose nucleic acid. *Nature* 1953;**171**(4356):737−8.

67. Watson JD, Crick FH. Genetical implications of the structure of deoxyribonucleic acid. *Nature* 1953;**171**(4361):964−7.

68. Cohen SC. *Alfred Ezra Mirsky. Dictionary of scientific biography.* New York, NY: Scribner's Sons; 1990. pp. 633−6.

69. Mirsky AE, Pauling L. On the structure of native, denatured, and coagulated proteins. *Proc Natl Acad Sci USA* 1936;**22**(7):439−47.

70. Boivin A. Directed mutation in colon bacilli, by inducing principle of desoxynucleic acid nature: its meaning for the general biochemistry of heredity. *Cold Spring Harb Symp Quant Biol* 1947;**12**:7−17.

71. Haddow A. Transformation of cells and viruses. *Nature* 1944;**154**(3902):194−9.

72. Gulland JM, Barker GR, Jordan DO. The chemistry of the nucleic acids and nucleoproteins. *Ann Rev Biochem* 1945;**14**:175−206.

73. Kalkar HM. The chemistry and metabolism of the compounds of phosphorus. *Ann Rev Biochem* 1945;**14**:283−308.

74. Woods DD. Bacterial metabolism. *Ann Rev Microbiol* 1947;**1**:115−40.

75. Chargaff E. On the nucleoproteins and nucleic acids of microorganisms. *Cold Spring Harb Symp Quant Biol* 1947;**12**:28−34.

76. Chargaff E. Preface to a grammar of biology. A hundred years of nucleic acid research. *Science* 1971;**172**(3984):637−42.

77. Chargaff E, Vischer E, et al. The composition of the desoxypentose nucleic acids of thymus and spleen. *J Biol Chem* 1949;**177**(1):405−16.

78. Watson JD. *The annotated and illustrated double helix*. New York, NY: Simon & Schuster; 2012.

79. Burnet M. *Changing patterns; an atypical autobiography*. Melbourne, Vic.: Heinemann; 1968.

80. Dale H. Address of the President Sir Henry Dale, O.M., G.B.E., at the anniversary meeting, 30 November 1945. *Proc R Soc* 1946;**185A**(1001):127−43.

81. Beadle GW. Genes and biological enigmas. *Am Sci* 1948;**36**(1):69−74.

82. Beadle GW. Physiological aspects of genetics. *Annu Rev Biochem* 1948;**17**:727−52.

83. Beadle GW. *Genetic control of metabolism: the first R.E. Dyer lecture [US PHS Publ. No. 142]*. Washington, DC: US Gov. Printing Office; 1952.

84. Beadle GW. The role of the nucleus in heredity. In: McElroy WD, Glass B, editors. *A symposium on the chemical basis of heredity*. Baltimore, MD: The Johns Hopkins Press; 1956. pp. 3−22.

85. Marshak A, Walker AC. Mitosis in regenerating liver. *Science* 1945;**101**(2613):94−5.

86. Murphy JB. Experimental approach to the cancer problem. I. Four important phases of cancer research. II. Avian tumors in relation to the general problem of malignancy. *Bull Johns Hopkins Hosp* 1935;**56**(1):1−6.

87. Stanley WM. *Biochemistry and biophysics of viruses. Handbuch der Virusforschung*. Wien: Springer; 1938. pp. 447−545.

88. Knight CA. The nucleoproteins in virus reproduction. *Cold Spring Harb Symp Quant Biol* 1947;**12**:104−14.

89. Stahl FW, editor. *We can sleep later: Alfred D. Hershey and the origins of molecular biology*. New York, NY: Cold Spring Harbor Laboratory Press; 2000.

90. Pollister AW, Swift H, Alfert M. Studies on the desoxypentose nucleic acid content of animal nuclei. *J Cell Physiol Suppl* 1951;**38**(Suppl. 1):101−19.

91. McCarty M, Taylor HE, Avery OT. Biochemical studies of environmental factors essential in transformation of pneumococcal types. *Cold Spring Harb Symp Quant Biol* 1946;**11**:177−83.

92. Fischer FG, Böttger I, Lehman-Echternacht H. Über die thymo-polynucleotidase aus pankreas nucleinsäure V. *Z Physiol Chem* 1941;**271**:246−64.

93. McCarty M. Reversible inactivation of the substance inducing transformation of pneumococcal types. *J Exp Med* 1945;**81**(5):501−14.

94. McCarty M, Avery OT. Studies on the chemical nature of the substance inducing transformation of pneumococcal types; II. Effect of desoxyribonuclease on the biological activity of the transforming substance. *J Exp Med* 1946;**83**(2):89−96.

95. McCarty M, Avery OT. Studies on the chemical nature of the substance inducing transformation on pneumococcal types; III. An improved method for the isolation of the transforming substance and its application to Pneumococcus Types II, III, and VI. *J Exp Med* 1946;**83**(2):97−104.

96. McCarty M. Purification and properties of desoxyribonuclease isolated from beef pancreas. *J Gen Physiol* 1946;**29**(3):123−39.

97. Dochez AR, Avery OT. Antiblastic immunity. *J Exp Med* 1916;**23**(1):61−8.

98. Blake FG. Studies on antiblastic immunity. *J Exp Med* 1917;**26**(4):563−80.

99. Barber MA. Antiblastic phenomena in active acquired immunity and in natural immunity to pneumococcus. *J Exp Med* 1919;**30**(6):589−96.

100. Barber MA. A study by the single cell method of the influence of homologous antipneumococcic serum on the growth rate of pneumococcus. *J Exp Med* 1919;**30**(6):569−87.

101. Award of the Gold Medal of the New York Academy of Medicine. *Science* 1944;**100**(2598):328−9.

102. The Lasker Awards for 1947. *Am J Pub Health Nations Health* 1947;**37**(12):1612−16.

103. Judson HF. *The eighth day of creation*. Expanded ed. New York, NY: Cold Spring Harbor Laboratory Press; 1996.

104. Hotchkiss RD. Ètud chimiques sur le facteur transformant du pneumocoque. *Colloq Intern Centre Natl Rechereche Sci (Paris)* 1959;57−65.

105. Hotchkiss RD. The biological nature of the bacterial transforming factors. *Exp Cell Res Suppl* 1952;**2**:383−9.

106. Hotchkiss RD. Growing up into our long genes. In: Stahl FW, editor. *We can sleep later*. New York, NY: Cold Spring Harbor Press; 2000. pp. 33−43.

107. Cohen JS, Portugal FH. The search for the chemical structure of DNA. *Conn Med* 1974;**38**(10):551−2 554−7.

108. Boivin A, Vendrely R, Lehoult Y. L'acide thymonucléique hautement polymérisé, prineipe capable de conditionner la spécificité sérologique et léquipement enzymatique des bactéries. Conséquence pour la biochimie de l'hérédité. *C R Acad Sci Paris* 1945;**221**:646−7.

109. Boivin A, Vendrely R. Rôle de l'acide désoxy-ribonucléique hautement polymérisé dans le déterminisme des caractéres héréditaires des bactéries. Signification pour la biochimie générale de l'hérédité. *Helvet Chim Acta* 1946;**29**(5):1338−44.

110. Olby R. Avery in retrospect. *Nature* 1972;**239**(5370):295−6.

111. Mandel M, Higa A. Calcium-dependent bacteriophage DNA infection. *J Mol Biol* 1970;**53**(1):159−62.

112. Cohen SN, Chang AC, Hsu L. Nonchromosomal antibiotic resistance in bacteria: genetic transformation of *Escherichia coli* by R-factor DNA. *Proc Natl Acad Sci USA* 1972;**69**(8):2110−14.

113. Obituary Notice: Oswald Theodore Avery, 1877−1955. *J Gen Microbiol* 1957;**17**(3):539−49.

114. Dochez AR. *Oswald Theodore Avery*. National Academy of Sciences Biographical Memoirs. Washington, DC: National Academy of Sciences Press; 1958. pp. 31−49.

Chapter 4

Hershey and Chase Clinched the Role of DNA as the Genetic Material

Phage Studies Propelled the Birth of Molecular Biology

Chapter Outline

M. Fry: Landmark Experiments in Molecular Biology. DOI: http://dx.doi.org/10.1016/B978-0-12-802074-6.00004-7
111

4.1   THE LATENT PERIOD 1944–52: ALTHOUGH DNA WAS IDENTIFIED AS THE PNEUMOCOCCAL-TRANSFORMING FACTOR IT WAS NOT UNIVERSALLY RECOGNIZED AS THE GENETIC MATERIAL

The 1944 announcement by Avery, MacLeod, and McCarty that DNA was the "transforming principle" of *pneumococcus*, that is, its genetic material,[1] was received with considerable amount of skepticism. To a large extent the resistance to the proposition that DNA dictated phenotypic properties of cells was due to an entrenched belief that genes were comprised of protein whereas nucleic acids lacked the necessary structural complexity to serve as genetic material. Few prominent scientists, such as Macfarlane Burnet (1899–1985), Erwin Chargaff (1905–2002), and Joshua Lederberg (1925–2008), were convinced by the *pneumococcus* transformation results that genes were made of DNA. However, the majority of the scientific community of that time held on to the idea that the genetic material consisted of protein and that DNA played an undefined structural role in the organization of chromatin. This conception persisted after 1944 despite the introduction of additional evidence in 1946/47 that strengthened the case of DNA as the *pneumococcus* transforming material. McCarty presented at a 1946 Cold Spring Harbor meeting[2] newer data that reinforced the 1944 identification of DNA as the pneumococcal transforming principle.[1] Shortly thereafter two formal papers by McCarty and Avery further buttressed the case for DNA as the bearer of genetic information in *pneumococcus*.[3,4] Nearly two decades after the 1946 Cold Spring Harbor meeting, the biologist H. Bentley Glass (1906–2005) described the circumspect reception of McCarty's new data[5]:

> [...] The demonstration that the transforming principle is DNA is very strong, although purification is not so complete that everyone is convinced that some protein does not remain in the preparation [...] We must recognize, it was said that only a single gene need be transferred to transform the rough cells to smooth, and the presence of a single protein gene in the partially purified material cannot be firmly excluded.

Even among scientists who accepted DNA as the transforming material, its capacity to induce phenotypic change in *pneumococcus* cells was regarded as a specific case, restricted perhaps only to bacteria. The consensus in the early 1950s was, however, that nucleic acids were "orphan" molecules with no known biological function. As already cited in Chapter 3, "Avery, MacLeod, and McCarty Identified DNA as the Genetic Material," this view was articulated by Pollister, Swift, and Alfert in their 1951 summary of a 1950 Oak Ridge Conference on Nucleic Acids: "with one exception (bacterial transformation), there has been no proof of a specific biological function mediated by one of these polynucleotides."[6]

The belief that genes were made of proteins and that nucleic acids lacked a known role was radically transformed by a single elegant experiment that Alfred Hershey (1908−97) and Martha Chase (1927−2003) conducted in 1952.[7] This experiment, which used a bacterial virus (named bacteriophage or phage), as model "organism" triggered historic shift that robbed proteins of their ascribed role in heredity and established DNA as the genetic material.

The Hershey and Chase experiment capped a string of basic discoveries that gainfully used bacteriophages as model systems for the investigation of fundamental questions in biology. Because of the central place of these studies in the history of molecular biology, this chapter first describes some earlier junctures in the deployment of bacteriophages to solve biological problems and only then the Hershey and Chase experiment itself is presented.

4.2 THE DISCOVERY OF BACTERIAL VIRUSES

4.2.1 Frederick Twort Discovered in 1915 a Transferable "Bacteriolytic Agent"

The British physician Frederick Twort (Box 4.1) was the first to report in 1915 that a transferrable factor could cause lysis of *Staphylococcus* cells.[10]

Working with his brother, Twort original naïve aim was to grow vaccinia virus in a synthetic medium. His attempts to culture the virus were repeatedly aborted because of frequent contamination of the growth medium by colonies of *Staphylococcus*. Twort noticed, however, that in some cases the staphylococcal colonies became transparent as a result of spontaneous cell lysis. He next passed the bacterial lysates through a fine filter and showed that the bacteria-free filtrate could cause lysis of fresh cells. In a 1915 *Lancet* paper Twort interpreted these results as indicative of the release from lysed cells of an element, which he termed *Bacteriolytic agent*, that could recurrently lyse new cells.[10] Unfortunately, Twort's investigations were interrupted by his military service during World War I and after his return from the war he never went back to his studies on the lysing agent.

BOX 4.1 Frederick Twort and Felix d'Herelle

Frederick Twort studied medicine at St Thomas's Hospital, London. He was first trained in pathological techniques at the Clinical Laboratory of St Thomas Hospital and then worked as bacteriologist in the London Hospital. In 1909 Twort was designated superintendent of the Brown Institution, a pathology research center, where he remained for the duration of his career. In the course of his work he had made some pioneering and perhaps premature contributions. He reported the adaptation of bacteria to changed carbohydrate source and succeeded in culturing leprosy bacteria by using killed tubercle bacilli as source of growth factors. The pinnacle of his scientific accomplishments was the identification of the staphylococci lytic agent that was later named bacteriophage. Diminishing financial support for his research and the destruction of his laboratory in 1944 by a bomb forced Twort to retire in 1949. His life and scientific contributions were described in a biography authored by his son Antony.[8]

Felix d'Herelle, son to French immigrants, was born in Canada and returned to Paris as a young child. d'Herelle who had only high school education became a self-taught microbiologist. Moving back to Canada at age 24, he built a home laboratory at which he started doing experiments. After publishing his first scientific paper, he worked as bacteriologist in Guatemala, Mexico, France, and Argentina. At later times he also worked in India, Egypt, the United States, and the Soviet Union. Although his contributions to microbiology were many, d'Herelle's key achievement was his discovery of phages and his pioneering attempts to use them as therapeutic agents. For a comprehensive personal and scientific biography of d'Herelle, see Ref. 9 (Fig. 4.1).

Frederick William Twort
(1877–1950)

Felix d'Herelle
(1873–1949)

FIGURE 4.1 Frederick Twort and Felix d'Herelle.

Explaining in 1930 his decision to abandon this line of investigation Twort wrote[11]:

> [...] it was sometime after the end of the war before I was really free again to continue the investigation, but at that time most of the additional details of the phenomenon had been published by other workers under the title of 'the bacteriophage'.

By "other workers," Twort most likely alluded to the French-Canadian bacteriologist Felix d'Herelle (Box 4.1) who made in 1917 a supposedly independent discovery of bacteria lysing agents.

4.2.2 Felix d'Herelle's Independent 1917 Discovery of Bacteriophages

d'Herelle had claimed to have initially observed lysis by a filterable agent of locust-infecting *coccobacilli* without having been aware of Twort's report of the staphylococcal bacteriolytic agent. Some historians believe, however, that he was familiar with the work of Twort and that in order to assert priority, he fashioned the story of his independent discovery of a lysing agent of locust-infecting bacteria. Be it what it may, after d'Herelle moved to the Pasteur Institute in Paris to investigate "Shiga dysentery" (*Shigellosis*), he isolated an analogous lytic agent of *Shigella* bacteria. Reporting his results in 1917, he named the infective agent "bacteriophage" (meaning "bacteria devourer").[12] The claim for the existence of bacteriophages was not readily accepted. The immunologist and 1919 Nobel laureate Jules Bordet (1870−1961) argued in a 1930 Croonian Lecture that the "*the invisible virus of Felix d'Herelle*" (whom he labeled "a self-taught microbiologist") did not exist and that the bacteria themselves produced the lytic factor. Soon thereafter Bordet was proven wrong when d'Herelle successfully employed bacteriophages as therapeutic agents to fight bacterial infections. Being sympathetic to communism, d'Herelle was instrumental in establishing in the 1930s a bacteriophage institute in Soviet Georgia that productively amassed a large collection of bacterial strain−specific phages. Although deployment of phages as therapeutic agents did not achieve medical usability, the growing current threat of antibiotic-resistant bacteria recently rekindled an interest in the potential use of phages as instruments to fight bacterial diseases.[13] Historically, however, the most important contribution of phages was their use in the early phases of molecular biology as superbly useful simple model systems. Interestingly, d'Herelle himself, who remained a life-long Lamarckian, never displayed interest in the field of phage genetics, which became a central part of molecular biology. To an extent, the life and work of d'Herelle inspired the American author and the 1930 Nobel laureate Sinclair Lewis to write his influential novel *Arrowsmith* (Box 4.2).

BOX 4.2 Sinclair Lewis and His Novel *Arrowsmith*

d'Herelle's colorful life and his pursuit of phages as antibacterial therapeutic agents served in part as basis for the novel *Arrowsmith* by Sinclair Lewis.[14] This book tells the story of Martin Arrowsmith, a provincial physician who evolves into a leading scientist. Arrowsmith discovers phages while working in a research institute modeled after the Rockefeller Institute. Later he pioneers a clinical trial of phage therapy to control a bacterial epidemic.[15] Lewis, who won and turned down the 1926 Pulitzer Prize for *Arrowsmith*, went on to be the first American writer to be awarded in 1930 a Nobel Prize in Literature. In writing *Arrowsmith* he was tutored on the early chapters in the study of phages and on the birth of basic biomedical research by Paul de Kruif, the microbiologist author of the famed *Microbe Hunters* book.[16] *Arrowsmith* inspired generations of young people (among them James Watson (see Chapter 5: "Discovery of the Structure of DNA") and Seymour Benzer (see Chapter 7: "Defining the Genetic Code")) to become scientists. For a description of the scientific background to the creation of *Arrowsmith* by Lewis and de Kruif, see Ref. 17 (Fig. 4.2).

Sinclair Lewis   Published 1925
1885–1950   Pulitzer Prize 1926

FIGURE 4.2   Lewis and his novel *Arrowsmith*.

Subsequent to the discovery of phages, basic research that was carried out between the late 1930s and the 1950s in various laboratories was focused mostly on T-family bacteriophages that specifically attack *Escherichia coli*. The fine structure of these bacterial viruses was revealed by electron microscopy (Fig. 4.3), and their kinetics of replication, genetics, and biochemistry were gradually uncovered. Knowledge that was thus gathered was critical for the use of phages as model systems in the early period of molecular biology.

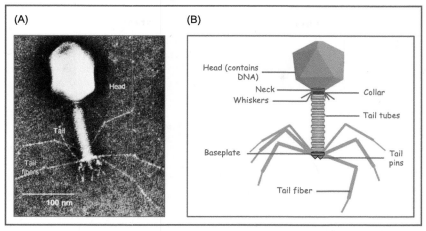

FIGURE 4.3 Anatomy of *Escherichia coli* T-family bacteriophage. (A) Electron micrograph of a phage. (B) Scheme of the fine anatomy of the phage.

4.3 INTRODUCTION OF BACTERIOPHAGE AS MODEL SYSTEM

4.3.1 Max Delbrück Pioneered the Use of Phage as the Simplest System to Investigate Basic Principles of Biology

Max Delbrück (Box 4.3), the influential physicist-turned-biologist, pioneered the use of bacteriophage as the most basic experimental system to probe fundamental principles of biology.

At the beginning of his scientific career, Delbrück studied quantum mechanics under leaders of the field: Max Born (1882−1970) in Göttingen, Niels Bohr (1885−1962) in Copenhagen, and Wolfgang Pauli (1900−58) in Zürich. Although they and others valued his intellectual potential, Delbrück did not distinguish himself as an exceptionally innovative or productive physicist.[20] Searching for a field that was open for pioneering research, Delbrück accepted an offered position at the Berlin Kaiser Wilhelm Institute as theoretician in the group of Lise Meitner (1878−1968). He chose this position in part because of the opportunity to interact with biologists in that institution.[20] After attending in 1932 a lecture by Bohr on "Light and Life," he became fully converted to the study of biology.[20−22] Discussing with biologists the outstanding problems of the day in the field, Delbrück became convinced that genetics was the most appealing and promising direction of research. Taking this course of investigation, he collaborated with the geneticist Nikolay Timofeev-Ressovsky (1900−81) and with the radiation biologist Karl Zimmer (1911−88) to predict in 1935 that X radiation induced mutations by rearranging atoms in molecules that

BOX 4.3 Max Delbrück, Salvador Luria, and Alfred (Al) Hershey

Max Delbrück, a German-American physicist-turned-biologist, was born into a family of prominent scholars. He earned a doctoral degree in theoretical physics at the University of Göttingen and did physics in Denmark, England, and Switzerland before turning to biology. After his 1937 move to the United States, Delbrück initially studied the genetics of *Drosophila*. Deeming *Drosophila* too complex a system, he chose bacteriophages as the simplest available model organisms for the exploration of fundamental laws of life. Following this line of research, he conducted with Salvador Luria their celebrated Luria–Delbrück experiment that won Luria and him the 1969 Nobel Prize for Medicine and Physiology. Equally important was his role as facilitator and leader of the Phage Group, a consortium of bacteriophage researchers whose collective efforts greatly advanced molecular biology and genetics.

After completing his studies of medicine in Turin, Salvador Luria fled fascist Italy in 1938; first to France and then in 1940 to the United States. Settling in 1943 in Indiana University, he performed the famous Luria–Delbrück experiment. Luria then moved in 1950 to the University of Illinois at Urbana-Champaign where he discovered the restriction system for host strain specificity of phage propagation. From 1959 he was on the faculty of the Massachusetts Institute of Technology (MIT) where he studies the mechanism of action of bacteriocins. From 1972 he served as Chair of the MIT Center for Cancer Research. A comprehensive assemblage of documents, correspondence, and images pertaining to Luria's life and career are to be found in an online National Library of Medicine Profiles in Science collection: http://profiles.nlm.nih.gov/QL/.

Alfred (Al) Hershey earned in 1930 a PhD degree in microbiology from Michigan State University. From 1934 he was on the faculty of Washington University where he initiated in 1940 his studies on phage genetics. In 1950 he joined the Carnegie Institution of Washington's Department of Genetics in Cold Spring Harbor where he and Martha Chase performed in 1952 their historical experiment. For this experiment, which confirmed the role of DNA as the bearer of genetic information, he shared with Delbrück and Luria the 1969 Nobel Prize for Medicine and Physiology. Much of Hershey's scientific contributions and personality were covered by Stahl[18] and more broadly in a book dedicated to him[19] (Fig. 4.4).

Max Delbrück
(1906–1981)

Salvador E. Luria
(1921–1991)

Alfred D. Hershey
(1908–1997)

FIGURE 4.4   Max Delbrück, Salvador Luria, and Alfred Hershey.

determined phenotypic properties of an organism.[23] With the rise of the Nazi regime and his diminished ability to freely conduct scientific research, Delbrück moved in 1937 to the United States. There he joined the Caltech laboratory of the eminent *Drosophila* geneticist Thomas Hunt Morgan (1866–1945). However, he soon deserted *Drosophila*, regarding it as too complex to serve as model system. He thus joined forces in the late 1930s with Emory Ellis (1906–2003) to study the simplest available system of bacteriophages. Delbrück and Ellis developed methods of counting phages and tracking their growth. They found that unlike the exponential multiplication of bacteria, phages displayed one-step growth.[24]

4.3.2 Establishment of the Phage Group

Owing to his scientific acumen and considerable charisma, Delbrück, who initially was on the faculty of Vanderbilt University and later at Caltech, founded and led the Phage Group, a loose consortium of researchers who used phage as their model system. The Group started in 1940, when Delbrück and Salvador Luria (Box 4.3) who had just recently arrived from Italy to temporarily settle in Columbia University, met at a physics conference in Philadelphia. After completing his studies of medicine, Luria was exposed to rigorous scientific thinking when he spent a year with the physics group of Enrico Fermi (1901–54) in Rome.[25] While still in Italy, he also learned about bacteriophages. Stating that his affair with phages was "love at first sight," Luria performed early experiments with the viruses that set him on a lifetime path of research.[25] Luria was highly impressed by the Timofeev-Ressovsky, Zimmer, and Delbrück paper[23] and had shared interest with Delbrück in phages.[20] It was thus fitting that their 1940 meeting led to a fruitful collaboration and to the eventual formation of the Phage Group.

The third founding father of the Phage Group was Alfred (Al) Hershey who was then working at Washington University (Box 4.3). In 1944, Delbrück introduced and promoted the "Phage Treaty," an accord between phage researchers to focus on a limited number of *E. coli*-infecting T-family phages and on a few bacterial strains, and to employ standardized experimental conditions. Abiding by this accord, different laboratories were able to compare and replicate results and thus to systematize the fields of phage and later of bacterial genetics. The Group included at different times several prominent pioneers of molecular biology such as James Watson (1928–), Seymour Benzer (1921–2007), Franklin Stahl (1929–), and Gunther Stent (1924–2008). Apart from the shared systematic methodology and the establishment of direct collaborations, a major legacy of the Group was the founding by Delbrück of an annual phage summer course at the Cold Spring Harbor Laboratory. Beginning in 1945 and continuing uninterrupted for successive 26 years, Delbrück and others used the course to teach younger biologists the fundamentals of phage biology and experimentation and to instill the Phage Group distinctive mathematics and physics-oriented approach to biology. Many alumni of the phage course became in later times leaders and productive contributors to molecular biology.

4.4 LURIA AND ANDERSON HAD A FIRST GLIMPSE OF BACTERIA-INFECTING PHAGE

The physical chemist Thomas Anderson (1911–91) pioneered the use of the first functioning electron microscope that RCA fabricated in 1940. In collaboration with Luria, he produced in 1942 the first high-resolution images of several strains of E. coli and Staphylococcus-specific bacteriophages and of their attachment to the infected bacteria. The electron microscopic pictures also revealed that incubation of the infected bacteria for about 20 minutes resulted in increase in the number of observed phage particles and damage to the cells. The assumption was that the added number of viruses reflected progeny phages that destroyed the host cells. Most interestingly, however, the bodies of the original infecting phages did not penetrate the bacterial cells and remained adsorbed to their outer wall.[26] These observations raised high interest among phage researchers and were the basis for awarding Luria a Guggenheim Fellowship (Ref. 20, p. 129). Using this award, Luria joined Delbrück in Vanderbilt in the fall of 1942 to initiate a landmark experiment that was completed after Luria got a position in Indiana University.

4.5 THE LURIA–DELBRÜCK EXPERIMENT

The crowning contribution of Delbrück and Luria was their 1943 "fluctuation test," which was later named the Luria–Delbrück experiment.[27] This experiment, which won the two scientists the 1969 Nobel Prize for Medicine and Physiology, demonstrated that phage-resistance mutations in E. coli were not induced by the virus but rather and that they arose spontaneously and at random prior to exposure of the cells to the phage. More generally, this experiment established that it is not a changed environment that induces mutations in bacteria but that they occur spontaneously and at random prior to the introduction of a selective element (in this case, a virus). To illustrate the status of science in 1942 and the challenge that Luria and Delbrück faced, it should be pointed out that at that time the influential 1956 Nobel Prize winner Sir Cyril Hinshelwood (1897–1967) argued that bacteria had no genes and that their resistance to phage was due to "change in chemical balance" (Ref. 20, p. 128). How then should one demonstrate that mutations arise at random? Luria got the idea after he already moved to the University of Indiana. There he observed a slot machine (which later appeared in the title of his autobiography[25]) that commonly yielded small or no return but that occasionally (and randomly) hit the jackpot. In analogy, if phage-resistance mutations in bacteria appeared at random, their distribution among separately grown multiple cultures should greatly vary. This idea was translated into a historic experiment to which (as indicated in the paper) Delbrück contributed the theory (calculation of expected fluctuations) and Luria provided the experimental work. *Escherichia coli* B cells were seeded in parallel onto multiple plates that contained T1 phages in their growth medium. Following incubation, phage-resistant bacterial colonies were scored in each plate and their distribution was plotted (Fig. 4.5).

If resistance to phage infection was induced in some of the cells by exposure of the bacterial populations to phage, it could be expected that the number of resistant colonies in replicate plates would have obeyed normal (Poisson) distribution. However, if mutations arose spontaneously prior to exposure to the phage, the number of resistant colonies would vary in different plates, with higher numbers of resistant colonies observed for mutations that occurred at earlier generations (Fig. 4.5A). Thus, plots of the number of resistant cells against the frequency of their incidence were expected to be exponential for spontaneous mutations whereas if mutations were induced they would follow Poisson distribution (Fig. 4.5B). In practice, Delbrück and Luria found that the distribution of phage-resistant mutant bacteria was exponential. This result indicated that mutations were not induced by the virus but rather that they were generated spontaneously prior to exposure of the cells to phages.[27]

4.6 DNA WAS ASSIGNED ONLY A SECONDARY ROLE IN EARLY THINKING ON THE CHEMICAL NATURE OF PHAGE GENETIC MATERIAL

Until the early 1950s most members of the phage research community did not consider the possibility that DNA might be the viral genetic material. In retrospect, one missed lead was the observation that UV-irradiated phages were efficiently reactivated by super-infection with nonirradiated phages. However, Luria interpreted this result as a reflection of the recombination of independently multiplying "genetic subunits."[28,29] In his thinking, the proteinaceous hereditary subunits contained very little DNA that was required for their final and unspecified process of "baking" (see Hershey's retrospective analysis of this model in Ref. 30, pp. 100–102). This perception was abandoned only after the 1952 demonstration by Hershey and Chase that the actual hereditary material of phages was DNA. Curiously, prior to that discovery Hershey did not openly contemplate the question of the chemical nature of the genetic material of phages. Before his celebrated experiment with Chase, his work was purely genetic, focusing on recombination of T-phage markers. Still, although he did not express at that earlier stage an explicit interest in the chemical composition of the genetic material of the virus, Hershey was attentive to new experimental evidence by others and to emerging ideas on a possible genetic function of DNA. With this

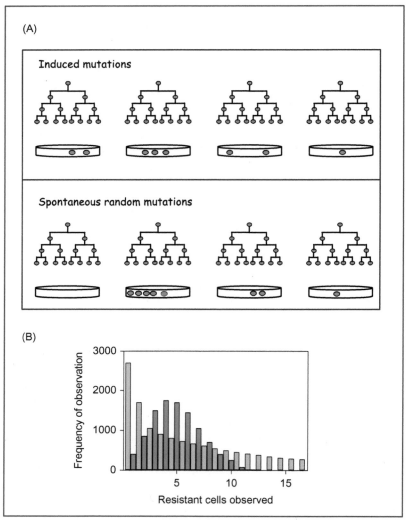

FIGURE 4.5   Schematic representation of the Luria–Delbrück experiment. (A) Scheme of the experiment: if phage-resistance mutations (*red circles*) were induced only after cells were exposed to phage, the number of mutant colonies would be normally distributed among replicate plates. However, if mutations occurred spontaneously prior to and independently of exposure to phage, replicate plates will contain variable numbers of mutants such that the earlier the generation at which a mutation occurred, the higher will be the number of resistant colonies. (B) Plots of expected distributions of the number of phage-resistant mutants. *Induced mutations*: Poisson distribution (*pink columns*). *Spontaneous mutations*: exponential distribution (*light blue columns*).

awareness in the background, Hershey's thinking and experimental approach evolved until it culminated in the critical experiment that he performed with Martha Chase.

4.7   DESPITE HIS ACTIVE STUDIES ON PHAGE GENETICS, HERSHEY SHOWED AT FIRST ONLY TACIT INTEREST IN THE CHEMICAL NATURE OF THE GENETIC MATERIAL

As an early student of bacteriophages and one of the three founding fathers of the Phage Group, Hershey's initial interest was in phage genetics. He first identified two independently segregating mutations in T2 phages: *h* (host range) and *r* (rapid lysis). Next he showed that bacteria that were infected with a mixture of the two mutants yielded recombinant viruses that carried both mutations.[31,32] Hershey next extended this observation by demonstrating similar recombination of three independently segregating markers.[33] Interestingly, however, the papers that documented these findings did not include any hint of an interest that Hershey might have had in the chemical nature of the studied genes. The record reveals, however, that despite this lack of explicit interest in the chemical composition of genes, Hershey was familiar with Avery's identification of DNA as the *pneumococcus* transforming principle and that he was attentive to the debate that this finding had raised. In an Apr. 18, 1949, letter to Rollin Hotchkiss (1911−2004) of the Rockefeller Institute, Hershey inquired (for facsimiles of the Hershey−Hotchkiss correspondence, see Ref. 19, p. 36):

> [. . .] *I hear from Adams and Delbrück that you have convinced yourself that Avery's stuff is really DNA. I would like to be able to say something about this at a round table on nucleic acids at the SAB* [Society of American Bacteriologists] *meeting. If you are willing I would appreciate hearing something about what you are doing.*

In reply, Hotchkiss sent to Hershey his unpublished data that showed that the level of protein in Avery's pneumococcal transforming DNA was negligible (at most 0.02%) (see Chapter 3: "Avery, MacLeod, and McCarty Identified DNA as the Genetic Material" for description of Hotchkiss' findings). Hershey's response in a May 3, 1949, letter reflected his perception that the Hotchkiss data strongly reinforced the claim that DNA was the true genetic material:

> . . .*The experiments are very beautiful.* [. . .] *My own feeling is that you cleared up most of the doubts.*

Evidently, therefore, although Hershey had not engaged yet in 1949 in direct efforts to identify the chemical composition of genes, he tended to accept the argument that DNA was the pneumococcal transforming principle and hence was its genetic material.

4.8 FIRST STEP TOWARD THE IDENTIFICATION OF DNA AS THE GENETIC MATERIAL OF PHAGES: VIRUS KILLING BY RADIOACTIVE PHOSPHOROUS IMPLIED A VITAL ROLE FOR NUCLEIC ACIDS

Hershey and associates obtained in 1951 an early experimental clue for the central role of nucleic acids in the ability of T bacteriophages to replicate.[34] As described at length by the historian Angela Creager,[35] this and subsequent experiments became feasible only after the radioactive ^{32}P isotope was put into use as a tool to follow the metabolism of DNA in phages. In his early experiment, Hershey propagated phages in bacteria that grew in a medium that contained ^{32}P at different specific activities. Calculable numbers of atoms of the radioactive phosphorous were incorporated into the phage particles and became integral components of the virus.[34] Scoring the viability of phages that incorporated different amounts of the ^{32}P isotope revealed that viruses were killed by the ^{32}P β radiation and that the loss of viability was proportional to the specific activity of the incorporated ^{32}P. In discussing these results, Hershey et al. concluded[34]:

> The finding that the nuclear reaction $P^{32} \rightarrow S^{32}$ kills radioactive phage with an efficiency of about 0.086 per atom transformed presumably means that at least this fraction of the phosphorous atoms of the phage particle is situated in vital structures, and therefore that the vital structures contain nucleic acid.

This inference was a first formal statement that reflected Hershey's evolving perception that nucleic acid (DNA) was vital for phage reproduction.

4.9 LAYING THE GROUNDWORK FOR THE HERSHEY AND CHASE EXPERIMENT: PHAGE DNA AND PROTEIN WERE SHOWN TO BE SEPARABLE

Prior to Hershey and Chase's critical experiment, other investigators gained first inkling about the different fates of the phage protein and DNA subsequent to the viral infection of bacteria.

4.9.1 DNA Was Evicted From Osmotically Shocked Phages Leaving Behind DNA-Free "Ghosts"

While Hershey was conducting his analysis of phage killing by ^{32}P β radiation, reports from several other laboratories laid crucial foundations for the Hershey and Chase 1952 landmark experiment. Thomas Anderson subjected in 1950 T-phages to osmotic shock by first suspending the viruses in high concentration of NaCl and then diluting them in water. Electron microscopy

BOX 4.4 Roger Herriot

Herriot had earned MSc and PhD degrees in chemistry from Columbia University and did his postdoctoral studies on pepsin action with the Nobel laureate John Northrop at the Rockefeller Institute in Princeton. After studying the effects of mustard gas during the Second World War, he joined Johns Hopkins University in 1948 at which he served for 27 years as Head of the Biochemistry Department at the School of Hygiene and Public Health. His most prominent scientific achievements were the separation of phage DNA from protein ghosts and his perceptive prognostication of how phages might propagate by injecting their DNA into the infected bacteria (Fig. 4.6).

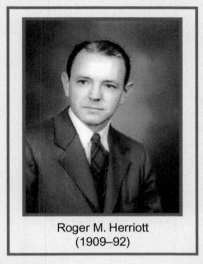

Roger M. Herriott
(1909–92)

FIGURE 4.6   Roger M. Herriott.

revealed that the osmotically shocked phages released their DNA, leaving behind DNA-free protein "ghosts."[36] About a year later, Roger Herriot at Johns Hopkins University (Box 4.4) showed that after the productive infection of bacteria by T2 phage, DNA-free "ghosts" of the phage remained absorbed to the walls of the infected bacteria.

The viral ghosts could be pelleted by centrifugation leaving nearly pure DNA in the supernatant.[37] The biophysicist and 1946 Nobel Prize laureate John H. Northrop (1891–1987) conducted in the University of California Berkeley similar studies on a phage of *Bacillus megatherium*. Based on his results, he made in 1951 the following farsighted comment[38]:

*The nucleic acid may be essential, autocatalytic part of the molecule, as in the case of the transforming principle of pneumococcus (*Avery, MacLeod and McCarty, 1944*), and the protein portion may be necessary only to allow entrance to the host cell.*

4.9.2 Roger Herriot Astutely Hypothesized That by Acting as a Hypodermic Needle, the Phage Injected Its DNA Into the Infected Bacterium

Most perceptively, Roger Herriot daringly postulated a mechanism of phage infection. In a Nov. 16, 1951, letter to Hershey (cited by Olby,[39] pp. 317–318), he wrote:

> *I've been thinking—and perhaps you have, too—that the virus may act like a little hypodermic needle full of transforming principle; that the virus as such never enters the cell; that only the tail contacts the host and perhaps enzymatically cuts a small hole through the outer membrane and then the nucleic acid of the virus head flows into the cell.*

Yet, Herriot struggled with the design of an experiment that might put his prediction to a test:

> *[...] If this is so, then two experiments suggest themselves (a) one should be able to get virus formed by the nucleic acid alone if one only knew how to get into the cell—and that of course, is the $64 problem, and although I have some idea on how to approach it, I'm not very proud of them. The other thing is that if the above notion is correct, then one should find the ghost in the cell debris after lysis. The latter shouldn't be hard to determine. I got ^{35}S to do it and heard recently that you have done some experiments very much like this. If you are working along this line, I'll work on something else for there are plenty of things to be done. If you are on a different trail entirely, I'd like to answer the problem or idea posted above.*

In a letter to Herriot dated Nov. 20, 1951 (cited in Ref. 40, p. 318), Hershey responded:

> *I have arrived at your notion the hard way, namely by doing the ^{35}S experiments you planned [...] There is little or no ^{35}S in the phage progeny, because we are able to separate it from them sometimes. I believe we now have a reliable method for doing this [...]*
>
> *Your idea about the nucleic acid is an intriguing one. I haven't any notion how to prove it. At present we can say that the intracellular prophage is either very small or very fragile [...]*
>
> *I might add my opinion, for what it is worth, that the intracellular phage cannot be soluble nucleic acid, but is more likely to be a small, highly organized nucleus. The surprising thing is that it probably contains little protein.*

Two points of interest emerge from Hershey's letter: (1) although he might not had shared at the outset Herriot's insight that it was the phage DNA that penetrated the bacteria, he was alerted to this possibility after he found that the phage progeny did not inherit the parental ^{35}S-labeled proteins. (2) In writing "I believe we now have a reliable method for doing this

[. . .]" Hershey most likely alluded to his use of a blender (dubbed then "Waring blendor" for the manufacturing Waring corporation) to separate phage ghosts from the bacteria early after their infection. This technique was at the crux of the Hershey and Chase experiment (*vide infra*).

4.10 THE HERSHEY AND CHASE EXPERIMENT IDENTIFIED DNA AS THE GENETIC MATERIAL OF PHAGE

In an Apr. 9, 1952, paper in the *Journal of General Physiology*,[7] Hershey and Chase (Fig. 4.7A; Box 4.5) provided definitive experimental evidence that the genetic material of T2 phages was their DNA.

4.10.1 Demonstration of the Separability of the Phage DNA From Its Protein Coat

Prior to the decisive experiment that identified DNA as the viral genetic material, Hershey and Chase conducted several preparatory experiments that confirmed the separability of phage DNA and protein, defined conditions for phage adsorption to bacteria, and determined the timing of the penetration of the viral DNA into the infected bacteria. In the first of these pilot experiments, Hershey and Chase confirmed the observations of Anderson[36] and of Herriot[37] on the separability of the DNA and protein components of T2 bacteriophage. Phage DNA or proteins were, respectively, labeled with ^{32}P or ^{35}S. Comparisons were then made of the sensitivity of intact or osmotically shocked viruses to DNase digestion, of

(A)

Martha Chase and Alfred Hershey
around the time of their 1952 experiment

(B)

The Waring blender that
Hershey and Chase used

FIGURE 4.7 (A) Martha Chase and Al Hershey circa the time of their experiment. (B) The blender ("Waring blendor") that was used by Hershey and Chase to detach adsorbed phage ghosts after the viral DNA was injected into the infected bacteria. The blender is kept in the Cold Spring Harbor Laboratory Archives.

Martha Chase
(1928–2003)

FIGURE 4.8 Martha Chase.

their ability to absorb to bacteria, and of their precipitation by specific antiphage antibodies. As illustrated in Fig. 4.9, whereas DNA of intact T2 phages was inaccessible to digestion by DNase, DNA that was released from the osmotically shocked viruses became exposed to the enzyme and was hydrolyzed.

Also, ghosts of the osmotically shocked phages retained almost full capacity to absorb to sensitive bacteria, whereas the released viral DNA did not adsorb to bacteria. Finally, antiphage antibodies precipitated the viral protein but not its DNA (Table I in Fig. 4.9). A related exploratory experiment revealed that phage DNA was rendered DNase-sensitive when the viruses were adsorbed to heat-killed bacteria.[7]

In a second preparatory experiment, Hershey and Chase demonstrated that upon infection phages release their DNA *into* the bacteria. At the time

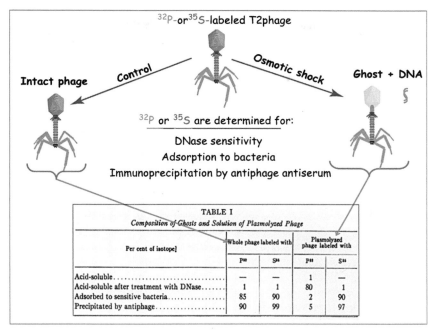

The whole figure including:

^{32}P-or^{35}S-labeled T2phage

Intact phage — Control — Osmotic shock — Ghost + DNA

^{32}P or ^{35}S are determined for:

DNase sensitivity
Adsorption to bacteria
Immunoprecipitation by antiphage antiserum

TABLE I
Composition of Ghosts and Solution of Plasmolyzed Phage

Per cent of isotope]	Whole phage labeled with		Plasmolyzed phage labeled with	
	P^{32}	S^{35}	P^{32}	S^{35}
Acid-soluble..............................	—	—	1	—
Acid-soluble after treatment with DNase.......	1	1	80	1
Adsorbed to sensitive bacteria.................	85	90	2	90
Precipitated by antiphage....................	90	99	5	97

FIGURE 4.9 Osmotic shock separated the phage DNA from its protein coat. Whereas the DNA of intact viruses resisted DNase digestion, viral DNA that was released by osmotic shock was rendered nuclease-sensitive as was assessed by the generation of acid-soluble products of the hydrolysis of DNA. Antiphage antibodies precipitated the DNA-free ghosts but not the DNA itself and only the protein ghosts, but not the viral DNA, could be absorbed onto sensitive bacteria. *Table I is from Hershey AD, Chase M. Independent functions of viral protein and nucleic acid in growth of bacteriophage. J Gen Physiol 1952;36(1):39–56.*

of that experiment not much was known about events during the interval between the adsorption of the phage to the bacterium and the appearance of newly produced progeny viruses. This limited extent of knowledge was summarized retrospectively in 1966 by Hershey (Ref. 30, p. 101):

After phage infection, the bacterial metabolic system falls under the control of the nuclear apparatus of the phage, producing mainly phage-specific materials (Cohen, 1949). The infecting phage particle does not itself survive the infection, losing at least its ability to infect another bacterium (Doermann, 1948). Thus intracellular phage growth is characterized by a phase of eclipse, lasting about half the life of the infected bacterium, during which the cell does not contain demonstrable phage particles. During the phase of eclipse, genetic recombination takes place, apparently preceded by replication of genetic determinants of the phage. Phage growth therefore consists of two stages: replication of some noninfective form of virus, followed by conversion of the products of replication back into finished phage particles which do not themselves participate in reproduction.

In addressing the matter of molecular events during the eclipse period, Hershey and Chase found that the phage DNA was internalized at an early stage into the infected bacterium. This was shown in an experiment in which the phage DNA was labeled with ^{32}P and the viruses were adsorbed onto sensitive bacteria for 5 minutes. After completion of the absorption phase, freezing and thawing were employed to permeabilize the bacterial cell walls and the bacteria were fixed with formaldehyde. Control bacteria were not permeabilized and the intact cells were fixed by formaldehyde. Subsequently, cells and supernatant were separated by centrifugation and DNA in each fraction was examined for its sensitivity to DNase digestion (Fig. 4.10). Results of this experiment (Table III in Fig. 4.10) revealed that although some of the viral DNA leached to the supernatant, it mostly remained confined within the permeabilized and fixed cells.

Whereas phage DNA in control cells that were not permeabilized was inaccessible to DNase, a large part of the DNA in the permeabilized and fixed cells was digested by the nuclease. These results indicated that internalization of phage DNA was an early event in the eclipse period. This finding, together with the previous report that phage protein ghosts remained outside the cells,[37] implied that upon infection the viral DNA was injected into the bacteria (or as Hershey and Chase originally put it[7]: "*[. . .] adsorption is followed by ejection of the phage DNA from its protective coat.*").

To additionally demonstrate the different fates of the phage protein coat and of its DNA, T2 or T4 phages whose protein or DNA were respectively labeled with ^{35}S or ^{32}P were mixed with bacterial debris (comprising mostly of disrupted cell walls) and the amounts of the adsorbed radioactive isotopes were measured.[7] Although the results of these experiments were instructive about the specificity of phage adsorption to the bacterial cell wall, they were inconclusive as to the release and fate of the viral DNA. As described later, this problem was ultimately tackled by the clever use of a blender to separate the infected bacterial cells and empty phage ghosts that remained after the viral DNA was injected into the bacteria.

4.10.2   The Hershey and Chase Blender Experiment Suggested That the Phage Genetic Material Was Its DNA

Hershey and Chase devised an ingenious experiment to address directly the question of the parts that the viral protein and DNA played in phage infection and replication. Based on the electron microscopic observations of Anderson,[36] Hershey and Chase specified the rationale and method of their experiment[7]:

Anderson (1951) has obtained electron micrographs indicating that phage T2 attaches to bacteria by its tail. If this precarious attachment is preserved during the progress of the infection, and if the conclusions reached above are correct, it ought to be a simple matter to break the empty phage membranes off the infected bacteria, leaving the phage DNA inside the cells.

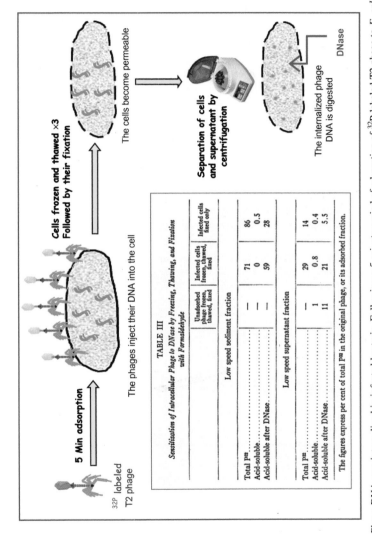

5 Min adsorption

³²P labeled T2 phage

The phages inject their DNA into the cell

Cells frozen and thawed ×3
Followed by their fixation

The cells become permeable

Separation of cells and supernatant by centrifugation

The internalized phage DNA is digested

DNase

TABLE III

Sensitization of Intracellular Phage to DNase by Freezing, Thawing, and Fixation with Formaldehyde

	Unadsorbed phage frozen, thawed, fixed	Infected cells frozen, thawed, fixed	Infected cells fixed only
Low speed sediment fraction			
Total P³²........................	—	71	86
Acid-soluble....................	—	0	0.5
Acid-soluble after DNase...	—	59	28
Low speed supernatant fraction			
Total P³²........................	—	29	14
Acid-soluble....................	1	0.8	0.4
Acid-soluble after DNase...	11	21	5.5

The figures express per cent of total P³² in the original phage, or its adsorbed fraction.

FIGURE 4.10 Phage DNA was internalized in infected bacteria. Following a 5-minute period of adsorption of ³²P-labeled T2 phage to *E. coli* cells, the bacteria were permeabilized by freezing and thawing and then fixed by formaldehyde. Two controls were unadsorbed, frozen-thawed and fixed free phages and formaldehyde-fixed intact infected cells. After the cells and supernatant were separated by centrifugation, the various fractions were treated with DNase and hydrolysis of the DNA was monitored by the release of acid-soluble ³²P nucleotides. *Table III is from Hershey AD, Chase M. Independent functions of viral protein and nucleic acid in growth of bacteriophage. J Gen Physiol 1952;36(1):39–56.*

In practice, sensitive bacteria were infected by phages whose DNA or proteins were radioactively labeled. Following internalization of the viral DNA into the cells, phage ghosts (or as dubbed by Hershey and Chase; *empty phage membranes*) were severed off the bacteria and the fates of the labeled viral DNA or protein were monitored. As Fig. 4.11 schematically illustrates, live *E. coli* cells were infected with ^{32}P- or ^{35}S-labeled T2 bacteriophages and after a brief adsorption period, the infected bacterial suspensions were subjected to agitation in a blender for different periods of time (Fig. 4.7B).

This treatment cut offs the adsorbed viral ghosts from the bacteria. The supernatant fraction that contained the phage ghosts was next separated by centrifugation from the bacteria that contained the internalized viral DNA and ^{35}S and ^{32}P radioactivity was measured in the supernatant. The effect of agitation in the blender on cell viability was determined by titration with antiphage antiserum of the number of bacteria that were capable of yielding phage particles.

Results of this experiment (Fig. 4.12) revealed that agitation of the bacteria in the blender for up to 8 minutes did not affect their viability to a significant extent. Without agitation in the blender only small respective proportions of 5% and 15% of the ^{32}P DNA and ^{35}S-labeled phage protein leached spontaneously into the supernatant. However, agitation of the bacteria in the blender released different proportions of ^{32}P- or ^{35}S-labeled phages. Whereas up to only ~30% of the ^{32}P-labeled DNA was freed from the bacteria, about 75−80% of the ^{35}S was stripped from the cells (Fig. 4.12).

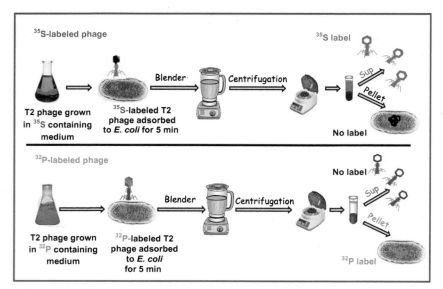

FIGURE 4.11 Scheme of the Hershey and Chase blender experiment. See text for details of the experimental procedure.

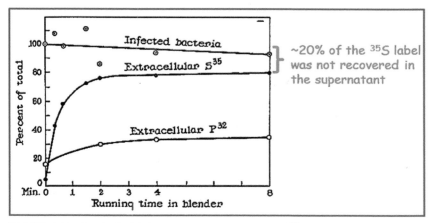

FIGURE 4.12 Survival of infected bacteria and release of ^{35}S-labeled phage protein and ^{32}P-labeled phage DNA from bacteria in the course of agitation in blender. *Graph is from Hershey AD, Chase M. Independent functions of viral protein and nucleic acid in growth of bacteriophage. J Gen Physiol 1952;36(1):39–56.*

This result was consistent with the idea that most of the ^{32}P-labeled viral DNA was internalized within the bacteria, whereas the ^{35}S-labeled phage protein remained outside and was thus detached from the cells by the blades of the blender. However, although the blender stripped off the bulk of the ^{35}S-labeled phage protein from the infected cells, about 20% of the radioactivity of this isotope remained unaccounted for (Fig. 4.12).

The possibility that this missing fraction represented internalized, and possibly heritable, protein was discounted in an experiment that measured the amount of ^{35}S protein that persisted in first-generation progeny phage. Essentially, cells were infected by ^{35}S-labeled phages and then were either subjected to agitation in the blender to strip phage ghosts off the bacteria or were left untreated (Fig. 4.13A).

Results of this experiment (Fig. 4.13B) showed that control cells that were not agitated in the blender released progeny phages at a yield of nearly 90%. These recovered viruses carried over 9.4% of the initial ^{35}S protein radioactivity. By contrast, a lower yield of nearly 60% of progeny phages released from the blender-agitated cells notwithstanding, these viruses contained only 0.7% of the initial ^{35}S parental radioactivity. This negligible amount of labeled protein strongly suggested that parental phage protein was not inherited and it was thus unlikely that it functioned as the genetic material. Yet, it should be pointed out that the measured relative amount of contaminating protein in the progeny phage was greater than the amount of residual protein in Avery's *pneumococcus* transforming principle (see Chapter 3: "Avery, MacLeod, and McCarty Identified DNA as the Genetic Material"). This had left some lingering doubts in the minds of

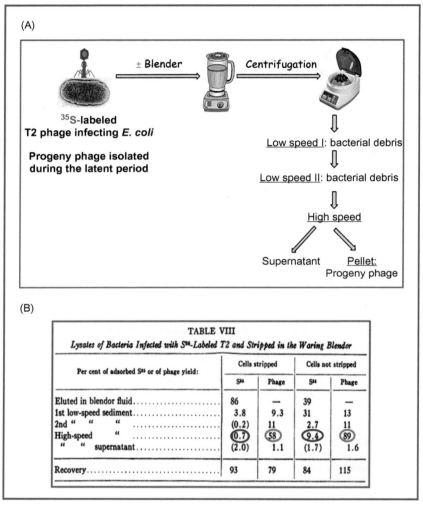

FIGURE 4.13   Progeny phages do not carry over parental ^{35}S-labeled proteins. (A) Scheme of the experimental procedure (see text). (B) Experimental results indicate that whereas nearly 60% of the parental ^{32}P DNA was recovered in the progeny phages (*red circle*) their content of ^{35}S-labeled protein was only 0.7% (*blue circle*). *Table VIII is from Hershey AD, Chase M. Independent functions of viral protein and nucleic acid in growth of bacteriophage. J Gen Physiol 1952;36(1):39–56.*

Hershey and Chase on whether or not the residual protein was heritable. The summarizing statement of their paper reflected their caution in interpreting the experimental results[7]:

> [...] *the following questions remain unanswered. (1) Does any sulfur-free phage material other than DNA enter the cell? (2) If so, is it transferred to the phage progeny? (3) Is the transfer of phosphorus (or hypothetical other*

*substance) to progeny direct--that is, does it remain at all times in a form
specifically identifiable as phage substance or indirect?*

*Our experiments show clearly that a physical separation of the phage T2 into
genetic and non-genetic parts is possible. [...] The chemical identification of the
genetic part must wait, however, until some of the questions asked above have
been answered.*

Hershey's judicious position can perhaps be explained by his awareness of
the criticism that was provoked by Avery's claim that DNA was the transforming principle of *pneumococcus*.[1] A principal argument against that finding
was that rather than DNA, it was minute amount of associated pneumococcal
protein that could have acted as the transforming material (see Chapter 3:
"Avery, MacLeod, and McCarty Identified DNA as the Genetic Material").
Hershey, therefore, was probably aware that the presence of very small but measurable amount of viral protein in first-generation progeny phages could raise
similar disapproval. Surprisingly, however, despite Hershey and Chase's cautious proclamations, their results were quite readily accepted as a bona fide demonstration that phage DNA and not its protein was the viral genetic material.

4.10.3 The Problem of Residual Protein in Progeny Phages Was Solved Years After DNA Had Already Been Universally Recognized as the Genetic Material of Phage

Notably, several years after the results of the Hershey and Chase experiment
had already been universally accepted as proof that phage DNA and not its
protein was the genetic material, the technical problem of residual protein in
the viral DNA was solved. The cell wall of *E. coli* can be removed by treat-
ing the bacteria with lysozyme in the presence of the cation-binding chelator
ethylenediaminetetraacetic acid (EDTA). Unlike intact bacteria, the resulting
so-called protoplasts, which are cells that were divested of their wall, become
permeable to DNA. In 1957, the Indiana University group of Charles A.
Thomas transfected *E. coli* protoplasts with purified DNA of T2 bacteriophage
(Fig. 4.14A). Protoplasts that were transfected with the T2 DNA yielded progeny of nearly 100 phage particles per bacterium (Fig. 4.14B).

The failure to produce progeny viruses by incubating under transfection
conditions phage DNA or uninfected protoplasts alone (Fig. 4.14B) indicated
that transfection of protoplasts with purified DNA represented bona fide multiplication of the viruses within the protoplasts.[41]

At about the same time John Spizizen (1917−2010) of Case Western
University reported that protoplasts of bacteria that were normally resistant
to infection by intact T2 bacteriophage yielded progeny viruses when they
were transfected with disrupted T2 phages. Although for some technical reason transfection of the protoplasts by purified DNA did not produce progeny
viruses, the effective transfection with disrupted phages was nevertheless

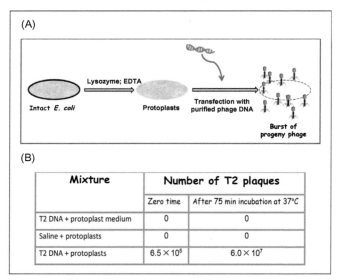

FIGURE 4.14 Purified phage DNA effectively mediated phage propagation. (A) Schematic representation of infection of *E. coli* protoplasts by isolated T2 phage DNA.[41] (B) Representative results of the experiment. *Panel (B) is from Fraser D, Mahler HR, Shug AL, Thomas CA. The infection of sub-cellular Escherichia coli, strain B, with a DNA preparation from T2 bacteriophage. Proc Natl Acad Sci USA 1957;43(11):939–47.*

interpreted as an indication that the viral DNA was responsible for propagation of the phages.[42]

4.11 EARLY REACTION TO THE HERSHEY AND CHASE RESULTS

Whereas the 1944 identification by Avery, MacLeod, and McCarty of DNA as the *pneumococcus* transforming principle was met with lingering skepticism, the evidence of Hershey and Chase that phage DNA was its genetic material was received more readily and approvingly by most, though not by all, of the interested scholars. Reacting to Hershey's account of his results prior to their formal publication, the Danish microbiologist Ole Maaløe (1915–1988) wrote to him on Apr. 2, 1952: "Very beautiful piece of work, and it gives us a lot to think about" (this one and the following letters are cited in Ref. 39, p. 319). On the same day, Max Delbrück also wrote "The best paper you ever wrote, as to substance, I mean. For once I am really envious [...] Weidel's paper had some very suggestive evidence that bacterial membrane can do something to the phages, namely cause a phage substance to be released. He swore it was not DNA, but I think it must have been DNA." In contrast to these positive reactions, Luria expressed in an earlier Mar. 2, 1952, letter to Hershey some reservations. Being under the impression that 30–50% of the parental ^{35}S-label was found in progeny viruses, he wrote: "I must say, your story sounds convincing and plausible [...] There are, however, several facts that make me keep an open mind as to whether DNA or protein rules reproduction."

The Hershey and Chase result had perhaps its historically most significant impact on James Watson who was then a postdoctoral fellow at the University of Cambridge. As was the case, Luria was invited to present the work of the American phage groups to attendants of an Apr. 1952 meeting of the Society of General Microbiology in Oxford, England. However, due to his left-leaning politics at a time of swelling paranoia in the United States about the threat of communism, the State Department denied Luria a passport and his travel to England was barred. Watson, who was Luria's first graduate student, was asked, therefore, to speak in his place. Just before the meeting Watson received a letter from Hershey with a detailed report of the results of the Hershey and Chase experiment. The experiment and its outcome convinced Watson that the phage DNA was its genetic material. In his talk in the Oxford meeting, Watson described the experiment and underscored its significance as evidence that DNA is the likely material that genes were made from. At the conclusion of his presentation, he stated that: "It is tempting to conclude that the virus protein functions largely as a protective coat for the DNA and that the perpetuation of genetic specificity is largely or entirely a function of the DNA."[43] Almost 50 years later, Watson retrospectively described the impact that Hershey's results had on him in 1952 (Ref. 19, pp. xi–xii):

> [...] Although there was already good evidence from the 1944 announcement of Avery, MacLeod, and McCarty at the Rockefeller Institute that DNA could genetically alter the surface properties of bacteria, its broader significance was unknown. The Hershey-Chase experiment had a much stronger impact than most confirmatory announcements and made me ever more certain that finding the three-dimensional structure of DNA was biology's next most important objective. The finding of the double helix by Francis Crick and me came only 11 months after my receipt of a long Hershey letter describing his blender experiment results. Soon afterward, I brought it to Oxford to excitedly read aloud before a large April meeting on viral multiplication.

Yet, as Watson pointed out in his book "The Double Helix" (Ref. 44, pp. 121–122), his excitement was definitely not shared by the other attendants of the Oxford meeting:

> [...] Their [Hershey and Chase's] experiment was thus a powerful new proof that DNA is the primary genetic material. Nonetheless, almost no one in the audience of over 400 microbiologists seemed interested as I read long sections of the letter. Moreover, when it came out that I was an American, my uncut hair provided no assurance that my scientific judgment was not equally bizarre.

It should also be stressed that the lack of immediate general acceptance of the Hershey and Chase result as a conclusive proof of the role of DNA as the genetic material of the phage was arguably also due to Hershey's own restrained interpretation of his data. Indeed, as he had already cautioned in the original paper,[7] Hershey went on to state later in a 1953 Pasteur Institute symposium[40]:

> *Parental DNA components are, and parental membrane components are not, materially conserved during reproduction. Whether this result has any fundamental significance is not yet clear.*

Fortunately, however, the importance of the Hershey and Chase result did not have to wait for acceptance for as long a period as it took for the Avery, MacLeod, and McCarty discovery to be recognized. Less than a year after the Hershey and Chase experiment, the role of DNA as the genetic material became evident with the 1953 publication of the Watson and Crick model of the double-stranded structure of DNA and the inherent suitability of each of its strands to be replicated (see Chapter 5: "Discovery of the Structure of DNA").

4.12 THE CONTRIBUTIONS OF AVERY, MacLEOD, AND McCARTY, OF HERSHEY AND CHASE, AND OF WATSON AND CRICK TO THE IDENTIFICATION OF DNA AS THE GENETIC MATERIAL

Despite the higher proportion of contaminating protein in the phage progeny (Fig. 4.13) than in the *pneumococcus* transforming principle (see Chapter 3: "Avery, MacLeod, and McCarty Identified DNA as the Genetic Material"), the results of Hershey and Chase were accepted with less reluctance than those of Avery, MacLeod, and McCarty. Several factors contributed to the more welcoming reception of the phage data. (1) Whereas the *pneumococcus* transformation principle induced in recipient cells change in the expression of only one or two phenotypic properties, the introduction of phage DNA into infected bacteria resulted in the reproduction of complete viruses. (2) Whereas the blender experiment was highly reproducible, *pneumococcus* transformation was harder to duplicate. (3) The image of phage particles acting as syringes that inject their DNA into the infected cells while leaving the protein coat outside of the bacteria could be easily and intuitively envisaged. By contrast, the intricate system of *pneumococcus* transformation was less understood and was harder to grasp. (4) It was argued by Stent that the discovery of Avery, MacLeod, and McCarty was premature.[45] This assertion, which was strongly contested by Hotchkiss[46], argued that at the time of the Avery discovery there was little or no interest in the chemical nature of the genetic material. By contrast, the Hershey and Chase result fell on more fertile ground by having been announced at a time of emerging interest in the biochemistry of inheritance. It is illuminating to read in this context Hershey's own comments on the importance and shortcomings of the *pneumococcus* system. In a 1994 symposium at the Rockefeller University to celebrate the 50th anniversary of the Avery, MacLeod, and McCarty paper,[1] Hershey assessed their work and commented on its subdued reception (cited in Ref. 19, p. 105):

> *Two things that struck me about the work of Avery/MacLeod/McCarty. First of all it's wonderful. It was wonderful and still is wonderful. But second, it had so little influence, and that's has been brought out here. Now why is that? Well, first of all, as long as you're thinking about inheritance, who gives a damn what the substance is—it's irrelevant. Now, once you know the genetic material is DNA, there is only one inference: you should study DNA! And Dr. Chargaff did, but few people did seriously, until of course Watson and Crick. Furthermore, another reason for the work of Avery and his colleagues not attracting quite as much attention as they might was that they were too modest. They refused to advertise. A third thing, and this is really not an explanation, but a curiosity. This was an extremely awkward system to work with, the* pneumococcus, *and a close homolog of course was the tobacco mosaic virus which was the first vehicle for demonstrating the role of nucleic acid in viral infection. These are the last systems you would have chosen, if you'd been looking for material to study from this point of view.*

On that very same occasion Hershey also made the important point of the cardinal place that the discovery of the structure of DNA by Watson and Crick had on its ultimate identification as the genetic material:

> *Well, so, some 10 years had elapsed before the structure of DNA was elucidated. The first fortunate thing about that discovery was that it did have some meaning. As Jim Watson once said, the structure of DNA could have been completely uninteresting. But it was not. But the elucidation of the structure had another effect. It made the work of Avery/Macleod/McCarty, and Hershey, and many other people completely unneeded, superfluous. Because once you have that structure, it has to be genetic material!*

4.13   EPILOGUE: HERSHEY'S PECULIAR LATE VERSION OF THE SOURCE OF HIS EXPERIMENT WITH CHASE

As noted, up to 1952 Hershey studied phage genetics without explicit regard to the question of the chemical nature of the genetic material. How then did he come about an experiment that was aimed at finding whether it was phage protein or DNA that carried the viral genetic information? As already discussed in this chapter, evidence shows that Hershey was familiar with and interested in the *pneumococcus* transformation story and that he was aware of the debate about the role of residual protein in the bacterial

DNA preparations. Additionally, he concluded from results of his 1951 study on ^{32}P-induced killing of bacteriophage that *"vital structures [of the virus] contain nucleic acid."*[34] Finally, Hershey was closely familiar with the observations of Anderson[36] and of Herriot[37] on the separation of viral protein and DNA in osmotically shocked phages. In fact, his cited correspondence with Herriot from late 1951 indicates that at that time he was already engaged in the work that culminated in his celebrated experiment with Chase. Despite all these pieces of evidence, a question still remains of the exact path that Hershey took from abstract genetics to the discovery of DNA as the genetic material of phage. The prominent virologist Norton Zinder (1928−2012) posed this question directly to Hershey during the aforementioned 1994 Rockefeller University symposium. It is left to reader, however, to decide whether Hershey's reply and his version of the events was a serious rendition of a true turn of events or a tongue-in-cheek response (Ref. 19, p. 106):

> *Zinder: I'd like to ask Al Hershey a question. In 1951 Al Hershey was writing a paper on Cold Spring Harbor Symposium on phage heterozygotes. In it the recombination of bacteriophage was performed. Yet, in 1952, you published the Hershey-Chase experiment, which was a biochemical discovery. Can you tell us how—what made you give up the phage genetics and go against the grain to biochemistry?*
>
> *Hershey: [. . .] I'd gone to considerable trouble to set up the equipment necessary for working with isotopes, following the work of Seymour Cohen, who first used isotopes to study bacteriophage. Having done this all by myself, buying the equipment and learning how to use it, more or less, I had to—now to ask what shall I do with it? Well, that answers your question.*

REFERENCES

1. Avery OT, Macleod CM, McCarty M. Studies on the chemical nature of the substance inducing transformation of pneumococcal types: induction of transformation by a desoxyribonucleic acid fraction isolated from pneumococcus type III. *J Exp Med* 1944;**79**(2): 137−58.
2. McCarty M, Taylor HE, Avery OT. Biochemical studies of environmental factors essential in transformation of pneumococcal types. *Cold Spring Harbor Symp Quant Biol* 1946;**11**: 177−83.
3. McCarty M, Avery OT. Studies on the chemical nature of the substance inducing transformation of pneumococcal types; effect of desoxyribonuclease on the biological activity of the transforming substance. *J Exp Med* 1946;**83**(2):89−96.
4. McCarty M, Avery OT. Studies on the chemical nature of the substance inducing transformation on pneumococcal types; an improved method for the isolation of the transforming substance and its application to Pneumococcus Types II, III, and VI. *J Exp Med* 1946;**83**(2):97−104.
5. Glass B. The long neglect of genetic discoveries and the criterion of prematurity. *J Hist Biol* 1974;**7**(1):101−10.

6. Pollister AW, Swift H, Alfert M. Studies on the desoxypentose nucleic acid content of animal nuclei. *J Cell Physiol* 1951;**38**(Suppl. 1):101−19.

7. Hershey AD, Chase M. Independent functions of viral protein and nucleic acid in growth of bacteriophage. *J Gen Physiol* 1952;**36**(1):39−56.

8. Twort A. *In focus, out of step: a biography of Frederick William Twort, F.R.S. 1877−1950.* Alan Sutton Pub. Ltd.; 1993.

9. Summers WC. *Felix d'herelle and the origins of molecular biology.* Yale University Press; 1999.

10. Twort FW. An investigation on the nature of ultra-microscopic viruses. *The Lancet* 1915;**186**(4814):1241−3.

11. Twort FW. Filter-passing transmissible bacteriolytic agents (bacteriophage). *The Lancet* 1930;**216**(5594):1064−7.

12. d'Herelle F. Sur un microbe invisible antagoniste des bacilles dysentèrique. *Compt Rend Acad Sci (Paris)* 1917;**165**(1):373−5.

13. Abedon ST, Kuhl SJ, Blasdel BG, Kutter EM. Phage treatment of human infections. *Bacteriophage* 2011;**1**(2):66−85.

14. Lewis S. *Arrowsmith.* New York, NY: Signet Classics; 1961.

15. Löwy I. Martin Arrowsmith's clinical trial: scientific precision and heroic medicine. *J R Soc Med* 2010;**103**(11):461−6.

16. de Kruif P. *Microbe Hunters.* 3rd ed. San Diego, CA: Mariner Books; 2002.

17. Summers WC. On the origins of the science in Arrowsmith: Paul de Kruif, Felix d'Herelle, and phage. *J Hist Med Allied Sci* 1991;**46**(3):315−32.

18. Stahl FW. Hershey. *Genetics* 1998;**149**(1):1−6.

19. Stahl FW, editor. *We can sleep later: Alfred D. Hershey and the origins of molecular biology.* Cold Spring Harbor, New York, NY: Cold Spring Harbor Laboratory Press; 2000.

20. Segrè G. *Ordinary geniuses: how two mavericks shaped modern science.* New York, NY: Penguin; 2011.

21. Kay LE. Conceptual models and analytical tools: the biology of physicist Max Delbruck. *J Hist Biol* 1985;**18**(2):207−46.

22. McKaughan DJ. The influence of Niels Bohr on Max Delbruck: revisiting the hopes inspired by "light and life". *Isis* 2005;**96**(4):507−29.

23. Timofeeff-Ressovky NW, Zimmer KG, Delbrück M. Über die natur der genmutation und der genstruktur. *Nachr Ges Wiss Göttingen* 1935;**1**(13):189−245.

24. Ellis EL, Delbruck M. The growth of bacteriophage. *J Gen Physiol* 1939;**22**(3):365−84.

25. Luria SE. (Alfred P. Sloan Foundation series) *A slot machine, a broken test tube: an autobiography.* Harpercollins; 1984

26. Luria SE, Anderson TF. The identification and characterization of bacteriophages with the electron microscope. *Proc Natl Acad Sci USA* 1942;**28**(4):127−31.

27. Luria SE, Delbruck M. Mutations of bacteria from virus sensitivity to virus resistance. *Genetics* 1943;**28**(6):491−511.

28. Luria SE. Reactivation of irradiated bacteriophage by transfer of self-reproducing units. *Proc Natl Acad Sci USA* 1947;**33**(9):253−64.

29. Luria SE. Bacteriophage: an essay on virus reproduction. *Science* 1950;**111**(2889):507−11.

30. Cairns J, Stent GR, Watson JD, editors. *Phage and the origins of molecular biology.* Cold Spring Harbor, NY: Cold Spring Harbor Laboratory; 1966.

31. Hershey AD. Mutation of bacteriophage with respect to type of plaque. *Genetics* 1946;**31**(6):620−40.

32. Hershey AD, Rotman R. Genetic recombination between host-range and plaque-type mutants of bacteriophage in single bacterial cells. *Genetics* 1949;**34**(1):44—71.

33. Hershey AD, Rotman R. Linkage among genes controlling inhibition of lysis in a bacterial virus. *Proc Natl Acad Sci USA* 1948;**34**(3):89—96.

34. Hershey AD, Kamen MD, Kennedy JW, Gest H. The mortality of bacteriophage containing assimilated radioactive phosphorus. *J Gen Physiol* 1951;**34**(3):305—19.

35. Creager AN. Phosphorus-32 in the phage group: radioisotopes as historical tracers of molecular biology. *Stud Hist Philos Biol Biomed Sci* 2009;**40**(1):29—42.

36. Anderson TF. Destruction of bacterial viruses by osmotic shock. *J Appl Phys* 1950;**21**(1):70.

37. Herriott RM. Nucleic-acid-free T2 virus "ghosts" with specific biological action. *J Bact* 1951;**61**(6):752—4.

38. Northrop JH. Growth and phage production of lysogenic *B. megatherium*. *J Gen Physiol* 1951;**34**(5):715—35.

39. Olby R. *The path to the double helix: the discovery of DNA*. New York, NY: Dover Publications; 1994.

40. Hershey AD. Intracellular phases in the reproductive cycle of bacteriophage T2. *Ann Inst Pasteur (Paris)* 1953;**85**(1):99—112.

41. Fraser D, Mahler HR, Shug AL, Thomas CA. The infection of sub-cellular *Escherichia coli*, strain B, with a DNA preparation from T2 bacteriophage. *Proc Natl Acad Sci USA* 1957;**43**(11):939—47.

42. Spizizen J. Infection of protoplasts by disrupted T2 Virus. *Proc Natl Acad Sci USA* 1957;**43**(8):694—701.

43. Watson JD. The nature of virus multiplication (contribution to the discussion). *Symp Soc Gen Microbiol* 1952;**2**:113—16.

44. Watson JD. *The annotated and illustrated double helix*. New York, NY: Simon & Schuster; 2012.

45. Stent GS. Prematurity and uniqueness in scientific discovery. *Sci Am* 1972;**227**(6):84—93.

46. Hotchkiss RD. The identification of nucleic acids as genetic determinants. *Ann N Y Acad Sci* 1979;**325**(1):321—44.

Chapter 5

Discovery of the Structure of DNA

The Most Famous Discovery of 20th Century Biology

Chapter Outline

M. Fry: Landmark Experiments in Molecular Biology. DOI: http://dx.doi.org/10.1016/B978-0-12-802074-6.00005-9
143

Of the many landmark advances that biology has made in the 20th century, the discovery of the DNA double helix by James Watson and Francis Crick[1] is arguably the one that attracted the greatest amount of attention among scientists, historians of science, and the general public. Only very few discoveries in modern biology, and in science at large, have been scrutinized with as much attentiveness to the minutest technical, historical, and personal details. Primary material of original correspondence, interviews, and images of scientists who played central roles in this discovery is deposited in several archives. Various facets of the double helix story were explored and analyzed in books by scientists who were actively involved in the discovery, by historians of science and biographers. Whereas some authors focused on the immediate experimental and modeling work that produced the double helix model of DNA, others put this groundbreaking accomplishment in a broader context of the history of the isolation and chemical analysis of DNA, its identification as the genetic material and the unsuccessful early attempts to elucidate its structure. Another group of books and articles placed the discovery of the structure of DNA within an even broader framework of the birth and development of molecular biology. Last, the story of the determination of the structure of DNA was also told as part of autobiographies and biographies of the main protagonists, Watson, Crick, Maurice Wilkins, Rosalind

Franklin, Linus Pauling, and Erwin Chargaff. It would be futile to attempt to cover here the enormous literature on the myriad aspects of the discovery of the DNA double helix. The henceforth-cited sources represent, therefore, just a sample of a much wider body of scholarship.

Archives containing wide-ranging primary material on the scientific contributions and lives of leading DNA researchers are listed in Table 5.1. As noted, in addition to the wealth of scholarly articles that deal with various scientific, historical, and philosophical aspects of the discovery of the double helical structure of DNA, a significant number of books were written on the subject. Prime among books on the history of the discovery of the double helix is Robert Olby's classical *"The Path to the Double Helix"*[2] which is arguably the weightiest exploration of the subject. Similar grounds were also

TABLE 5.1 Archival Sources on the Life and Scientific Work of Leading Investigators of the Structure of DNA

Scientist	Archive	URL
Francis F.H. Crick (1916–2004)	National Library of Medicine "Profiles in Science"*	http://profiles.nlm.nih.gov/SC/ Views/Exhibit/narrative/ doublehelix.html
	Wellcome Library Collection*	http://wellcomelibrary.org/ collections/digital-collections/ makers-of-modern-genetics/ digitised-archives/francis-crick/
James D. Watson (1928–)	Archives at Cold Spring Harbor Laboratory*	http://library.cshl.edu/personal-collections/james-d-watson
Rosalind Franklin (1920–58)	Wellcome Library Collection*	http://wellcomelibrary.org/ collections/digital-collections/ makers-of-modern-genetics/ digitised-archives/rosalind-franklin/
	ASM—Anne Sayre Collection of Rosalind Franklin Materials	http://www.asm.org/index.php/ choma3/71-membership/ archives/8230-anne-sayre-collection-of-rosalind-franklin-materials
	The Churchill Archives Centre, Churchill College, Cambridge	http://janus.lib.cam.ac.uk/db/ node.xsp?id = EAD%2FGBR% 2F0014%2FFRKN
	National Library of Medicine "Profiles in Science"*	http://profiles.nlm.nih.gov/ps/ retrieve/Narrative/KR/p-nid/187/ p-docs/true

(Continued)

TABLE 5.1 (Continued)

Scientist	Archive	URL
Maurice Wilkins (1916–2004)	Wellcome Library Collection*	http://wellcomelibrary.org/collections/digital-collections/makers-of-modern-genetics/digitised-archives/maurice-wilkins-mrc-biophysics-unit-archive/
William T. Astbury (1898–1961)	The National Archives—Leeds University Library, Special Collections	http://discovery.nationalarchives.gov.uk/browse/r/56f82c6b-9f37-4e21-874d-db6c9e2d6861/next/56f82c6b-9f37-4e21-874d-db6c9e2d6861/378662f2-3b0e-4808-9126-85bc8378866f
Linus Pauling (1901–94)	National Library of Medicine "Profiles in Science"*	http://profiles.nlm.nih.gov/MM
	Oregon State online archive "Linus Pauling and the Race for DNA"*ᵃ	http://scarc.library.oregonstate.edu/coll/pauling/dna/index.html

Material in collections marked by red asterisk "*" is available online.
ᵃArchival material on Pauling's pursuit of the structure of DNA. This collection of drafts, manuscripts, letters, and images is part of the much larger Oregon State University Linus Online Archive (http://pauling.library.oregonstate.edu/) that encompasses the full range of Pauling's contributions to science and to public life.

covered in a book by Portugal and Cohen[3] and in a short volume by Lagerkvist.[4] A substantial portion of Judson's *"The Eighth Day of Creation"*[5] is dedicated to description of the contributions of Watson and Crick, Wilkins, Franklin, and Pauling.

Because of its prime significance, the deciphering of the three-dimensional structure of DNA is also the subject of dedicated chapters in books on the history of molecular biology.[6,7] In their autobiographical books, key contributors to the elucidation of the structure of DNA described their part of the story. The most famous is Watson's highly personal *"The Double Helix"* which was originally published in 1968 and was reprinted in 2012 with added enlightening annotations.[8] Crick told more concisely his version of the story in his 1988 book *"What a Mad Pursuit."*[9] Other autobiographical accounts of the discovery were divulged by Wilkins[10] and Chargaff.[11] Different aspects of the discovery of the structure of DNA were also covered in monographs and biographies of Astbury,[12,13] Watson,[14−17] Crick,[16,18,19] Franklin,[20−22] Wilkins,[23] and Pauling.[24,25] In addition, there are reviews and memoirs of witnesses to the discovery.[26−28] The National Center for Biotechnology Education (NCBE) at the University of Reading posted online a useful timeline of the road to the double helix: http://www.ncbe.reading.ac. uk/menu.html.

It is impossible to write on "Landmark Experiments in Molecular Biology" without including the tale of the elucidation of the double helical structure of DNA. Yet, considering the massive scholarship that this milestone achievement had engendered, it is hard to introduce a fresh viewpoint to the story. Another difficulty is that whereas the present volume is dedicated to landmark *experiments*, Watson and Crick made their discovery not by performing direct experiments but rather by building a model of DNA. However, their model was built on the basis of X-ray diffraction data that were collected by Rosalind Franklin, Maurice Wilkins, and their doctoral student Raymond Gosling (1926−) and on biochemical results of Erwin Chargaff (1905−2002). Thus, true to the title of this volume, the main focus of this chapter is on the *experiments* that formed the basis for the Watson and Crick model of the DNA double helix. Much of the described experimental work entailed analysis of the diffraction of X-rays in DNA. The solution of the structure of DNA by X-ray diffraction analysis is, therefore, reviewed here in the broader context of the history of the discovery of X-rays and of their use to solve the structure of simple compounds and later on of complex biological molecules.

Essentially, regularly arranged atoms in a crystal or a crystalline fiber diffract X-rays. Waves that are diffracted off a structure that they traverse are recorded on a detection medium (ie, film) as sets of spots of different intensity. Each of these spots corresponds to a wave diffracted from orbiting electrons of atoms that are situated at a particular plane in the three-dimensional molecule.

Mathematical tools are then applied to reconstruct the three-dimensional arrangement of atoms in the molecule from knowledge of the phase of the diffracted waves (relative position of the peaks and troughs of the wave), and their amplitude (the strength of the wave as measured by the intensity of spots). A detailed exposition of the mathematical processing of the experimental data is much beyond the scope and focus of this volume. To fully comprehend the analytical−mathematical part of some of the described experiments, interested readers should consult textbooks and technical manuscripts on the subject.

5.1 EARLY HISTORY OF X-RAY CRYSTALLOGRAPHY

The three-dimensional structures of numerous biological molecules—polysaccharides, proteins, and nucleic acids—were deduced from the patterns of the diffraction of X-rays through the tightly spaced lattice of atoms in fibrous and mostly in crystallized forms of the molecules. Historically, the Watson and Crick model of a DNA double helix was based in large part on the diffraction of X-rays in DNA fibers. Electromagnetic X-radiation (comprised of

X-rays) has wavelengths ranging between 0.01 and 10 nm $(10^{-12}-10^{-8}\,\mathrm{m})$, and frequencies of $3 \times 10^{16}-3 \times 10^{19}\,\mathrm{Hz}$. Because the wavelengths of X-rays are shorter than the interatomic spaces in fibers or crystals, they collide with and are scattered by orbiting electrons of the arrayed atoms. As is described later, the obtained diffraction patterns of the X-rays serve to reconstruct arrangement of atoms within the fiber or crystal. Before introducing the actual X-ray diffraction experiments that led to the solution of the three-dimensional structure of DNA, it is necessary to briefly describe the history of the discovery of X-rays by Wilhelm Röntgen (Box 5.1) and of the development of X-ray crystallography. For a comprehensive review of the subject, interested reader should consult Andrè Authier's book on the early history of X-ray crystallography.[29]

BOX 5.1 Wilhelm Conrad Röntgen and Max von Laue

Röntgen was born in Rhenish Prussia and in his youth his family moved to Holland. Starting his studies at the University of Utrecht, he was expelled from that institution because of a prank that another student had made and was banned from every other school in the Netherlands. Eventually he studied mechanical engineering in the Federal Polytechnic Institute in Zurich (now ETH) and received a PhD degree from the University of Zurich. After serving on the faculties of the German universities of Strasburg, Hohenheim, and Giessen, he was appointed Professor of Physics in the University of Würzburg where he performed the experiments that led to the discovery of X-rays (also named "Röntgen rays"). In 1900 he moved to his last academic station at the University of Munich. For his discovery of X-rays, Röntgen was awarded in 1901 the first Nobel Prize in Physics (Fig. 5.1).

Max von Laue studied physics at the Universities of Strasbourg, Göttingen, and Munich. He received a PhD degree in physics from the University of Berlin in 1903. After working as an assistant to Max Planck at the Institute for Physics in Berlin, Laue was from 1909 to 1912 Privatdozent (a rank that allowed him to conduct independent research), at the Munich Institute for Theoretical Physics, which was headed by Arnold Sommerfeld. There he came up with the idea that due to their short wavelengths, X-rays should be diffracted in crystals. With the help of the experimentalists Friedrich and Knipping, he demonstrated in 1911 diffraction of X-rays in crystals. For this achievement, von Laue was awarded the 1914 Nobel Prize in Physics. Serving later as Professor of Theoretical Physics in Swiss and German universities, von Laue made significant contributions to the theory of superconductivity. Laue opposed National Socialism and its promotion of "Deutsche Physik" (German physics) that rejected Einstein's theory of relativity as "Jüdische Physik" (Jewish physics). Laue openly resisted Nazi policies and he and Otto Hahn helped exiled Jewish scientists to find positions in other countries. Laue was killed at age 80 in a road accident in Berlin.

(Continued)

BOX 5.1 (Continued)

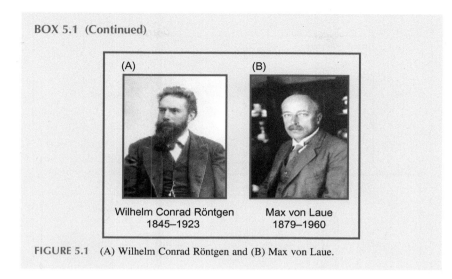

(A)   (B)

Wilhelm Conrad Röntgen   Max von Laue
1845–1923   1879–1960

FIGURE 5.1   (A) Wilhelm Conrad Röntgen and (B) Max von Laue.

5.1.1   The Discovery of X-Radiation by Wilhelm Röntgen

Several leading physicists of the second half of the 19th century conducted experiments with electrical discharge tubes. These devices, first the earlier Geissler tube and later Crookes-type tubes, consisted of a partly evacuated glass vessel that had at their two respective ends two metal electrodes: a cathode and an anode (Fig. 5.2A).

High voltage that was applied between the electrodes caused ionization of residual molecules of air in the tube. The electric field in the tube accelerated the positively charged ions toward the negative cathode. Because of their high velocity, the accelerated ions hit the surface of the metal cathode with great momentum causing it to release a large number of negatively charged electrons that then traveled at high speed toward the positive anode. This stream of electrons was dubbed "cathode rays." Because of their great speed, some electrons passed through the anode and hit the glass wall behind it. Colliding with atoms in the glass, the cathode rays bumped electrons of these atoms into higher orbits and energy levels. Upon returning to their original energy levels, the bumped electrons emitted fluorescence that made the glass glow (Fig. 5.2A). In a later development of electrical discharge tubes, their back glass wall was smeared with phosphor or other fluorescent agents to intensify the emitted fluorescence.

Historically, investigations of cathode rays have led to the discovery of electrons. Studies by the British physicist and 1906 Nobel laureate Joseph John (J.J.) Thomson (1856–1940) and by the German Nobel laureate of 1905 (and a devoted Nazi and anti-Semite) Philipp von Lenard (1862–1947)

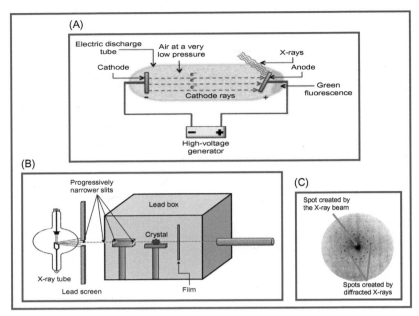

FIGURE 5.2 (A) Generation of cathode rays and X-rays in an electric discharge tube. High voltage between a cathode and an anode at the opposite ends of a partly evacuated glass tube ionizes the residual air molecules. The positively charged ions collide at high speed with the negative cathode and release electrons from its surface that stream toward the anode. Due to their great momentum, some electrons pass the anode and collide with the glass wall. The streaming electrons ("cathode rays") bump electrons of the glass atoms into higher orbits and energy levels. When these electrons return to their original orbits, they emit fluorescence that makes the glass glow. Röntgen discovered that in addition to the fluorescing cathode rays, a new and different type of radiation ("X-rays") was also emitted from the electric discharge tube. (B) Scheme of the apparatus that Friedrich and Knniping constructed to detect diffraction of X-rays by crystals. X-rays passed through progressively narrower slits (collimators) in lead screens, hit the crystal, and their diffraction was recorded on film. (C) The 1912 documentation by Friedrich, Knniping, and Laue of the diffraction of X-ray pattern of a zinc sulfate crystal. *For (B): Reproduction of the original scheme of the Friedrich, Knniping, and Laue apparatus[30] was republished in Ref. 185. For (C): From Friedrich W, Knipping P, Laue M. Interferenzerscheinungen bei Röntgenstrahlen.* Bayerische Akad Wiss München, Sitzungsber math-phys Kl 1912:303−22.

showed that cathode rays consisted of negatively charged particles which could be bent by a magnetic field. Calculations revealed that these particles were much smaller than atoms and had a very large charge-to-mass ratio. Realizing that electrons were components of the atom, Lenard rightly argued that atoms consisted of mostly empty space.

  Like other investigators of the time, the University of Würzburg physicist Wilhelm Conrad Röntgen (Box 5.1) studied cathode rays. Although the ultimate objective of his experiments is unknown, it is clear that he was examining at the time various types of cathode tubes and investigated excitation of fluorescence. In the summer of 1895, he was using a cathode tube that von Lenard devised. That tube was furnished with an aluminum window that permitted exit of the cathode rays. To protect the aluminum from damage by the strong electrostatic field, Röntgen placed a cardboard cover that blocked escaping visible light. Despite the presence of this light-obstructing cardboard, Röntgen discovered that when cathode rays were emitted, a screen painted with barium platinocyanide $(BaPt(CN)_4)$ that was placed near to the aluminum window fluoresced. Repeating the experiment with light tight Crookes-type tube that had thicker glass walls, he found that the barium platinocyanide screen fluoresced even when it was placed at a relatively large distance from the cathode tube. Speculating that he was observing a new type of radiation that could penetrate both the thick glass and the light-blocking cardboard, he named it "X-rays," using the mathematical designation "X" for an unknown entity (Fig. 5.2A). In Germany the rays were named in his honor "*Röntgen strahlung*" (Röntgen radiation). Röntgen's early experiments distinguished between X-rays and cathode rays. Whereas the cathode rays that were found by Lenart to be deflected by a magnet, the progression of X-rays was not affected by the magnetic field. In studying the traversal of X-rays through different media, he first took pictures of brass weights that were walled within a wooden box. Most significantly, however, Röntgen was the literal pioneer of X-ray medical imaging when, just 2 weeks after he first detected X-radiation, he took a picture of the hand of his wife Anna Bertha. It was claimed that upon seeing her skeletal bones, his wife cried out: "*I have seen my death.*"

Between March 1896 and March 1897, Röntgen summarized his observations and measurements in three concise communications that covered *in toto* just 31 pages. These reports were promptly translated into English and reprinted in the journals *Nature* and *Science*. The discovery of X-rays was immediately recognized as an exceptional achievement and the reticent Röntgen was ordered by Kaiser Wilhelm II to demonstrate his discovery in the Berlin palace.[32] An immediate outcome of Röntgen's historic discovery was the mushrooming application of X-rays for medical imaging. The focus of this chapter, however, is not on the highly beneficial use of X-radiation in medicine but rather on the application of the diffraction of X-rays in crystals and fibers for the determination of the three-dimensional atomic structure of materials.

5.1.2   Max von Laue's Discovery of X-Ray Diffraction

In his third communication, the cautious and exacting experimenter Röntgen reported on his repeated failures to observe diffraction of X-rays (cited in Ref. 33, p. 5):

> *Ever since I began working on X-rays, I have repeatedly sought to obtain diffraction with these rays; several times, using narrow slits, I observed phenomena*

which looked very much like diffraction. But in each case a change of experimental conditions, undertaken for testing the correctness of the explanation, failed to confirm it, and in many cases I was able directly to show that the phenomena had arisen in an entirely different way than by diffraction. I have not succeeded to register a single experiment from which I could gain the conviction of the existence of diffraction of X-rays with a certainty which satisfied me.

It was ultimately the German theoretical physicist Max von Laue (Box 5.1) who demonstrated some 15 years after the discovery of X-rays that they were diffracted in crystals. In his 1914 Nobel lecture, which he delivered only in 1920 after the end of World War I, von Laue retrospectively maintained that the search for diffraction was based on the presupposition that X-rays were waves[34] (http://www.nobelprize.org/nobel_prizes/physics/laureates/1914/laue-lecture.pdf):

In the case of X-rays their discoverer [Röntgen] had already made efforts to locate diffraction or interference phenomena in order to solve the question of whether or not they represent a wave phenomenon or the ejection of any small particles. But in this quest his research, which had otherwise been so successful, met with failure.

Indeed, the hypothesis that X-rays were waves received support from several outstanding physicists of the time among whom was the 1911 Nobel laureate Wilhelm Wien (1864−1928) who estimated their wavelengths to range between 10^{-10} and 10^{-9} cm. A second consideration that formed a basis for von Laue's discovery was the space lattice hypothesis. That theory contended that atoms are organized in crystals in a space lattice, or as von Laue put it in his Nobel lecture: "[*crystals are comprised of*] *similar molecules similarly situated.*"

Laue himself and some of his contemporaries offered on various occasions retrospective reconstruction of the status of theories on the wave nature of X-rays and on crystal space lattices that existed when von Laue came up with the idea of demonstrating diffraction. The reconstructed versions of the basic assumptions that stood behind the discovery of X-ray diffraction and the circumstances of the actual experiments are a matter of controversy among physicists and historians of physics.[35−37] The consensus lore, however, is that the hypothesis of diffraction originated during a conversation that von Laue had in late 1911 with the physicist Paul Peter Ewald (1888−1985) (for a very readable short history of the discovery of X-ray diffraction in crystals, see Ref. 38). Ewald was then writing his doctoral thesis on crystal optics. He attempted to develop a theory on the scattering of light by regularly arranged elements ("resonators") in a crystal. However, Ewald was concerned that wavelengths in the visible range of the spectrum were much larger than the spacing between the resonators in a crystal. During their casual walk in the park, Laue came up with the idea that because spaces between elements of the crystal were greater than the conjectured wavelengths of X-rays, atoms in the crystal lattice should diffract the X-ray waves. The argument was, therefore, that if observed, diffraction of a beam of X-ray in a crystal would provide evidence for both the three-dimensional lattice arrangement of resonators in crystals and the wave nature of X-rays. When he broached this idea with Arnold Sommerfeld, some doubts were raised. One was that thermal motion of atoms within the crystal would blur any possible interference pattern. Sommerfeld was also of the opinion that the crystals would not diffract the primary beam of X-rays but rather would act as emitters of secondary, so-called characteristic, X-rays. Nevertheless, Laue prevailed and having been a theoretician, he recruited to do the critical experiment Sommerfeld's assistant Walther Friedrich (1883–1968) who had hands-on experience with X-rays. Also joining the experiment was Röntgen's doctoral student Paul Knipping (1883–1935). These two investigators constructed an apparatus to capture images of diffracted X-rays that were targeted at crystals of copper sulfate ($CuSO_4 \cdot 5H_2O$) and later of zinc sulfide (ZnS; Zincblende) (Fig. 5.2B). The experiment was initially set up to document "characteristic X-rays" such that photographic plates were placed left, right, and back of the crystal.[35,39] When no interference was recorded, Friedrich and Knipping positioned the plate behind the crystal. Finally, after exposure of the plate for 7 or 10 days, diffraction spots were sighted around a central spot (a sample image from the original Friedrich, Knipping, and Laue 1912 paper[31] is shown in Fig. 5.2C). The results suggested that each of the X-rays-generated "Laue spots" corresponded to some lattice constant and wavelength. When the original 1912 paper by Friedrich, Knipping, and Laue[31] was republished a year later in the *Annalen der Physik*,[30] Laue appended a note in which he proposed a crystal that contained several atoms per cell and formulated what was later named the "structure factor." All in all, Laue suggested that a crystal acts as three-dimensional diffraction gratings for X-rays. However, assuming that the ZnS crystal had primitive cubic lattice, Laue could not explain why some spots that corresponded to particular indices were generated whereas others with narrowly close indices were missing. In an attempt to explain this phenomenon, Laue hypothesized that diffraction patterns could arise by fluorescence that the X-rays induced in excited atoms of the crystal. An alternative speculation was that the crystal separated five main wavelengths of the primary beam to generate the various diffraction spots. This quandary was solved shortly thereafter by the English father and son team of William Henry and William Lawrence Bragg.

5.1.3 The Contribution of William Henry and William Lawrence Bragg to the Analysis of Crystal Structure by X-Ray Diffraction

The Leeds University physicist William Henry Bragg (Box 5.2) strongly believed that X-rays were particles. He had held to this notion for a long period of time.

BOX 5.2 William Henry Bragg and William Lawrence Bragg

William Henry Bragg (Sir William) was educated at Cambridge University where he excelled in mathematics. In 1886 at the age of 23 he was appointed Professor of Mathematics and Physics in the then tiny South Australian University of Adelaide. There he developed interests in electromagnetism, radioactivity, and instrument building. In 1908 he returned to England as the Cavendish Chair of Physics in the University of Leeds where he invented the X-ray spectrometer. Together with his son Lawrence, who was then a graduate student at Cambridge, he laid the foundations to the science of X-ray crystallography by developing the analysis of crystal structure by X-ray diffraction. For this achievement, father and son were jointly awarded the Nobel Prize in Physics in 1915. Later Sir William was appointed Quain Professor of Physics at University College London and in 1923 he became Fullerian Professor of chemistry at the Royal Institution (Fig. 5.3).

William Lawrence Bragg (Sir Lawrence) was the first born son of William Henry and his wife Gwendoline Todd (there were also a daughter, Gwendolen, and another son, Robert, who was killed in the Battle of Gallipoli). Entering the University of Adelaide at age 16, he studied mathematics, chemistry, and physics and graduated in 1908. When the family moved back to England, he was elected a Fellow at Trinity College, Cambridge. In 1912, during his first year as research student, he developed Bragg's law, which allowed calculation of the positions of atoms in a crystal from the diffraction of an X-ray beam by the crystal lattice. Together with his father he successfully applied the law to solve the crystal structure of many inorganic compounds. Shortly after his brother Robert was killed in the war, it was announced that the father and his 25-year-old son had won the 1915 Nobel Prize in Physics. From 1919 to 1937, he was the Langworthy Professor of Physics at Victoria University in Manchester. In 1938, he replaced Ernest (Lord) Rutherford as Cavendish Professor and Head of the Cavendish Laboratory in the University of Cambridge. There he promoted the application of X-ray diffraction analysis to the elucidation of the structure of proteins and later of DNA. Since 1953 Sir Lawrence served in different capacities at the Royal Institution in London and from 1966 until his death in 1971 he was Emeritus Professor at that Institution.

(A) William Henry Bragg
(1862–1942)

(B) William Lawrence Bragg
(1890–1971)

FIGURE 5.3   (A) William Henry Bragg and (B) William Lawrence Bragg.

He defended this view on the pages of *Nature* in arguments with the English physicist and future (1917) Nobel laureate Charles Barkla (1887−1944) who believed in the wave nature of X-rays. When Laue's diffraction patterns were published in 1912, they were interpreted by Bragg as reflection of the traveling of the "X-particles" along "avenues" between the crystal atoms. Bragg also discussed Laue's report with his son William Lawrence (Box 5.2) who was then a first year graduate student at Cambridge.

Although both Braggs rejected von Laue's fluorescence idea, unlike his father, Lawrence Bragg accepted Laue's premise of the wave nature of X-rays and his proposal that resonators within the crystal diffracted those waves. Yet, he found Laue's speculations about the selective presence of certain spots to be lacking. While he was struggling with the problem, the younger Bragg was cognizant of a model of crystal structure that the English chemists William Barlow (1845−1934) and William Jackson Pope (1870−1939) have developed. The Barlow−Pope model visualized atoms in crystals as closely packed "balls" that were of the same diameter in an element and of two different sizes in binary compounds such as KCl or NaCl.[40,41] With this model in mind and by his intuition, Bragg successfully linked X-ray diffraction to the atomic organization of a crystal. At the root of his insight were the different shapes of spots that Laue obtained in two plates that were placed at different distances from the crystal. Whereas a plate that was in close proximity to the crystal had round spots, the spots on a more distantly positioned plate were elliptical (Fig. 5.4A).

Bragg's intuition was that the observed focusing effect was due to reflection of the X-rays by successive sheets of atoms in the crystal. In essence, he predicted that information on the structural organization of materials could

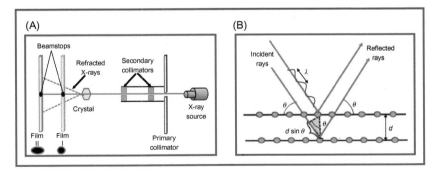

FIGURE 5.4 (A) Scheme of the experiment that drove Lawrence Bragg to the conclusion that X-ray spots were generated by their reflection from organized sheets of atoms in a crystal. He gained this insight from the different shapes of spots on films that were placed at different distances from the refracting crystal. Spots on the closer photographic plate I were round whereas spots on the more distant plate II were elliptical. (B) Scheme of the collision of X-rays with parallel planes of atoms within a crystal. Parameters of the Bragg equation are indicated: λ, X-ray wavelength; d, inter-planar distance; θ, angle of reflection.

be deduced from the intensity distribution of the diffraction pattern. Pope, who was at the time professor of chemistry in Cambridge, was delighted with this interpretation and suggested that crystals of simple molecules such as alkali halides should be tested for their diffraction patterns. Following this advice, the senior Bragg obtained simpler diffraction patterns for compounds such as NaCl, KCl, and KBr. This result convinced him of the wave nature of X-rays.[42] In the course of these analyses, William Bragg advanced the experimental work by constructing a precise X-ray spectrometer and an ionization chamber that allowed quantification of X-rays intensities.

Based on their accumulated experimental results and on the theory that Lawrence Bragg developed, the Braggs proposed that crystals contained well-defined sets of surfaces (lattice planes). Each set of these planes consisted of a large number of parallel planes that were separated from one another by fixed distance. This distance between two parallel planes of similar type (an inter-planar distance) had been marked d. Importantly, both the wavelengths of X-rays, λ, and the inter-planar distances were of atomic dimensions. Thus, when X-rays hit a crystal, they generated a diffraction pattern that was characteristic of the crystal. Fig. 5.4B illustrates the geometry of X-ray diffraction. The lower of the two rays has to travel a longer distance than the upper one. If the added distance is an exact integer number of wavelengths (1, 2, 3, . . .), then the two reflected X-rays emerge in-phase from the crystal. However, when the distance is not an integer number of wavelengths, the two out-of-phase rays will be annulled by destructive interference. Thus, each Laue spot would consist of several in-phase harmonics of a specific wavelength that was selected by its correspondence to integral multiples of the inter-planar distance in the crystal. To make the rays leave in-phase a crystal with a given inter-planar d spacing, it may be tilted relative to the source of X-radiation such that the angle of the incident rays is changed. A different wavelength of X-rays should then diffract at each relative position of the crystal. Based on their gathered insights, William and Lawrence Bragg formulated in a joint 1913 paper[43] the "Bragg equation" which states that a diffracted beam is obtained if: $n\lambda = 2d \sin \theta$, where λ is the X-ray wavelength, d the inter-planar distance, θ the angle of reflection (also named "Bragg angle"), and n is an integer number (1, 2, 3, . . .) (Fig. 5.4B).

5.1.4   Application of Mathematical Tools Was Necessary to Reconstruct the Atomic Structure of Materials From Their Patterns of X-Ray Diffraction

Usage of mathematical tools was crucial for the elucidation of the atomic structure of materials from their patterns of X-ray diffraction. Mathematical manipulation was first applied for the structural analyses of relatively simple materials and then for fibers of organic polymers and later for proteins and

DNA. Prime among those tools was the Fourier transform that decomposes periodic functions into their sinusoidal components. Fourier transform was introduced in the 19th century in a different context and was first applied to the problem of diffraction of light in 1906.[44] Because different waves of X-rays are superimposed on one another during diffraction, it becomes a problem to isolate the contribution of each diffraction event such that the lattice structure of a crystal can be determined. Without going into technical details, performing Fourier synthesis enables the separation of individual frequencies and relative strengths of diffracted X-rays from mixtures of superimposed waves of different amplitudes, frequency, and phase. After William Bragg first applied in 1915 Fourier summation to determine the density of scattered X-rays in a crystal,[45] Lawrence Bragg used two-dimensional summation to draw electron density map of diopside ($MgCaSi_2O_6$). Others applied Fourier synthesis in the 1920s and 1930s to determine the structure of more complex materials including organic polymers.

Three pieces of information are required to calculate electron densities from an X-ray diffraction experiment: (1) indices of reflections, (2) intensities of the reflections, and (3) the phase angles of the reflections. Indices of a reflection can be determined by the symmetry of the crystal and intensities are measured directly by photon counters. However, because the exact same intensity is obtained by combination of wave peaks (positive phase) or troughs (negative phase), the phase cannot be directly deduced from the X-ray diffraction data. Since knowledge of the phase is essential for the calculation of electron density within a crystal, several methods have been developed to overcome the so-called *phase problem*. An early phasing method that was initiated by the work of Cork on alums[46] was *isomorphous replacement*. Under this method, diffraction patterns are compared in two almost identical crystals that differ by a heavy atom that replaces a single native atom in the molecule (see Section 5.2.2). Another approach is the application of the *Patterson function*. This mathematical tool was devised by Arthur Patterson (1902−66) in 1934 and 1935[47,48] and was promptly recognized as a valuable tool such that Dorothy Hodgkin, for instance, adopted it already in 1935 for her analysis of the crystal structure of insulin (Ref. 49, p. 116). The Patterson function yields a map of the distances and direction of all the possible vectors between the atoms in an investigated crystal and of the height of every peak. Despite the ample information that Patterson analysis produced, it was not trivial to derive from it an accurate arrangement of the atoms in a crystal. If a unit cell in a crystal contains n atoms, than the number of vectors is n^2 and thus with a growing number of atoms the calculations are apt to become increasingly difficult and even impossible. The Patterson function is usually applied, therefore, when the crystal contains heavy atoms or when a substantial part of the structure had already been solved. More recent additional methods for the solution of the phase problem were reviewed in depth elsewhere.[50]

5.2 THE EMERGENCE OF X-RAY CRYSTALLOGRAPHY OF BIOLOGICAL MOLECULES

Beginning in the 1920s, laboratories in Germany and then in England applied X-ray diffraction techniques to solve the atomic structure of biomolecules. These efforts encompassed studies of cellulose fibers, smaller molecules such as sterols, B vitamins, penicillin, and increasingly of proteins. Because these studies provided a fertile ground on which the structural analysis of DNA by X-ray diffraction developed, they are surveyed here first. Early attempts to solve the structure of DNA are described later in this chapter.

5.2.1 Diffraction of X-Rays in Fibers

The different arrangement of molecules in fibers and crystals prescribed different reading of their patterns of diffracted X-rays. In general, fibers consist of extended molecules that are arranged in parallel to one another along the axis of the fiber. Relative to crystals, fibers scatter X-rays more weakly and thus require longer exposure time for the development of readable images of the diffraction pattern. The pattern of X-rays that are diffracted off a fiber appears as a series of lines, termed layer lines, which are perpendicular to the axis of the fiber. The distance between these lines is correlated to the periodicity of elements alongside the axis of the fiber (see Section 5.3.1).

5.2.2 Early X-Ray Analyses of the Atomic Structure of Fibers

The discovery that patterns of the diffraction of X-rays could reveal the atomic structure of materials opened up new fields of research in physics, chemistry, and material science. After the successful use of X-ray diffraction to elucidate in 1914 the atomic structure of simple minerals and diamonds, subsequent studies made X-ray crystallography a powerful tool for the elucidation of the structure and explication of the properties of diverse materials such as silicates, metals, and alloys. X-ray diffraction analysis of the atomic structure of biological molecules developed more slowly. Fiber, powder, or liquid forms of biomaterials were initially considered to lack crystalline organization that may act as diffraction gratings for X-rays. This misconception was strengthened when Friedrich, Knipping, and Laue reported in 1912 that powder which they produced by grinding up whole crystals lacked diffraction pattern.[31] Yet, first X-ray diffraction diagrams of powder and fiber were obtained as early as 1913.[51,52] Evidence that microcrystalline powders and fibers also possessed three-dimensional atomic structure was obtained shortly thereafter when more powerfully evacuated X-ray tubes were developed by the Swiss Paul Scherrer (1890−1969) and by the American William Coolidge (1873−1975). Using an improved tube and cylindrical diffraction camera, Scherrer and the renowned University of Göttingen physicist Peter Debye (1884−1966) documented in

1916 diffraction lines in powder of lithium fluoride (LiF).[53] Using monochromatic X-rays that were generated in a Coolidge tube, Reginald Herzog (1878−1935) and his assistant Willie Jancke in the Kaiser Wilhelm Institute of Fiber Chemistry (*Faserstoffchemie*) in Berlin-Dahlem, irradiated in 1920 cellulose fibers of ramie (China grass) and other sources. These fibers yielded a pattern of 26 smeared diffraction spots that were symmetrically positioned in groups of four around two mirrors planes.[54] Similar albeit less clear, patterns of X-ray diffraction in cellulose were also produced in the same year by Scherrer. Herzog asked the young Hungarian (and later English) polymath Michael Polanyi (1891−1976) to interpret the diffraction pattern of the cellulose fibers. Within a week or two, Polanyi came up with the conclusion that the images of four spots reflected arrangement of parallel crystals around the axis of the fiber. Polanyi also concluded that all together, the fiber had rotational symmetry around its axis.[55] The chemist Kurt Meyer (1883−1952), head of a research laboratory of Badische Anilin-und Soda-Fabrik (BASF) (a division of the IG Farben industrial concern), carried out in the 1920s studies on the absorption of dyes by natural fibers. In the course of these investigations, he became interested in the structure of the fibers themselves. This interest led to his collaboration with Herman Mark (1895−1992), a young Austrian crystallographer who taught X-ray diffraction analysis to Linus Pauling and Max Perutz. Based on their analysis of diffraction by fibers, Meyer and Mark proposed a model for the structure of cellulose. In their model, strongly bonded (which they dubbed "main valence") units formed long molecules that bundled into micelles by laterally adhering to one another through weaker bonds.[56] Expanding their analyses to the structure of rubber and silk,[57] Meyer and Mark constructed molecular models to test alternative modes of polymer organization. These studies led them to conclude that cellulose, rubber, and silk shared similar general structural characteristics. In summarizing their ideas in 1930, they introduced a ribbon chain model for cellulose that had a screw axis of symmetry and in which all the folds of the ribbon remained in the same plane[58] (for a detailed discussion of their model, see Olby,[2] pp. 35−37). Yet, as was indicated (Ref. 2, p. 38), despite these and other advances, no fiber structure had been solved to the extent that the positions of atoms in the molecule were conclusively determined. Atomic-level resolution of the structure of biomolecules was attained within the next two to three decades, mostly through the efforts of the British school of crystallography.

5.2.3 The British School of X-Ray Crystallography of Biomolecules

Although studies of the atomic structure of biomolecules began in Germany, English researchers in the universities of Leeds, Cambridge, London, and Oxford dominated this area of research from the late 1920s and on. Examples of achievements of the English school of crystallographers include the solution of the structures of vitamin B_{12}, cholesterol, penicillin, and insulin by Dorothy Crowfoot Hodgkin (1910–94), of hemoglobin by Max Perutz (1914–2002), of myoglobin by John Kendrew (1917–97), and of DNA by Rosalind Franklin (1920–58), Maurice Wilkins (1916–2004), Francis Crick (1916–2004), and James Watson (1928–). This explosion of talent and discoveries flourished on the fertile soil of crucial earlier work by leading figures of British crystallography such as Lawrence Bragg, John Desmond Bernal (1901–71), William Astbury (1898–1961) (Box 5.3), and John Turton Randall (1905–84) (see Box 5.5 later in this chapter). The limited space of this chapter allows for only a superficial survey of the main developments in England in structural analysis of molecules of biological significance. A book by de Chadarevian[7] is recommended as an excellent source on the development after World War II of structural biology in England, and particularly at Cambridge University.

BOX 5.3 William (Bill) Astbury and John Desmond Bernal

As one of seven sons of a potter, William (Bill) Astbury's initial prospects of an academic career were dim. Yet, owing to his outstanding academic achievements, he won a scholarship to Cambridge from which he obtained (after disruption by World War I) a degree in physics. Subsequent to his graduation, Astbury worked with William Bragg in University College and at the Royal Institution. In 1928 he was appointed Lecturer in Textile Physics at the University of Leeds and in 1946 he became Professor of Biomolecular Structure and remained the incumbent of that chair until his death in 1961. Preceding the discovery by Pauling of the α-helix and β-sheet structural elements of proteins, Astbury was the first to suggest on the basis of his crystallographic studies of the fibrous proteins keratin and collagen that native proteins possess secondary structure. He later pioneered fiber crystallography of DNA. Although Astbury's interpretation of the structure of proteins was inaccurate and his model of DNA was wrong, he is acclaimed as a pioneer of structural biology and is commemorated in the Astbury Centre for Structural Molecular Biology in the University of Leeds (Fig. 5.5).

Bernal was Irish-Catholic of Sephardic-Jewish origin on his father's side and American-Catholic on his mother's. Winning a scholarship to Cambridge, he got in 1922 a BA degree in mathematics and physics. He later did his doctoral work on X-ray crystallography under William Bragg at the Royal Institution. Returning to Cambridge in 1927, Bernal was later appointed Assistant Director of the Cavendish Laboratory at Cambridge University. Having been denied tenure in Cambridge by Ernest Rutherford, he was appointed in 1938 Professor of Physics at Birkbeck College of the University of London at which he established after World War II a Biomolecular Research Laboratory. Bernal did pioneering crystallographic studies of proteins and viruses. He taught, inspired, and worked closely with Dorothy Crowfoot Hodgkin, the 1964 Nobel Prize laureate for her work on the structure of vitamin B_{12} and other biological molecules. Bernal was also the doctoral adviser of Max Perutz, the 1962 Nobel Prize laureate for his solution of the structure of hemoglobin. A true polymath, Bernal was nicknamed "Sage" by

(Continued)

BOX 5.3 (Continued)

his students and peers "because he knew everything." Beside his stature as a highly influential scientist, he was also a lifelong communist and laureate of the Lenin Prize for Peace.

(A) William Astbury   John Desmond Bernal
1898–1961   1901–1971

FIGURE 5.5   (A) William Astbury and (B) John Desmond Bernal.

This section focuses on the pioneering experiments to solve the three-dimensional structure of proteins. While it took decades (from the 1930s to the 1950s and up to the 1980s) to elucidate the atomic structure of proteins such as insulin or hemoglobin, present-day X-ray crystallography is so well advanced that by 2015 it solved the three-dimensional structures of nearly 100,000 proteins (http://www.rcsb.org/pdb/statistics/contentGrowthChart.do?content=explMethod-xray&seqid = 100).

Protein Crystallography at the University of Leeds

The Cambridge University educated physicist William Astbury (Box 5.3) was trained in X-ray crystallography by William Bragg at University College and at the Royal Institution. There he worked side by side with fellow trainees who later became luminaries in the field: Bernal and Dame Kathleen Yardley Lonsdale (1903–71). Losing to Bernal in competition for a position at Cambridge University, Astbury was appointed in 1928 Lecturer in the University of Leeds. The largest department of this university (formerly named Yorkshire College) was at the time the department of textile industries. Historically, the principal backers of this department were the local textile industries that also prevailed in determining its teaching curriculum and its research activities (for a history of Leeds University and its department of textile industries before and at the time of Astbury's recruitment, see Ref. 2, pp. 41–46). A lecturer (later professor) in the department J.B. Speakman (1896–1969) alerted the new Vice Chancellor of the university, the moral philosopher James Baillie (1872−1940) to the unsatisfactory quality of the students at the department of textile industries and to the need for more rigorous teaching and research in mathematics, physics, and chemistry. Rising to the challenge and searching for new faculty, Baillie received from Bragg a letter recommending Astbury as "*[. . .] a very brilliant man [. . .] very energetic and persevering, has imagination, and in fact he has the research spirit*" (cited in Ref. 59). On the strength of this recommendation, Astbury was appointed in 1928 Lecturer in Textile Physics. Although Astbury was initially despondent about his prospects of doing serious research in a department of textile industries, soon after his arrival in Leeds, he had nevertheless launched vigorous research activity. Working with an assistant and with several research students, the brusque, passionate, idealistic, and creative Astbury (who was also regarded by some peers to be amateurish) approached the study of wool fibers with gusto (for a lively description of the early days in Leeds, see Hall's biography of Astbury[12]). Using a newly purchased X-rays generator and camera, he took multiple X-ray diffraction pictures of unstretched and maximally stretched moist wool or hair protein (keratin) fibers. In a series of papers, Astbury and his students showed in the first half of the 1930s that the unstretched keratin fibers possessed a coiled structure with a characteristic repeat of 5.1 Å.[60−62] Based on these data, Astbury suggested that the unstretched protein molecules formed a helix, which he named α-form that upon stretching became uncoiled into a different β-form. Although incorrect in detail, Astbury's model of hair protein was fundamentally right in being an early inaccurate version of the α-helix and the β-sheet structures of polypeptides that Pauling Corey and Branson correctly modeled in 1951.[63] Soon after the model was published, its details were disputed. Thus, for instance, the protein biochemist Hans Neurath (1909−2002) pointed out that the side chains and the hydrogen atoms could not be accommodated in α-keratin helix of the specified dimensions (for a discussion of the opposition to Astbury's model, see Ref. 2, pp. 62−64). It should be noted, however, that despite these criticisms, Astbury's work on keratin stands out as an early demonstration of the effect of changed conformation of a protein on its mechanical properties.

Together with the Imperial College biochemist and his friend Kenneth Baily (1909−63), Astbury ventured beyond fibrous protein to study the globular protein edestin. Their biochemical and X-ray diffraction analyses correctly described the transition from the soluble form of the native protein to its insoluble denatured form whose X-ray diffraction pattern was not unlike that of fibrous keratin.[64] As will be detailed later in this chapter, Astbury also conducted X-ray diffraction analyses of DNA fibers (see Section 5.3.1). Although his interpretation of the obtained results and the model of DNA that he constructed were wrong, and although these data later misled Pauling to propose a different and also incorrect model of DNA, Astbury is justifiably regarded as a pioneer in the study of the structure of DNA.

Structural Analysis of Biomolecules by Bernal and Hodgkin
John Desmond Bernal

John Desmond Bernal (Box 5.3) was trained in X-ray crystallography by William Bragg at the Royal Institution. After he successfully determined the structure of graphite, Bernal competed with Astbury to win in 1934 a position in the University of Cambridge. There he was first a lecturer and then Assistant Director of the Cavendish Laboratory. As was described by Brown in an excellent biography of Bernal[65] (pp. 80−83), his interest in the structure of biomolecules was aroused when he visited the BASF laboratory of Herman Mark and Kurt Meyer and learned of their work on the structure of cellulose fibers (see Section 5.5.2). Their interest in the structure of biomolecules made him think about the use of X-ray diffraction analyses to solve structures of proteins. Together with his doctoral student at Cambridge, the future (1964) Nobel laureate in Chemistry Dorothy Crowfoot (later Hodgkin), they first studied the X-ray diffraction patterns of vitamin B_1 and adenine.[66] Moving on to X-ray diffraction analysis of a protein, Bernal and Hodgkin first used polarized light microscopy to show that crystals of the protease pepsin were birefringent when they were bathed in their mother liquor. Loss of birefringence in dried out crystals suggested that they have lost an internal order. Based on this observation, Bernal constructed a sealed thin-walled glass capillary that kept the pepsin crystals bathed in liquid. Using this contraption, he and Hodgkin showed that wet pepsin crystals diffracted X-rays to high resolution.[67] This observation, which indicated that all the pepsin molecules shared an identical structure, contrasted the then prevalent colloid theories, which claimed that proteins lacked distinct structure. As they astutely stated in this early work, Bernal and Hodgkin reached the opposite conclusion that a protein possesses a definite structure[67]:

> *From the intensity of the spots near the centre, we can infer that the protein molecules are relatively dense globular bodies, perhaps joined together by valency bridges, but in any event separated by relatively large spaces which contain water. From the intensity of the more distant spots, it can be inferred that the arrangement of atoms inside the protein molecule is also of a perfectly definite kind, although without the periodicities characterising the fibrous proteins. [...] At this stage, such ideas are merely speculative, but now that a crystalline protein has been made to give X-ray photographs, it is clear that we have the means of checking them and, by examining the structure of all crystalline proteins, arriving at a far more detailed conclusion about protein structure than previous physical or chemical methods have been able to give.*

Another outstanding doctoral student of Bernal was the future (1962) Nobel laureate in Chemistry Max Perutz. His ambitious study of the structure of hemoglobin commenced when he worked under Bernal.[68] It then took him 25 more years of work to bring this study to conclusion (see the next subsection). After his move in 1938 to Birkbeck College, Bernal initiated with his postdoctoral trainee Isidore Fankuchen (1905–64) X-ray crystallographic studies of the structure of plant viruses.[69–72] These investigations continued after World War II and involved, among others, Rosalind Franklin, who after leaving King's College for Birkbeck College did X-ray analysis of the tobacco mosaic virus (TMV).[73] For a comprehensive scientific biography of Bernal and a full bibliography of his writings, the reader is directed to Dorothy Hodgkin's captivating memoir.[74]

Dorothy Crowfoot Hodgkin

After completing her doctoral thesis under Bernal, Dorothy Hodgkin left Cambridge for Oxford University at which she worked for the duration of her career. There she solved the structure of biologically important molecules such as cholesterol, penicillin, vitamin B_{12}, and insulin (see http://www. google.co.il/url?sa=t&rct=j&q=&esrc=s&source=web&cd=5&ved=0C DcQFjAE&url=http%3A%2F%2Frsbm.royalsocietypublishing.org%2Fhig hwire%2Ffilestream%2F3490%2Ffield_highwire_adjunct_files%2F0%2Fr sbm20020011.doc&ei=IPwHVZ-eAYP8aPnwgqgM&usg=AFQjCNFGCF ZbrmNP8bGnfCnE-EbtUtFGVg for a full bibliography of Hodgkin's sizeable body of contributions). Notably, soon after Hodgkin solved the structure of vitamin B_{12}, the leading crystallographer Peter Ewald wrote to Bernal to say: "*I think if you did nothing else in science but train this woman, your name would last forever*" (cited in Ref. 75). Because this section deals with the analysis of the structure of proteins, only the solution of the three-dimensional structure of insulin is discussed here. Hodgkin's study of insulin had started already in 1935 when she took X-ray photographs of rhombohedral crystals of insulin.[76] Although it was expected that the structure of insulin would be readily solved, the challenge proved to be much more difficult than was initially thought. Practices of the time were not advanced enough to solve the structure of a protein (insulin) of 5780 Da. Because of the complexity of the problem, Hodgkin and her associates proceeded stepwise improving their techniques until in 1969 the three-dimensional structure of rhombohedral 2 zinc insulin was solved at a resolution of 2.8 Å.[77] Ultimately, more than 50 years after she started her pursuit of the structure of the hormone, Hodgkin and her team solved in 1988 the structure of pig insulin at a resolution of 1.5 Å.[78] Hodgkin's contributions to the solution of the atomic structure of biomolecules and her life have been described in a fine biography by Georgina Ferry[49] and in shorter articles.[79,80]

X-Crystallography of Proteins at the University of Cambridge Cavendish Laboratory

Protein crystallography in Cambridge had two prominent triumphs: the solutions of the three-dimensional structures of hemoglobin by Max Perutz and of myoglobin by John Kendrew. Because these achievements had direct effect on the work of Crick and Watson on DNA, and because of the prominence of these discoveries in the history of structural molecular biology they are described here at some length. Attention is particularly paid to the work of Perutz on hemoglobin.

The Elucidation of the Structure of Hemoglobin by Max Perutz

After obtaining a degree in chemistry in his native Austria, the 22-year-old Perutz (Box 5.4) arrived in Cambridge in 1936.

He started his graduate studies in the Cavendish Laboratory in late 1936 under the supervision of Bernal. Despite his lack of prior knowledge of X-ray crystallography, he rapidly learned the methodology and started looking for a specific subject for research. As he later noted, it was Bernal who inspired him to use X-ray diffraction to elucidate the structure of proteins[81]:

BOX 5.4 Max Ferdinand Perutz and John Cowdery Kendrew

Max Perutz was born in Vienna to a family of Jewish descent that converted to the Catholic faith. After receiving in 1936 a degree in chemistry from the University of Vienna and despite his total ignorance of X-ray crystallography he was accepted by Bernal as a doctoral student in his University of Cambridge group. Having been encouraged by Bernal to study the structure of proteins he chose to investigate horse hemoglobin, which remained his focus of research for the length of his career. After Bernal had left Cambridge Perutz completed his doctoral work under Lawrence Bragg, who continued to encourage and support his long and often frustrating quest of the structure of hemoglobin. When World War II broke Perutz was first incarcerated as an alien in England and Canada and then, on the recommendation of Bernal, was recruited to do war-related research. When the war ended Perutz resumed his attempts to elucidate structure of hemoglobin. After many failures he adapted the technique of isomorphous replacement to the study of hemoglobin, which after 6 more years yielded in 1960 a solution of the three-dimensional structure of the tetrameric hemoglobin. For this achievement, Perutz shared with Kendrew the 1962 Nobel Prize in Chemistry. In subsequent studies, Perutz elucidated the molecular bases for the cooperative oxygen binding by hemoglobin and for its defective activity in various hemoglobinopathies. Perutz was also an inspired writer of articles for the general public on science-related subjects (Fig. 5.6).

(Continued)

BOX 5.4 (Continued)

John Kendrew was born in Oxford to an academic family. In 1939 he graduated in chemistry from the University of Cambridge. After a short period of research on reaction kinetics, he joined the war effort working first on radar at the Air Ministry Research Establishment and then doing operational research at the Royal Air Force at which he received an honorary rank of Wing Commander. Conversations with Bernal during the war aroused his interest in protein structure and after the war he joined Perutz in crystallographic study of fetal hemoglobin. In 1949 he embarked on an independent project to elucidate the structure of myoglobin by X-ray diffraction analysis. This effort culminated in 1958 with a report of the three-dimensional structure of myoglobin. This was the first reported solution of the tertiary structure of any protein. In recognition of this achievement, Kendrew shared with Perutz the 1962 Nobel Prize in Chemistry. Ending at that time his active laboratory research, Kendrew was among the founders of the European Molecular Biology Organization (EMBO) and its first president. He also founded and was the Editor in Chief of the *Journal of Molecular Biology*.

(A)   (B)

Max Ferdinand Perutz   John Cowdery Kendrew
1914–2002   1917–1997

FIGURE 5.6   (A) Max Ferdinand Perutz and (B) John Cowdery Kendrew.

He taught me that the riddle of life was hidden in the structure of proteins, and that X-ray crystallography was the only method of solving it, and I became his disciple.

Perutz choice to pick hemoglobin as his protein of choice was somewhat inadvertent. On a visit to Prague in 1937 his relative by marriage, the biochemist Felix Haurowitz (1896–1987)[82] persuaded him to purify and crystallize hemoglobin. Upon his return to Cambridge, Perutz was given some hemoglobin crystals that yielded in his hand rich X-ray diffraction patterns. Though proud of this result, Perutz did not know what to make of it (Ref. 83, pp. XVIII–XIX):

> [...] it was not until 1937 that the physiologist Gilbert Adair gave me some beautiful crystals of horse hemoglobin from which I obtained rich X-ray diffraction patterns. I was hooked! I proudly showed my X-ray photographs to all my friends, but when asked what they meant I changed the subject, because I had no idea.

By late 1938, Bernal had left Cambridge to assume his responsibilities as Professor of Physics at Birkbeck College in London. At the same time, Lawrence Bragg arrived in Cambridge to replace Ernest Rutherford as the Cavendish Professor and Director of the Cavendish Laboratory. Seeing Perutz' X-ray diffraction images of hemoglobin crystals, Bragg who took over Bernal's duty as the supervisor of Perutz' doctoral thesis encouraged him to continue the pursuit of the atomic structure of hemoglobin. It should be pointed out that at that time the task of reconstructing a three-dimensional arrangement of the 10,000 atoms of the hemoglobin tetramer (molecular weight 64,000) appeared to be insurmountable. However, the tenacity of Perutz and the unwavering support that Bragg gave him brought this enterprise into completion 25 years after it had started.

The crystallographic analysis of the structure of hemoglobin was interrupted by the outbreak of World War II. At first Perutz was interned as an enemy alien in England and then in Canada. After having been freed, he was recruited by Bernal, who did prominent military research under General Earl Mountbatten, to participate in a top-secret research project. This venture that had never been realized was the construction of a floating airfield (code-named "Habakkuk") that was to be built from a mixture of wood pulp and ice. Descriptions of this imaginative but totally impractical project were provided by Perutz himself[84] and more concisely by David Eisenberg.[85] Despite his being preoccupied by this and other war-related activities, Perutz used his additional duty of fire watching at nights to measure the intensities of ~ 7000 X-ray reflection spots. Following Hodgkin's methodology of three-dimensional Fourier summation, he then performed tedious calculations that suggested that the most conspicuous facet of hemoglobin was the parallel arrangement of rods that were packed 10.5 Å apart with strong signals at intervals of 5.1 Å.[86] Drawing on similarities between his and Astbury's diffraction patterns of keratin, Perutz deduced that hemoglobin consisted of tightly packed keratin-like polypeptide chains (Ref. 83, pp. 35–40). Francis Crick who joined Perutz in 1948 as a PhD student insightfully criticized this model. His recalculations indicated that only approximately one-third of the molecule comprised of parallel keratin-like polypeptide chains (Ref. 83, p. 40). After the structure of hemoglobin was solved, it became clear that the rods were actually α-helices that were only a minor structural element of the globin molecule. Bragg suspected that the observed 5.1 Å repeat was a feature of a helical polypeptide chain and he, Perutz, and John Kendrew published in 1950 a fourfold polypeptide helix model that did not take into account the planarity of the peptide bond.[87] A year later Pauling and associates published their model of the α-helix structure of polypeptide chains.[63] Their hypothetical α-helix had planar amide groups and each carbonyl group was engaged in a hydrogen bond. Pauling also proposed that the pitch of the helix was 5.4 Å and not 5.1 Å, as was suggested by the Cambridge team (the 5.1 Å periodicity in keratin was later shown to be due to two intertwined helices). Perutz realized that if the pitch was indeed 5.4 Å, then peptide bonds should progress by 1.5 Å along the helix axis. On the very same day that he learned of the Pauling's α-helix model, Perutz took an X-ray diffraction picture of horsehair and discovered strong reflection signals at spacing of 1.5 Å. This measurement provided an immediate experimental support for Pauling's model.[88] Famously, Perutz and Bragg were frustrated by not having been the first to discover the α-helix. Understandably, therefore, Bragg was hard-pressed to conceal his delight when 2 years later Watson and Crick won the race with Pauling to solve the structure of DNA (see Section 5.7.1).

By 1953 the unsolved phase problem hindered any real advance toward a solution of the actual structure of hemoglobin. After spending years visually estimating the intensities of thousands of diffractions spots in hemoglobin crystals at different stages of swelling and shrinkage, Perutz succeeded in determining the phases of just 7 out of 7000 reflections. At that point even Crick regarded the phase problem to be hopeless.[89] A glimmer of a solution appeared when Perutz noticed that the intensities of rays diffracted by hemoglobin crystals were much weaker than the original incident rays. Reasoning that the weakened intensities were due to negative interference of the scattered rays, Perutz had the idea of applying the isomorphous replacement strategy that was first used in 1936 by the crystallographer John Monteath Robertson (1900−89). To solve the phase problem for the small molecule of phthalocyanine, Robertson bound a heavy atom to it. He was then able to determine the phases by measuring the change in reflections of the X-rays that the heavy atom produced.[90] This approach appeared, however, inapplicable to hemoglobin that contained ~5000 atoms compared with the 40 atoms of phthalocyanine. Nevertheless, Perutz dared to try to detect intensity differences by substituting atoms in hemoglobin with heavier electron-dense mercury atoms. Austen Riggs (1924−), then at Harvard University, reported in 1952 that the oxygen binding properties of hemoglobin did not change when its sulfhydryl groups were blocked with the mercury-containing compound *p*-chloromercuribenzoate.[91] Following this clue, Perutz replaced two atoms in hemoglobin by mercury atoms and compared the intensities of diffraction spots in the unsubstituted and in mercury-containing hemoglobin molecules. Prominent intensities of mercury-generated peaks allowed determination of the mercury−mercury vector and revealed the exact positions of the heavy atoms. Based on this information, the phases of most of the reflections could be deduced and thus the signs of most of the reflections were determined with certainty.[92] Using this approach, Perutz presented in a series of papers in the *Proceedings of the Royal Society* a preliminary picture of hemoglobin. However, although these studies revealed the outline of the molecule and pinpointed the positions of residues that contained a sulfhydryl group, essential data such as the location of the heme groups or of the iron atoms remained unknown. Moreover, no information could be gleaned on the three-dimensional structure of the molecule. This was mostly because the 63-Å-thick molecule along the axis of projection corresponded to ∼40 atomic diameters and thus various details of the molecule were superimposed on one other, making the diffraction pattern uninterpretable. Also, as often happens in experimental work, added to this fundamental difficulty, an unexpected affliction of the crystals muddled the lattices and rendered them useless for phase determination.[89] After 3 years of frustration, Perutz put the lattices back in order by modifying the pH of the solution in which the hemoglobin crystals were bathed. Aided by a method that Michael Rossmann (1930−) devised for determining the relative positions of two heavy atoms in a unit cell,[93] Perutz and colleagues painstakingly collected multiple diffraction patterns with three or four different isomorphous replacements in the same unit cell. Based on 40,000 measurements, they completed in the summer of 1959 a model that was published in early 1960[94] of the tertiary and quaternary structure of hemoglobin at a resolution of 5.5 Å. After he received the 1962 Nobel Prize in Chemistry in recognition of this achievement, Perutz went on to solve the structure of hemoglobin at greater resolutions, first of 2.8 Å[95] and then 1.7 Å.[96] Perutz never forgot having been told by Felix Haurowitz already in 1937 that hemoglobin changes its structure when it binds oxygen. Thus, after he unlocked the structure of hemoglobin, he was ready to elucidate the molecular details of this conformational change. Comparison of the structures of oxyhemoglobin (the oxygenated form of hemoglobin) and deoxyhemoglobin (the reduced form of hemoglobin) provided atomic pictures of the different conformations of these two forms and offered molecular explanation to the cooperative modes of binding and release of oxygen. Also, by studying the effects of different mutations on the properties of hemoglobin, the molecular causes for some hemoglobinopathies were revealed.[97] A beautifully illustrated concise review of the findings up to 1983 on structure, physiology, and pathology of hemoglobins can be found in a book by Dickerson and Geis.[98]

John Kendrew Solved the Structure of Myoglobin

Kendrew met with John Bernal in Ceylon (today's Sri Lanka) during the war when the two scientists were involved in war-related research. Bernal's burning passion for the determination of the structure of biomolecules aroused

Kendrew's interest in the structure of proteins. When the war ended, Kendrew returned to England via California where he met with Linus Pauling at his Caltech laboratory. Pauling's deep insight of structural chemistry and his ideas on the elucidation of the structure of biomolecules strengthened Kendrew's determination to pursue studies of protein structure. Thus, when he arrived in Cambridge in 1945 and became Perutz' first PhD student, he started by performing X-ray diffraction analysis of sheep fetal hemoglobin. In 1949 Kendrew carved his own independent line of investigation by choosing to study the structure of myoglobin, the oxygen binding protein of muscle tissue. Unlike the $\sim 64,000$ Da tetrameric hemoglobin which contains some 10,000 atoms (or ~ 5000 atoms without hydrogens), the $\sim 17,000$ Da monomeric myoglobin had only ~ 2500 atoms (or ~ 1200 atoms excluding hydrogens). The relatively small size of myoglobin and its lower complexity relative to hemoglobin made it a preferred subject for X-ray analysis. Yet, these advantages were relative only because at the time that Kendrew embarked on his study of sperm whale myoglobin, the most complex molecule whose structure was fully solved was vitamin B_{12} that contains only 93 atoms. It is understandable, therefore, that as in the case of hemoglobin, the limited power of the tools and methods that were available in the 1950s made the elucidation of the structure of myoglobin a nearly impossible mission. Indeed, early efforts to solve the structure of myoglobin were unsuccessful. A promise of solution surfaced only after Perutz adopted for hemoglobin the strategy of isomorphous replacement by heavy mercury atoms. However, unlike hemoglobin, myoglobin did not contain free sulfhydryl groups and Kendrew and his associates had to introduce other heavy atoms molecule by co-crystallization of the protein with silver or mercury-containing compounds. Ultimately, metal-containing ligands have been found to combine at five specific sites in sperm whale myoglobin. The Kendrew group then performed a large series of X-ray diffraction analyses on crystals that had heavy atom in a unit cell that comprised of two myoglobin molecules (for a detailed technical description of the experiments, see Kendrew's review in the *Nature Physics* portal: http://www.nature.com/physics/looking-back/kendrew/index.html). Interestingly, whereas Perutz initially distrusted the newly introduced electronic computers and although the calculations could be done in the customary way, Kendrew pioneered the usage of computers by processing the diffraction data in the early British computer Electronic Delay Storage Automatic Calculator (EDSAC) Mark I at Cambridge and in the first commercially available British computer Digital Electronic Universal Computing Engine (DEUCE) at the National Physical Laboratory (for a discussion of the early use of computers in X-ray crystallography at Cambridge, see Ref. 7, pp. 107–135). The computed results yielded in 1958 a low-resolution three-dimensional structure of myoglobin.[99] A more finely resolved structure (at 2 Å) which was based on 200,000 measurements was published in 1960[100] and the later use of more powerful computers allowed an increase of the resolution to 1.4 Å.[101] In general, the single myoglobin polypeptide was found to fold into a flat disk within which short nonhelical regions linked eight separate right-handed α-helices. The most striking finding was that despite their greatly different amino acid compositions, myoglobin and each of the globin chains in hemoglobin folded into closely similar three-dimensional structures.

X-Ray Crystallography at King's College

The experimental physicist John Turton Randall, who spent much of his career doing basic and applied research on luminescence, won great acclaim when he invented in the early years of World War II together with Henry Booth (1917–83) the cavity magnetron, a vital component of microwave radar (Box 5.5).

BOX 5.5 John Turton Randall

Born at Newton-le-Willows, Lancashire, Randall received BSc and MSc degrees in physics from the University of Manchester. From 1926 and for 11 years he conducted research on X-ray diffraction and luminescence in the General Electric laboratories at Wembley, London. After winning a Royal Society fellowship in 1937, he moved to the University of Birmingham to work on luminescence. With the outbreak of World War II, he was assigned to develop better radar. At the start of the war, the British radar systems used meter-wave transmitters that enabled detection of echoes from approaching aircraft at a limited distance of 100 miles. Another major drawback of that system was its inability to detect low-flying objects. Thus, German aircraft that crossed the channel at low altitude reached shore without being detected. Given the task of building a high-power generator of microwaves, Randall and the engineer Henry Boot (1917–83) invented within just 3 months the cavity magnetron in which electrons traveled in a circle and generated microwaves when they passed by resonators. The new magnetron was 10^3 times more powerful than the hitherto available meter-wave generator and its use in radar increased its range and resolution. From 1941 and on, the new microwave radar systems were installed and deployed by the British and American forces. These radars proved to be vital instruments in the ultimate defeat of Germany. This scientific triumph made Randall famous. In 1944 he was appointed to the Chair of Natural Philosophy at the University of St. Andrews from which he submitted a proposal to establish a Biophysics Research Unit. This proposal won the support of the Medical Research Council (MRC) and Randall was appointed director of a new Biophysics Research Unit (presently named the Randall Division of Cell and Molecular Biophysics) at King's College, London. Believing that DNA was the genetic material, Randall encouraged and supported the X-ray structural analysis of DNA that was conducted by members of the Unit: Maurice Wilkins, Rosalind Franklin, and Raymond Gosling (Fig. 5.7).

(Continued)

BOX 5.5 (Continued)

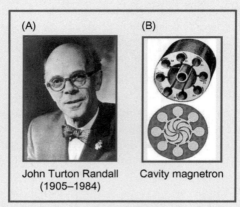

FIGURE 5.7 (A) John Turton Randall and (B) the cavity magnetron.

Although his education and research activities up to the end of the war were in physics, he was always interested in biology and specifically in applying physics to biological questions. When he worked in the General Electric Laboratories, Randall tried his hand in biophysics by conducting X-ray diffraction experiments on cellulose fibers and by attempting to assess the effect of X-rays on seeds.[102] He continued to maintain academic interest in biological problems after he moved to the University of Birmingham and later to the University of St. Andrews in Scotland at which he was the Chair of Natural Philosophy. When the war ended on May 8, 1945, Randall began to make plans for biological research. In September of the same year, he recruited to his St. Andrews department Maurice Wilkins (Box 5.6) who was his PhD student in Birmingham.

BOX 5.6 Maurice Hugh Frederick Wilkins

Wilkins was born in New Zealand where his father was a physician. At age 6 he moved with his family to Birmingham, England. After receiving a BA degree in physics from the University of Cambridge, Wilkins started a doctoral thesis on phosphorescence and electron traps under John Randall at the University of Birmingham. World War II interrupted his studies and he first worked in Birmingham on improving radar screens and later joined the Manhattan Project to work in 1944—45 on isotope separation in the University of California, Berkeley. Only after his return to England in 1945 did Wilkin receive a PhD degree. After a short period in the University of St. Andrews, he moved with Randall to King's College where he served as Assistant Director in Randall's

(Continued)

BOX 5.6 (Continued)

Biophysics Research Unit. In addition to overseeing the various research activities in the Unit, Wilkins had his own projects: first development of new optical microscope and later X-ray crystallography of DNA. Wilkins' first diffraction image of DNA that was taken in 1950 indicated that the DNA had ordered atomic structure. To improve the quality of the diffraction pictures, he installed better equipment and requested that Rosalind Franklin, who was originally recruited to do protein crystallography, would instead work on DNA. Because of Randall's faulty management of the project, and due to clash of personalities, Wilkins and Franklin's work relationships deteriorated and in March 1953 Franklin had left for Birkbeck College. When Watson visited King's College in early 1953, Wilkins showed him Franklin and Gosling's famous high-quality "photo 51" of X-ray diffraction pattern of B-form DNA. The information in this photograph was instrumental in the construction of the double helix model of DNA by Watson and Crick. Because of this episode and other circumstances, the relative roles of Franklin and Wilkins in the discovery of the structure of DNA became a matter of ongoing debate among historians and scientists. After Watson and Crick published their model of the double helix in April 1953, Wilkins and his associates conducted experiments to substantiate the details of the helical structure of DNA. Wilkins' contribution to the discovery of the structure of DNA was recognized in 1962 when he shared with Watson and Crick the Nobel Prize in Physiology and Medicine (Fig. 5.8).

**Maurice Hugh Frederick Wilkins
1916–2004**

FIGURE 5.8   Maurice Hugh Frederick Wilkins.

Similarly to Randall, Wilkins was an experimental physicist who engaged during the war in research on radar and later on isotope separation under the Manhattan Project. After atomic bombs were dropped on Hiroshima and Nagasaki, Wilkins became tormented by a sense of guilt and thus abandoned nuclear physics for biophysics. His interest in biology developed when he read Schrödinger's book "*What Is Life*."[103] He was also impressed by the studies of Astbury on the diffraction of X-rays in biomolecules (Ref. 10, p. 90). Deciding to switch to biophysics, he intended to investigate chromosomes by rupturing them with ultrasonic waves.[102]

In late 1945, Randall applied to the Physical Secretary of the Royal Society for a grant of £800–1000 per annum for 5 years. This sum was supposed to fund research on the physics of mitosis and on "long-range" forces between proteins. The Biological Secretary of the Royal Society deemed the application too modest and by the end of the year, Randall's revised and more ambitious application was approved (for a highly detailed account of this historical episode, see de Chadarevian,[7] pp. 55–61, and Olby,[2] pp. 326–331). In early 1946, Randall was invited to serve as distinguished Wheatstone Chair of Physics at King's College, London. He accepted the invitation mostly because the Royal Society considered St. Andrews to be too small to sustain a viable biophysics program and held that such activity could thrive only in Cambridge or in London. After rebuilding the King's College physics laboratories, which had been destroyed during the war by a bomb, Randall established and directed a new Biophysics Research Unit in the Department of Physics and had appointed Wilkins to be his Assistant Director. The stated mission of the new Unit was (to conduct) *"an interdisciplinary attack on the secrets of chromosomes and their environment"* or, as was Randall's motto: *"to bring the logi of physics to the graphi of biology."*[104] The staff of the Unit, which grew in the two decades since its establishment in 1946 from 40 people to 120 (Ref. 2, p. 329), engaged in wide-ranging biophysical research projects that were funded by both the MRC and the Rockefeller Foundation (for the role of Rockefeller Foundation in supporting the rising discipline of molecular biology in general and specifically in the funding of Randall's Unit, see Ref. 105). As is described in Section 5.5, Wilkins performed in 1950 early X-ray diffraction analyses on samples of high molecular size DNA that he received from the Swiss organic chemist Rudolph Signer (1903–90). Although the obtained photographs indicated that DNA possessed crystalline structure, Wilkins realized that in order to reveal more details, their quality had to be improved. He thus got a microfocus X-ray generator that had been developed and donated by Werner Ehrenberg (1901–75) and Walter Spear (1921–2008) of the Bernal group at Birkbeck College[27] and a coupled microcamera.[106] He also asked that Rosalind Franklin, who was to join the Unit on a 3-year research fellowship, would work on crystallography of DNA rather than of protein as was originally planned. The details of their experiments and their consequences are described in Section 5.5.

5.3   EARLY STUDIES OF DIFFRACTION OF X-RAYS BY DNA FIBERS

Attempts to deduce the structure of DNA from diffracted X-rays patterns began in the Leeds laboratory of Astbury and were continued by Wilkins, Franklin, and Gosling in King's College. Unlike later studies of crystallized nucleic acids, these early X-ray diffraction analyses were done with DNA fibers. In general, deciphering X-ray diffraction in fibers that possess crystalline structure poses a specific challenge because although the fibers are organized in a roughly parallel arrangement, their relative orientations are different. Thus, an experimentally obtained diffraction pattern of a fiber reflects the average of patterns that would be produced by different orientations. Using the known chemical structure of the monomers, the structure of a polymer can be reconstructed by comparison of calculated diffraction patterns in the model with the experimentally documented ones. The model is then repeatedly corrected to fit the diffraction intensities in the actual pattern. This model "building" approach was taken by Watson and Crick when they constructed the double helix model of DNA.

Before delving into a description of the actual experiments and model building, the basic parameters of helical fibers should be introduced first.

5.3.1   X-Ray Diffraction by Helical Fibers

The defining parameters of a helical molecule include the repeat (c) of the helix, its pitch (p) and the pitch angel (β), its rise (h), and radius (r) (Fig. 5.9A).

To conduct X-ray diffraction analysis of a DNA fiber, it is pulled out from a concentrated solution, stretched to be aligned along the fiber axis, and placed in an X-ray beam. The scattering of the beam by the fiber is recorded on detection medium such as film, which is placed behind the irradiated fiber. As pointed out (Section 5.2.1), unlike crystals, fibers present a particular challenge. Although the elongated molecules are arranged in a roughly parallel orientation to one another along the axis of the fiber, not all the molecules are rotationally oriented relative to one another in a consistent fashion. Bands in the obtained X-ray image represent, therefore, rotational average of patterns that are produced by different orientations around the fiber axis. Shown in Fig. 5.9B is a sketch of diffraction pattern of the schematic helix in Fig. 5.9A. Reflections appear as a series of equidistant lines, named *layer lines*, which are perpendicular to the fiber axis. The scattering along each layer line is made up from Bessel functions, named after the German astronomer Friedrich Wilhelm Bessel (1784−1846) who originally used them to make accurate calculations of planetary orbits. In diffraction by helices, Bessel function replaces the sines and cosines that are used for crystals. A hallmark of a helix is a cross-shaped pattern that the layer lines form (see Section 5.3.2). The spacing that separates the layer lines is inversely proportional to the repeat (c) of the helix. The repetition of the cross-pattern every four layer lines in the scheme in Fig. 5.9B indicates that there are four residues per turn and thus that the rise of the helix is one-fourth of the repeat ($c/4$). Even though in actual cases the X-ray patterns of DNA fibers are more complex than the simplified scheme in this figure, they include the basic parameters of the helix. It should be noted, however, that the patterns do not yield the precise arrangement of atoms within each residue. A recommended comprehensive treatment of the analysis of X-ray diffraction by helices can be found in Ref. 107, Chapter 16, "Diffraction by Helical Structure."

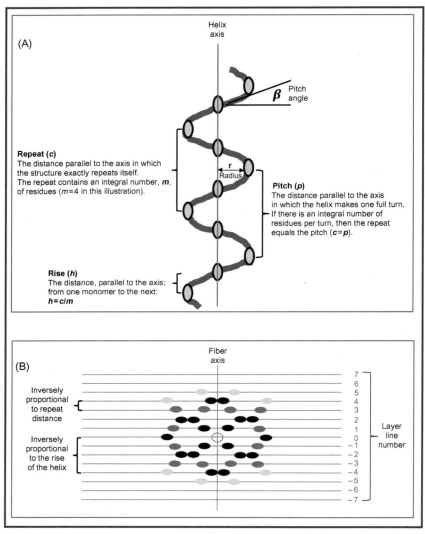

FIGURE 5.9 (A) Scheme of a helical molecule and its defining parameters. In the illustrated helix the four residues repeat equals the pitch. (B) Scheme of X-ray diffraction from a fiber.

5.3.2   Astbury Pioneered X-Ray Diffraction Analysis of the Structure of DNA Fibers

Reginald Herzog and Willie Jancke of the Kaiser Wilhelm Institute of Fiber Chemistry in Berlin-Dahlem produced in 1920 the first X-ray diffraction images of DNA fibers. Unfortunately, because their DNA was mostly degraded, the X-ray diffraction patterns that they produced were uninterpretable. Some 10 years later the Freiburg University crystallographer Friedrich Rinne (1863−1933) also obtained similarly low-quality images of DNA. The first interpretable patterns of diffraction of X-rays by DNA fibers were generated in the Leeds University laboratory of William Astbury. Although less known now, at his time Astbury was an internationally renowned scientist whose Leeds laboratory was acclaimed by Perutz as "*the X-ray Vatican*" (cited in Ref. 12). Because of his reputation as an expert on the structure of fibers, other researchers sent him various types of fibers for his inspection. In 1935 a zoologist from the Justus-Liebig University in Giessen, Wilhelm Joseph Schmidt (1884−1974) sent him birefringent fibers of DNA. However, when examined for their diffraction of X-rays, these fibers did not yield informational diagrams. Things changed in 1937 when Florence Bell (later Sawyer) joined the Astbury laboratory as a PhD student. Bell had a perfect background: Girton College, Cambridge, then a short period of work with Bernal at the Cavendish Laboratories, and last working under Lawrence Bragg in the Physics Department of the University of Manchester. Her PhD thesis in Leeds dealt with X-ray diffraction by a wide variety of biomolecules: collagens, globular proteins, and nucleic acids. Of Bell's various lines of study, only her studies on DNA are of enduring historical interest. Bell had relative success in obtaining readable X-ray diffraction images that Astbury used to build the first (erroneous) model of the structure of DNA. Although she first tried her hand at nucleic acids from different sources such as yeast, pancreas, and TMV, the key to Bell's relatively legible images was the use of high-molecular-weight DNA from calf thymus. This DNA was isolated in the Karolinska Institute by Torbjörn Caspersson (1910−97) and Einar Hammersten (1889−1968) who developed an extraction method that minimized the breakage of the DNA. Rudolph Signer inspected their DNA under polarized light and reached the conclusion that the molecule was a long rod of $0.5−1.0 \times 10^6$ Da.[108] When Bell sent an X-ray beam through this high-molecular-weight DNA, she was not aware that the conformation of that DNA changes with increasing humidity from an A-form into a different B-form (see below). Because the humidity of the DNA sample was not controlled, her DNA had the two forms superimposed on one another and the obtained X-ray diffraction images were therefore blurred and hard to interpret (Fig. 5.10A).

Yet, Astbury and Bell deduced from the diffraction patterns that the DNA molecule comprised of nucleotides stacked on one another in what they likened to "*a pile of pennies.*" Moreover, taking into account the spacing between the rings in the diffraction images and the dimension of the molecules, they calculated that the distance that separated any two neighboring bases along the axis of the molecule was 3.3 Å (Refs. 109,110; see also a facsimile of Florence Bell's doctoral thesis at http://www.leeds.ac.uk/

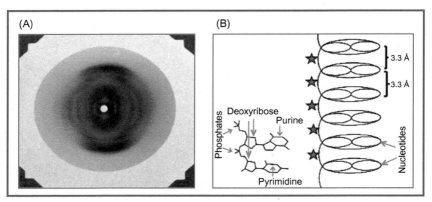

FIGURE 5.10 (A) Florence Bell's X-ray diffraction patterns of thymus DNA. (B) A model of the structure of DNA that Astbury and Bell proposed on the basis of their X-ray diffraction data. The two main elements of the model were: (a) DNA was described as a single-chain helix. (b) The bases were stacked on one another and separated by inter-nucleotide distance of 3.3 Å. Red stars mark the phosphodiester backbone. The scheme of stacked pyrimidine and purine on the bottom-left corner indicate that Astbury thought that the base and sugar moieties were coplanar and positioned perpendicularly to the phosphodiester backbone. *For (A): The image is from her 1939 PhD thesis: http://www.leeds.ac.uk/heritage/Astbury/Bell_Thesis/index.html. For (B): Modified from Astbury WT, Bell FO. Some recent developments in the X-ray study of proteins and related structures.* Cold Spring Harbor Symp Quant Biol *1938;6:109–21.*

heritage/Astbury/Bell_Thesis/index.html). Later Astbury tended to think that a more exact inter-nucleotide spacing was 3.4 Å (Ref. 2, p. 324). Based on these data, Astbury and Bell constructed a single-chain helix model of DNA in which the bases that were linked by a phosphodiester backbone were stacked on one another at spacing of 3.3 Å (Fig. 5.10B).

In his original model, Astbury deemed the base and deoxyribose moieties to be coplanar. He changed his mind only in 1942. A model that his assistant at the time Mansel Davies (1913–95) had built convinced him that the sugar and base were positioned at an angle to one another. Based on a perceived similarity between the measured inter-nucleotide spacing in DNA and the distance between adjacent amino acids in a fully extended polypeptide chain, Astbury came up with an idea about the functional implication of the deduced structure of DNA[109]:

> *The significance of these findings for chromosome structure and behavior will be obvious, for we cannot fail to be struck by the fact that the* spacing of successive nucleotides is almost exactly equal to the spacing of successive side-chains in a fully extended polypeptide chain. *It is difficult to believe that the agreement is not more than coincidence: rather it is a stimulating thought that probably the interplay of proteins and nucleic acids in the chromosomes is largely based on this very circumstance.*

And:

The idea is equivalent to saying that the molecule of thymonucleic acid fits per-
fectly on side-chain pattern of a fully extended polypeptide chain that interac-
tion should take place almost without any steric hindrance whatsoever; most
easily between the basic side-chains and the phosphoric acid groups, but pre-
sumably, too, between the acid side-chains and the basic groups of the nucleo-
tides. Furthermore, the products of combinations should also be fibrous, like
the two original constituents.

With the assumption that duplicating genes were proteins, Astbury interpreted the similar spacing of amino acids in protein and nucleotides in DNA
as reflecting the role of DNA in protein duplication. Evoking in 1967
Astbury's leap of imagination in the late 1930s, Florence Bell said that bas-
ing his thinking only on the similar inter-residue spacing in DNA and in
extended polypeptide, he came up with the daring idea that DNA provides
structural support for the replication of protein genes (cited in Ref. 2, p. 67):

We were 'ecstatic' to find a spacing identity—but not really surprised, because
we had been hoping to find some relationship. Astbury considered that the
nucleic acids were templates for protein duplication and organization, and held
the polypeptide chains stretched and parallel for division process. This was one
time when his 'vox diabolica' (as Astbury labelled me) gave little arguments.

It must be pointed out, however, that Astbury's views on the nucleotide
composition of DNA and the potential basis for correspondence between
DNA and protein became more nuanced in time. He later began to believe
that the gene was a nucleoprotein complex and that the DNA component of
the complex was more variable than posited by Levene's tetranucleotide
hypothesis (Section 2.4). Historical evidence indicates that particularly after
Avery identified DNA as the genetic material of *pneumococcus* (see
Chapter 3: "Avery, MacLeod, and McCarty Identified DNA as the Genetic
Material"), Astbury's thinking evolved to impart on DNA a role beyond
mere mechanical support for protein genes. Space here does not permit full
discussion of this matter and the interested reader should consult Hall's comprehensive treatment of the subject.[12,59]

In 1951 Astbury returned for the last time to X-ray analysis of DNA. At
this instance the examined DNA sample, which was contributed by the
Columbia University biochemist Erwin Chargaff (1905−2002), yielded a diffraction pattern much superior to the images that Florence Bell had produced. As is described in Section 5.4, it was Chargaff who provided in 1950
unambiguous biochemical evidence that the molar ratios of the four nucleotides differed in DNA from different species.[111] This finding put an end to
the long-standing tetranucleotide theory of Levene that professed universal
1:1:1:1 molar ratios for the four nucleotides in DNA (see Section 2.4).
Astbury was highly impressed with this report because it matched his earlier

prior idea that varying nucleotide compositions may dictate different shapes of different DNA regions. He thus wrote a letter to Chargaff asking for DNA samples that he could examine by X-rays in an attempt to detect their different shapes. Chargaff promptly sent a sample of calf thymus DNA and Astbury assigned a member of the laboratory, Elwyn Beighton, to examine its structure by X-ray diffraction analysis. Beighton joined the Astbury laboratory in 1939, first as "lab-boy" who did menial jobs around the laboratory and later as technician. After serving in the military during World War II, Beighton returned to the laboratory to do a PhD thesis under the supervision of Astbury. It was at that stage that he performed X-ray diffraction analysis of Chargaff's DNA. Beighton's experiment was never published and the following description of its technical details is based on Olby's historical research (Ref. 2, p. 379). Because Chargaff was worried that the low water content of the DNA could impede the analysis, Beighton dripped water onto the fiber while stretching it by threefold. He then passed through the moistened DNA an X-ray beam that emanated from an X-ray generator, which was equipped with a rotating anode that reduced damage by the radiation to the DNA. The image that Beighton produced was much clearer and more revealing that the X-ray diffraction patterns that Bell generated in 1938. First, we retrospectively recognize that whereas Bell analyzed a mixture of the A- and B-forms of DNA, Beighton's wetting of the DNA fiber ensured that it adopted a pure B-form conformation. Second, and most clearly, the image had at its center a cross-shaped pattern of spots (Fig. 5.11).

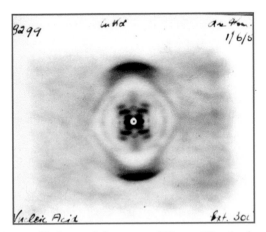

FIGURE 5.11   Photographic image of the pattern of X-rays diffraction by calf thymus DNA. Elwyn Beighton produced this image on June 1, 1951, in the University of Leeds laboratory of William Astbury. *Image from the University of Leeds Museum of the History of Science, Technology and Medicine website: http://www.leeds.ac.uk/heritage/Astbury/Beighton_photo/ index.html.*

Unbeknown to Astbury, this pattern held key information of a helical structure of the examined DNA. Based on their early images of the diffraction of X-rays by DNA fibers (see Section 5.5.1), Maurice Wilkins suspected that the DNA might be a helical molecule, mostly because the molecule did not diffract at all along its length. He thus asked in 1950 his King's College colleague, the theoretical physicist Alexander Stokes (1919−2003), to develop a mathematical solution to the question of what type of diffraction pattern does a helical molecule produce. Using Bessel function, Stokes promptly came up with the solution that an X-ray diagram of a helix should have a centrally positioned cross-shaped configuration of diffraction spots. Without going into the mathematical reasoning for this conclusion, it should be noted that the X-shaped pattern of diffraction is generated by the scattering of the X-rays perpendicularly to the "zigzag" arrangement of the DNA molecule. Stokes shared his new insight with Wilkins and it also became known to Crick who was at the time a research student at the Cambridge Cavendish Laboratory. Stokes had never published his theory but it was corroborated by an independent study in Cambridge. In 1951 William Cochran (1922−2003), then a lecturer in the Cavendish, collaborated with Crick and with the Czech crystallographer Vladimir Vand (1922−62) in Glasgow to develop a theory that explains the X-ray diffraction pattern of the synthetic peptide poly-γ-methyl-L-glutamate. Their solution indicated that a cross-shaped diffraction pattern reflected helical structure of the molecule. This result was published in 1952 with the remark that it *"was also derived independently and almost simultaneously by Dr A.R. Stokes."*[112]

Unaware of both the Stokes theory and the work of Cochran, Crick, and Vand,[112] Astbury could not grasp the significance of the cross-shaped pattern in Beighton's image. He considered Florence Bell's pictures of 1938 to be superior to Beighton's because they contained more spots and were thus thought to represent a purer form of the DNA (Ref. 2, p. 379; also see Ref. 12, pp. 145−148, 151−161). At that point in time Astbury gave up his studies on DNA and had never returned again to the subject. Kersten Hall, Astbury's biographer, pointed out that he may have left DNA because he felt defeated by the progress that was made by the better funded and strong group in King's College (Ref. 12, pp. 133−134). Hall also argued that in his later years Astbury became disheartened and was questioning the worth of his career-long efforts (Ref. 12, pp. 147−148). Be it what it may, it was at that point that Astbury abandoned and never returned to his studies of the structure of DNA.

In May 1952, almost a year after Beighton produced his image of cross-shaped pattern of diffraction by B-form DNA, Franklin and Gosling produced in King's College their famous "photo 51" of B-form DNA that had at its center a very similar cross-shaped configuration of diffraction spots (see Figs. 5.18A and 5.27). After Watson saw photo 51 and conveyed its salient features to Crick, it became clear to the two that DNA must have had helical structure.

5.3.3 Sven Furberg's Proposal of Helical DNA

The Norwegian crystallographer Sven Verner Furberg (1920–83) came to Birkbeck College in 1947 to work with Bernal. Harry Carlisle, who later inherited Bernal's chair at Birkbeck, turned Furberg's attention to the emerging importance of nucleic acids and put him to solve the three-dimensional structure of crystallized cytidine. Considering the technical limitations of the time, this was not a trivial undertaking. Yet, Furberg succeeded in determining that the deoxyribose moiety was almost perpendicular to the base.[113] Using his structure of cytidine, he went on to construct two models of DNA of which only one is described here. Based on Astbury's crystallographic data, Furberg's model proposed that the bases that were oriented perpendicularly to the deoxyribose ring were separated from one another by 3.4 Å and formed a helix (a "spiral" by his term) that had eight nucleotides in a repeat of 27 Å (Ref. 114 and Furberg's PhD thesis cited in Ref. 2, pp. 336–339). This was the first model of DNA that placed the bases on the inside of the helix. However, additionally to its incorrect dimensions, the Furberg model was of a single-strand helix. It is retrospectively clear that although his model was published in 1952, he was not familiar with the Chargaff 1950 report of purine–pyrimidine complementarity[111] (Section 5.4). This experimental finding became later a key element in the construction by Watson and Crick of the double helix model of DNA.

5.4 ERWIN CHARGAFF'S DISCOVERY OF BASE COMPLEMENTARITY IN DNA

Erwin Chargaff (Box 5.7), biochemist, linguist, and intellectual of the old European tradition, was working in Columbia University when he read in 1944 the landmark paper of Avery, Macleod, and McCarty.[115]

BOX 5.7 Erwin Chargaff

Chargaff was born in 1905 in the Austro-Hungarian town of Czernowitz. He was a gifted linguist who was reputed to have perfect command of 15 different languages. He thus hesitated when it came to choosing a career and finally elected to study chemistry rather than linguistics. After obtaining a PhD degree in 1928 from the University of Vienna, he conducted research work at Yale University for 2 years. Not liking America, he returned to Europe to spend 3 more years at the University of Berlin. Being Jewish and facing the new Nazi policies of 1933 that excluded Jews from academic positions, Chargaff had left Germany for the Pasteur Institute in Paris and shortly thereafter for the United States. He settled in 1935 at Columbia University where he stayed on until his retirement in 1978. In the earlier part of his career, he focused mostly on the biochemistry of bacterial lipids (specifically, lipids of *Mycobacterium tuberculosis*). However, exposure to

(Continued)

BOX 5.7 (Continued)

Avery's 1944 paper on DNA as the *pneumococcus* transforming principle led Chargaff to change course and concentrate on the biochemistry of nucleic acids, which he understood to be the hereditary material. His experiments disproved Levene's tetranucleotide hypothesis of the composition of DNA. More importantly, he discovered the so-called Chargaff Rules of base complementarity in DNA that became a cornerstone of the Watson and Crick model of the double helix. At later times Chargaff became bitter, feeling that his contribution to the discovery of the structure of DNA was not adequately recognized. He also developed a strong antipathy toward the new discipline of molecular biology. His views on science and culture were laid out in his 1978 autobiography *"Heraclitean Fire"*[11,21] (Fig. 5.12).

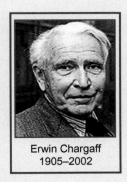

Erwin Chargaff
1905–2002

FIGURE 5.12   Erwin Chargaff.

In this article DNA was identified as the substance that induced phenotypic changes in the bacterium *pneumococcus* (see Chapter 3: "Avery, MacLeod, and McCarty Identified DNA as the Genetic Material"). The inference of this report that DNA was the genetic material enthralled Chargaff so much that he changed his area of research from the biochemistry of lipids in the bacterium *Mycobacterium tuberculosis* to the biochemistry of nucleic acids. Describing this turn of heart he wrote in 1971[116]:

> *This* [Avery's] *discovery, almost abruptly, appeared to foreshadow a chemistry of heredity and, moreover, made probable the nucleic acid character of the gene. It certainly made an impression on a few, not on many, but probably on nobody a more profound one than on me. For I saw before me in dark contours the beginning of a grammar of biology [...] Consequently, I decided to relinquish all that we had been working on or to bring it to a quick conclusion [...]*

Chargaff started his studies of nucleic acids by assessing their nucleotide compositions. He became interested in this question after reading a 1945 review article by the Nottingham University nucleic acid chemist John

Masson Gulland (1898–1947) and his associates, G.R. Barker and D.O. Jordan.[117] These British investigators revisited Phoebus Levene's tetranucleotide hypothesis which contended that DNA consisted of repeating units of the four nucleotides: $(ATCG)_n$ (Section 2.4). Without disputing the equal molar ratios of the nucleotides, Gulland and associates argued that instead of regular repeat in DNA of the same order of four bases (a so-called structural tetranucleotide arrangement: ACTG ACTG ACTG···), the bases might be distributed randomly in quartets without losing their equimolarity (a "statistical tetranucleotide" arrangement: TAGC GCAT TCGA···).

Based on this idea, Chargaff pondered whether the assembly of nucleotides in different orders might be a key to the capacity of DNA to dictate specific traits in living organisms. In pursuit of this notion, he dedicated the later years of the 1940s to the quantification of nucleotides in DNA and RNA. The technical basis for his analysis was laid in a 1944 paper by investigators from the Wool Industries Research Association in Leeds that described the use of paper chromatography to separate amino acids.[118] In this method, a mixture of organic molecules (in that early case, amino acids) was separated into its constituent components. Briefly, the mixture was loaded onto one end of a sheet of filter paper that was then soaked at its origin in an empirically defined mixture of organic solvents. The liquid slowly advanced by gravity along the paper carrying with it the mixture of amino acids. While the solvent was progressing, the components of the mixture were gradually separated from one another because of their different partition coefficients between paper and solvent. Reading that paper, Chargaff realized that by finding the right solvent and the conditions of chromatography, he could adapt the method for the resolution of the four nucleotides in hydrolyzates of DNA or RNA. Moreover, at the end of the chromatographic run, the separated nucleotides could be excised and extracted from the paper and their relative amounts in the extracts could be determined by measurement of the ultraviolet adsorption (at 250 mμ). Indeed, in 1948, Chargaff developed and refined procedures for the chromatographic separation, extraction, and quantification of nucleotides in hydrolysates of DNA and RNA.[119–121] He next showed that the molar proportions of the four nucleotides were the same in DNA from calf spleen and thymus.[122] He also found that unlike the unchanged ratios of nucleotides in DNA from different tissues of the same species, the relative amounts of the four nucleotides differed in DNA from *Mycobacterium tuberculosis*, from yeast[123] and from human cells.[124] In 1950, Chargaff published in the respected journal *Experientia* (now defunct) his most comprehensive survey of the base compositions of DNA and RNA of different species and cell types.[111] The main experimental steps that Chargaff took are charted in Fig. 5.13A.

Three main principal points emerged from the quantitative analysis of the four nucleotides in hydrolysates of DNA from bovine, human, bacterial, and yeast cells (Fig. 5.13B). (1) DNA from different tissues in any one species

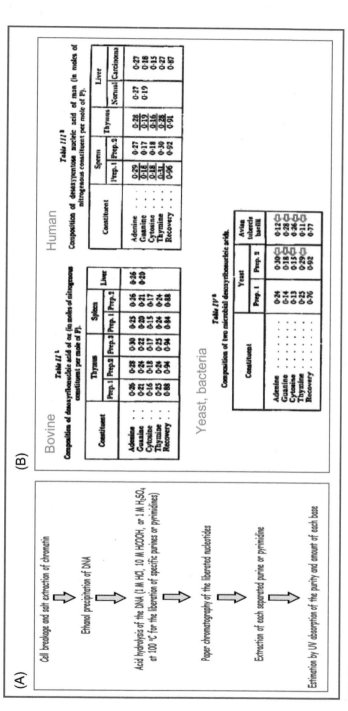

FIGURE 5.13 (A) Chargaff's procedure of acid hydrolysis of DNA, separation of the four nucleotides by paper chromatography, their isolation, extraction, and quantification. (B) Molar ratios of the constituent bases of DNA from four different species. Relative values of the content of nucleotides DNA are underlined or marked by arrows in representative cases. Same color underlines or arrows highlight the nearly identical relative contents of adenine and thymine and of guanine and cytosine. *For (B): Tables from Chargaff E. Chemical specificity of nucleic acids and mechanism of their enzymatic degradation. Experientia 1950;***6***(6):201–09.*

(bovine or human) had practically the same relative contents of purines (adenine and guanine) and pyrimidines (cytosine and thymidine). For instance, the average relative amounts of pyrimidines (cytosine + thymine) in the bovine tissues thymus and spleen were, respectively, 42% and 41%. Likewise, the relative content of purines (adenine + guanine) was also the same in different tissues of any one species (Fig. 5.13B). (2) DNA from different species had prominently different relative content of pyrimidines and purines. For example, numbers in Fig. 5.13B indicated that the relative content of pyrimidines (cytosine + thymine) in bovine spleen, human sperm, avian tubercle bacilli, and in yeast were 40%, 48.5%, 37%, and 41%, respectively. Another way to appreciate the different nucleotide compositions of DNA from different species was to compare the ratios of adenine to guanine or of thymine to cytosine. For example, the adenine to guanine ratios in DNA from human sperm and from tubercle bacilli were, respectively, 1.6 and 0.42 and their respective ratios of thymine to cytosine were 1.7 and 0.39. (3) Chargaff's measurements showed that in all the examined species and tissue types, the relative content of guanine always equaled, within the margin of experimental error, that of cytosine and the content of adenine was the same as that of thymine. For instance, adenine comprised 29% and thymine 31% of human sperm DNA and these two respective nucleotides comprised 12% and 11% of the DNA of tubercle bacilli. In parallel, guanine and cytosine consisted 18% each of human sperm DNA and 28% and 26% of the DNA of tubercle bacilli (Fig. 5.13B). A common way to present these results was to give a value of 1 to the molar ratios of adenine to thymine and of guanine to cytosine.

In a series of additional experiments, Chargaff next substantiated these three fundamental observations and also expanded his measurements to RNA. First, multiple measurements of the ratios of adenine to guanine and of thymine to cytosine in DNA from four species demonstrated the sizeable differences in their nucleotide compositions (Fig. 5.14A).

By contrast, determination of the relative amounts of nucleotides in DNA of different bovine or human tissues validated the earlier conclusion that they were identical in different tissues of any single species had the same relative content of the four nucleotides (Fig. 5.14B). Last, quantification of the nucleotide composition of RNA from tissues of five different species revealed a fundamental difference between the base compositions of DNA and RNA. On the one hand, the nucleotide contents of both RNA and DNA exhibited similar species-specific diversity. Thus, the ratio of purines to pyrimidines in RNA from five examined species ranged from 1.1 to 2.5 (Fig. 5.14C). Yet, unlike DNA, the ratios of guanylic to cytidylic acid or of adenylic to uridylic acid were not 1 in every examined RNA samples. For instance, the ratio of guanylic to cytidylic acid in RNA from pig pancreas was 2.29 and the ratio of adenylic to uridylic acid was 2.17 (respective arrows in Table VIII, Fig. 5.14C). Analyses of the RNA from the other four species revealed similar unequal contents of guanine and cytidine and of

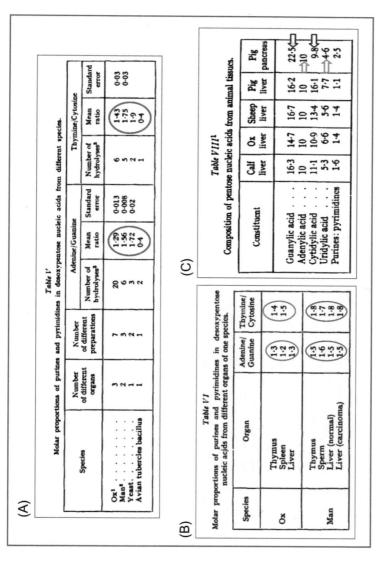

FIGURE 5.14 (A) DNA samples from different species had very different base composition as expressed by their ratios of adenine to guanine and of thymine to cytosine. The different ratios are encircled by red or blue ellipses. (B) DNA samples from different tissues of the same species had practically identical base compositions. Similar ratios in each species are encircled by blue ellipses. (C) Unlike DNA, RNA samples from different species did not maintain a ratio of 1 of guanine to cytosine or of adenine to uridine. White and red arrows point at the different contents of guanylic and cytidylic acid and of adenylic and uridylic acid, respectively. All three Tables are from Ref. 111. *Tables in (A), (B), and (C) are from Chargaff E. Chemical specificity of nucleic acids and mechanism of their enzymatic degradation. Experientia 1950;6(6):201−9.*

adenine and uridine. Thus, unlike DNA, and except for one accidental case of the guanine to cytosine ratio in pig liver RNA, the ratios of guanine to cytosine and of adenine to uridine differed in RNA of different species and were not 1 (Fig. 5.14C).

Returning to the origin of Chargaff's set of experiments, it should be remembered that the accepted wisdom at their inception was Phoebus Levene's model of DNA as a polymer of repeating (ATCG) four-base units. This theory imposed molar ratios of 1:1:1:1 of the four bases in DNA (25% each). Chargaff's finding that these ratios were not 1:1:1:1 and that they differed in different species convincingly repudiated the tetranucleotide theory. Because Levene's theory was for decades an unchallenged paradigm, Chargaff's delight at its refutation was understandable. He opened his discussion of the results with a philosophical comment on the advantages and perils of generalizations in science. Then, after recapping his results, he made his own generalization by declaring the demise of the tetranucleotide hypothesis and its replacement by a new premise[124]:

Generalizations in science are both necessary and hazardous; they carry a semblance of finality which conceals their essentially provisional character; they drive forward, as they retard; they add, but they also take away. Keeping in mind all these reservations, we arrive at the following conclusions. The deoxypentose nucleic acids [DNA] from animal and microbial cells contain varying proportions of the same four nitrogenous constituents, namely adenine, guanine, cytosine, thymine. Their composition appears to be characteristic of the species, but not the tissue.

And:

The results serve to disprove the tetranucleotide hypothesis.

The dispute with the tetranucleotide model of DNA is now mostly forgotten. By contrast, the "Chargaff Rule" of the equimolarity of guanine and cytosine and of adenine and thymine in duplex DNA from every species that will ever endure. Curiously, however, Chargaff made in 1950 only the following cautious and terse comment about this most important discovery[124]:

It is, however, noteworthy—whether this is more than accidental, cannot yet be said—that in all deoxypentose nucleic acids [DNA] examined thus far the molar ratios of total purines to total pyrimidines, and also of adenine to thymine and guanine to cytosine, were not far from 1.

Returning to the matter of the heterogeneous nucleotide content of DNA from diverse species, Chargaff made a naïve assessment of the size of bovine DNA and yet he was farsighted about the potential importance of DNA base sequence (my emphasis in a italic and bold font)[124]:

[...] a decision as to the identity of natural high polymers often still is beyond the means at our disposal. **This will be particularly true of substances that**

differ from each other only in the sequence, not in the proportions, of the constituents. *The number of possible nucleic acids having the same analytical compositions exhibiting the same molar proportions of individual purines and pyrimidines as the deoxyribonucleic acid of the ox is more than 10^{56}, if the nucleic acid is assumed to consist of only 100 nucleotides; if it consists of 2,500 nucleotides, which probably is much nearer the truth, then the number of possible 'isomers' is not far from 10^{1500}.*

A few years after Chargaff discovered the equimolarity of guanine and cytosine and of adenine and thymine in DNA, it became a key part of the double helix model of DNA. Chargaff's direct contribution to this landmark discovery is described later in the context of the work of Watson and Crick (Section 5.6.4).

5.5 STUDIES AT KING'S COLLEGE ON THE DIFFRACTION OF X-RAYS IN DNA FIBERS

Maurice Wilkins, Rosalind Franklin, and Raymond Gosling of King's College performed in 1950−52 X-ray diffraction studies of DNA fibers. Some of their data as well as their direct interaction with James Watson and Francis Crick at the Cavendish Laboratory of Cambridge University were essential to the construction by Watson and Crick of the double helix model of DNA. This section describes the experimental work that was accomplished at King's College.

5.5.1 Wilkins' Early Study of X-Ray Diffraction by DNA Fibers

After a period of exploration of diverse directions of research, Randall's Biophysics Research Unit at King's College narrowed its experimental work to the analyses of DNA, nucleoproteins, TMV, muscle, and collagen using ultraviolet light, dichroism, and X-ray diffraction. According to Raymond Gosling (see Box 5.8) who began his work at King's College as a PhD student of Randall and Wilkins and later of Franklin, Randall was persuaded by the work of Avery to think that DNA was a key to cell division.[104,125] He thus steered part of the research in his Unit to studies of DNA. Maurice Wilkins, who was Randall's Assistant Director, initiated the employment of X-ray diffraction to explore the structure of DNA fibers. Fortune was on his side when he attended in May 12, 1950, a meeting in London of the Faraday Society. One attendant at the meeting was the Bern University biochemist Rudolph Signer (Fig. 5.15A, first panel) who was invited to present his new method for the isolation of high-molecular-weight DNA.

As was noted in Section 5.3.2, Signer collaborated in the late 1930s with Caspersson and Hammersten and used polarized light to determine the size and contours of DNA.[108] These experiments convinced him that DNA was a

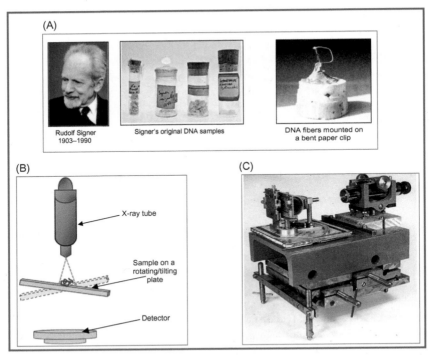

FIGURE 5.15 (A) Rudolph Signer, his vials of freeze-dried high-molecular-weight DNA, and a bent paper clip that held the DNA fibers for X-ray analysis at King's College. (B) Scheme of the main components of an X-ray imaging apparatus. (C). The actual tilting microcamera and stand that were used by the King's College DNA crystallographers. *Images in panels (A) and (C) are from a King's College online exhibition: http://www.kingscollections.org/exhibitions/archives/dna/.*

long and highly fragile molecule that readily broke during purification. To minimize DNA breakage, Signer and his student H. Schwander developed in the late 1940s a modification of the procedure of Mirsky and Pollister[126] for the isolation of high-molecular-size DNA.[127] When he was invited to the Faraday Society meeting, Signer took with him vials of freeze-dried sodium salt of calf thymus DNA that he purified by his new method and which was later shown to have a size of $\sim 7 \times 10^6 \, \text{Da}$[128,129] (Fig. 5.15A, second panel). By the end of his talk, he offered these DNA samples as a gift to fellow attendees (for an engaging history of this episode and of Signer's career, see Ref. 130). One who received this valuable present was the Harvard University chemist and molecular biologist Paul Doty (1920−2011) who later went on to determine the molecular size of this DNA[128] and another was Maurice Wilkins. Wilkins studies of Signer's DNA started with measurements of its ultraviolet absorption curves and determination of its dichroic ratios by achromatic reflecting microscope.[131] He next reported that stretching the DNA fiber reduced its diameter and caused base pair inclination, suggesting that the DNA was an extensible molecule.[132] Most importantly, he found that in an atmosphere of saturated water vapors, the fibers were mostly microcrystalline as was reflected by their negative birefringence, ultraviolet absorption, and infrared dichroism. However, the same measurements indicated that at humidity of 50%, the crystallinity of the DNA was mostly lost.[132] In the course of his experiments, Wilkins noticed that the DNA could be spun into a thin fiber by touching it with the tip of a glass rod and then drawing the rod away. In time he became so adept at spinning out the DNA fibers that Raymond Gosling described him as *"being like a wonderful spider."* The uniform appearance of the fibers suggested that they must have been aligned in parallel orientation to one another. Also, polarized light microscopy of the DNA indicated that the fibers were likely to consist of well-organized crystals. Based on these preliminary clues, Wilkins undertook to perform diffraction of X-rays by his DNA fibers. Joined by Raymond Gosling, he used an outmoded Raymax X-ray apparatus in the basement of the Chemistry Department. Because of the soft focus X-ray tube of this apparatus, it was ill suited for work with a single fiber. Thus, to obtain a specimen of a size that would detectably diffract X-rays, Wilkins had to bundle together 35 fibers that were packed together with the aid of small blobs of glue.[104] The clumped fibers were mounted on a make-shift holder made of a bent paper clip (Fig. 5.15A, third panel) and were subjected to X-irradiation. The first X-ray diffraction image of a bundle of 35 DNA fibers was ill defined. Randall pointed out that the fuzziness was likely due to the similar atomic content of nitrogen and oxygen in both the fibers and the air within the camera. He thus instructed Gosling to replace the air in the apparatus with hydrogen. To monitor the amount of streamed hydrogen and to avoid filling the room with the gas, it was bubbled through water. According to Gosling, this procedure also serendipitously wetted the DNA and caused microcrystallites to form within each fiber.[125] Wilkins had a different memory of the circumstances of the wetting of the fibers. According to his version, he and Gosling kept the DNA fibers moist because Bernal had reported before that protein crystals yielded better diffraction images when they were wet.[133] Be it what it may, the diffraction pattern of the dampened fiber bundles had many well-defined spots, suggesting that the DNA adopted a crystalline form (Fig. 5.16).

Both Gosling and Wilkins felt at the moment that the obtained pattern showed that they had successfully *"crystalized genes."*[104] Gosling with Alex Stokes' help determined the repeat values of the unit cell by measuring the line separation in the image. They also defined the unit cell as belonging to the monoclinic C2 space group (http://metafysica.nl/monoclinic_1.html), whose dyad axis was at right angles to the fiber axis.[125]

Some months later, Randall was invited to a conference in Naples on "Submicroscopical Structure of Protoplasm." Unable to attend, he asked

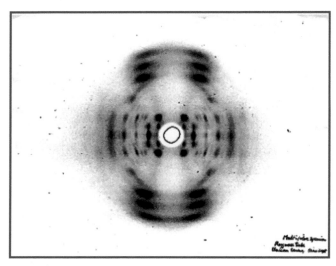

FIGURE 5.16 An image of X-rays diffraction by a bundle of 35 DNA fibers that Wilkins and Gosling produced in the spring of 1950.

Wilkins to go in his place. In his talk at that May 1951 meeting, Wilkins presented the DNA diffraction image that Gosling and he had produced (for a detailed description of the Wilkins talk at the conference, see Ref. 2, pp. 338–340. Also in the same reference, see citation from the talk on p. 355). This evidence for a crystalline structure of the DNA impressed James Watson who was in the audience. Much later he wrote of the impact that the Wilkins and Gosling image made on him (Ref. 8, pp. 25–26):

> *Suddenly I was excited about chemistry. Before Maurice's talk I had worried about the possibility that the gene might be fantastically irregular. Now, however, I knew that gene could crystallize; hence they must have a regular structure that could be solved in a straightforward fashion.*

In fact, Watson intended to ask Wilkins if he could join his laboratory but according to him, Wilkins vanished at the end of the talk and he could not make his request (Ref. 8, p. 26). As remembered by Gosling, however, he was actually turned down *"because Wilkins was afraid of him. He's quite scary old Jim, on full flight."*[104] Instead, Watson applied to Lawrence Bragg and was accepted to the Cavendish Laboratory where his collaboration with Crick culminated in the construction of double helix model of DNA.

After his return from Naples, Wilkins began to think that in analogy with the just discovered α-helix in proteins, the DNA molecule might also be helical. He thus asked Alex Stokes in July 1951 to determine mathematically what X-ray diffraction pattern was to be expected from a helical molecule. As mentioned, Stokes' analysis indicated that a helix should produce a central cross-shaped pattern.

5.5.2   Rosalind Franklin and the Discovery of Humidity-Dependent Forms A and B of DNA

Although Wilkins and Gosling produced X-ray diffraction images that were indicative of crystalline structure of DNA, Randall was not sure that they were up to the task of solving the atomic structure of the DNA. He decided, therefore, to recruit for this undertaking a more experienced crystallographer.[104] Impressed by the X-ray diffraction analysis of carbon that Rosalind Franklin (Box 5.8) performed at the Central Government Laboratory in Paris, Randall invited her to work in his Unit. The original research plan was that she would conduct crystallographic analysis of proteins in solution.

However, after Franklin had already been granted a 3-year Turner and Newall fellowship, Randall asked her to change the course of her intended research from proteins to DNA. This he did in a December 4, 1950, letter to Franklin (a facsimile of this letter is in the National Library of Medicine "Profiles in Science" "The Rosalind Franklin Papers" (http://profiles.nlm.nih.gov/ps/access/KRBBBB.pdf) :

After very careful consideration and discussion with the senior people concerned, it now seems that it would be a good deal more important for you to

---

BOX 5.8 Rosalind E. Franklin and Raymond G. Gosling

Franklin was born in London to a well-to-do English-Jewish family. After studying chemistry and physics at the University of Cambridge, she started in 1942 a 4 years' period of research at the British Coal Utilization Research Association at which she used helium absorption to assess the porosity of coal. After receiving in 1945 a PhD degree from Cambridge, she worked from 1947 until 1951 at the Central Government Laboratory for Chemistry in Paris. That was the happiest period of her life during which she did X-ray diffraction analysis of carbons. In 1951, Randall recruited her to his King's College Unit. Although she was originally assigned to do crystallography of proteins in solution, Randall changed her direction of research to X-ray diffraction analysis of DNA and put Gosling to be her PhD student. Unfortunately, Wilkins, who had initiated X-ray studies of DNA with Gosling, was not informed of this development. As a result, his professional and personal relationship with Franklin deteriorated. Franklin and Gosling produced the first clear pictures of A- and B-forms of DNA, most prominently the famous "photo 51" of B-form DNA that Wilkins showed to Watson. This image has proved to be crucial for the construction by Watson and Crick of the double helix model of DNA. Because of her friction with Wilkins and general dissatisfaction with King's College, Franklin moved to the laboratory of Bernal at Birkbeck College where she conducted X-ray analyses of TMV. Rosalind Franklin died in 1958 of ovarian cancer. She was only 37 years old at the time of her death. Her life, scientific career, and role in the discovery of the structure of DNA were the subject of biographies[20,21] and a memoir[134] (Fig. 5.17).

(Continued)

BOX 5.8 (Continued)

Gosling studied physics at University College, London. After working for 2 years as a hospital physicist, he became in 1949 research student in King's College. In his early work under Wilkins, he produced the first clear image of X-ray diffraction by a bundle of DNA fibers. In his later work with Franklin, he obtained under conditions of controlled humidity clear images of the B- and A-forms of DNA. After receiving a PhD degree in 1954, he stayed for a while at King's College and then moved first to University of St. Andrews and then to the University of the West Indies in Jamaica. In 1967, he returned to London and worked until the end of his career at Guy's Hospital Medical School. After leaving King's College, Gosling had left crystallography for the development of ultrasound devices for the diagnosis and treatment of atherosclerosis.

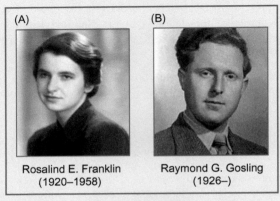

(A) | (B)

Rosalind E. Franklin (1920–1958) | Raymond G. Gosling (1926–)

FIGURE 5.17   (A) Rosalind E. Franklin and (B) Raymond G. Gosling.

investigate the structure of certain biological fibres in which we are interested, both by low and high angle diffraction, rather than to continue with the original project of work on solutions as the major one. [...] Gosling, working in conjunction with Wilkins, has already found that fibres of desoxyribose nucleic acid derived from material provided by Professor Signer of Bern gives remarkably good fibre diagrams. [...] As you no doubt know, nucleic acid is an extremely important constituent of cells and it seems to us that it would be very valuable if this could be followed up in detail. [...] I hope you will understand that I am not in this way suggesting that we should give up all thought of work on solutions, but we do feel that the work on fibres would be more immediately profitable and, perhaps, fundamental.

Franklin and Wilkins Separated Their Ways

Franklin started her work as a research associate in the King's College Biophysics Unit in January of 1951. Famously, as soon as she started to work on DNA, contacts between her and Wilkins were severed and each worked separately without sharing results, techniques, or reagents. Many wrote about the clash of personalities between Wilkins and Franklin. However, Randall's mismanagement of the DNA project was also a major contributing factor to the detrimental schism that opened between Franklin and Wilkins. First, Wilkins had not seen Randall's December 4, 1950, letter to Franklin. He remained uninformed about its substance for a long time and learned of it only many years later. Second, immediately after Franklin's arrival in London, Randall reassigned Gosling to work with her and decided that Signer's calf thymus DNA would be exclusively hers to work on. Remarkably, Wilkins was neither consulted nor informed about these decisions. This is what Gosling remembered in 2013 about his first meeting with Franklin in Randall's office[104]:

> *The whole trouble was that there was a meeting in Randall's office where Rosalind turned up and Alex Stokes and myself were invited along to meet her, and Wilkins was somewhere else. In his autobiography, I think he says that he was away in Wales with a new girlfriend. But that was the key to what followed, he wasn't there. [...] It was a very curious thing. Randall actually wrote to Rosalind saying that she would be asked to direct the X-ray crystallographic work on the Signer DNA material, and I didn't know that he'd done that.*

Gosling believed that Randall's move was intentional[104]:

> *He definitely subscribed to the divide and rule principle, as lots of people did. He thought it would make them competitive and improve their work.*

Another more generous interpretation was that Randall wanted to avert a direct confrontation with Wilkins and air the doubts that he had about his ability to successfully accomplish X-ray analysis of the atomic structure of DNA. Based on Randall's word, Franklin was convinced that the X-ray analysis of calf thymus DNA was her exclusive domain. Thus, after Wilkins presented in a July 1951 meeting in Cambridge the idea of a helical DNA in conjunction with Stokes' predicted pattern X-ray diffraction in a helix, Franklin demanded that he would leave the diffraction work to her and return to his optical studies. Although Wilkins accepted that the calf thymus DNA would be left to Franklin, he did not abandon his work on DNA. On a visit to the United States, he received from Chargaff samples of DNA from *Escherichia coli*, wheat germ, and pig thymus which, upon his return to London, he had put to analysis by X-ray diffraction (Ref. 2, p. 347). Franklin was displeased by his continued studies of DNA and her professional and personal relations with Wilkins deteriorated to the point that they worked in total separation from one another. Having had their offices two floors apart also contributed to their disconnection. Thus, the situation was that Signer's calf thymus DNA was analyzed only by Franklin and Gosling whereas

Wilkins and his postdoctoral fellow Rees Wilson (1929–2008) studied DNA from *E. coli*, wheat germ, and pig thymus.[26] Yet, whereas the calf thymus DNA could be manipulated to yield well-orientated single fibers, the DNA preparations that Wilkins and Wilson had used proved to be more difficult to manipulate and their X-ray diffraction images were less crisp and harder to interpret.

Franklin and Gosling Identified the A- and B-Forms of DNA

Taking off from the point that Wilkins had left, Franklin found it was difficult to maintain bundled fibers in perfect parallel alignment. She realized, therefore, that a better alternative was to work with single fibers whose diameter was <40 μm. The use of thinner fibers increased the extent of alignment of the DNA molecules in the examined fiber. Working with Signer's calf thymus DNA, Franklin and Gosling mastered the technique of isolating single fibers.[135]

> *Fibres were prepared by the method of Wilkins: sufficient distilled water is added to a small piece of the fibrous solid to form a stiff gel and a needle-point is then placed in the gel and slowly withdrawn. By suitably varying the speed of withdrawal and the water-content of the gel, fibres of diameter from about 100 μ to less than 1 μ can be obtained at will. In general, the smaller the diameter of the fibre the greater the degree of orientation obtained.*

Like Wilkins before her, Franklin recognized that the resolving power of the hitherto used equipment was insufficient for the analysis of small single-fiber specimens. She thus replaced the unsatisfactory Raymax X-ray tube that Wilkins had used with a more advanced fine-focus X-ray tube that Ehrenberg and Spear of Birkbeck College loaned in May 1950 to King's College.[136] Also, the previously used Unicam camera was put aside for a more advanced Philips microcamera. Prior to exposing the DNA fiber to X-rays, Franklin and Gosling replaced the air in the camera by hydrogen that was bubbled through a saturated salt solution. A key to the eventual discovery of the A- and B-forms of DNA was Franklin and Gosling's careful control of the levels of relative humidity of the examined DNA samples. When she measured the uptake of water by the DNA fibers by determining their weight at different relative humidity, Franklin noticed a distinct transition of the DNA from one form to another at a relative humidity of 75%[135] (for a facsimile of Franklin's hand-drawn results, see Fig. 14 in Ref. 28). X-ray analysis revealed that when the heavily hydrated DNA soaked water it became longer and adopted a "paracrystalline" structure whose diffraction image had at its center a telltale cross-shaped pattern of spots (Fig. 5.18A). Franklin and Gosling named this form of DNA "structure B" (this term was later changed to B-form DNA). At a relative humidity of <75%, the "dry" molecule was shorter by 25% compared to the "wet" DNA and its diffraction

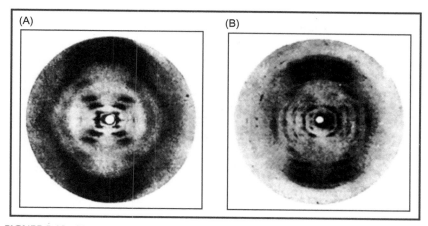

FIGURE 5.18 X-ray diffraction images of single fibers of calf thymus DNA that Franklin and Gosling produced in 1952 (published in April[137] and September[135] of 1953). (A). Fiber of 50 μm at a relative humidity of 75%. This image, taken on May 2, 1952, became known as "photo 51." (B) Fiber of 120 μm at a relative humidity of 56%. *For (B): The images are from Franklin RE, Gosling RG. The structure of sodium thymonucleate fibres. I. The influence of water content.* Acta Cryst *1953;6(8−9):673−7.*

image lacked a cross-shaped central pattern of spots. This conformer of the DNA was named "structure A" (termed later A-form DNA) (Fig. 5.18B).

Back and forth changes of the relative humidity of the DNA induced repeated transformations of form A to B and vice versa. However, when the DNA was subjected to prolonged drying, it sustained an A structure even at a relative humidity of 92%.[135] After learning of these results and seeing that that the "wet" DNA yielded a diffraction picture that matched the pattern that Stokes projected for a helix, Wilkins proposed that he and Stokes would collaborate with Franklin and Gosling. When his offer was turned down, his relations with Franklin became even more strained than before. Yet, at Randall's urging, it was agreed that Franklin would focus on the A-form of Signer's DNA whereas Wilkins would work the B-form of the Chargaff DNA. Unfortunately, Wilkins could not make much progress because the DNA preparations that Chargaff contributed broke down during purification and were inadequate for X-ray diffraction analysis.

The Different Crystalline Structures and Geometries of the A- and B-Forms of DNA

Before continuing with the historical story of the quest to solve the structure of DNA, it is appropriate to pause here and clarify the basis for the different X-ray diffraction patterns of its A- and B-forms. The X-ray diffraction image of the "wet" B-form DNA (Fig. 5.18A) had a distinctive cross-shaped pattern of bands, which were organized as a set of layer lines. Based on the unpublished theory of Stokes and on the similar model of Cochran, Crick, and Vand,[112] this cross-shaped pattern was a characteristic feature of a helix. By contrast, the diffraction pattern of A-form DNA was more complex and was devoid of a central cross-shaped pattern of streaks (Fig. 5.18B). The observed relatively simple diffraction pattern of B-form DNA represents an average of the different orientation of the molecules around the fiber axis. In contrast to the random rotational orientations of molecules in B-form DNA, the molecules in A-form DNA are organized in small crystalline modules. In crystalline A-form DNA, the fibrous molecules pack microcrystals that are arranged around a common axis. However, unlike a crystal, microcrystals in the fiber have different orientations of their constituent molecules (Fig. 5.19A).

X-ray reflections from such a fiber are grouped along layer lines that result from the repeating structure along the axis. However, at high resolution, the reflections fall on top of one another and the three-dimensional X-ray data appear as a muddled pattern in two dimensions (Fig. 5.18B).

Both the A- and B-forms of DNA were eventually recognized to be double helices with different geometrical parameters (see Fig. 5.9B). It was later deduced from the layer lines in the diffraction pattern of B-form DNA that the pitch of the helix was 34 Å and that the X-ray reflection on the meridian corresponded to a rise of 3.4 Å. The repeat was thus 34/3.4 = 10. A-form DNA was later recognized to be a more tightly coiled compact helix with a rise of 2.56 Å, pitch of 28 Å, and a repeat of 11. Also, the bases in A-form DNA are askew of the axis and it was this tilting that obscured the characteristic cross-shaped pattern of spots of a simple helix (Fig. 5.19B).

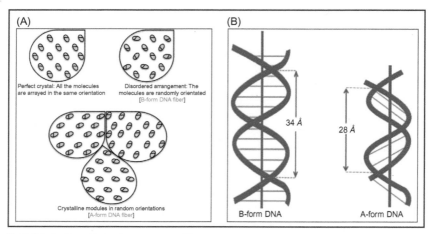

FIGURE 5.19   (A) Molecules are differently packed in A- and B-forms DNA fibers. (B) Schemes of B-form and A-form DNA helices.

Franklin Oscillated Between Helical and Nonhelical Structure of A-Form DNA

In her attempts to solve the structure of A-form DNA, Franklin chose to collect data and analyze it with the expectation that a structure of the DNA would eventually materialize from the processed empirical results. This approach contrasted the strategy of model building that was taken by Linus Pauling at Caltech and by Watson and Crick in Cambridge. Thus, Franklin based her (changing) ideas on the likely structure of A-DNA on her analyses of images of the diffraction of X-rays by this DNA. Specifically, she conducted Patterson analysis of the A-form in an effort to solve its structure.

At a relatively early time, Franklin thought that A-DNA and also B-DNA were helical molecules. In November 21, 1951, she presented this idea in a colloquium at King's College. One of the attendees in that colloquium was James Watson who came over from Cambridge. The notes that Franklin prepared for the talk were preserved by her colleague and personal friend, the future (1982) Nobel laureate Aaron Klug (1926−), and are stored in the archives of Cambridge University Churchill College. They are also available online at the Wellcome Library website: http://wellcomelibrary.org/player/b19831249#?asi=0&ai=12&z=-0.4657%2C0%2C1.9315%2C1.2133. As those notes show, at that point in time, Franklin deduced from the diffraction patterns that the DNA molecules were cylindrical and were likely to be helical:

> Evidence for spiral [helical in today's term] structure. Straight chain untwisted is highly improbable. Absence of reflections on meridian in xtlline form suggests spiral structure [. . .] Nucleotides in equivalent positions occur only at intervals of 27 Å [matching] the length of turn of the spiral.

Moreover, after listing her evidence for a spiral structure of the DNA, Frankiln ended her notes of 1951 with the following comment (cited in full by Judson,[138] p. 415):

> Near-hexagonal packing suggests that there is only one helix (containing possibly >1 chain) per lattice point. Density measurements (24 residues/27 A) suggest > 1 chain.

This comment (" >1 chain") serves to indicate that as early as 1951, Franklin had already the idea that the DNA molecule is composed of more than a single-helical chain.

At that point in time, Franklin hypothesized that A-form DNA was a bundle of two or three strands that had outside-facing phosphate groups and inside turned nucleotides. According to this premise, these bundled chains associated to form a fiber by weak links that were mediated by sodium ions and water (Fig. 5.20A).

Her notes also reveal that Franklin thought that upon the transition from A-form to B-form DNA, the water that engulfs the DNA disrupts the weak

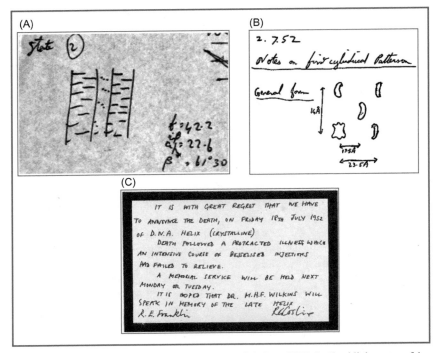

FIGURE 5.20 (A) Scheme of possible structure of A-form DNA in Franklin's notes of her November 1951 King's College colloquium presentation. At that point in time, Franklin considered the DNA to be a bundle of two or three strands that had the bases on the inside and the phosphates on the outside. The bundles formed fibers by associating with one another through weak links provided by water and sodium ions. Franklin depicted in this scheme two DNA molecules of two chains each linked by ionic bonds (dotted lines). (B) A drawing from Franklin's notebook dated July 2, 1952. The headline says, *"Notes on first cylindrical Patterson."* Underneath the sketch Franklin wrote: *"There is no indication of a helix of diameter 11A."* (C) Franklin's handwritten note of July 18, 1952, announcing the death of the DNA helix. *For (A): Image from The Rosalind Franklin Papers, Cambridge University Churchill College Archives. Available online: Papers of Rosalind Franklin, Wellcome Library: http://wellcomelibrary.org/ player/b19831249#?asi = 0&ai = 12&z = -0.4657%2C0%2C1.9315%2C1.2133. For (B): Image from "Profiles in Science" The Rosalind Franklin Papers: http://profiles.nlm.nih.gov/ps/access/ KRBBJF.pdf. For (C): Facsimile in the Library of Medicine "Profiles in Science" The Francis Crick papers: http://profiles.nlm.nih.gov/ps/retrieve/ResourceMetadata/SCBBXL.*

linkages between the bundles without affecting their parallel orientation or helical structures. Her belief at the time that B-form DNA was also a helix, albeit with dimensions that were different from those of the A-form, was directly expressed in the notes (bold mine):

Helical structure *in the* [wet] *form cannot be the same as in the* [dry] *because of large increase in length.*

Yet, as time passed, Franklin changed her mind and began to regard A-form DNA as nonhelical. As explained by Klug,[28] this new view was the outcome of one faulty X-ray diffraction image that was produced in the spring of 1952 and had left the impression that the orientation of molecules in crystallites of A-form DNA were not totally random (Fig. 5.20A). A greater recurrence of some orientations over others misled Franklin to think that the DNA did not have a cylindrical symmetry after all and that it was, therefore, unlikely to be a helix. Guided by her Patterson analysis, she tended to think that rather than a helix, A-DNA was rod-like, a sheet, or a "figure of eight" (Fig. 5.20B). Contrasting her image of a joyless and stern person, Franklin posted in July 1952 a black-bordered humorous invitation to a memorial service for the helix (Fig. 5.20C):

> It is with great regret that we have to announce the death, on Friday 18th July 1952 of D.N.A. Helix (crystalline).
>
> Death followed a protracted illness which an intensive course of besselised injections had failed to relieve.
>
> A memorial service will be held next Monday or Tuesday.
> It is hoped that Dr. M.H.F. Wilkins will speak in memory of the late helix
> — R.E. Franklin, R. Gosling

Although Franklin's attempts to solve the structure of A-form DNA by Patterson analysis have failed, one outcome of her efforts turned out to become a crucial factor in the construction of the double helix model of DNA by Watson and Crick. Her analysis of the crystal symmetry of this DNA revealed that it was of the C2 face-centered monoclinic space group. As is described later (Section 5.6.8), when this essential piece of information became known to Crick, he immediately recognized that A-form DNA comprised of two coaxial strands that ran in opposite (antiparallel) directions: $5' \rightarrow 3'$ and $3' \rightarrow 5'$.

Franklin's Return to B-DNA

After her efforts to solve the structure of A-DNA had hit the wall, Franklin returned in early 1953 to B-form DNA. As her notes of the period indicate, she had inside of the helix whereas the phosphates faced outward. As Klug pointed out,[28] by March 17, 1953, 1 day before she had learned of the Watson and Crick's model, she had already prepared a draft of a paper with Gosling that presented these conclusions. This article was published in the April 25 issue of *Nature*, back-to-back with the articles of Watson and Crick[1] and of Wilkins, Stokes, and Wilson.[139]

After 2 Years, Franklin Had Left King's College

Franklin was miserable in King's College. In addition to her strained relationship with Wilkins, her early biographer Sayre wrote that she suffered from the disparaging attitude toward women that was common in the period as manifested, for instance, by their exclusion from the faculty common rooms.[20] Her upper crust upbringing and accent were also loathed by most of her colleagues.[21] At the same time, based on his interviews with Franklin's female coworkers at King's, Judson argued that her rigidity and isolation from others in the laboratory were largely responsible for her situation.[138] Whatever were the circumstances, Franklin planned early on to leave King's College for Bernal's laboratory at Birkbeck College. After Bernal had agreed to accept her, Franklin wrote to him in June 19, 1952, to make arrangements for the move (facsimile of the letter is in the National Library of Medicine "Profiles in Science": The Rosalind Franklin Papers: http://profiles.nlm.nih.gov/ps/access/KRBBDN.pdf):

> *I have now spoken to Professor Randall about the possibility of moving to your laboratory and applying to the Fellowship Committee for permission to take the remains of my Turner and Newall Fellowship with me. Professor Randall has no objection to my doing this I hope this arrangement is still agreeable to you. If it is, would you suggest a time when I could come to see you to discuss things in more detail?*

Despite these early arrangements it was only on March 14, 1953, that Franklin had actually left King's for Birkbeck College. Shortly afterward Randall demanded in an April 17, 1953, letter that she should discontinue all work on DNA (which she did) and cease advising Gosling (which she did not) (facsimile of the letter is in the National Library of Medicine: "Profiles in Science" The Rosalind Franklin Papers: http://profiles.nlm.nih.gov/ps/retrieve/ResourceMetadata/KRBBHZ#transcript):

> *You will no doubt remember that when we discussed the question of your leaving my laboratory you agreed that it would be better for you to cease to work on the nucleic acid problem and take up something else. I appreciate that it is difficult to stop thinking immediately about a subject on which you have been so deeply engaged, but I should be grateful if you could now clear up, or write up, the work to the appropriate stage. A very real point about which I am a little troubled is that it is obviously not right that Gosling should be supervised by someone not specifically resident in this laboratory. You will realise that the necessary reorganisation for this purpose which arises from your departure cannot really proceed while you remain, in an intellectual sense, a member of the laboratory.*

During her early time at Birkbeck College, Franklin completed the writing with Gosling of three manuscripts that summarized their X-ray analysis of DNA fibers.[135,137,140] After this she had never returned to active study of DNA and her years at Birkbeck and later in Cambridge were dedicated to crystallography of viruses (for an exhaustive study of Franklin's scientific work after DNA, see Ref. 73). Nevertheless, as is described later

(Section 5.6.2), Franklin's insight about the orientations of bases and phosphates in DNA turned Watson and Crick away from their erroneous early model of triple helix DNA. More importantly, the famous "photo 51" of B-form DNA that Gosling took in May 1952 (Figs. 5.18A and 5.27) and Franklin's discovery that the crystal symmetry of A-form DNA was a C2 face-centered monoclinic were crucial underpinnings of the Watson and Crick model of the DNA double helix.

5.6 FIRST MODELS OF DNA WERE BUILT BY WATSON AND CRICK AND BY LINUS PAULING

Historically, two different strategies were taken in the quest for the structure of DNA. One was an empirical approach that was pioneered by Astbury in the late 1930s and was continued by Wilkins, Franklin, and Gosling in the early 1950s. This strategy entailed production and analysis of patterns of X-ray diffraction by DNA fibers with the expectation that the mathematically processed experimental data would reveal the atomic structure of the DNA. Watson and Crick at Cambridge University and the eminent Caltech chemist and two-times (1954 and 1962) Nobel Prize laureate Linus Pauling (1901−94) took an alternative approach of model building. In principle, this line of attack uses known data of unit cell dimensions, interatomic distances, bond lengths, water content, etc., to build alternative models of possible crystal structures. The models are next compared with the X-ray diffraction data and corrected to fit them. After several cycles of corrections and refinements of the model in accord with the actual experimental data, a molecular structure is ultimately determined when it fully corresponds to the diffraction results.

Much had been written about the Pauling and Corey triple-stranded model of DNA[141] and on the Watson and Crick double helix model.[1] Since the model-building phase was exhaustively covered and because it did not involve direct experiments, which is the theme of this book, this section just outlines the building of the double helix model of DNA.

5.6.1 Watson and Crick—An Unlikely Team of "an Adolescent Postdoc and an Elderly Graduate Student"

After obtaining a BSc degree from the University of Chicago at the age of 19, James Watson (Box 5.9) intended to pursue in graduate school ornithology, which had been his passion since childhood.

However, because of circumstances that were described by Olby (Ref. 2, p. 298), and the inspiration that he drew from Schrödinger's "*What Is Life*"[103] and from Sinclair Lewis' "*Arrowsmith*,"[143] he chose instead to focus on genetics. He elected to attend Indiana University in Bloomington, which was home to the renowned geneticist and 1946 Nobel Prize laureate Hermann Muller (1890−1967) and to the biologist and geneticist Tracy

BOX 5.9 James Dewey Watson

Watson was born and got his early education in Chicago. He enrolled in the University of Chicago at the age of 15 and received a BSc degree in zoology in 1947. He next did his graduate studies in the University of Indiana at Bloomington and conducted PhD research on X-ray inactivation of bacteriophage under the guidance of Salvador Luria. After winning a National Research Council (NRC) and Merck & Co postdoctoral fellowship, Watson joined in 1950 the laboratory of Herman Kalckar at the University of Copenhagen. Finding Kalckar's work on nucleotide metabolism irrelevant to his interest in genetics, he collaborated instead with the Caltech-trained microbiologist Ole Maaløe (1915–88). After seeing Wilkins' early X-ray diffraction patterns of DNA in a May 1951 conference in Naples, Watson became excited about the possibility of using crystallography to solve the structure of nucleic acids. With the help of Luria, he transferred his NRC fellowship to the Cavendish Laboratory at Cambridge University with the intention of learning crystallography from Perutz and Kendrew. There he shared a room with Crick who was doing his PhD research under Perutz. Learning of the Hershey and Chase experiment (see Chapter 4: "Hershey and Chase Clinched the Role of DNA as the Genetic Material"), Watson became convinced that DNA was the key to heredity. Although they were assigned to do crystallography of proteins and viruses, Crick and Watson began building a model of DNA. After one failed attempt, they constructed a double helix model of DNA on the basis of information gleaned from Chargaff (base complementarity), Franklin via Wilkins ("photo 51"), and Franklin via Perutz (crystal symmetry). In recognition of their discovery of the double helix, Watson, Crick, and Wilkins shared the 1962 Nobel Prize in Physiology and Medicine. After his sojourn in Cambridge, Watson did postdoctoral work at Caltech. There he tried without success to solve the crystal structure of RNA. In 1956 he joined the Biology Department at Harvard University where he contributed to the discovery of mRNA (see Chapter 9: "The Discovery and Rediscovery of Prokaryotic Messenger RNA"). He also urged a change of focus of the department to molecular biology and authored the first textbook on the subject "The Molecular Biology of the Gene"[142] (which had its 7th edition in 2013). In 1968 Watson became President of the Cold Spring Harbor Laboratory (CSHL). During his 35 years' tenure, there he transformed the laboratory into an internationally renowned education and research institution. From 1990 to 1992, Watson served as the Head of the Human Genome Project at the National Institutes of Health. In 2007, he retired from CSHL under a cloud of criticism of his views on race and intelligence and became its Chancellor Emeritus (Fig. 5.21).

James Dewey Watson
(1928–)

FIGURE 5.21   James Dewey Watson.

Sonneborn (1905−81). There he conducted research on X-ray inactivation of bacteriophages under the guidance of the future (1969) Nobel Prize laureate Salvador Luria (1912−91) (for a detailed description of Watson's graduate studies and thesis, see Ref. 2, pp. 299−305).

In the course of his graduate work, Watson came to know Max Delbrück and other members of the Phage Group and became attracted to their style of thinking and experimenting, which eventually evolved into molecular biology (Ref. 17, pp. 7−15). During that time, he also developed a growing interest in nucleic acids and after winning an NRC postdoctoral fellowship, he chose to work with Herman Kalckar (1908−91) at the University of Copenhagen. Watson soon found Kalckar's studies on nucleotide metabolism to be irrelevant to his interest in *"the secrets of the gene"* and instead he collaborated with the microbial physiologist Ole Maaløe (1915−88) who had returned to Denmark from Delbrück's Caltech laboratory.

The two scientists reported in 1951 that 30−50% of radioactive phosphorous was transferred from parental to progeny phage whereas only 10−20% of the labeled protein was transferred[144] (see additional references to this work in Ref. 2, p. 306). This study was to an extent precursor to the famous Hershey and Chase 1952 experiment[145] that demonstrated that DNA was the genetic material of the virus (see Chapter 4: "Hershey and Chase Clinched the Role of DNA as the Genetic Material"). In May of 1951, Watson traveled with Kalckar to the aforementioned meeting in Naples on *"Submicroscopical Structure of Protoplasm."* As described (Section 5.5.1), it was at that meeting that Maurice Wilkins showed an X-ray diffraction image of DNA (Fig. 5.16). This picture was indicative that DNA adopted a crystalline structure. The arrangement of DNA in an ordered crystal structure mitigated Watson's initial worry that the gene might be an incredibly complex structure. Years later he wrote about the transformative effect that Wilkins' talk had on him (Ref. 146, p. 8)

> I had become monomaniacal about DNA only in 1951 when I had just turned 23 and as a postdoctoral fellow was temporarily in Naples attending a small May meeting on biologically important macromolecules.

After his request to join Wilkins at King's College was not answered, he was accepted by Bragg to the Cambridge University Cavendish Laboratory with the purpose of working with John Kendrew on X-ray crystallography of proteins (for details of the complicated process of the transfer of Watson's NRC fellowship to Cambridge, see Ref. 8, pp. 273−181, and Ref. 2, pp. 307−310).

The Watson and Crick Collaboration

Once Watson arrived in Cambridge and was placed in the same room with Francis Crick (Box 5.10), their collaboration took off.

BOX 5.10 Francis Harry Compton Crick

Crick was born and raised in Weston Favell, a village near Northampton. At age 14, he was accepted on a scholarship to Mill Hill School in London where he excelled in chemistry, physics, and mathematics. In 1937, he received a BSc degree from University College London and began to do graduate research project in its Physics Department on the viscosity of water. Interrupted by World War II, he was recruited to the Admiralty Research Laboratory to design magnetic and acoustic mines. After the war, he made the decision to leave physics for biology. Having received an MRC studentship, he spent almost 2 years studying biology and working at the Cambridge University Strangeways Laboratory. In 1947, he joined the Cavendish Laboratory as a PhD student of Max Perutz. Since he was assigned to work on the solution of the structure of hemoglobin, he gained expertise in X-ray crystallography. When Watson arrived in Cambridge in 1951, he and Crick became close collaborators despite their 12 years' difference in age. Watson's drive, Crick's deep understanding of X-ray crystallography, and the empirical data that Franklin, Gosling, and Wilkins garnered all culminated in the triumphant double helix model of DNA that Watson and Crick published in April 1953.[1] Crick's brilliance as a theoretician who possessed thorough familiarity and understanding of the experimental data continued to be manifested at several critical junctions in molecular biology. He thus played a central role in the definition of the genetic code (see Chapter 7: "Defining the Genetic Code"), came up with the adaptor hypothesis that predicted the existence of transfer RNA (tRNA) (see Chapter 8: "The Adaptor Hypothesis and the Discovery of transfer RNA"), and formulated the central dogma of molecular biology: DNA → RNA → Protein. Crick was also at the cradle of the discovery of mRNA (see Chapter 9: "The Discovery and Rediscovery of Prokaryotic Messenger RNA") and was the originator of the wobble hypothesis that explained the ability of one tRNA species to recognize more than one codon (see Chapter 10: "The Deciphering of the Genetic Code"). With others he also raised speculations on the origin of life on earth and on the origin of the genetic code and protein synthesis. In 1977, Crick left Cambridge University for the Salk Institute in La Jolla, California. After teaching himself various areas of neurobiology, he deserted molecular biology in 1980 to fully concentrate on theoretical neuroscience. Until his death in 2004, Crick explored questions of consciousness, sleep, and memory. In addition to the 1962 Nobel Prize in Physiology and Medicine that he had shared with Watson and Wilkins, he was the recipient of many honors and awards. Most of all, he was universally appreciated for his incisive analytical powers and brilliance (Fig. 5.22).

**Francis Harry Compton Crick
(1916–2004)**

FIGURE 5.22   Francis Harry Compton Crick.

On the face of it this was an unlikely partnership of a 23-year-old American with a 35-year-old English graduate student. In addition, the two had different backgrounds; Crick was a physicist by training and a crystallographer whereas Watson was a biologist with some experience in phage and bacterial genetics. Watson's original task in Cambridge was to crystallize myoglobin for Kendrew's X-ray analysis. Since this experimental venture was unsuccessful, he and Crick spent several hours a day talking. In a December 9, 1951 letter to Delbrück, Watson described the impression that Crick had made on him (the letter was cited in Ref. 8, p. 13):

> *The most interesting member of the group is a research student named Francis Crick [...] He is no doubt the brightest person I have ever worked with and the nearest approach to Pauling I've ever seen [...] He never stops talking or thinking and since I spend my spare time in his house (he has a very charming French wife* [Odile Crick] *who is an excellent cook I find myself in a state of suspended stimulation.*

At the time of their meeting Crick was deeply involved in theoretical problems that came up during the pursuit of the crystal structure of hemoglobin. However, with Watson talking incessantly about DNA, Crick became increasingly attentive to the possibility of building a model of DNA. It should be noted, however, that Crick and Watson had initially different motivations to solve the structure of DNA. Crick thought that just because the structure of DNA was not known, solving it would be a worthy crystallographic challenge. In addition, he considered the problem interesting because DNA was part of the nucleoprotein complex whose proteinaceous component was thought at the time to be the genetic material. Primed by his experience with phages, Watson, however, suspected that DNA might possibly be the carrier of genetic information. This intuition became a conviction after he read a copy of the Hershey and Chase paper that demonstrated that the phage DNA and not its protein was the likely genetic material of the virus.[145] Although Watson's presentation of this experiment at the April 1952 Oxford meeting of the Society of General Microbiology was met with indifference (for a description of this episode, see Section 4.11), his absolute dedication to DNA sustained his commitment to the solution of its structure. As a matter of record, it was not just Watson himself who admitted to have been monomaniacal about DNA. Pauling was also cited as having remembered Watson as *"something of a monomaniac"* (where DNA was concerned).

Once Crick and Watson began to think about the modeling of DNA, they became an efficient cross-correcting team. As Crick later wrote (cited in Ref. 2, p. 316), the secret to their success was their absolutely open and blunt evaluation of each other's ideas.

5.6.2   Watson and Crick's Initial Three-Strand Model of DNA

Crick taught Watson the principles of X-ray crystallography such that after a month and a half in Cambridge, it was felt that he could interpret diffraction data. Therefore, Crick decided that Watson would attend a King's College colloquium in November of 1951 at which Franklin presented her data (see Section 5.5.2). Unfortunately, at that point, Watson's understanding of crystallography was still insufficient and he could not convey much information upon his return to the Cavendish. Importantly, he misremembered the water content of DNA that Franklin determined. Despite all of that, less than a week later, Crick and Watson constructed their first model of DNA. The elements of the model were summarized in an eight-pages unpublished manuscript that was reproduced by Olby (Ref. 2, pp. 357−359). The main features of this early model are illustrated in Fig. 5.23: (1) Based on the available diffraction data and on the just hatched theory of Cochran, Crick,

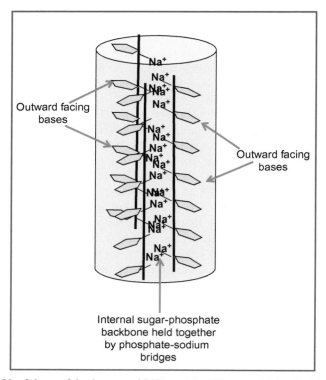

FIGURE 5.23   Scheme of the three-strand DNA model of Watson and Crick. In this model, the cylindrical helix consisted of three chains whose inside-facing phosphodiester backbones were neutralized by sodium ions whereas the bases faced outward. For simplification, the helicity of the DNA is not shown.

and Vand,[112] the DNA was proposed to be a cylindrical helix. (2) With the X-ray data having been compatible with any number from two to four DNA strands per molecule, Watson and Crick proposed a three-strand DNA. (3) In their model, the phosphodiester backbone faced the inside of the molecule whereas the bases faced outward. The fiber diffraction pattern indicated that DNA had a regular structure. Thus, the rationale for placing the bases outwardly was that because of their different sizes and shapes and their varying order along the molecule, if placed at the core of the molecule, the purines and pyrimidines were likely to distort the regular structure of the DNA.

Crick and Watson reasoned that the regular structure of the DNA was due to the regularity of the phosphodiester backbones at the core of the triple helix. They also assumed that by jutting out, the bases gained access to protein molecules in the nucleoprotein complex. As to the inside-facing phosphodiester backbone, the assumption was that the charged phosphates were neutralized by sodium ions. One aspect that the model did not take into account was the content and role of water molecules. Watson and Crick were pleased with their model and on the advice of Kendrew who felt that the model should be shared with their King's College counterparts, they invited Wilkins, Franklin, Gosling, and William (Bill) Seeds to see it. Arriving in Cambridge by train on the next day, the London scientists were introduced to the triple helix by Crick. Franklin was quick to play the killjoy. First, she insisted that the then available X-ray diffraction data did not favor any specific structure and thus it was too early to decide on a helix. Second, she had determined the content of water to be eight molecules per nucleotide, a value that was much higher than what Watson remembered. The large number of water molecules would enfold and neutralize the sodium ions and render them incapable of counteracting the phosphates in DNA. Last, Franklin pointed out that DNA was a "thirsty" molecule that soaked up to 10-folds more water than the three-strand model allowed. She argued, therefore, that because the nucleotides were nonpolar, the excessive binding of water by the DNA must have been due to the positioning of the phosphates on the surface of the molecule where they could be encased in a shell of water. In interviews with Olby, Crick attributed in hindsight his and Watson's erroneous three-strand model to their insufficient knowledge of chemistry at the time (Ref. 2, pp. 360−361).

Following Franklin's cutting criticism and realizing that they were on the wrong track, Watson and Crick proposed collaboration with the King's College team and were promptly rebuffed. The unsuccessful meeting between the London and Cambridge scientists had immediate negative repercussions. Randall and Bragg discussed the matter and came to an agreement that only the King's College team would continue to study the structure of DNA. Crick was then ordered to get back to his work of the diffraction of X-rays by polypeptides and proteins. Watson was assigned to work with the plant pathologist Roy Markham (1916−79) on X-ray diffraction by TMV.

Indeed, despite his continued passion for DNA, Watson did do X-ray crystallography of TMV which he later published.[147] Intriguingly, he and Crick continued to maintain an interest in the structure of viruses for several years after the publication of their double helix model of DNA (see Section 9.2.3). Nevertheless, despite Bragg's decree, rather than give up on the DNA problem, Crick and Watson took it underground, talking it over quietly in their office or over drinks at a local pub. It was not until Pauling came up with a (wrong) model of his own that Bragg had lifted his ban and Watson and Crick could return full-steam ahead to the construction of what proved to be the ultimately successful double helix model.

5.6.3 Linus Pauling's Attempt to Solve the Structure of DNA

Linus Pauling (Box 5.11), the most celebrated chemist of his time and one of the greatest scientists of the 20th century, was mostly known in the early 1950s for his pioneering work in quantum chemistry.

BOX 5.11 Linus Carl Pauling

Pauling was born in Portland, Oregon, and despite the modest circumstances of his family, he excelled academically showing an early interest in chemistry. In 1916, at age 15, he was admitted to Oregon State University (named then Oregon Agricultural College). Already during his undergraduate studies, he committed himself to the study of the atomic basis of the physical−chemical properties of substances. After obtaining a BSc degree in 1924, he did graduate research at Caltech on X-ray diffraction by minerals. In 1925, he received a PhD in physical chemistry and had left for 2 years in Europe where he was exposed to quantum mechanics by masters of the field: Sommerfeld, Bohr, and Schrödinger. In 1927, he returned to Caltech as an Assistant Professor and was promoted to Professor already in 1930. During the late 1920 and throughout the 1930s, Pauling used quantum mechanics to develop his theory of the chemical bond. His large body of research, which explained covalent, ionic, and hydrogen bonds, was laid out in his magnum opus textbook "*The Nature of the Chemical Bond.*" For this landmark achievement, Pauling received the 1954 Nobel Prize in Chemistry. Since the 1930s Pauling maintained an active interest in biomolecules, particularly proteins. His discovery of the abnormal structure of hemoglobin in sickle cell anemia (which he named a "molecular disease"[148]) was preamble to molecular genetics and to molecular medicine. Years before the structure of any entire protein was solved, Pauling discovered α-helix and β-sheet, the two major structural elements of proteins. Pauling became politically active after World War II. He joined the Emergency Committee of Atomic Scientists, which informed the public of the dangers of nuclear weapons and preached nuclear disarmament. During the dark years of Senator McCarthy, the US State Department denied him in 1952 a passport, then issued a "limited

(Continued)

BOX 5.11 (Continued)

passport" until finally his passport was returned in 1954 when he traveled to Stockholm to receive his Nobel Prize. His efforts to stop nuclear testing were recognized in 1962 by a Nobel Peace Prize. Incensed by the lukewarm reaction at Caltech to this Prize, he left the institution and its Department of Chemistry, of which he was the head for many years. In his later life, he controversially advocated the use of mega doses of vitamin C to treat cancer (Fig. 5.24).

Linus Carl Pauling
(1901–1994)

Pauling's "The Nature of
the Chemical Bond"

FIGURE 5.24   (A) Linus Carl Pauling and (B) his *"Nature of the Chemical Bond."*

After a period of training in Europe with leaders of the new field of quantum mechanics—Sommerfeld, Schrödinger, and Bohr—he was appointed in 1927 Assistant Professor of theoretical chemistry in Caltech. There he began in the late 1920s his seminal work that applied principles of quantum mechanics to describe chemical bonds. This body of work was expounded in his 1939 book *"The Nature of the Chemical Bond"* (Box 5.11) that was then and still is the most influential chemistry book. Pauling's deep understanding of structural chemistry was also informed by his X-ray crystallographic analysis of the structure of various compounds. Starting in the 1930s, he had developed a strong interest in proteins and their structure. Preceding by almost a decade the first X-ray crystallographic solutions of the structures of protein, Pauling with his longstanding collaborator Robert Corey (1897–1971) and with Herman Branson (1914–95) modeled in 1951 the two major structural elements of proteins: the α-helix and β-sheet.[63] Pauling's theoretical insight was guided by his knowledge of the structure of amino acids and peptides and of the planar nature of the peptide bond. A concise history of this landmark discovery which cannot be detailed here is available in Ref. 149. A treasure trove of original material on Pauling's work on proteins can be found in the Oregon State University *"Linus Pauling*

Online" website: http://scarc.library.oregonstate.edu/coll/pauling/proteins/materials/index.html.

Although proteins were Pauling's major interest, he also turned his attention in the early 1950s to nucleic acids. His motivation to look into the structure of DNA is not completely clear. By his own admission, the impetus to solve the structure of DNA was certainly not the recognition of its function as the hereditary material. Retrospectively describing his state of mind in 1951, Pauling wrote (http://scarc.library.oregonstate.edu/coll/pauling/dna/narrative/page5.html):

> *I knew the contention that DNA was the hereditary material, but I didn't accept it. I was so pleased with proteins, you know, that I thought that proteins probably are the hereditary material rather than nucleic acids—but that of course nucleic acids played a part. In whatever I wrote about nucleic acids, I mentioned nucleoproteins, and I was thinking more of the protein than of the nucleic acids.*

A reasonable possibility was that he regarded the problem of the structure of DNA to be a worthy challenge for a structural chemist. This scenario was favored by Pauling's son Peter[150]:

> *To my father nucleic acids were interesting chemicals just as sodium chloride is an interesting chemical and both presented interesting structural problems.*

Another possible motive was Pauling's fierce competitiveness. Learning from Delbrück about Watson's work on DNA and hearing that Wilkins had nice diffraction images of DNA (Ref. 2, p. 378), he was perhaps tempted to join the race. Be it what it may, by 1952, Pauling started to think seriously about the problem of the structure of DNA.

Pauling's Winding Path to the Formalization of His Model of DNA

After his discovery in the spring of 1951 of the α-helix and β-sheet structures of proteins, Pauling attention was turned to DNA. To try and solve its structure, he needed X-ray diffraction patterns of DNA fibers. Although X-ray diffraction images have been produced at Caltech, Pauling and Corey considered them to be inferior to the previously published patterns of Astbury and Bell.[110] Having heard that Wilkins at King's College obtained excellent pictures of DNA fibers, he asked to have a look at these data but was turned down.

In late 1951, Pauling was invited to a special meeting of the Royal Society at which he was to field questions from British researchers about his α-helix and β-sheet structures. The meeting was scheduled for May 1, 1952, and Pauling was planning to take the opportunity to visit Wilkins and have a look at his X-ray diffraction images of DNA. In January 1952, he applied to renew his passport. Having been a prominent peace activist and suspected to be a member of the communist party (which he was not), Pauling's application was denied. It was Mrs Ruth Shipley, head of the State Department's passport division, and a fervent anti-Communist, who rejected his request. As a matter of routine at the time, Shipley refused passports to anyone that she, the State Department's security personnel, or the FBI suspected of being too far left and too voluble about it. In her letter to Pauling Shipley, she wrote (Fig. 5.25A):

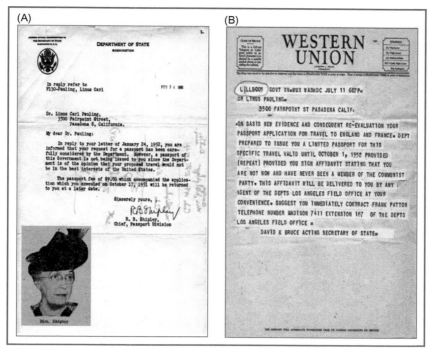

FIGURE 5.25   (A) Facsimile of a February 14, 1952, letter to Pauling from Ruth Shipley of the State Department. In this letter, his application for a passport was denied. The facsimile of the letter is from the Oregon State University online archive *"Linus Pauling and the Race for DNA."* In the lower left corner is a picture of Ruth Shipley from the same archive. (B) Facsimile of a July 11, 1952, telegram from the Acting Secretary of State to Pauling stating that he was granted a limited passport. *For (A): From http://scarc.library.oregonstate.edu/coll/pauling/dna/corr/ bio2.002.5-shipley-lp-19520214.html; http://scarc.library.oregonstate.edu/coll/pauling/dna/pic- tures/1952n.31-shipley.html. For (B): From http://scarc.library.oregonstate.edu/coll/pauling/dna/ corr/bio2.003.2-bruce-lp-19520711.html.*

> In reply to your letter of January 24, 1952, you are informed that your request for a passport has been carefully considered by the Department. However, a passport of this Government is not being issued to you since the Department is of the opinion that your proposed travel would not be in the best interests of the United States.
>
> The passport fee of $9.00 which accompanied the application which you executed on October 17, 1951 will be returned to you at a later date.

Pauling's British hosts were outraged that their guest of honor was denied a passport and the affair also raised significant public protest in America itself. Probably because of these pressures, David K. Bruce, Acting Secretary of State, informed Pauling just 10 weeks later that he was granted a limited passport to travel to only England and France (Fig. 5.25B).

Surprisingly, during his month-long visit in England, Pauling never spoke with anyone, including Wilkins, about DNA. He later said that this was due to his preoccupation with protein structure and to his suspicion that Wilkins would not let him see his X-ray diffraction images. It should be remembered that just when Pauling was in London, Franklin and Gosling produced their excellent images of A- and B-form DNA (Fig. 5.18). It is reasonable to assume that had she been asked, Franklin would have shown him those pictures. Wilkins later said that had Pauling visited his lab he would present him with the pictures of which he was so proud.

To Pauling's experienced eye, the cross-shaped central pattern of B-form DNA would have spelled a helix and the twofold symmetry would exclude a three-strand structure. It should also be pointed out that at that time Pauling was totally unaware of the unpublished excellent image of B-form DNA that Beighton produced in Astbury's laboratory (Fig. 5.11). Thus, with no fresher and better images available to him, when Pauling returned to California to build his model of DNA, he had to do with the inferior X-ray diffraction patterns of mixed A- and B-forms of DNA that Astbury and Bell had produced in the late 1930.[110]

Pauling Constructed a Model of Three-Stranded DNA

On November 25, 1952, 3 months after his return from England, Pauling finally launched his attempt to solve the structure of DNA. An immediate impetus for this project was a seminar given at Caltech by the Berkeley virologist and biophysicist Robley Williams (1908–95). In his talk, Williams presented electron microscope images of small biological structures. One picture showed strands of sodium salt of RNA that was metal-shaded to display three-dimensional details. To Pauling the strands appeared cylindrical and after examining the image he conjectured that similarly to the RNA, DNA was also likely to be helical. Indeed Williams later published electron micrographs of DNA[151] from which Pauling deduced that the diameter of DNA was 15−20 Å, a range that was also independently approximated by others.[152−154] Pauling figured that only a helix could fit both Astbury and Bell's X-ray patterns and Williams' electron microscope images. In constructing his model of DNA, Pauling also relied on recent discoveries on the stereochemistry of nucleotides that were made in the Oxford University laboratory of the future (1957) Nobel Prize laureate Alexander (Lord) Todd (1907−97).[155,156] He also consulted Furberg's X-ray determined structure of cytidine[114] (Section 5.3.3). Using the images of Astbury and Bell and their inaccurate calculation of the density of the DNA, he started with a double-stranded helix but concluded that having just two strands would leave too much free space and he thus switched to a more tightly packed three-stranded structure (for a detailed step-by-step description of Pauling's considerations in the building of his model, see Ref. 2, pp. 379−383).

In December 19, 1952, Pauling wrote to Henry Allen Moe of the Guggenheim Foundation to tell him (rather proudly it seems) that he solved the structure of nucleic acids (for a facsimile of the letter, see http://scarc. library.oregonstate.edu/coll/pauling/dna/corr/sci14.014.7-lp-moe-19521219-01-large.html):

> I have now discovered, I believe, the structure of the nucleic acids themselves. Biologists probably consider that the problem of the structure of nucleic acid is fully as important as the structure of proteins. I think that Dr. Corey and I will probably send in a note on the discovery of the structure of nucleic acids next month; we want to check up on our present structure a bit more before announcing it.

After Robert Corey had made some important corrections to Pauling's calculations, Pauling and he published a short announcement of their model in the February 21, 1953, issue of *Nature* without giving much detail.[157] The full model was submitted in December 1952 and was published in the March 1953 issue of the *Proceedings of the National Academy*.[141]

The Main Features of the Pauling and Corey Model of Three-Stranded DNA

Relying on Astbury and Bell's old diffraction images and on their measurements, Pauling and Corey reasoned that DNA must be three-stranded[141]:

> The x-ray photographs show a very strong meridional reflection, with spacing about 3.40 A. This reflection corresponds to a distance along the fiber axis equal to three times the distance per residue. Accordingly, the reflection is to be attributed to a unit consisting of three residues. [...argument to preclude a single-strand helix...] The alternative explanation of the x-ray data is that the cylindrical molecule is formed of three chains, which are coiled about one

another. The structure that we propose is a three-chain structure, each chain being a helix with fundamental translation equal to 3.4 A, and the three chains being related to one another (except for differences in the nitrogen bases) by the operations of a threefold axis.

Addressing the question of what constituted the core of the molecule, Pauling and Corey argued that stereochemical constraints precluded the packing of the purines and pyrimidines or the sugar moieties into a regularly organized core. An evenly packed core could form, however, if the phosphate groups were situated internally along the axis of the DNA triple helix (Fig 5.26)[141]:

A close-packed core of phosphoric acid residues, HPO_4^-, can easily be constructed. At each level along the fiber axis there are three phosphate groups. These are packed together [...]. Six oxygen atoms, two from each tetrahedral phosphate group, form an octahedron, the trigonal axis of which is the axis of the three-chain helical molecule. A similar complex of three phosphate tetrahedra can be superimposed on this one, with translation by 3.4 A along the fiber axis, and only a small change in azimuth.

They also argued that hydrogen bonds held together the phosphates groups at the core[141]:

Accordingly we assume that hydrogen bonds are formed between the oxygen atoms of the phosphate groups in the same basal plane, along outer edges of the octahedron [...]

Notably, Pauling and Corey did not take into consideration the place of sodium ions in the DNA. Having been unaware of the existence of forms

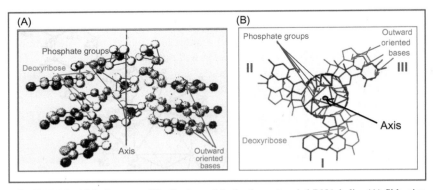

FIGURE 5.26 Main features of Pauling's model of a three-stranded DNA helix. (A) Side view (only two strands are shown). Note that the phosphates were placed along the axis and at the center of the molecule whereas the bases protruded outward. (B) View from above: Note the helical arrangement of the outward-facing stacked bases in the three stands (marked I, II, and III) and the packing of the phosphate groups at the center of the molecule. *(A) and (B) adapted from Pauling L, Corey RB. A proposed structure for the nucleic acids.* Proc Natl Acad Sci USA 1953;**39**(2):84−97.

A and B of DNA, they also did not consider the content and role of water molecules. As pointed out by Olby (Ref. 2, p. 381), Pauling was primarily concerned with the stereochemical fit of the model and he was satisfied with its agreement with both the Astbury and Bell X-ray diffraction data[110] and the Cochran, Crick, and Vand[112] theory of helix. When errors in the van der Waals distances were later discovered, they were quickly corrected without changing the general features of the model. On the whole, Pauling was pleased with the structure that he and Corey modeled (Ref. 2, p. 383).

The Pauling and Corey Model Had Some Glaring Faults

With the benefit of hindsight it is easy now to discern the conspicuous errors of the Pauling and Corey model. However, even without our present-day knowledge, it is surprising that Pauling and Corey did not detect some of their mistakes already when they constructed their model.

First, there was the chemo-physical error of ignoring the effect of the physiological pH on the ionization of phosphate groups. As cited, Pauling and Corey proposed that the phosphate backbone core of the triple helix was held together by hydrogen bonds between $O \cdots H \cdots O$ groups of opposing phosphates. Their mistake was, however, that the oxygen atoms become protonated only at pH < 2.0 whereas at the physiological pH ~ 7.0 all the hydrogen ions dissociate imparting DNA with its acidity and leaving behind mutually repellant O^- groups which would tear the DNA core apart.

A second mistake was that although Pauling was well aware that DNA was isolated as sodium salt, sodium ions were not even mentioned in his and Corey's paper and their model did not leave space and exposed negative charges to accommodate them. The third, and in hindsight the most important shortcoming of the model, was that it did not take into account the Chargaff Rules of base pairing and the biologically fundamental principle of replication of DNA was ignored. Instead, the potential of the bases to form hydrogen bonds was regarded as a device for the DNA to interact with other (unspecified) molecules. It was pointed out that since the model allowed the DNA to have any sequence of nucleotides, it could have a practically infinite capacity to specifically interact with diverse molecules[141]:

> It is interesting to note that the purine and pyrimidine groups, on the periphery of the molecule, occupy positions such that their hydrogen-bond forming groups are directed radially. This would permit the nucleic acid molecule to interact vigorously with other molecules. Moreover, there is enough room in the region of each nitrogen base to permit the arbitrary choice of any one of the alternative groups; steric hindrance would not interfere with the arbitrary ordering of the residues. The proposed structure accordingly permits the maximum number of nucleic acids to be constructed, providing the possibility of high specificity.

Two relevant historical happenings should be mentioned at this juncture. First, it is hard to understand why Pauling ignored the problem of replication when he and Delbrück proposed already in 1940 the principle of complementarity.[158] Moreover, in a Sir Jesse Boot Foundation Lecture in England in May 18, 1948 (5 years before he proposed the triple-stranded model of DNA), he lucidly explained a template-replica model of gene duplication of *two* complementing molecules (for a facsimile of the lecture, see http://scarc. library.oregonstate.edu/coll/pauling/dna/papers/1948p.13.html):

> *If the structure that serves as a template (the gene or virus molecule) consists of say, two parts, which are themselves complementary in structure, then each of these parts can serve as the mould for the production of a replica of the other part, and the production of duplicates of itself. In some cases the two complementary parts might be very close together in space, and in other cases more distant from one another—they might constitute individual molecules, able to move about within the cell.*

Without providing an explanation for overlooking his own farsighted prediction, Pauling wrote to Olby in a March 13, 1967 (Ref. 2, p. 380):

> [I would have preferred] *two chains, because I had decided several years earlier that the gene involved two complementary molecules, each of which could act as a template for the synthesis of a replica of the other.*

Another unfortunate circumstance was that Pauling could have learnt of the Chargaff Rules had he been more attentive. When they shared a shipboard in 1947, Chargaff told Pauling of his findings. However, because he found Chargaff annoying, Pauling ignored this crucial information.

The various flaws of the triple helix model were perhaps due to an extent to the speed at which Pauling worked. In an interview with Olby, the crystallographer Verner Schomaker (1914−97) described Pauling's style (Ref. 2, p. 383):

> *The typical thing with Pauling is that he works and thinks enormously hard on something and makes a breakthrough on it. Then he works enormously fast and prepares marvelous seminar that tells all about it.*

5.6.4 Back in Cambridge: Watson and Crick's Auspicious Meeting With Chargaff

As mentioned (Section 5.6.2), despite Bragg's ban, Watson and Crick continued to work on the DNA problem. Crick began to think of the possibility that DNA might adopt a structure of two paired chains that could act each as template for a complementary strand. The idea of template was "in the air" for years and whereas the German Physicist Pascual Jordan (1902−80) promoted a view of like-attracts-like, Pauling and Delbrück argued for the principle of complementarity, that is, formation of negative image by a positive one and vice versa.[158] In thinking of the chemistry of the bonding of the two chains, Crick initially rejected the possibility that hydrogen bonds could be the attractive force because he thought that bases adopted their two tautomeric forms (enol and keto) in the same molecule (see Section 5.6.10). Instead, he thought that the two strands could be held together by electrostatic attraction between the surfaces of stacked bases of two interleaved helical DNA chains (Ref. 8, p. 134). To assess whether such an attraction was feasible, Crick sought the help of a young Cambridge mathematician, John Griffith, a nephew of Fred Griffith of the *Pneumococcus* transformation fame (see Chapter 3: "Avery, MacLeod, and McCarty Identified DNA as the Genetic Material"). Griffith's quantum mechanical calculations suggested, however, that a double-stranded structure might form if the bases would pair through hydrogen bonds. He also deduced that the linking would be between adenine and thymine and guanine with cytosine. In interviews that Crick granted to Olby in 1968 and 1972, he recalled his exchange with Griffith (Ref. 2, p. 387):

> Have you done the calculations?" and Griffith said: "Yes, and I find that adenine attracts thymine and guanine attracts cytosine" Now I did not say: "This explains Chargaff's rules," which any sensible person would have done if they ever heard of them. What I said immediately, before he had time to say anything was: "Well that is all right, that is perfectly O.K., A goes and makes B and B goes and makes A, you just have complementary replication.

This revealing episode points to Griffith's early insight of hydrogen-bonded base pairs and to Crick's instantaneous perception that such pairing may be the basis to replication. However, the incident also shows that Crick was unacquainted at the time with the Chargaff Rules. It should be noted, however, that Watson knew of Chargaff's results and he claimed to "have muttered to Crick" about them (Ref. 8, p. 135). Whatever was the case, the full or partial lack of familiarity with Chargaff's Rules were gone after Erwin Chargaff himself had visited Cambridge July of 1952 and met there with Watson and Crick. Kendrew arranged for the two to meet in his room with the American guest. The meeting was an unpleasant but constructive clash between very different scientists. This is how Judson conveyed Crick's impressions of that occasion (Ref. 5, p. 120):

> By Crick's account of the meeting, he and Watson put Chargaff 'slightly on the defensive'. Thinking about structures, 'We were saying to him as protein boys, What has all this work on nucleic acid led to? It hasn't told us anything we want to know'. Chargaff answered, Crick said, 'Well of course there's the one-to-one ratios.' So I said, What's that? So he said, 'Well, it's all published.' Course, I'd never read the literature, so I wouldn't know. So he told me. Well, the effect was electric, this is why I remember it. I suddenly realized, by God, if you have complementary replication, you can expect to get one-to-one ratios.

Unlike Crick, who admitted to his ignorance of the Chargaff Rules, Watson did not divulge at the meeting whether or not he was familiar with them. This, however, did not make him more agreeable to Chargaff, as was described by Watson himself (Ref. 8, p. 136):

> *Chargaff, as one of the world's experts on DNA, was at first not amused by dark horses trying to win the race. Only when John [Kendrew] reassured him by mentioning that I was not a typical American did he realize that he was about to listen to a nut. Seeing me quickly reinforced his intuition. Immediately he derided my hair and accent, for since I came from Chicago I had no right to act otherwise. Blandly telling him that I kept my hair long to avoid confusion with American Air Force personnel proved my mental instability.*

More importantly, however, Chargaff was irritated by the pair's ignorance of biochemistry:

> *The high point in Chargaff's scorn came when he led Francis into admitting that he did not remember the chemical differences among the four bases. The faux pas slipped out when Francis mentioned Griffith's calculations. Not remembering which of the bases had amino groups, he could not qualitatively describe the quantum-mechanical argument until he asked Chargaff to write out their formulas. Francis' subsequent retort that he could always look them up got nowhere in persuading Chargaff that we knew where we were going and how to get there.*

Last, this is how Chargaff described the meeting in an interview with Judson (Ref. 5, p. 119):

> *They impressed me by their extreme ignorance. Watson makes that clear! I never met two men who knew so little—and aspired to so much. They were going about it in a roguish, jocular manner, very bright young people who didn't know much. They didn't seem to know of my work, not even the structure and chemistry of the purines and pyrimidines. But they wanted to construct a helix' a polynucleotide to rival Pauling's alpha helix. They talked so much about 'pitch' that I wrote down afterwards, 'Two pitchmen in search of a helix'.*
>
> *I explained to them the observations on the regularities in DNA, and told them adenine is complementary to thymine, guanine to cytosine, purines to pyrimidines; that any structure would have to take account of our complementary ratios. It struck me as a typical British intellectual atmosphere, little work and lots of talk.*

As Watson told it, the noxious atmosphere of the meeting did not extinguish Crick's excitement about complementarity and soon after it ended he rushed to see Griffith and then to the library to make sure that he knew which base complemented which (Ref. 8, p. 136).

5.6.5   Pauling's Triple Helix Model of DNA Reached Cambridge

Linus Pauling's son Peter was a doctoral student of John Kendrew at the Cavendish Laboratory and had shared an office there with Crick and Watson. Sometime in the middle of December of 1952, Peter Pauling showed his roommates a letter that his father had sent that said that he now had a structure of DNA. Crick and Watson were frustrated by the absence in the letter of any details of the model and were mostly distressed by the likely prospect of having been scooped by Pauling. Considering Pauling's eminence, it is easy to see why the young postdoctoral fellow Watson, and the graduate student Crick, felt so threatened by his interest in the structure of DNA. The awe that Pauling's exceptional scientific achievements and intellectual prowess aroused among his contemporaries is illustrated in an episode that Alexander Rich (1924−2015) recounted[159]:

> Pauling was widely recognized for his outstanding contributions to humankind in many different arenas. The esteem with which he was regarded was illustrated to me in a vivid way in 1951, when, as a postdoctoral fellow of Pauling's, I visited Albert Einstein in Princeton. Einstein's comment to me was 'Ah, that man is a real genius!'

Peter wrote back to his father requesting a copy of the DNA paper. Two copies of the article arrived in Cambridge in early February of 1953: one was addressed to Bragg who put it aside, and one to Peter. As told by Watson, before he or Crick have had a chance to read the paper, Peter told them that his father's model was of a three-strand DNA with the bases were facing outward (Ref. 8, pp. 172−173). This sounded suspiciously like the failed triplex DNA model of Watson and Crick. Most importantly, however, when Watson read the manuscript he realized that with DNA being acidic at a physiological pH, the phosphate groups at the core could not bind hydrogen atoms that were claimed by the model to mediate $O \cdots H \cdots O$ bonds. At first he and Crick were incredulous that Pauling has made such a basic mistake but consultation with several chemists assured them that DNA was indeed an acid and that Pauling's model was incorrect.

5.6.6   Photo 51 Was Shown to Watson

On Friday, January 30, 1953, Watson paid a visit to King's College with the intention of bringing the Pauling and Corey paper to the attention of Rosalind Franklin. Their meeting did not end well. First, Franklin who was a practicing X-ray crystallographer did not regard Watson, who was not one, as a true peer. She also felt that the experimental evidence did not justify Pauling's claim that DNA was a helix. Last, she was irritated by the fact that Watson got the Pauling and Corey paper whereas she, who amiably corresponded with Corey, had not received it (Ref. 2, p. 396).

As Watson told it, the meeting ended badly and he had to retreat (Ref. 8, p. 178). However, his subsequent meeting with Maurice Wilkins was friendlier and it proved to be crucial for the development of the double helix model. On this well-documented occasion, Wilkins showed to Watson the famous "photo 51" of the "wet" B-form of DNA that Gosling and Franklin took in May 1952 (Fig. 5.27; the same image was also reproduced in Fig. 5.18A).

Watson immediately noticed that unlike the more confusing picture of A-form DNA, photo 51 of the B-form had at its center a revealing cross-shaped pattern of streaks. Having been familiar with the theoretical unpublished work of Stokes and of Cochran, Crick, and Vand,[112] he recognized that this pattern reflected the zig and the zag of a helix. Also, the absence of streaks between the zero and ten layer along the meridian suggested that the structure of the DNA was helical. In addition, there were spots along ten-layer lines, suggesting that there were an integral number of 10 residues (nucleotides) per repeat and the repeat and pitch were the same (Fig. 5.9). Describing later his reaction to this clear evidence for a helix, Watson wrote: "...*my mouth fell open and my pulse began to race*" (Ref. 8, p. 181). Although it was believed for some time that Wilkins had shown photo 51 to Watson without Franklin's permission, the truth was that he had legitimate possession of the image. In his 2003 autobiography, Wilkins

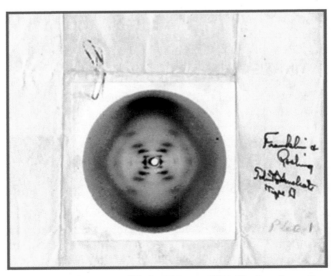

FIGURE 5.27 Photo 51 of "structure B" of calf thymus DNA ("sodium deoxyribose nucleate") taken by Franklin and Gosling on May 2, 1952. The handwritten annotation at right is Linus Pauling's. *From Oregon State University "Linus Pauling and the Race for DNA" online archive: http://scarc.library.oregonstate.edu/coll/pauling/dna/pictures/sci9.001.5.html.*

described the circumstances of his receiving of the photograph (Ref. 10, pp. 197−198):

> One day in January 1953, Raymond [Gosling] met me in the corridor and handed me an excellent B pattern that Rosalind and he had taken. For me to be shown raw data in this way was quite without precedent and, even more extraordinary, Raymond made it clear that I was to keep the photograph [...] Raymond gave me to understand that Rosalind was handing the pattern over to me to use as I wished.

This version was corroborated in an interview with Gosling that was cited in Ref. 8, p. 182. According to him, after putting the finishing touches to their *Acta Crystallographica* paper (Ref. 135) that included the photo 51 image of B-form DNA, and in view of Franklin's imminent leave for Birkbeck College:

> ...She [Franklin] therefore decided to make a 'present' to Maurice [Wilkins] of the original film of our best structure B diffraction pattern, the 51st exposure in our series of X-rays of single fiber specimens held at various steady humidities. Accordingly, I went down the corridor to Maurice's lab/office, sometime in January 1953 and gave him this beautiful negative. He was very surprised and wanted reassurance that Rosalind was actually saying that he could make whatever use he wished with this interesting data.

In their January 30, 1953, meeting, Wilkins also told Watson that his King's College colleague Robert Donald Bruce Fraser (1924−) used Astbury's data to construct a model of triple-stranded DNA helix that had its phosphate groups on the outside and the bases facing inward at a distance of 3.4 Å from one another. In response to Watson's question, Wilkins said that Fraser based the number of strands in his model (three) on the density of dried DNA (1.63 g/mL) that Astbury determined. Although Watson and Crick cited in their April 1953 *Nature* paper the Fraser model as a manuscript "in press," it had never been published.

On his train ride back to Cambridge and with the strong impression of photo 51 still fresh in his mind, Watson sketched what he could recall in a blank margin of a newspaper. Thinking of the problem of the number of chains in the DNA molecule, he came to the conclusion that the Fraser model of triplex DNA was not unassailable because it included an estimate of the water content of the DNA that could be erroneous. He thus made up his mind to bet on a two-chain model. He retrospectively described this decision (Ref. 8, p. 184):

> Thus by the time I had cycled back to college and climbed over the back gate, I had decided to build two-chain models. Frances would have to agree. Even though he was a physicist, he knew that important biological objects come in pairs.

5.6.7 Bragg Gave Watson and Crick a Green Light to Return to DNA

According to Watson (Ref. 8, p. 185), he had met with Bragg in Perutz' office on the morning after his return from London. He told Bragg that, based on photo 51, DNA appeared to be a helix which had the bases perpendicularly oriented to its axis and stacked upon one another, separated by a distance of 3.4 Å and forming a repeat of 34 Å. Watson alerted Bragg to the possibility that once Pauling realized that his triplex DNA model was wrong, he would immediately launch another attempt to solve the structure of DNA and would have a good chance of beating the Cavendish on such second attempt. He therefore intended to ask the Cavendish machine shop to make models of purines and pyrimidines so that Crick and he could start their modeling work. Bragg immediately allowed Watson and Crick to return to the DNA problem. The change in his attitude was most likely motivated by a fear that, as has happened with the discovery of the α-helix of proteins, Pauling would once again defeat the Cavendish in the contest to solve the structure of DNA.

5.6.8 Another Crucial Piece of Experimental Evidence Reached Watson and Crick

Crick did not accept Watson's assertion that the DNA had to be two-stranded. He argued that on the basis of its diameter, density, and the X-ray evidence, it could be either a double or triple helix. Watson nevertheless began building models of two-stranded DNA. Since the machine shop was still working on the models of purines and pyrimidine, Watson had built a model of only the phosphodiester backbone. He started by keeping the backbone at the center of the helix but since such model did not come close to the experimental data, he placed the backbone on the outside of the molecule. After some tinkering, the configuration of this model appeared to agree with Franklin's stereochemical measurements but it was still without the bases.

At this juncture Watson and Crick came upon another crucial piece of experimental evidence. Max Perutz was appointed to serve as member of an MRC committee that was assigned to review the work at the King's College Biophysics Research Unit. Under the instruction of Randall, members of the Unit prepared in advance of the committee's visit comprehensive reviews of their research. After receiving a copy of the report, Perutz showed Watson and Crick the sections in which Wilkins and Franklin summarized their unpublished results. Crick had gleaned two important pieces of evidence from these summaries. First, he was satisfied that after his visit to King's, Watson reported accurately the essential features of B-form DNA (Ref. 8, p. 194). More importantly, however, he found that Franklin defined the A-form

DNA as belonging to the face-centered C2 monoclinic space group (Section 5.5.1). In addition to being a keen crystallographer, Crick was especially familiar with the C2 space group because it was also a feature of hemoglobin crystals that he was working on at the time. He was thus quick to deduce from this symmetry that the two chains of DNA ran in opposite directions. It is interesting to note that Franklin had not garnered from her data the antiparallel orientation of the two DNA strands. Years later, after the double helix model had already been established, she told her friend Aaron Klug that she could have kicked herself for this oversight.[28]

After Watson wrote in his *"Double Helix"* book[8] about the disclosure of the King's College report by Perutz, Chargaff wrote in *Science* that divulging to Crick and Watson the content of a confidential report was unethical.[160] Perutz responded by publishing the relevant parts of the original King's College report and arguing that when they saw it Watson and Crick have already been familiar with photo 51 of B-form DNA that Wilkins had shown to Watson.[161] As to the C2 space group of A-form DNA, he pointed out that Franklin presented that information long before, in a 1951 Kings' colloquium that Watson had attended (Section 5.5.2). Yet, Perutz wrote that although the report was not classified as confidential, courtesy demanded that he should have asked for Randall's permission before showing it to Watson and Crick. Perutz clarification was accompanied by separate comments on the episode of Wilkins and Watson himself.[161]

5.6.9   Back to the Model: How Should the Bases Be Positioned at the Center of the Helix?

Once Watson chose to place the bases at the center of the helix, he had to arrange the differently shaped purines and pyrimidines at the core of the DNA molecule in a way that a perfectly regular outward-facing backbone would be maintained. Consulting Davidson's standard text on nucleic acids,[162] he drew on paper and cut out pictures of the bases and organized them in pairs that could be linked by hydrogen bonds. After some failures, he reached the idea of arranging like-with-like base pairs: adenine with adenine, cytosine with cytosine, guanine with guanine, and thymine with thymine. In thinking about the possible type of bonds that could link bases to form pairs, he relied on papers by Gulland and Jordan[163] and by Jordan[164] which maintained that hydrogen bonds could link together two adjacent chains of DNA. Watson probably also consulted a PhD thesis and papers[165,166] by June Broomhead who was working at the time at the Cavendish. Studying the structure of crystals of adenine and guanine, she had identified regular patterns of hydrogen bonds in crystals of pure adenine or of pure guanine. Watson conclusion was, therefore (Ref. 8, p. 198):

> *[...] I was drawing the fused rings of adenine on paper. Suddenly I realized the profound implications of a DNA structure in which the adenine residue*

formed hydrogen bonds similar to those found in crystals of pure adenine. If DNA was like this, each adenine residue would form two hydrogen bonds to an adenine residue related to it by a 180-degree rotation.

Indeed, when he rotated one base by 180° against another identical base, he spotted two hydrogen bonds that could link pairs of each of the four bases and it thus appeared that four like-with-like base pairs could be formed (Fig. 5.28A).

One feature of the like-with-like arrangement of the pairs within the core of the double helix was that because the size of the purine pairs

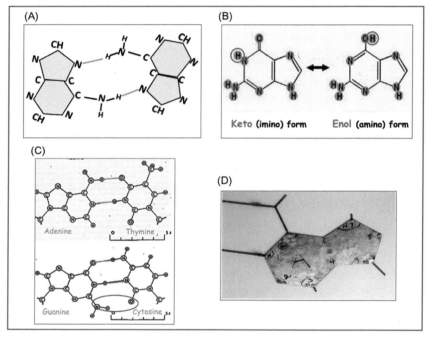

FIGURE 5.28   (A) Example of Watson's early idea of like-with-like base pairing. Here one adenine residue was rotated by 180° relative to another adenine to form two hydrogen bonds (red lines) between the two identical bases. Similar manipulations enabled hydrogen bonding between like-with-like pairs of guanine, thymine, and cytosine. (B) Enol and keto forms of guanine. Positions of the migrating hydrogen are circled. (C) Watson and Crick's original schemes of purine−pyrimidine base pairs. It was proposed that the two bases were linked by two hydrogen bonds (red lines) between adenine and thymine and guanine and cytosine. In this early scheme, Watson and Crick overlooked a third hydrogen bond between guanine and cytosine (encircled in an ellipse) which was later noticed by Pauling and Corey. (D) Aluminum-made template of adenine that the Cavendish machine shop made for the Crick and Watson 1953 model of DNA. *For (C): Image adapted from Watson JD, Crick FH. Genetical implications of the structure of deoxyribonucleic acid.* Nature *1953;171(4361):964−7. For (D): From the Science Museum London, Science and Society Picture Library.*

(guanine—guanine and adenine—adenine) was greater than the size of pyrimidine pairs (thymine—thymine and cytosine—cytosine), the outer backbone protruded or caved in at the respective sites of purine or pyrimidine pairs. Another worrisome aspect of the like-with-like pairing was that it was incongruous with the Chargaff Rules. However, despite these difficulties, Watson was very pleased with the model because it appeared to provide a basis for replication. Thus, once the two DNA chains separated, each could serve as template for the synthesis of a daughter chain that would be an exact replica of the parental strand. In fact, Watson was so satisfied with his like-with-like model that, without disclosing any specific details, he wrote in February 20, 1953, to Delbrück: *"I have a very pretty model, which is so pretty that I am surprised that no-one ever thought of it before."*

5.6.10   Learning of Base Tautomers, Watson Came Up With Purine—Pyrimidine Base Pairing

Watson's satisfaction with his like-with-like base pairing was short-lived. Returning to the laboratory on the next day, he showed his scheme to the Caltech crystallographer Jerry Donohue (1920—85) who shared a room with him and Crick during his 6 months' visit to Cambridge. Donohue immediately noticed that Watson copied from the Davidson text and used in his scheme enol tautomers of guanine and thymine. He explained to Watson that organic chemists habitually drew bases in their enol forms although it was much more likely that the bases adopted in DNA their keto forms (Fig. 5.28B). The conversion from enol to keto tautomers bode badly for Watson's base pairing model. Because of the shifted positions of hydrogen atoms, hydrogen bonds of like-with-like pairs could be maintained only if the bases were rotated in ways that were incompatible with the crystallographic data (Ref. 8, pp. 202—205).

With the collapse of the like-with-like model, Watson literally returned to the drawing board. He cut out stiff cardboard models of the keto forms of the bases and tried to form hydrogen-bonded base pairs. While attempting unsuccessfully to once again create like-with-like pairs, he noticed that by bonding adenine to thymine, the formed base pair was identical in shape and size to a guanine with cytosine pair. Moreover, unlike like-with-like pairing that required tilting of bases relative to one another, purine—pyrimidine pairs were formed smoothly without a need for rotation (Fig. 5.28C). Also, as a consequence of the same size of the two purine—pyrimidine pairs, the outward-facing backbone remained regular and undistorted. Most importantly, the pairing of purines with pyrimidines was in line with, and provided an elegant molecular basis for the 1:1 ratios between adenine and thymine and guanine and cytosine that Chargaff documented[111] (Fig. 5.14). Scrutinizing the new scheme of purine—pyrimidine pairing, Crick found no fault with the pairing of the cardboard base models. Yet, it still remained to be seen whether such base pairs could be fitted into a full model of a double helix. Crick also noticed that the two glycosidic bonds between base and deoxyribose were related by dyad axis that stood at an angle of 90° to the axis of the helix. Therefore, the complementing purine and pyrimidine in the two strands could be flipped while their glycosidic bonds maintained their original orientation. This served to indicate that each chain could contain both purines and pyrimidines (Ref. 8, p. 208).

5.6.11   Building a Double Helix Model of DNA

Because the machine shop was still working on the production of metal models of the bases, Crick and Watson had erected an interim model of a double helix that consisted of only the outer shell of the phosphodiester backbone. After several more days, the aluminum models of the bases (Fig. 5.28D) have arrived and purine—pyrimidine pairs were placed in the center of the double helix in a manner that was compatible with the stereochemical constraints and the X-ray data. Crick and Watson then rapidly completed a model of intertwined double helix with the purine—pyrimidine pairs fitted into its center and the phosphodiester backbone on the outside. Bragg who was the first to see the model was impressed but cautioned Crick and Watson to also seek the opinion of the leading authority on nucleotides, Alexander Todd. In time Todd and his associates came over to examine the model and gave it their seal of approval. Wilkins arrived from London and after having seen the model went back with a promise to examine its correspondence with the X-ray data. Two days later he called to say that the model was in full agreement with his and Franklin's crystallographic results. Moreover, to Watson and Crick's surprise, Franklin readily accepted their contention that the DNA was helical. In retrospect, however, this acknowledgment was not unexpected since Franklin's laboratory notes of the time show that she had independently reached the conclusion that B-form DNA was helical.[28] In discussions with Wilkins and Franklin, it was then decided that the Watson and Crick's double helix model would be submitted to *Nature* side by side with two supporting experimental papers: one by Wilkins, Stokes, and Wilson, and another by Franklin and Gosling.

While all this was happening, Watson's fears of being scooped by Pauling were put to rest when a letter from Delbrück arrived. Apparently, Pauling's triple helix model contained a stereochemical error, which Corey and Schomaker were trying to correct. But even the corrected model was received with little enthusiasm. When Pauling presented his triplex DNA model in a seminar at Caltech, it got a cool reception. Afterward Delbrück told Schomaker that he thought Pauling's model was not convincing. He had also mentioned in this conversation that Watson wrote to him that Pauling's structure contained "*some very bad mistakes*" and that he himself had "*a very pretty model [...] that I am surprised that no-one ever thought of it before*." Watson's disclosure had reached Pauling who had acted upon it (see Section 5.7.1). While they were submitting their paper to *Nature*, Watson wrote a letter to Delbrück in March 12, 1953, that described in detail the double helix model. For a facsimile of the letter, see http://scarc.library.oregonstate.edu/coll/pauling/dna/corr/corr432.1-watson-delbruck-19530312.html. At the bottom of that letter he added:

> *P.S. We would prefer your not mentioning this letter to Pauling. When our letter to* Nature *is completed we shall send him a copy. We should like to send him coordinates.*

Knowing that Pauling was planning a visit to Cambridge on his way to a Solvay Conference in Belgium, Watson and Crick wrote to him directly on March 21, 1953, to inform of the impending publication of their paper (a copy of which they appended) and to express their wish to discuss the model during his visit in England (for a facsimile of the letter, see http://scarc.library.oregonstate.edu/coll/pauling/dna/corr/sci9.001.32-watsoncrick-lp-19530321.html).

5.6.12   The Double Helix Model of DNA Was Formally Announced

Drafts of the manuscripts of Watson and Crick and of the King's College investigators were passed back and forth between the authors until they were finally agreed upon. Then Bragg submitted the three papers to *Nature* with a strong letter of support. All three articles were published back-to-back in the April 25, 1953, issue of *Nature*.

Crick's original intention was to discuss in his and Watson's paper the biological implications of the double helix model. It was ultimately decided, however, that the first article[1] would just outline the principal features of the model and that a separate article would be dedicated to its genetic consequences. That second paper was eventually published just 5 weeks later in the May 30, 1953, issue of *Nature*.[167] Only after the two articles have been published and just before he had left Cambridge for Caltech did Watson submit in August 1953 a technical paper to the *Proceedings of the Royal Society* that had the coordinates of the principal atoms in the diffracted DNA fiber. That paper,[168] which was published in April 1954, remained relatively unacknowledged.

5.6.13   The Watson and Crick First Paper

The landmark announcement of the double helix model of DNA spanned a single page of the journal, including an illustration of the double helix that Odile Crick drew (Fig. 5.29).

This famous diagram was drawn by Crick's wife Odile

Crick and Watson after the publication of their letter to *Nature*

FIGURE 5.29 Facsimile of the historical announcement in *Nature* of the double helix model of DNA.[1] At right: The authors at the time of the publication of their paper.

The major points that Watson and Crick had made in this article[1] were: (1) In contrast to Pauling's three-strand helix, they proposed a two-stranded helix. (2) The two intertwined DNA chains in the model were right-handed helices running in opposite (antiparallel) directions. (3) There was a residue (nucleotide) every 3.4 Å with a 36° angle between two subsequent bases such that a repeat of the helix had 10 bases. (4) The bases were placed at the center of the helix whereas the phosphates faced outward. (5) Hydrogen bonds held together the helix by pairing adenine with thymidine and guanine with cytosine. In their model, Watson and Crick specified two hydrogen bonds per each of the two base pairs (Fig. 5.28C). It was only later that Pauling and Corey pointed out that guanine was linked to cytosine by three hydrogen bonds (see Section 5.7.1). Although a broader discussion of the biological implications of the model was postponed for the next paper, Crick had added at the end of the first article a sentence that became in time the most famous understated comment in the annals of molecular biology[1]:

> *It has not escaped our notice that the specific pairing we have postulated suggests a copying mechanism for the genetic material.*

The Wilkins, Stokes, and Wilson Paper[139]

These King's College investigators presented experimental results and made the following main points: (1) The X-ray diffraction of paracrystalline (fiber) DNA suggested "*a helical structure with axis parallel to fiber length.*" (2) Their results were "*not in conflict*" with the double helix model that Watson and Crick proposed in their adjoining paper. (3) DNA from *E. coli*, sperm, T2 bacteriophage and an active "transforming principle" (source unspecified), all yielded roughly similar X-ray diffraction patterns, suggesting a universality of the structure of DNA.

The Franklin and Gosling Paper[137]

Franklin and Gosling presented in the article their famous "photo 51" of B-form DNA. This and additional experimental data led to the following major conclusions: (1) Considering the X-ray diffraction pattern of B-form DNA, they stated that: "*a helical structure is highly probable.*" Notably, however, Franklin and Gosling did not suggest at that point that the helix was double-stranded. Several months later they published an additional paper that presented convincing evidence for the two-strand structure of DNA.[140] (2) The diffraction pattern suggested that the DNA had a rise of 3.4 Å, repeat of 36 Å, and thus 10 residues per turn. (3) The calculated distance between the phosphorous atoms fit a fully extended polymer indicating that the phosphate groups were on the outside of the helix and were not compactly packaged at its center as Pauling proposed.

Watson and Crick's Second Paper[167]

Exactly 5 weeks after their first paper appeared in print, Watson and Crick published a second paper on the genetic implications of their double helix model.[167] The basic point of that article was that since DNA was the carrier of genetic information, it should be capable of duplication. In this article, Watson and Crick made the following principal points: (1) Steric restrictions impose full complementarity of the pairing of free nucleotides with a template strand. (2) Regarding the mechanism of the joining of the complementing free nucleotides into a polymer of a daughter strand, they stated: *"Whether a special enzyme is required to carry out the polymerization or whether a single-helical chain already formed acts effectively as an enzyme, remains to be seen."* (3) Since the two DNA chains are intertwined, their separation would necessitate untwisting of the helix. Watson and Crick suggested that DNA-associated proteins mediate the unwinding of the double helix.

Citing Chargaff's biochemical data, they stressed hydrogen bonding and pairing of adenine with thymine and guanine with cytosine and concluded that[167]:

> *If the actual order of the bases on one of the pair of chains were given, one could write down the exact order on the other one [. . .] It* (complementarity) *is this feature which suggests how the DNA molecule might duplicate itself.*

They then went on to propose a semiconservative mechanism of duplication, although they had not termed it yet as such:

> *Previous discussions of self-duplication have usually involved the concept of a template or mould. Either the template was supposed to copy itself directly or it was to produce a 'negative', which in its turn was to act as a template and produce the original 'positive' once again. In no case has it been explained in detail how it would do this in terms of atoms and molecules.*
>
> *Now our model for deoxyribonucleic acid is, in effect, a* pair *of templates, each of which is complementary to the other. We imagine that prior to duplication the hydrogen bonds are broken, and the two chains unwind and separate. Each chain then acts as a template for the formation on to itself of a new companion chain, so that eventually we shall have two pairs of chains, where we only had one before. Moreover, the sequence of the pairs of bases will have been duplicated exactly.*

This semiconservative mode of replication was contested in the next few years and competing dispersive and conservative mechanisms of replication have been proposed. Ultimately, it was the elegant decisive experiment of Meselson and Stahl[169] that demonstrated that DNA was replicated semiconservatively as originally hypothesized by Watson and Crick (see Chapter 6: "Meselson and Stahl Proved That DNA Is Replicated in a Semiconservative Fashion").

An additional succinct sentence that Watson and Crick have made in this 1953 paper[167] instigated a decade-long pursuit of the genetic code and its eventual deciphering:

It seems likely that the precise sequence of the bases is the code which carries the genetical information.

Watson and Crick concluded their second article with a careful and yet rather confident statement[167]:

...the general scheme we have proposed for the reproduction of deoxyribonucleic acid must be regarded as speculative. Even if it is correct, it is clear from what we have said that much remains to be discovered before the picture of genetic duplication can be described in detail. What are the polynucleotide precursors? What makes the pair of chains unwind and separate? What is the precise role of the protein? Is the chromosome one long pair of deoxyribonucleic acid chains, or does it consist of patches of the acid joined together by protein?

Despite these uncertainties we feel that our proposed structure for deoxyribonucleic acid may help to solve one of the fundamental biological problems—the molecular basis of the template needed for genetic replication.

5.7   RECEPTION OF THE DOUBLE HELIX MODEL OF DNA

Watson and Crick's immediate colleagues in Cambridge have accepted the model of a DNA double helix and its exciting biological consequences with general enthusiasm. Curiously, however, Watson himself had doubts whether or not the model was correct. In March 22, 1953, he confessed in a letter to Delbrück (cited in Ref. 2, p. 421):

I have a rather strange feeling about our DNA structure. If it is correct, we should obviously follow it up at a rapid rate. On the other hand it will at the same time be difficult to avoid the desire to forget completely about nucleic acid and to concentrate on other aspects of life.

In May 21 of the same year he wrote again to Delbrück (*ibid.*):

[. . .] I have not infrequent spells of seriously worrying about whether it [the double helix model] *is correct or whether it will turn out to be Watson's folly.*

At that point in time the model got a life of itself as it was presented and examined by ever widening audiences.

5.7.1   Pauling Accepted That Watson and Crick Had the Correct Model of DNA

Pauling learned in advance that Watson and Crick were working on the structure of DNA. As mentioned (Section 5.6.8), Watson had written to

Delbrück about his pretty model of DNA (the like-with-like erroneous model) and this news reached Pauling. Without really knowing what the pretty model was, Pauling was quick to invite Watson to present his model in a meeting on proteins that he was organizing. At a later time, after they already had figured out the correct purine–pyrimidine complementation and had completed their double helix model, Watson and Crick wrote to Pauling about the impeding publication of their model and sent him a preprint of their *Nature* paper.[1] In response, Pauling wrote to the two in March 27, 1953 (for a facsimile of the letter, see http://scarc.library.oregonstate.edu/coll/pauling/dna/corr/sci9.001.34-lp-watsoncrick-19530327.html):

> *I think that it is fine that there are now two proposed structures for nucleic acid and I am looking forward to finding out what the decision will be as to which is incorrect. Without doubt the King's College data will eliminate one or the other.*

Shortly thereafter Pauling arrived in Cambridge in the first week of April 1953. After spending the night with his son Peter, he walked into the office that Crick and Watson occupied. There he saw for the first time their model of the DNA double helix and had a look at Franklin's photo 51. Crick talked about the features of the double helix while Pauling was scrutinizing the wire and metal model. Watson and Crick waited and then, as was remembered by Watson (Ref. 8, p. 236): "...*gracefully, he gave his opinion that we had the answer.*"

It was a jubilant moment for Watson and Crick and a less happy one for Pauling. He was astonished that such an unlikely team of "*an adolescent post-doc and an elderly graduate student*" had beat him by finding such elegant solution to the structure of a key biomolecule. Nevertheless, he has accepted the fact that the two Cavendish scientists were almost certainly right.

A day or two later, Bragg and Pauling went together to the Solvay Conference on proteins, which took place in Brussels. It was there that Bragg had made the first public announcement of the double helix. Generously seconding Bragg's assertions, Pauling said[170]:

> *Although it is only two months since Professor Corey and I published our proposed structure for nucleic acid, I think that we must admit that it is probably wrong. Although some refinement might be made, I feel that it is very likely that the Watson–Crick structure is essentially correct. [...] I think that the formulation of their structure by Watson and Crick may turn out to be the greatest development in the field of molecular genetics in recent years.*

Pauling reiterated his public proclamation in letters to his wife Ava Helen that he had sent from the Solvay Conference (available at http://scarc.library.oregonstate.edu/coll/pauling/dna/corr/safe1.021.3.html and http://scarc.library.oregonstate.edu/coll/pauling/dna/corr/safe1.021.4.html). On April 6, 1953, he wrote: "*I have seen the King's College nucleic acid pictures, and talked with Watson and Crick, and I think that our structure is probably wrong, and theirs right.*" In an April 7 letter he added: "*I have thought more about the nucleic acid structure of Crick and Watson, and I think that it is probably right.*" Then, on April 20, Pauling wrote to Delbrück (http://scarc. library.oregonstate.edu/coll/pauling/dna/corr/sci9.001.39-lp-delbruck-19530420-01.html):

> *I was very deeply impressed by the Watson–Crick structure, I do not know whether you know what put Corey and me off on the wrong track. The x-ray photographs that we had, which had been made by Dr. Rich, and which are essentially identical with those obtained some years ago by Astbury and Bell, are really the superposition of two patterns, due to two different modifications of sodium thymonucloates. This had been discovered a year or more ago by the King's College people, but they had not announced it, and I did not know that this was so. Corey and I had tried to find the structure that accounted for one of the principal features of one pattern, and simultaneously for one of the principal features of the second pattern, Watson and Crick saw the x-ray photographs made in King's College a couple of months ago, when they attended a seminar there, and they immediately began work on the problem. The King's College people had already derived one conclusion from the photographs, as to the nature of the helical structure. Watson and Crick amplified this by the idea of complementariness between purine and pyrimidine residues, and formulated their structure. While there is still a chance that their stricture is wrong, I think that it is highly probable that it is right. It has very important implications, as you mention. I think that it is the most significant step forward that has been taken for a long time.*

A footnote to Pauling's involvement in the DNA story was his February 18, 1963, letter to Crick. Written a short time after Watson, Crick, and Wilkins had received the 1962 Nobel Prize in Physiology and Medicine, the letter alerted Crick to the missing hydrogen bond in their scheme of a guanine–cytosine base pair (http://scarc.library.oregonstate.edu/coll/pauling/dna/corr/sci9.001.50-lp-crick-19630218-01.html):

> *I am writing about the matter of the three hydrogen bonds between guanine and cytosine. A man giving a seminar on the nucleic acids here a few days ago used the structures with two hydrogen bonds, and said that he supposed that you were still supporting it, as shown, for example, by your article in the Scientific American, which was published after the publication of the paper by Professor Corey and me, in which we pointed out that guanine and cytosine form three hydrogen bonds with one another. He asked me if it did not seem, from your publications, that you and Watson had doubt about the third hydrogen bond.*
>
> *I trust that you are going to introduce the third hydrogen bond in your published Nobel lecture. I am writing just to be sure that, through oversight, you do not continue to refer to guanine and cytosine as forming two hydrogen bonds with one another.*

5.7.2 Watson's Presentation of the Double Helix Model at Cold Spring Harbor

The 18th annual CSHL Symposium on Quantitative Biology was taking place on the week of June 5–11, 1953. The topic of the conference, which was attended by 272 researchers, was "viruses" and its organizer was Max Delbrück who had understandably given strong emphasis to studies of bacteriophages. The audience included almost all the leaders of the field of what would become molecular biology. Among those attending were Françoise Jacob, Macfarlane Burnet, Renato Dulbecco, Alfred Hershey, Salvador Luria, André Lwoff, Barbara McClintock, George Palade, Edward Tatum, and other leading researchers. At the last moment, Delbrück invited Watson to present his and Crick's model of DNA. Prior to Watson's talk, Delbrück circulated among the attendees copies of the first Watson and Crick *Nature* paper[1] and of the adjoining articles of Wilkins, Stokes, and Wilson[139] and of Franklin and Gosling.[137] Watson's paper at the conference[171] (coauthored by Crick who was not attending the meeting) summarized the three papers and discussed the biological implications of the double helix model. It ended with a short discussion on mutations that Watson predicted to involve substitution of one base for another.[171] This is how Judson described Watson's manner at Cold Spring Harbor (Ref. 5, p. 239):

> *Watson wandered through the week tall and skinny and remote, dressed in loose cotton shorts and short-sleeved shirt, tails flapping. [...] In the lecture hall at Cold Spring Harbor, with the barber pole drawing of the structure projected on a high screen, Watson pointed to the wide groove that allowed access to the inward lying base pairs.*

Jacob vividly described Watson's presentation and its reception[172]:

> *His manner more dazed than ever, his shirttails flying in the wind, his legs bare, his nose in the air, his eyes wide, underscoring the importance of his words, Watson gave a detailed explanation of the structure of the DNA molecule; breaking into his talk with short exclamations [about] the construction of atomic models to which he had devoted himself at Cambridge with Francis Crick; [...]*
>
> *For a moment the room remained silent. There were a few questions. How for example, during the replication of the double helix, could the two chains untwined around one another separate without breaking? But no criticism. No objections. This structure was of such simplicity, such perfection, such harmony, such beauty even, and biological advantages flowed from it with such vigor and clarity that one could not believe it to be untrue.*

Although the reactions to Watson's data and conclusions were overwhelmingly positive, they were not universally approving. One prominent doubter was no other than Al Hershey, who just a year before demonstrated with Martha Chase that DNA carried the genetic information of bacterio-phage.[145] Curiously, however, even after Watson's talk, Hershey doubted that DNA was the exclusive hereditary material. After listing in his talk in the same 1953 Cold Spring Harbor Symposium all the arguments in favor of the role of DNA in heredity he added[173]:

> *None of these, nor all together, forms a sufficient basis for scientific judgment concerning the genetic function of DNA. The evidence or this statement is that biologists (all of whom, being human, have an opinion) are about equally divided pro and con. My own guess is that DNA will not prove to be a unique determiner of genetic specificity, but that contributions to the question will be made in the future only by persons willing to entertain the contrary view.*

Indeed, some scientists maintained Hershey's skeptical attitude up until the late 1950s. Everyone became convinced of that DNA was the sole genetic material only after its function as the carrier of hereditary informa-tion was demonstrated in multicellular organisms.

5.8   AFTER THE DISCOVERY

The determination of the double helix structure of DNA is arguably among the few most important discoveries in biology and it certainly is the break-through finding that launched molecular biology as a full-blown discipline. This key discovery received its formal seal of approval from the scientific community when Crick, Watson, and Wilkins were awarded in December 1962 the Nobel Prizes in Physiology and Medicine for their DNA work. The restricted space of this chapter will not allow an appropriate discussion of the enormous consequences of the discovery of the structure of DNA. Interested readers may choose to consult any one of the books that were listed at the opening of this chapter or other sources. The chapter ends, there-fore, with just a brief mention of the turns that the careers and lives of some of the protagonists of the DNA story had taken after the heroic period of the early 1950s.

5.8.1   Francis Crick

After spending a short time as a postdoctoral fellow in New York, Crick returned to the MRC Unit in Cambridge where he made with others major contributions to molecular biology throughout the 1950s and 1960s. Chief among his achievements were the definition to the genetic code (see Chapter 7: "Defining the Genetic Code"), the adaptor hypothesis that predicted the existence and role of tRNA, and formulated the central dogma of molecular biology (see Box 5.10). In 1977 Crick moved to the Salk Institute in California and worked on the neuronal basis of vision and consciousness until his death in 2004.

5.8.2 James Watson

Watson had left Cambridge at the end of his 3 years' fellowship there and joined Delbrück's laboratory at Caltech. There he attempted without success to solve the structure of RNA. After he was appointed Assistant Professor at Harvard he codiscovered bacterial mRNA (see Chapter 9: "The Discovery and Rediscovery of Prokaryotic Messenger RNA") and authored a highly readable first textbook on molecular biology.[174] In 1968 he was appointed President of the CSHL where he was for 35 years an inspired and inspiring scientific administrator, writer, and facilitator of the research of others.

5.8.3 Maurice Wilkins

After the double helix model of DNA was proposed, Wilkins and his group conducted careful experiments to rigorously ascertain the model.[175–179] He also authored an autobiography that told his side of the double helix story.[10]

5.8.4 Erwin Chargaff

Chargaff was a leading, perhaps *the* leading, nucleic acid biochemist of his time. His achievements won appropriate recognition; he was awarded the National Science Medal and assorted other medals, was elected to several Academies of Science, received several honorary doctorates, and was invited to give distinguished lectures. Despite all these honors, he felt ignored and thought that his contribution to the solution of the structure of DNA was overlooked. "Yes and no" Chargaff replied when he was asked in 1972 by Horace Judson whether he had recognized at the time the consequences of his discovery of the 1:1 molar ratios of guanine to cytosine and of adenine to thymine.[138] In retrospect, however, a "no" is a more apt answer. Because Chargaff considered DNA to be a single strand, there was no way he could expound the equimolar ratios of purines and pyrimidines. In his hand this finding remained an unexplained empirical observation and it was only the insight of Watson and Crick that DNA was double-stranded that solved the puzzle of Chargaff's empirical finding.[138]

In his autobiography "*Heraclitean Fire—Sketches From a Life Before Nature,*"[11] he recalled his fateful meeting with the "*ill-matched pair*" of Watson and Crick whom he regarded as having (Ref. 11, p. 102): "[...] *enormous ambition and aggressiveness, coupled with almost complete ignorance of, and a contempt for, chemistry.*" He thought, however, that: "*the double-stranded model of DNA came about as a consequence of our conversation.*" Yet, he made the point that Watson and Crick did not acknowledge his help and failed to cite his 1950 and 1951 papers[111,124,180] in their historic 1953 *Nature* paper. Most likely because of this traumatic experience, he developed great hostility toward the birth and growth of molecular biology as a reductive solution to life. In an interview with Judson, he said (Ref. 5, p. 196):

I am against the over-explanation of science, because I think it impedes the flow of scientific imagination and associations. My main objection to molecular biology is that by its claim to be able to explain everything, it actually hinders the free flow of scientific ideas. But there is not a scientist I have met who would share my opinion.

Chargaff's time at Columbia University ended with a sad note. He was banished from his office and displaced to a distant building. When he was invited back by the new chairman of the Department of Biochemistry David Hirsh, Chargaff refused saying that in the event of his death; *"my simple request is: I do not want to be remembered by the university"* (cited in a June 30, 2002, obituary in the *New York Times*: http://www.nytimes.com/2002/06/30/obituaries/30CHAR.html).

5.8.5   Rosalind Franklin

As described, Franklin had left King's College for the more congenial setting of Birkbeck College. After she had published her work on DNA,[135,137,140] she completely left this topic and instead made important contributions to the X-ray crystallographic analysis of the structure of plant viruses and in particular of TMV[181–185] (for an in-depth study of Franklin's research after DNA, see Ref. 73). By the end of her life, she befriended Francis and Odile Crick and had moved her laboratory to Cambridge, where she undertook dangerous work on the poliovirus.

Franklin's role in the discovery of the double helix was extensively researched and written about. Watson's original condescending attitude toward her in his 1968 *"The Double Helix"* infuriated many readers and expert historians and scientists. Her untimely death, 4 years before Watson, Crick, and Wilkins received the 1962 Nobel Prizes, had raised speculations whether or not she would have been included among the awardees. Of the plethora of writings about Franklin (listed in the first pages of this chapter), it is interesting to read what Crick wrote about her role in the discovery of the double helix. In a December 31, 1961, letter to Jacque Monod (exactly 1 year before the 1962 Nobel Prizes), Crick evaluated the relative contributions of Wilkins and Franklin to the discovery of the double helix (bold mine) (facsimile of the letter in the National Library of Medicine, Profiles in Science, The Francis Crick papers: http://profiles.nlm.nih.gov/ps/retrieve/ResourceMetadata/SCBBFW):

On the matter of Maurice Wilkins. I think his contribution was twofold. He initiated the careful x-ray work on DNA, and since 1953 he has done numerous extensive, accurate and painstaking studies on it. It is true that he has worked rather slowly, but then hardly anybody else has done anything. **However, the data which really helped us to obtain the structure was mainly obtained by Rosalind Franklin who died a few years ago.** *"*

REFERENCES

1. Watson JD, Crick FH. Molecular structure of nucleic acids: a structure for deoxyribose nucleic acid. *Nature* 1953;**171**(4356):737−8.
2. Olby R. *The path to the double helix: the discovery of DNA.* New York, NY: Dover Publications; 1994.
3. Portugal FH, Cohen JS. *A century of DNA.* Cambridge, MA: MIT Press; 1977.
4. Lagerkvist U. *DNA pioneers and their legacy.* New Haven, CT: Yale University Press; 1998.
5. Judson HF. *The eighth day of creation.* Expanded ed. New York, NY: Cold Spring Harbor Laboratory Press; 1996.
6. Morange M. *A history of molecular biology.* Cambridge, MA: Harvard University Press; 1998.
7. de Chadarevian S. *Designs for life: molecular biology after World War II.* Cambridge, UK, and New York, NY: Cambridge University Press; 2002.
8. Watson JD. *The annotated and illustrated double helix.* New York, NY: Simon & Schuster; 2012.
9. Crick FH. *What mad pursuit: a personal view of scientific discovery.* New York, NY: Basic Books; 1988.
10. Wilkins M. *The third man of the double helix: the autobiography of Maurice Wilkins.* Oxford, UK: Oxford University Press; 2003.
11. Chargaff E. *Heraclitean fire: sketches from a life before nature.* New York, NY: Rockefeller University Press; 1978.
12. Hall KT. *The man in the monkeynut coat. William Astbury and the forgotten road to the double-helix.* Oxford, UK: Oxford University Press; 2014.
13. Bernal JD. William Thomas Astbury. 1898−1961. *Biogr Mems Fell R Soc: Royal Soc* 1963;1−35.
14. McElheny VK. *Watson & DNA: making a scientific revolution.* New York, NY: Basic Books; 2004.
15. Inglis JR, Sambrook J, Witkowski JA, editors. *Inspiring science: Jim Watson and the age of DNA.* Cold Spring Harbor, NY: Cold Spring Harbor Laboratory Press; 2003.
16. Edelson E. *Francis Crick and James Watson: and the building blocks of life.* New York, NY: Oxford University Press; 1998.
17. Watson JD. *A passion for DNA: genes, genomes, and society.* Cold Spring Harbor, NY: Cold Spring Harbor Laboratory Press; 2001.
18. Ridley M. *Francis Crick: discoverer of the genetic code.* New York, NY: Harper Perennial; 2006.
19. Olby R. *Francis Crick: hunter of life's secrets.* New York, NY: Cold Spring Harbor Laboratory Press; 2009.
20. Sayre A. *Rosalind Franklin and DNA.* New York, NY: W.W. Norton & Co; 1975.
21. Maddox B. *Rosalind Franklin: the dark lady of DNA.* New York, NY: Harper Perennial; 2003.

22. Glynn J. *My sister Rosalind Franklin: a family memoir*. Oxford: Oxford University Press; 2012.
23. Arnott S, Kibble TWB, Shallice T. Maurice Hugh Frederick Wilkins CBE. 15 December 1916−5 October 2004. *Biogr Mem Fell R Soc Lond* 2006;457−78.
24. Serafini A. *Linus Pauling: a man and his science*. New York, NY: Paragon House Publishers; 1991.
25. Hager T. *Force of nature: the life of Linus Pauling*. New York, NY: Simon & Schuster; 1995.
26. Wilson HR. The double helix and all that. *Trends Biochem Sci* 1988;**13**(7):275−8.
27. Wilson HR. Connections. *Trends Biochem Sci* 2001;**26**(5):334−7.
28. Klug A. The discovery of the DNA double helix. *J Mol Biol* 2004;**335**(1):3−26.
29. Authier A. *Early days of x-ray crystallography*. Oxford: IUC/Oxford University Press; 2013.
30. Friedrich W, Knipping P, Laue M. Interferenzerscheinungen bei Röntgenstrahlen. *Ann Physik* 1913;**346**(10):971−88.
31. Friedrich W, Knipping P, Laue M. Interferenzerscheinungen bei Röntgenstrahlen. *Bayerische Akad Wiss München, Sitzungsber math-phys Kl* 1912;303−22.
32. Glasser O. *Wilhelm Conrad Roentgen and the early history of the roentgen rays*. San Francisco, CA: Jeremy Norman Co.; 1992.
33. Ewald PP, editor. Fifty years of X-ray diffraction. Available from: http://www.iucr.org/publ/50yearsofxraydiffraction; 1962.
34. Laue M. *Concerning the detection of X-ray interferences. Nobel Lectures, Physics 1901−1921*. Amsterdam: Elsevier; 1967. pp. 347−55.
35. Forman P. The discovery of the diffraction of x-rays by crystals: a critique of the myth. *Arch Hist Exact Sci* 1969;**6**(1):28−71.
36. Ewald PP. The myth of myths: comments on P. Forman's paper on "the discovery of the diffraction of X-rays in crystals". *Arch Rational Mech* 1969;**6**(1):72−81.
37. Eckert M. Disputed discovery: the beginnings of X-ray diffraction in crystals in 1912 and its repercussions. *Acta Crystallogr A* 2012;**68**(Pt 1):30−9.
38. Hildebrandt G. The discovery of the diffraction of X-rays in crystals—a historical review. *Cryst Res Technol* 1993;**28**(6):747−66.
39. Eckert M. Max von Laue and the discovery of X-ray diffraction in 1912. *Ann Phys (Berlin)* 2012;**524**(5):A83−5.
40. Barlow W, Pope WJ. A development of the atomic theory which correlates chemical and crystalline structure and leads to a demonstration of the nature of valency. *J Chem Soc Transactions* 1906;**89**:1675−744.
41. Barlow W, Pope WJ. The relation between the crystalline form and the chemical constitution of simple inorganic substances. *J Chem Soc Trans* 1907;**91**:1150−214.
42. Bragg WL. The structure of some crystals as indicated by their diffraction of X-rays. *Proc R Soc* 1913;**A89**(610):248−77.
43. Bragg WH, Bragg WL. The reflection of X-rays by crystals. *Proc R Soc* 1913;**A 88**(605):428−38.
44. Porter AB. On the diffraction theory of microscopic vision. *Lond Edin Dublin Phil Mag J Sci* 1906;**11**:154−66.
45. Bragg WH. Bakerian Lecture: X-rays and crystal structure. *Phil Trans R Soc Lond A* 1915;**215**(523−537):253−74.
46. Cork JM. The crystal structure of some of the alums. *Philo Mag* 1927;**4**(23):688−98.

47. Patterson AL. A Fourier series method for the determination of the components of interatomic distances in crystals. *Phys Rev* 1934;**46**(5):372−6.

48. Patterson AL. A direct method for the determination of the components of interatomic distances in crystals. *Z Krist* 1935;**A90**(1):517−42.

49. Ferry G. *Dorothy Hodgkin a life*. Cold Spring Harbor, NY: Cold Spring Harbor Laboratory Press; 1998.

50. Taylor G. The phase problem. *Acta Cryst D* 2003;**59**(11):1881−90.

51. Friedrich W. Eine neue interfernzersheinung bei Röntgenstrahlen. *Phys Z* 1913;**14**:317−19.

52. Nishikawa S, Ono S. Transmission of X-rays through fibrous, lamellar and granular substances. *Proc Math Phys Soc Tokyo* 1913;**7**(8):131−8.

53. Debeye P, Scherrer P. Interferenzen an regellos orientierten Teilchen im Röntgenlicht. I. *Phys Z* 1916;**17**:277−83.

54. Herzog RO, Jancke W. Über den physikalischen aufbau einiger hochmolecularer organischer verbindungen. *Ber Dtch Chem Ges* 1920;**53**:2162−4.

55. Herzog RO, Jancke W, Polanyi M. Röntgenspektrographische beobachtungen an zellulose. II. *Z Physik* 1920;**3**:343−8.

56. Meyer KH, Mark H. Über den Bau des krystallisierten Anteils der Cellulose. *Ber Dtch Chem Ges* 1928;**61**(4):593−614.

57. Meyer KH, Mark H. Über den aufbau des seiden-fibroins. *Ber Dtch Chem Ges* 1928;**61**(8):1932−6.

58. Meyer HK, Mark H. *Der Aufbau der hochpolymeren organischen Naturstoffe auf Grund (aufgrund) molekular-morphologischer Betrachtungen*. Leipzig, Germany: Leipzig Akademische Verlagsgesellschaft 1930.

59. Hall K. William Astbury and the biological significance of nucleic acids, 1938−1951. *Stud Hist Philos Biol Biomed Sci* 2011;**42**(2):119−28.

60. Astbury WT, Street A. X-ray studies of the structure of hair, wool, and related fibres. I. General. *Phil Trans R Soc Lond Series A* 1932;**A 230**(681−693):75−101.

61. Astbury WT, Woods HJ. X-ray studies of the structure of hair, wool, and related fibres. II. The molecular structure and elastic properties of hair keratin. *Phil Trans R Soc Lond* 1934; **A 232**(707−720):333−94.

62. Astbury WT, Sisson WA. X-ray studies of the structure of hair, wool, and related fibres. III. The configuration of the keratin molecule and its orientation in the biological cell. *Phil Trans R Soc Lond A* 1935;**150**(871):533−51.

63. Pauling L, Corey RB, Branson HR. The structure of proteins: two hydrogen-bonded helical configurations of the polypeptide chain. *Proc Natl Acad Sci USA* 1951;**37**(4):205−11.

64. Astbury WT, Dickinson S, Bailey K. The X-ray interpretation of denaturation and the structure of the seed globulins. *Biochem J* 1935;**29**(10):2351−60 1.

65. Brown A. *J. D. Bernal: the sage of science*. Oxford: Oxford University Press; 2005.

66. Bernal JD, Crowfoot D. Crystal structure of vitamin B1 and of adenine hydrochloride. *Nature* 1933;**131**(3321):911−12.

67. Bernal JD, Crowfoot D. X-ray photographs of crystalline pepsin. *Nature* 1934;**133** (3369):794−5.

68. Bernal JD, Fankuchen I, Perutz MF. An X-ray study of chymotrypsin and haemoglobin. *Nature* 1938;**141**(3568):487−528.

69. Bawden FC, Pirie NW, Bernal JD, Fankuchen I. Liquid crystalline substances from virus-infected plants. *Nature* 1936;**138**(3503):1051−2.

70. Bernal JD, Fankuchen I. Structure types of protein crystals from virus-infected plants. *Nature* 1937;**139**(3526):923−4.

71. Bernal JD, Fankuchen I. X-ray and crystallographic studies of plant virus preparations: I. Introduction and preparation of specimens II. Modes of aggregation of the virus particles. *J Gen Physiol* 1941;**25**(1):111−46.

72. Bernal JD, Fankuchen I. X-ray and crystallographic studies of plant virus preparations. III. *J Gen Physiol* 1941;**25**(1):147−65.

73. Creager AN, Morgan GJ. After the double helix: Rosalind Franklin's research on tobacco mosaic virus. *Isis* 2008;**99**(2):239−72.

74. Hodgkin DMC. John Desmond Bernal, 1901−1971. *Biogr Mems Fell R Soc* 1980;**26**:17−84.

75. Brown AP. J D Bernal: the sage of science. *J Phys Conference Series; John Desmond Bernal: Science and Society* 2007;**57**:61−72.

76. Crowfoot D. X-ray single-crystal photographs of insulin. *Nature* 1935;**135**(3415):591−2.

77. Adams MJ, Blundell TL, Dodson EJ, et al. Structure of rhombohedral 2 zinc insulin crystals. *Nature* 1969;**224**(5218):491−5.

78. Baker EN, Blundell TL, Cutfield JF, et al. The structure of 2Zn pig insulin crystals at 1.5 A resolution. *Philos Trans R Soc Lond B* 1988;**319**(1195):369−456.

79. Glusker JP. Dorothy Crowfoot Hodgkin (1910−1994). *Prot Sci* 1994;**3**(12):2465−9.

80. Howard JA. Dorothy Hodgkin and her contributions to biochemistry. *Nat Rev Mol Cell Biol* 2003;**4**(11):891−6.

81. Perutz MF. Hemoglobin structure and respiratory transport. *Sci Am* 1978;**239**(6):92−125.

82. Putnam FW. *Felix Haurowitz 1896−1987. Biographical Memoirs National Academy of Sciences*. Washington, DC: National Academy of Science USA; 1994. pp. 134−63.

83. Perutz MF. *Science is not a quiet life: unravelling the atomic mechanism of haemoglobin*. Singapore: World Scientific Pub; 1998.

84. Perutz MF. *I wish i'd made you angry earlier: essays on science, scientists, and humanity*. New York, NY: Cold Spring Harbor Laboratory Press; 2003.

85. Eisenberg D. Max Perutz's achievements: how did he do it? *Prot Sci* 1994;**3**(10):1625−8.

86. Perutz MF. An X-ray study of horse methemoglobin. *Proc R Soc London A* 1949;**195** (1043):474−99.

87. Bragg L, Kendrew JC, Perutz MF. Polypeptide chain configurations in crystalline proteins. *Proc R Soc London A* 1950;**203**(1074):321−57.

88. Perutz MF. The 1.5-A. reflexion from proteins and polypeptides. *Nature* 1951;**168** (4276):653−4.

89. Perutz MF. Life with living molecules. *New Scientist* 1976;**72**(1023):144−7.

90. Robertson JM. 255. An X-ray study of the phthalocyanines. Part II. Quantitative structure determination of the metal-free compound. *J Chem Soc* 1936;**0**:1195−209.

91. Riggs AF. Sulfhydryl groups and the interaction between the hemes in hemoglobin. *J Gen Physiol* 1952;**36**(1):1−16.

92. Green DW, Ingram VM, Perutz MF. The structure of haemoglobin. IV. Sign determination by the isomorphous replacement method. *Proc R Soc London A* 1954;**225**(1162):287−307.

93. Rossmann MG. The accurate determination of the position and shape of heavy-atom replacement groups in proteins. *Acta Cryst D* 1960;**13**(3):221−6.

94. Perutz GMF, Rossmann MG, Cullis AF, Muirhead H, Will G, North ACT. Structure of hæmoglobin: a three-dimensional Fourier synthesis at 5.5-Å resolution, obtained by X-ray analysis. *Nature* 1960;**185**(4711):416−22.

95. Perutz MF, Muirhead H, Cox JM, Goaman LC. Three-dimensional Fourier synthesis of horse oxyhaemoglobin at 2.8 A resolution: the atomic model. *Nature* 1968;**219** (5150):131−9.

96. Fermi G, Perutz MF, Shaanan B, Fourme R. The crystal structure of human deoxyhaemoglobin at 1.74 A resolution. *J Mol Biol* 1984;**175**(2):159−74.

97. Perutz MF. Fundamental research in molecular biology: relevance to medicine. *Nature* 1976;**262**(5568):449−53.

98. Dickerson RE, Geis I. *Hemoglobin: structure, function, evolution, and pathology*. Menlo Park, Reading, London, Amsterdam, Ontarion, Sydney: Benjamin/Cummings Publishing Company; 1983.

99. Kendrew JC, Bodo G, Dintzis HM, Parrish RG, Wyckoff H, Phillips DC. A three-dimensional model of the myoglobin molecule obtained by x-ray analysis. *Nature* 1958;**181**(4610):662−6.

100. Kendrew JC, Dickerson RE, Strandberg BE, et al. Structure of myoglobin: a three-dimensional Fourier synthesis at 2 A resolution. *Nature* 1960;**185**(4711):422−7.

101. Kendrew JC. *Myoglobin and the structure of proteins. Nobel Lectures in Chemistry 1942−1962*. Singapore, London, and New Jersey: World Scientific Publishing Co; 1962.

102. Wilkins MHF. John Turton Randall. 23 March 1905−16 June 1984. *Biogr Mem Fell R Soc* 1987;**492**−535.

103. Schrödinger E. *What is life?* New York, NY: Cambridge University Press; 1944 (reprinted 1967)

104. Attar N. Raymond Gosling: the man who crystallized genes. *Genome Biol* 2013;**14** (4):402.

105. Abir-Am PG. The Rockefeller Foundation and the rise of molecular biology. *Nat Rev Mol Cell Biol* 2002;**3**(1):65−70.

106. Chesley FG. X-ray diffraction camera for microtechniques. *Rev Sci Instrum* 1947;**18** (6):422−4.

107. Sherwood D. *Crystals, X-rays and proteins*. London: Longmans; 1976.

108. Signer R, Caspersson T, Hammersten E. Molecular shape and size of thymonucleic acid. *Nature* 1938;**141**(3559):122.

109. Astbury WT, Bell FO. Some recent developments in the X-ray study of proteins and related structures. *Cold Spring Harbor Symp Quant Biol* 1938;**6**:109−21.

110. Astbury WT, Bell FO. X-ray study of thymonucleic acid. *Nature* 1938;**141**(3573):747−8.

111. Chargaff E. Chemical specificity of nucleic acids and mechanism of their enzymatic degradation. *Experientia* 1950;**6**(6):201−9.

112. Cochran W, Crick FH, Vand V. The structure of synthetic polypeptides. I. The transform of atoms on a helix. *Acta Cryst* 1952;**5**(5):581−6.

113. Furberg S. On the structure of nucleic acids. *Acta Chem Scand* 1952;**6**:634−40.

114. Furberg S. Crystal structure of cytidine. *Nature* 1949;**164**(4157):22.

115. Avery OT, Macleod CM, McCarty M. Studies on the chemical nature of the substance inducing transformation of pneumococcal types: induction of transformation by a desoxyribonucleic acid fraction isolated from pneumococcus type III. *J Exp Med* 1944;**79** (2):137−58.

116. Chargaff E. Preface to a grammar of biology. A hundred years of nucleic acid research. *Science* 1971;**172**(3984):637−42.

117. Gulland JM, Barker GR, Jordan DO. The chemistry of the nucleic acids and nucleoproteins. *Ann Rev Biochem* 1945;**14**:175−206.

118. Consden R, Gordon AH, Martin AJ. Qualitative analysis of proteins: a partition chromatographic method using paper. *Biochem J* 1944;**38**(3):224−32.

119. Chargaff E, Zamenhof S. The isolation of highly polymerized desoxypentosenucleic acid from yeast cells. *J Biol Chem* 1948;**173**(1):327−35.

120. Vischer E, Chargaff E. The separation and quantitative estimation of purines and pyrimidines in minute amounts. *J Biol Chem* 1948;**176**(2):703−14.

121. Vischer E, Chargaff E. The composition of the pentose nucleic acids of yeast and pancreas. *J Biol Chem* 1948;**176**(2):715−34.

122. Chargaff E, Vischer E, Doniger R, Green C, Misani F. The composition of the desoxypentose nucleic acids of thymus and spleen. *J Biol Chem* 1949;**177**(1):405−16.

123. Vischer E, Zamenhof S, Chargaff E. Microbial nucleic acids: the desoxypentose nucleic acids of avian tubercle bacilli and yeast. *J Biol Chem* 1949;**177**(1):429−38.

124. Chargaff E, Zamenhof S, Green C. Composition of human desoxypentose nucleic acid. *Nature* 1950;**165**(4202):756−7.

125. Gosling G. The genesis of discovery: first steps. *Genet Eng Biotechnol News* 2013;**33** (7):29−31.

126. Mirsky AE, Pollister AW. Nucleoproteins of cell nuclei. *Proc Natl Acad Sci USA* 1942;**28** (9):344−52.

127. Signer R, Schwander H. Isoierung hochmolecularer nukleinsäure aus kalbsthymus. *Helv Chim Acta* 1949;**32**(3):853−9.

128. Reichmann ME, Bunce BH, Doty P. The changes induced in sodium desoxyribonucleate by dilute acid. *J Polymer Sci* 1953;**10**(1):109−19.

129. Peterlin A. Modèle statistique des grosses molécules à chaînes courtes. V. Diffusion de la lumière. *J Polymer Sci* 1953;**10**(4):425−36.

130. Meili M. Signer's gift—Rudolf Signer and DNA. *Chimia* 2003;**57**(11):735−40.

131. Seeds WE, Wilkins MHF. Ultra-violet microspectrographic studies of nucleoproteins and crystals of biological interest. *Discuss Faraday Soc* 1950;**9**:417−23.

132. Wilkins MHF, Gosling RG, Seeds WE. Physical studies of nucleic acid: nucleic acid: an extensible molecule? *Nature* 1951;**167**(4254):759−60.

133. Wilkins M. *The molecular configuration of nucleic acids. Nobel Lectures in Physiology and Medicine 1942−1962*. Amsterdam: Elsevier; 1964. pp. 752−82.

134. Glynn J. Rosalind Franklin: 50 years on. *Notes Rec R Soc* 2008;**62**(2):253−5.

135. Franklin RE, Gosling RG. The structure of sodium thymonucleate fibres. I. The influence of water content. *Acta Cryst* 1953;**6**(8−9):673−7.

136. Ehrenberg W, Spear WE. An electrostatic focusing system and its application to a fine focus X-ray tube. *Proc Phys Soc B* 1951;**64**(1):67.

137. Franklin R, Gosling R. Molecular configuration in sodium thymonucleate. *Nature* 1953;**171**(4356):740−1.

138. Judson HF. Reflections on the historiography of molecular biology. *Minerva* 1980;**18** (3):369−421.

139. Wilkins M, Stokes A, Wilson H. Molecular structure of deoxypentose nucleic acids. *Nature* 1953;**171**(4356):738−40.

140. Franklin R, Gosling R. Evidence for 2-chain helix in crystalline structure of sodium deoxyribonucleate. *Nature* 1953;**172**(4369):156−7.

141. Pauling L, Corey RB. A proposed structure for the nucleic acids. *Proc Natl Acad Sci USA* 1953;**39**(2):84−97.

142. Pauling L, Itano HA, Singer SJ, Wells IC. Sickle cell anemia, a molecular disease. *Science* 1949;**110**(2865):543−8.

143. Lewis S. *Arrowsmith*. New York, NY: Signet Classics; 1961.

144. Maaloe O, Watson JD. The transfer of radioactive phosphorus from parental to progeny phage. *Proc Natl Acad Sci USA* 1951;**37**(8):507−13.

145. Hershey AD, Chase M. Independent functions of viral protein and nucleic acid in growth of bacteriophage. *J Gen Physiol* 1952;**36**(1):39–56.

146. Watson JD. *Genes, girls, and Gamow: after the double helix*. New York, NY: Vintage Books; 2001.

147. Watson JD. The structure of tobacco mosaic virus. I. X-ray evidence of a helical arrangement of sub-units around the longitudinal axis. *Biochim Biophys Acta* 1954;**13**(1):10–19.

148. Kubbinga H. Crystallography from Hauy to Laue: controversies on the molecular and atomistic nature of solids. *Acta Cryst Section A* 2012;**68**(Pt 1):3–29.

149. Eisenberg D. The discovery of the α-helix and β-sheet, the principal structural features of proteins. *Proc Natl Acad Sci USA* 2003;**100**(20):11207–10.

150. Pauling P. DNA—the race that never was? *New Scientist* 1973;**58**(8):558–60.

151. Williams RC. Electron microscopy of sodium desoxyribonucleate by use of a new freeze-drying method. *Biochim Biophys Acta* 1952;**9**(3):237–9.

152. Cecil R, Ogston AG. The sedimentation of thymus nucleic acid in the ultracentrifuge. *J Chem Soc* 1948;**0**:1382–6.

153. Kahler H. The diffusion and sedimentation of sodium thymonucleate. *J Phys Chem* 1948;**52**(4):676–89.

154. Kahler H, Lloyd Jr BJ. The electron microscopy of sodium desoxyribonucleate. *Biochim Biophys Acta* 1953;**10**(3):355–9.

155. Clark VM, Todd AR, Zussman J. Nucleotides. Part VIII. Cyclonucleoside salts. A novel rearrangement of some toluene-*p*-sulphonylnucleosides. *J Chem Soc* 1951;2952–8.

156. Brown DM, Todd AR. Nucleotides. Part X. Some observations on the structure and chemical behaviour of the nucleic acids. *J Chem Soc* 1952;52–8.

157. Pauling L, Corey RB. Structure of the nucleic acids. *Nature* 1953;**171**(4347):346.

158. Pauling L, Delbruck M. The nature of the intermolecular forces operative in biological processes. *Science* 1940;**92**(2378):77–9.

159. Rich A. Linus Pauling: chemist and molecular biologist. *Ann NY Acad Sci* 1995;**758** (1):74–82.

160. Chargaff E. The double helix. A personal account of the discovery of the structure of DNAp. *Science* 1968;**159**(3822):1448–9.

161. Perutz MF, Randall JT, Thomson L, Wilkins MH, Watson JD. DNA helix. *Science* 1969;**164**(3887):1537–9.

162. Davidson JN. *Biochemistry of the nucleic acids*. London: Methuen; 1950.

163. Gulland JM, Jordan DO. The macro-molecular behaviour of nucleic acids. *Symp Soc Exp Biol* 1947;**1**:56–65.

164. Jordan DO. Physicochemical properties of the nucleic acids. *Prog Biophys Biophys Chem* 1952;**2**:51–89.

165. Broomhead JM. The structure of pyrimidines and purines. II. A determination of the structure of adenine hydrochloride by X-ray methods. *Acta Cryst* 1948;**1**(6):324–9.

166. Broomhead JM. The structures of pyrimidines and purines. IV. The crystal structure of guanine hydrochloride and its relation to that of adenine hydrochloride. *Acta Cryst* 1951;**4** (2):92–100.

167. Watson JD, Crick FH. Genetical implications of the structure of deoxyribonucleic acid. *Nature* 1953;**171**(4361):964–7.

168. Crick FHC, Watson JD. The complementary structure of deoxyribonucleic acid. *Proc R Soc* 1954;**223**(1152):80–96.

169. Meselson M, Stahl FW. The replication of DNA in *Escherichia coli*. *Proc Natl Acad Sci USA* 1958;**44**(7):671–82.

170. Pauling L, Bragg L. Discussion des rapports de MM. L. Pauling et L. Bragg. *Rep Inst Intl Chimie Solvay* 1953;111−12.

171. Watson JD, Crick FH. The structure of DNA. *Cold Spring Harbor Symp Quant Biol* 1953;**18**:123−31.

172. Jacob F. *The statue within: an autobiography.* New York, NY: Cold Spring Harbor Laboratory Press; 1995.

173. Hershey AD. Functional differentiation within particles of bacteriophage T2. *Cold Spring Harb Symp Quant Biol* 1953;**18**:135−9.

174. Watson JD. *Molecular biology of the gene.* New York, NY: W.A. Benjamin Inc; 1965.

175. Wilkins MH. Physical studies of the molecular structure of deoxyribose nucleic acid and nucleoprotein. *Cold Spring Harb Symp Quant Biol* 1956;**21**:75−90.

176. Marvin DA, Spencer M, Wilkins MH, Hamilton LD. A new configuration of deoxyribonucleic acid. *Nature* 1958;**182**(4632):387−8.

177. Hamilton LD, Barclay RK, Wilkins MHF, et al. Similarity of the structure of DNA from a variety of sources. *J Biophys Biochem Cytol* 1959;**5**(3):397−404.

178. Marvin DA, Spencer M, Wilkins MH, Hamilton LD. The molecular configuration of deoxyribonucleic acid. III. X-ray diffraction study of the C form of the lithium salt. *J Mol Biol* 1961;**3**(5):547−65.

179. Arnott S, Wilkins MH. Fourier synthesis studies of lithium DNA. 3. Hoogsteen models. *J Mol Biol* 1965;**11**(2):391−402.

180. Chargaff E. Some recent studies on the composition and structure of nucleic acids. *J Cell Physiol* 1951;**38**(Suppl. 1):41−59.

181. Franklin RE, Commoner B. Abnormal protein associated with tobacco mosaic virus; x-ray diffraction by an abnormal protein (B8) associated with tobacco mosaic virus. *Nature* 1955;**175**(4468):1076−7.

182. Franklin RE. Structural resemblance between Schramm's repolymerised A-protein and tobacco mosaic virus. *Biochim Biophys Acta* 1955;**18**(2):313−14.

183. Franklin RE. Structure of tobacco mosaic virus: location of the ribonucleic acid in the tobacco mosaic virus particle. *Nature* 1956;**177**(4516):929.

184. Franklin RE. X-ray diffraction studies of cucumber virus 4 and three strains of tobacco mosaic virus. *Biochim Biophys Acta* 1956;**19**(2):203−11.

185. Franklin RE, Holmes KC. The helical arrangement of the protein subunits in tobacco mosaic virus. *Biochim Biophys Acta* 1956;**21**(2):405−6.

Chapter 6

Meselson and Stahl Proved That DNA Is Replicated in a Semiconservative Fashion

An Elegant Experiment Decided Among Three Competing Theoretical Models of the Mechanics of DNA Replication

Chapter Outline

M. Fry: Landmark Experiments in Molecular Biology. DOI: http://dx.doi.org/10.1016/B978-0-12-802074-6.00006-0
249

The groundbreaking discovery of the double-stranded helical structure of DNA by James Watson (1928−) and Francis Crick (1916−2004)[1] and their insight into the genetic implications of this structure[2] raised three fundamental questions. The first was the problem of the mechanics of separation and duplication of the paired chains in double-stranded DNA. A second concern was the nature of the code that enables translation of sequences of four bases in DNA into corresponding sequences of 20 different amino acids in proteins. The third issue was the mode of transfer of information from DNA to the protein-synthesizing machinery. Of these three problems, the earliest to have been solved was the question of the mode of DNA replication. In case rare for biological research, three competing and experimentally testable theoretical models of DNA replication were formulated before any experimental work was launched to identify the operational mode of replication. Framed between 1953 and 1955, the three rival models argued for semiconservative, dispersive, or conservative mode of DNA replication. These alternative models then guided experiments that were aimed at the identification of the actual mode of replication of bacteriophage or bacterial DNA. After a relatively short phase of inconclusive results, two young Caltech researchers, Matthew Meselson (1930−) and Franklin Stahl (1929−) performed an elegant experiment which convincingly demonstrated that bacterial DNA is replicated in a semiconservative manner.[3]

6.1 THEORETICAL CONSIDERATIONS BRED THREE COMPETING ABSTRACT MODELS OF DNA REPLICATION

6.1.1 Watson and Crick Proposed a Semiconservative Mode of Replication

At the time of the introduction of their double helix model of DNA, Watson and Crick also speculated on the mode of replication of the DNA strands. They hypothesized that DNA is replicated in a semiconservative fashion by which each DNA chain serves as a template for the synthesis of a complementing chain. The general principle of complementary replication had been raised before by others. The zoologist Nikolai Koltsov (1872−1940) formulated in the 1927 Third All-Russian Congress of Zoologist, Anatomist, and Histologists the principle *"omnis molecule ex molecule"* (every molecule arises from a molecule). Specifically, he proposed that genes are organized side-by-side along giant molecules that serve as templates for the reproduction of complementary molecules.[4] However, as was unanimously thought at the time, Koltsov believed that the genes in the giant molecules were made of protein, whereas nucleic acids functioned as "hard protective crust" of chromosomes.[4] The idea of complementary replication of a template has been raised again by Linus Pauling. As was described in Section 5.6.3, Pauling proposed in 1948 a template−replica mechanism for the duplication of genes in two complementing molecules (a facsimile of the Pauling's original lecture is in: http://scarc.library.oregonstate.edu/coll/pauling/dna/papers/1948p.13.html). Notably, Pauling too thought that the replicating genes were comprised of protein.

There is no evidence that Watson or Crick had been aware of the Koltsov or Pauling propositions of semiconservative duplication of the hereditary material. Clearly, however, their suggestion of hydrogen bond—mediated complementarity of the DNA bases put their model on firmer chemical ground than their predecessors. Watson and Crick alluded to the question of DNA replication already in their landmark 1953 paper on the double helix model of DNA. At the end of that article they famously stated: "It has not escaped our notice that the specific base pairing we have postulated immediately suggests a possible copying mechanism for the genetic material."[1] Behind this terse proclamation, there has been an already well-formed idea of semiconservative mode of replication. On Mar, 19, 1953, just a day or two before the submission of the paper to *Nature*,[1] Crick described Watson's and his model of DNA in a seven-page letter to his 12-years-old son Michael who was away in a boarding school. Referring in this letter to the replication question, Crick wrote[5]:

> [...] *You can now see how Nature makes copies of the genes. Because if the two chains unwind into two separate chains, and if each chain then makes another chain come together on it, then because A always goes with T, and G with C, we shall get two copies where we had one before. For example* [...] [here Crick inserted a hand-drawn scheme of semiconservative replication shown in Fig. 6.1]. *In other words we think we have found the basic copying mechanism by which life comes from life.*

Thus, Crick and Watson envisioned at the outset that DNA replication occurred in a semiconservative manner in which each of the two DNA strands served as a template for a complementary daughter strand (illustrated in Fig. 6.2A).

This view was discussed explicitly in a correspondence between Watson, then in Cambridge, England, and his mentor Max Delbrück (1906—1981) in Caltech. Writing to Delbrück 1 month before the publication of the double helix DNA model, Watson described the principal attributes of the proposed structure of DNA (Watson, Mar. 12, 1953, letter to Max Delbruck http://osu-library.orst.edu/specialcollections/coll/pauling/dna/corr/corr432.1-watson-del-bruck-19530312-03.html):

> [...] *The main features of the model are (1) The basic structure is helical—it consists of two intertwining helices—the core of the helix is occupied by the purine and pyrimidine bases—the phosphates groups are on the outside. (2) The helices are not identical but complementary so that if one helix contains a purine base, the other helix contains a pyrimidine—this feature is a result of*

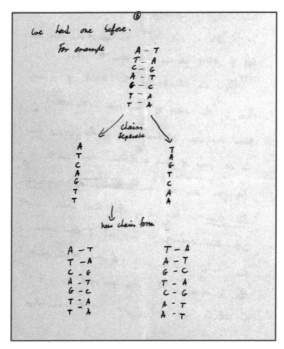

FIGURE 6.1   Crick's handwritten scheme of semiconservative DNA replication (page 6 of the seven-page-long Mar. 19, 1953, letter to his son Michael, *reproduced in Watson JD. The annotated and illustrated double helix. New York, NY: Simon & Schuster; 2012.*).

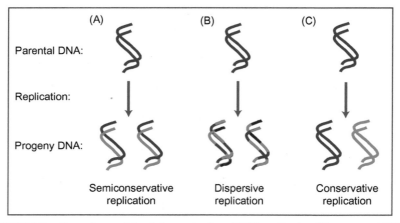

FIGURE 6.2   Alternative models of the mechanics of DNA replication. Blue: parental DNA. Red: progeny DNA.

our attempt to make the residues equivalent and at the same time put the purine and pyrimidine bases in the center. The pairing of the purine with pyrimidines is very exact and dictated by their desire to form hydrogen bonds—Adenine will pair with Thymine while Guanine will always pair with Cytosine.

Shortly thereafter, on Mar. 22, 1953, Watson sent to Delbrück a draft of the *Nature* article that included the statement: "The sequence of bases on a single chain does not appear to be restricted in any way. However, if only specific pairs of bases can be formed, it follows that if the sequence of bases on one chain is given, then the sequence on the other chain is automatically determined."[1] Drawing on this perception, Delbrück explicitly raised the possibility of semiconservative replication in an Apr. 14, 1953, letter to Watson (cited in Ref. 6, p. 14):

[...] In your model the DNA molecule consists of two threads each of which determines the other completely. One thinks of reproduction taking place by separation of the two threads, followed by the formation of a complementary thread by each one of them.

Indeed, this account faithfully reflected Watson's own vision of a semiconservative mode of replication as he asserted in his subsequent Apr. 25, 1953, reply to Delbrück (cited in Ref. 6, p. 17):

[...] we would also guess that reproduction takes place by separation of the two threads, followed by the formation of a complementary thread by each of them.

6.1.2 Delbrück Offered an Alternative Dispersive Mechanism of Replication

The Basis for Delbrück's Alternative Model

Deliberating the mode of replication, Delbrück concluded that intertwined DNA double helix ("plectonemic DNA," see Fig. 6.3) that was stipulated in the Watson and Crick model could not replicate in a semiconservative fashion. Delbrück argued in his letter to Watson that the number of turns needed to separate the interlocked DNA strands appeared implausible:

[...] For a DNA molecule of MW 3,000,000 there would be about 500 turns around each other. These would have to untwiddle to separate the threads.

Delbrück saw a way out of this seemingly improbable scenario by suggesting that the two DNA strands might be arranged in a side-by-side "paranemic" modus (Fig. 6.3).

Obviously, such arrangement could have enabled separation of the strands without uncoiling:

[...] A feasible way to do this would be to assume the existence of an alternate equilibrium state, in which the double helix is contracted. In contracting, it forms a superhelix (like chromosomes do), and at the same time the threads

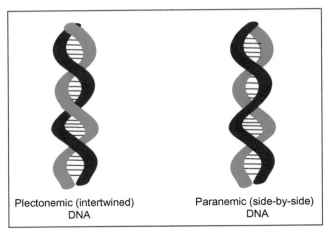

Plectonemic (intertwined)   Paranemic (side-by-side)
DNA   DNA

FIGURE 6.3 Plectonemic (interlocked) and paranemic (side-by-side, unlocked) DNA structures.

arrange themselves in a paranemic manner, i.e., such that for each turn of the superhelix the threads turn around each other in a compensating turn. In such a configuration the two threads can be pulled apart sideways without interlocking.

One month after the publication of their double helix model of DNA, Watson and Crick published in *Nature* in May 30, 1953, their theoretical article on the genetic implications of the structure of DNA. In this paper they explicitly expressed their view that DNA replicates semiconservatively[2]:

Now our model for deoxyribonucleic acid is, in effect, a pair of templates, each of which is complementary to the other. We imagine that prior to duplication the hydrogen bonds are broken, and the two chains unwind and separate. Each chain then acts as a template for the formation on itself of a new companion chain, so that eventually we shall have two pairs of chains, where we only had one before. Moreover, the sequence of the pairs of these bases will have been duplicated exactly.

Watson and Crick were mindful of Delbrück's point, namely, that in order to separate the two strands in plectonemic DNA the double strand must untwist multiple times. However, although they did not offer a mechanism for the untangling of the DNA, they felt that the problem of strand untwisting was not intractable. It should be noted in this context that 3 years later, Cyrus Levinthal (Box 6.1) and the physicist Horace Richard Crane (1907−2007), both of the University of Michigan, posited a theoretical model of Y-shaped replicating DNA whose vertical part is the parental duplex and the arms are the growing progeny. Their calculations for a DNA molecule with 6000 turns indicated that only a negligible increase in energy

BOX 6.1 Gunther Stent and Cyrus Levinthal

Born and raised in Berlin, Gunther Stent escaped Germany in 1938 and settled in Chicago. After obtaining in 1945 a Bachelor's degree in chemistry and in 1948 a PhD in physical chemistry, both from the University of Illinois, he worked in Max Delbrück's lab at Caltech studying phages and becoming a member of the Phage Group. From 1952 until his retirement he was on the faculty of UC Berkeley chairing at different times its departments of molecular biology and of cell and molecular biology. Although Stent was an early proponent and a practitioner of molecular biology, he asserted already in 1968 that it is a discipline in decline.[7] Switching therefore to neurobiology, he used the leech as a model organism for the study of neural connections and circuitry that underlie its behavior. Side-by-side with his work in molecular biology and neurobiology, Stent was also known for his studies and prolific writings on the history and philosophy of biology.

Cyrus Levinthal earned a PhD degree in physics from UC Berkeley in 1951 and soon switched to the emerging field of molecular biology. After a time in the University of Michigan, he joined in 1957 the MIT Department of Biology, moving in 1968 to Columbia University. In the early 1960s Levinthal demonstrated a direct relationship between genes and their encoded proteins and showed that bacterial mRNA is labile. He later pioneered the use of computers to represent and predict the three-dimensional structure of proteins and to trace three-dimensional images of brain cells (Fig. 6.4).

Gunther Stent
(1924–2008)

Cyrus Levinthal
(1922–1990)

FIGURE 6.4 Gunther Stent and Cyrus Levinthal.

was required for independent rotation against viscous drag of each branch of the Y around its own axis. Thus such geometry of the DNA could potentially enable unwinding of a long duplex DNA and allow its semiconservative replication.[8] This model challenged Delbrück's argument that unwinding of the whole duplex was physically unfeasible.

Delbrück Put Forward His Dispersive Model of Replication

Back in 1953 Delbrück responded critically to a preprint of the Watson and Crick second article.[2] In a May 12, 1953, letter to Watson (cited in Ref. 6, pp. 21−22), he expressed faith in the correctness of base complementarity as the basis for DNA replication. Nevertheless, Delbrück stressed his mistrust in the intertwined DNA model:

> *[...] I am willing to bet that the plectonemic coiling of the chains in your structure is radically wrong, because (1) The difficulty of untangling the chains do seem, after all, insuperable to me. (2) The X-ray data suggest only coiling but not specifically your kind of coiling. I would suggest, therefore, that your second publication de-emphasize the mode of coiling.*

Shortly thereafter Watson addressed the issue of the plectonemic coiling of the two DNA strands when he presented the double helix model of DNA in a Jun. 1953 Cold Spring Harbor Conference on Viruses. There he argued that the three-dimensional model of DNA that he and Crick have constructed indicated that a side-by-side paranemic coil could not sustain equivalent orientation of successive bases along each helix with regard to the helical axis.[9] Holmes in his history of the quest for the mechanics of replication proffered that this argumentation persuaded Delbrück to abandon his claim for paranemic DNA configuration and to accept that the DNA strands are intertwined.[10]

Although Delbrück accepted the plectonemic configuration of DNA, he held on to the idea that untwisting of the whole double helix was improbable. He thus introduced in late 1954 a purely theoretical model of *dispersive replication* that obviated the need for unwinding of the plectonemic double strand.[11] In an article in the *Proceedings of the National Academy*, he listed three possible ways of separating daughter duplexes from the parental DNA: (1) longitudinal slippage of the strands past one another; (2) resolution of the two duplexes from each other; (3) breakage and reunions of the duplexes. Delbrück's grounding in theoretical physics was evident in his rejection of the first two alternatives on the account of their inelegance and inefficiency. He thus considered the third model of breakage and rejoining to be the only credible one. Features of this model were described in both the 1954 paper[11] and in a subsequent review article.[12] The crux of Delbrück's argument was that separation of the two chains of a plectonemic duplex by lateral movement in opposite directions would encounter an interlock at each turn of the helix. Illustratively describing the barrier to replication in interlocked DNA, Delbrück and Stent wrote: *"[...] for any winding number greater than zero, the 'braid' consisting of two chains cannot be combed."*[12] To overcome this difficulty Delbrück proposed that as replication proceeds, the parental chains are broken at obstructing sites that are located at every half-turn of the helix. Then, following replication of the parental DNA segments, their lower terminals are rejoined in a criss-cross manner to open ends of daughter strands of equal polarity. DNA products of such dispersive replication should consist, therefore, of alternating segments of parental and progeny segments of average size of half a turn of the helix, that is, five nucleotides (Fig. 6.2B).

Although Delbrück's scheme was purely speculative, he also offered a thought experiment to decide between semiconservative and dispersive replication. In this proposed procedure, DNA was to be labeled with a radioactive precursor and just as replication was to begin, cells would be transferred to a medium containing an unlabeled precursor. Progeny DNA would then be analyzed for the distribution of the label among parental and daughter strands. As illustrated in Fig. 6.2A and B, the expected distribution of label among parental and daughter strands would be different if replication was semiconservative or dispersive.[11] Notably, however, Delbrück did not specify a method to resolve the parental and daughter strands.

6.1.3   David Bloch and John Butler Proposed Variant Models of Conservative Replication

Watson and Crick's and Delbrück's respective semiconservative and dispersive models of replication were challenged in 1955–56 by a third alternative hypothesis of a conservative mode of replication. This type of replication was proposed to yield wholly newly synthesized daughter duplex while maintaining fully conserved parental double helix (Fig. 6.2C). David Bloch (1926–86), then at Columbia University, offered a speculative model of conservative duplication by which histone proteins distorted the parental DNA strands, causing breakage of the interstrand hydrogen bonds in DNA and rotating the bases of each strand by about 180 degrees. The rotated bases became then accessible to bonding with incoming bases of growing daughter strands such that the entire parental duplex would serve as template.[13] Following replication, the hydrogen bonds between parental and daughter strands would break, allowing the bases to rotate back such that like-with-like pairing of parental and daughter strands would occur resulting in completely conservative duplication.[13]

John Butler (1899–1977) of the Chester Beatty Institute in London offered a variant model of conservative replication in which the two strands of the parental duplex were first unwound through the action of histones attached to one of their ends. Once each strand served as template for complementary chain, histone molecules would enact like-with-like rewinding and pairing of the two parental and the two daughter strands with an end result of the production of wholly newly made daughter duplex and a fully conserved parental double helix.[14]

To sum up, within just 1 year after Watson and Crick's proposal of semiconservative replication,[1,9] Delbrück came up with the distinctly different model of dispersive replication.[11] Soon thereafter alternative theoretical models for a third kind of conservative DNA replication were put forward.[13,14] It should be stressed that at the time of their introduction the three competing models of DNA replication were purely theoretical and no experimental evidence existed to favor one over the others. Only after 1955, when the groundwork of theoretical modeling was fully formulated, did several researchers commit to experiments that were aimed to determine which one of the three abstract models was consistent with the actual mechanics of DNA replication.

6.2 EARLY EXPERIMENTAL TESTING OF THE ALTERNATIVE THEORETICAL MODELS OF REPLICATION

As noted, Delbrück proposed to distinguish between dispersive and semiconservative replication by labeling replicating DNA with a radioactive precursor followed by examination of the distribution of label in the parental and progeny DNA duplexes.[11] However, in this thought experiment Delbrück did not specify how newly made duplexes could be resolved from parental DNA. In fact, the main challenge that experimentalists faced was to find a way to clearly distinguish between parental and daughter strands of DNA. Actual early experiments to decide between conservative, semiconservative, or dispersive modes of replication adhered to Delbrück's idea of determining the distribution of radioactively labeled replicating DNA among daughter cells. In these experiments Gunther Stent[15,16] and Cyrus Levinthal[17] (Box 6.1) employed different strategies to distinguish the labeled parental DNA from daughter DNA.

6.2.1 Gunther Stent's Experiments Yielded Inconclusive Results

In a first attempt Gunther Stent at Berkeley (Box 6.1) together with the eminent future immunologist and Nobel laureate Niels Jerne (1911−94) in Caltech tried to identify experimentally the mode of replication of phage DNA.

A year later Stent and his graduate student Clarence Fuerst (1928−2005) applied a similar approach to analyze the replication of bacterial DNA. Essentially, newly synthesized phage[15] or bacterial[16] DNA were labeled with radioactive ^{32}P isotope and the labeled cells were transferred to a medium with nonradioactive phosphorous. Then, loss of viability as a result of radioactivity-induced damage to the DNA was monitored through subsequent replication cycles. The objective was to identify the mode of replication by determining whether ^{32}P-DNA was equally distributed in all the viruses or bacteria or was confined to only some while others contained wholly unlabeled DNA. Employing this strategy, all the progeny bacteria was found to contain newly made labeled DNA intermingled with parental DNA with no detectable class of viable cells that had only unlabeled DNA.[16] The observation that *"[...] parental and newly assimilated DNA become intermingled with daughter nuclei"*[16] was inconsistent with a conservative mode of replication but semiconservative or dispersive replication could not be distinguished from one another.

6.2.2 An Experiment by Cyrus Levinthal Also Failed to Produce Decisive Conclusions

At about the time that Stent carried out his experiments, Cyrus Levinthal (Box 6.1) employed a different approach in an attempt to define the mechanics of DNA duplication. Replicating bacteriophage DNA was labeled with ^{32}P and the distribution of radioactivity among progeny viruses was determined by their exposure to electron-sensitive photographic emulsion. It was hoped that this technique would yield estimates of the radioactivity in a single molecule of DNA released from lysed virus.[17] Rather than being uniformly distributed in all the phages after the first and second replication cycles, radioactivity was found to be concentrated in few progeny viruses. This result was inconsistent with the dispersive replication model of breakage and reunion. Yet although Levinthal interpreted his results as consistent with semiconservative replication,[17] for multiple technical reasons he could not definitively identify the mode of replication of phage DNA.[6]

6.3 HERBERT TAYLOR'S EXAMINATION OF DUPLICATING CHROMOSOMES WAS CONSISTENT WITH A SEMICONSERVATIVE MODE OF REPLICATION

An early but still not fully conclusive direct indication that DNA was replicated in a semiconservative fashion was obtained by Herbert Taylor (Box 6.2), then at the Columbia University Department of Botany. Aided by Philip Wood and Walter Hughes of the Brookhaven National Laboratory, who prepared high-specific activity tritiated ^{3}H thymidine, he applied autoradiography to monitor the distribution of ^{3}H-labeled DNA in replicating *Vicia faba* chromosomes.[18]

BOX 6.2 Herbert Taylor

Herbert Taylor earned a PhD degree from the University of Virginia in 1944. After service with the US military in the Pacific theater (1943–46), he served on the faculties of the Universities of Oklahoma and Tennessee and of Columbia University. From 1964 until his retirement in 1990, he was Distinguished Professor at Florida State University. In a career that spanned more than 50 years, he studied with his wife and colleague Shirley Taylor the structure and replication timing of chromosomes (Fig. 6.5).

(Continued)

BOX 6.2 (Continued)

J. Herbert Taylor
(1916–1998)

FIGURE 6.5   J. Herbert Taylor.

Essentially, *V. faba* seedlings that were grown for one-third of the cell division time in ^{3}H thymidine-containing medium were transferred to a medium without labeled thymidine that contained colchicine that arrests cells at the anaphase stage of cell division. Autoradiography of successive metaphases revealed that after a single round of duplication all the chromosomes were evenly labeled. In cells that duplicated twice, only one of the two chromatids carried label and the other was unlabeled and in cells that duplicated three times one half of the chromosomes was devoid of label in either chromatid whereas the other half had one labeled and one nonlabeled chromatids[18] (Fig. 6.6). This observation, which was consistent "... at the microscopic level" with the semiconservative model of replication, raised much interest. In a letter Watson alerted Crick to Taylor's results (Watson, Sep. 23, 1956, letter to Crick http://profiles.nlm.nih.gov/ps/access/SCBBJH.pdf):

> *[...] 1st division - both chromosomes radioactive 2nd one hot—one cold. So chromosome is divided into two parts. Whether this is 2 fold polyteny and two stranded DNA is undecided. Nevertheless a very elegant result.*

Crick, Delbrück, and Alex Rich (1924–2015) paid visits to Taylor's laboratory to gain direct impression of his results.[6] Also, after listening to Taylor's description of his findings in a Caltech seminar, Delbrück briefly considered abandoning his breaks-and-reunions model of replication although on a second thought he chose to adhere to his dispersive model.[6] Indeed, Taylor's observations did not provide irrefutable evidence for semiconservative mode of replication. Because of the low resolution of the

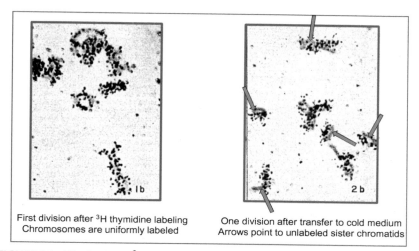

First division after ^{3}H thymidine labeling
Chromosomes are uniformly labeled

One division after transfer to cold medium
Arrows point to unlabeled sister chromatids

FIGURE 6.6   Distribution of ^{3}H label in *Vicia faba* sister chromatids before and after a single replication cycle. All the chromosomes were Feulgen-stained. *From Taylor JH, Woods PS, Hughes WL. The organization and duplication of chromosomes as revealed by autoradiographic studies using tritium-labeled thymidine.* Proc Natl Acad Sci USA *1957;43(1):122–8.*

autoradiography of whole chromosomes, a remaining possibility was that the distribution of label in the chromosomes could actually be non-uniform. Potential unequal dispersion of label would suggest that the observed semi-conservative scattering of label was restricted to only select segments of the DNA but not to its full complement. In addition, with its resolution being restricted to the chromosomal level, Taylor's analysis did not unveil the mode of DNA replication at the molecular level.

6.4 A CRITICAL EXPERIMENT BY MESELSON AND STAHL PROVIDED A PRACTICALLY UNEQUIVOCAL CORROBORATION OF THE SEMICONSERVATIVE MODEL OF REPLICATION

A practically definitive experimental demonstration that bacterial DNA is replicated in a semiconservative fashion was provided within a year after the Taylor report. The critical experiment was conducted by two young Caltech researchers; Matthew (Matt) Meselson who was then a doctoral student under Linus Pauling and Franklin (Frank) Stahl (Box 6.3; Fig. 6.8), a post-doctoral fellow with Giuseppe (Joe) Bertani (1924–2015).

Historical, personal, and scientific circumstances of the Meselson and Stahl experiment were described by Meselson himself[19] and in much greater detail by the historian Frederick Holmes in his comprehensive book.[6] Discussion here is limited to scientific and technical aspects of the experiment itself.

BOX 6.3 Meselson and Stahl

In the summer of 1954 Meselson, then a doctoral student at Caltech, came to the Woods Hole Marine Laboratory to serve as an assistant in a course on phage genetics. Stahl, then a PhD student at Rochester, was attending that course. As Stahl was sitting under a tree trying to solve a problem of phage mutation distribution, Meselson introduced himself and helped Stahl to the solution. In an ensuing conversation, Meselson introduced the problem of DNA replication and described his ideas for its solution. Eventually, he invited Stahl, who was planning postdoctoral work at Caltech, to collaborate with him on the problem (Fig. 6.7).

Matt Meselson by the
Beckman Model E ultracentrifuge

Frank Stahl speaking at the
1958 Cold Spring Harbor Symposium

FIGURE 6.7   Meselson and Stahl at around the time of their remarkable experiment.

Meselson (l), Stahl (r), and Martha Chase (bottom)
Cold Spring Harbor, 1950s

Meselson and Stahl in 1996 photographed by the tree at Woods Hole where they first met 42 years earlier

FIGURE 6.8   Meselson and Stahl at different times.

The key to the successful outcome of the Meselson and Stahl experiment was that rather than labeling DNA with radioactive phosphorous[15-17] they respectively marked parental and replicated progeny DNA with heavy ^{15}N and light ^{14}N isotopes of nitrogen. Ultracentrifugation was then applied to separate DNA molecules whose two strands were either heavy or light from DNA that had one heavy and one light strand.

6.4.1   Early Stages in the Evolution of the Experiment

After attending a Caltech research seminar by Jacque Monod (1910−76), Meselson conceived a general principle of solving biochemical riddles by the generation and then resolution of heavy and light molecules.[6] In his talk Monod posed the then unanswered question of whether enzyme induction represented activation of preexisting protein molecules or it reflected de novo synthesis of new molecules. Meselson came up with the idea of feeding cells with deuterium, a heavier isotope of hydrogen, which would be incorporated into enzyme molecules. Next, cells could be transferred into light hydrogen-containing medium and enzyme activity would be induced. If induction involved synthesis of new enzyme molecules, they would contain the lighter hydrogen. In principle, these newly made molecules could then be resolved from the preexistent heavier deuterium-containing molecules. On the other hand, if induction represented activation of preexisting enzyme, then only deuterium-containing protein would be detected. The principle of employing heavy and light isotopes was never put to practical testing to tackle the problem of enzyme induction. However, after Meselson was introduced by Delbrück to the question of the mechanics of DNA replication, he thought of applying the heavy−light separation approach to this problem. In principle, DNA was to be first labeled with a heavy constituent and then replicating cells would be transferred to a medium that contained a lighter precursor. Next, the heavy parental DNA was to be separated from light or mixed heavy/light progeny DNA by ultracentrifugation in a highly dense solution of salt such as cesium chloride (CsCl).

Originally, Meselson and Stahl chose to generate heavy parental DNA by incorporating into the DNA 5-bromouracil (5-BU), a heavier analog of thymidine. After transfer to thymidine-containing medium, the lighter newly synthesized DNA was separated from heavy parental DNA by centrifugation through a solution of concentrated (8.9 M) solution of CsCl.

Collaborating with Jerome Vinograd (Box 6.4), Meselson and Stahl mixed 5-BU-substituted heavy DNA and thymidine-containing lighter DNA in a solution of 7.0 M CsCl and subjected the mixture to analytical ultracentrifugation (Box 6.4).

The original expectation was that the heavy DNA would sink to form a pellet at the bottom of the tube, whereas the lighter DNA would float. To their surprise, Meselson, Stahl, and Vinograd discovered that ultracentrifugation generated a gradient of the CsCl itself and most importantly, that each DNA type settled as a narrow band at a region at which its density equaled the local

FIGURE 6.9   The analytical ultracentrifuge and Jerome Vinograd.

density of the CsCl. In addition, the molecular weight of the DNA could be derived from the standard deviation of the Gaussian distribution of its band.[20] Remarkably, apart from its original aim of separating old from new DNA, this work initiated the application of equilibrium density centrifugation to countless other research questions that required density separation of macromolecules.

While establishing the method of density equilibrium centrifugation, Meselson and Stahl also realized that using 5-BU to generate heavy DNA was problematic. First, it could not be ascertained that the analog was uniformly incorporated along the full length of the DNA. Second, being mutagenic, 5-BU could potentially damage the DNA. In searching for substitute for 5-BU, Meselson came upon the idea of forming heavy and light DNA by incorporating into it heavy or light isotopes of nitrogen. Among the 17 isotopes of nitrogen (most of which are artificial and short-lived), there are two naturally occurring stable ones: ^{14}N that comprises 99.63% of the nitrogen in nature and whose nucleus is composed of 7 protons and 7 neutrons, and a heavier ^{15}N isotope that constitutes 0.36% of the nitrogen in nature and has 7 protons and 8 neutrons.

To prove that ^{14}N- and ^{15}N-containing DNA were separable, Meselson and Stahl first formed heavy or light DNA by growing *Escherichia coli*

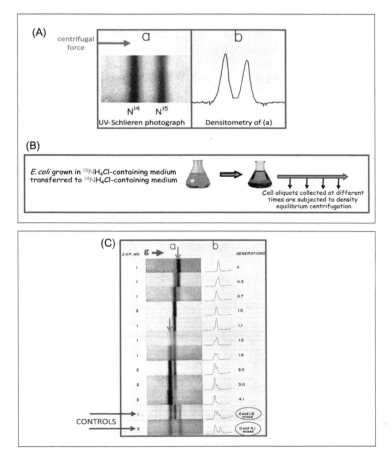

FIGURE 6.10   Density equilibrium centrifugal separation of 1:1 control mixture of *Escherichia coli*
^{15}N-DNA and ^{14}N-DNA. (A) (a) UV-Schlieren photograph of the ^{14}N and ^{15}N bands. (b)
Densitometry of the bands. (B) Scheme of the Meselson and Stahl experiment. (C) Density equilibrium centrifugal resolution of DNA at different times after transfer of the cells from ^{15}N-containing
medium to ^{14}N-containing medium. (a) UV-Schlieren photograph of the DNA bands. The red arrows
mark (from right to left) parental ^{15}N/^{15}N double-stranded heavy DNA; hybrid ^{15}N/^{14}N first-generation DNA; ^{14}N/^{14}N light progeny DNA in later generations. Controls were: (1) A mixture of DNA
from the 0 and 1.9 generations; (2) A mixture of DNA from the 0 and 4.1 generations. (b)
Densitometry tracings of bands separated in (a). *Panel (A): Adapted From Meselson M, Stahl FW.
The replication of DNA in* Escherichia coli. Proc Natl Acad Sci USA *1958;**44**(7):671−82. Panel
(C): From Watson JD, Crick FH. Genetical implications of the structure of deoxyribonucleic acid.*
Nature *1953;**171**(4361):964−7.*

in media that contained as source of nitrogen ^{15}NH$_4$Cl or ^{14}NH$_4$Cl, respectively. Next, a 1:1 mixture of ^{15}N- and ^{14}N-DNA was resolved by CsCl density equilibrium ultracentrifugation. Indeed, as illustrated in Fig. 6.10A, the
two DNA species migrated as two well-separated bands.

6.4.2 An Ultimate Elegant Experiment Provided Nearly Unambiguous Evidence for Semiconservative Mode of DNA Replication

In an experiment to determine the mode of replication (Fig. 6.10B and C), dividing *E. coli* cells were first grown for several generations in ^{15}N-containing medium and then transferred to a medium that contained the lighter ^{14}N isotope. Aliquots of the bacterial cultures were collected at different times after the transfer, the cells were lysed, mixed with concentrated CsCl, and subjected to density equilibrium centrifugation in analytical ultracentrifuge.[3] An observed drift from parental ^{15}N/^{15}N double-stranded DNA into ^{15}N/^{14}N hybrid duplex of parental/daughter strands and subsequent accumulation of ^{14}N/^{14}N light DNA in succeeding replication cycles was consistent with a semiconservative mode of replication. To prove that the lighter band that appeared after the first generation (single duplication) was indeed a double-stranded hybrid between a ^{15}N parental strand and ^{14}N daughter strand, bands were isolated and the DNA duplexes were heat-denatured. The centrifugal migration of the separated strands was compared with the migration of two controls; native and heat-denatured ^{15}N- or ^{14}N-DNA. Results showed unequivocally that the lighter band that appeared after a single round of replication was indeed comprised of one ^{15}N parental strand and one ^{14}N daughter strand.

At the time, the seemingly unambiguous results of this elegant experiment were promptly accepted as a definitive proof of semiconservative DNA replication. Rumor of the result spread rapidly and raised immediate excitement even before it came out in print. This was reflected for instance in a letter that Sydney Brenner wrote to Meselson more than 3 months before the actual formal publication of the paper (Brenner S. Feb. 18, 1958, letter to Meselson http://www.dnalc.org/view/16477-Gallery-21-Letter-from-Sydney-Brenner-to-Matt-Meselson.html):

Dear Matt: We were very excited to hear about your wonderful experiment with light and heavy DNA. As you say, perfect Watson-Crickery, and we eagerly await further news [...]

6.5 ALTERNATIVE INTERPRETATIONS OF THE MESELSON AND STAHL RESULTS WERE REFUTED

Despite its elegance and seeming finality, the Meselson and Stahl result was soon subjected to alternative interpretations that did not subscribe to semiconservative replication. First, Liebe F. Cavaliery (1919–2013) of the Sloan Kettering Institute argued that although the conserved DNA represented double strands, the heavy–light hybrids could have consisted of four-stranded complexes of two heavy parental and two light daughter strands[21] (Fig. 6.11A).

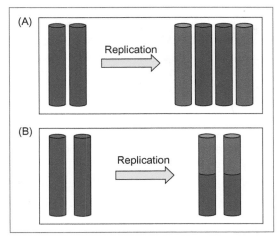

FIGURE 6.11   Schemes of alternative interpretations of the Meselson and Stahl result. (A) The $^{15}N/^{14}N$-DNA hybrids represent four-stranded DNA.[21] (B) The $^{15}N/^{14}N$-DNA hybrids represent end-to-end association of parental and daughter strands.

This possibility was ruled out by Meselson's graduate student John Menninger who used X-ray scattering to show that the hybrids comprised of two and not four strands.[22] Another alternative interpretation to the Meselson and Stahl result was that the hybrid bands could represent end-to-end association of parental and daughter strand rather than laterally copied one parental and one daughter strand (Fig. 6.11B).

Ronald Rolfe, who was also a graduate student of Meselson, discounted this possibility by showing that hybrid density was maintained in small fragments of ultrasonicated DNA and thus that the DNA strands were uniformly heavy or light along their full length.[23] The experimental rejection of alternative interpretations to the Meselson and Stahl result ultimately led in the early 1960s to consensual acceptance of the semiconservative mode as the way by which DNA is replicated in vivo.

6.6   POSTSCRIPT: THE DISTINCTIVE STATUS OF THE MESELSON AND STAHL EXPERIMENT

The quest for the mechanics of DNA replication that was practically concluded with the Meselson and Stahl result occupies a rather special place in the history of molecular biology. It represented a rare case in biological research in which well-defined and experimentally testable models were built purely on the basis of theoretical consideration and without prior experimental data. This case was also exceptional in that only after the competing abstract models were formulated were they put to critical experimental examination. Specifically, the three mutually irreconcilable models of semi-conservative, dispersive,[11,12] and conservative[13,14] replication were fully delineated between 1953 and 1955. Only then were they tested experimentally, first with inconclusive results,[15–17] later with suggestive but not definitive conclusion,[18] and lastly with the identification of the semiconservative mode as the actual mechanics of DNA replication.[3] Another uncommon aspect of the Meselson and Stahl endeavor was that it solved a difficult question in a single decisive experiment. Indeed the clean elegant quality of this experiment stands out in the annals of molecular biology. The status of this experiment as an exemplary standard-setter was best spelled out by Stahl himself (cited in Ref. 6, p. 446):

It set a standard for me in science which I don't know I can ever achieve again, but it's worth shooting for. Anytime I write a paper I remember that one, and say...come as close as you can. That's been awfully important. [...]

REFERENCES

1. Watson JD, Crick FH. Molecular structure of nucleic acids: a structure for deoxyribose nucleic acid. *Nature* 1953;**171**(4356):737–8.
2. Watson JD, Crick FH. Genetical implications of the structure of deoxyribonucleic acid. *Nature* 1953;**171**(4361):964–7.
3. Meselson M, Stahl FW. The replication of DNA in *Escherichia coli. Proc Natl Acad Sci USA* 1958;**44**(7):671–82.
4. Rubtsov NB. Organization of eukaryotic chromosomes: from Kol'tsov's studies up to present day. *Russ J Genet* 2013;**49**(1):10–22.
5. Watson JD. *The annotated and illustrated double helix.* New York, NY: Simon & Schuster; 2012.
6. Holmes FL. *Meselson, Stahl and the replication of DNA: a history of "the most beautiful experiment in biology".* New Haven & London: Yale University Press; 2001.
7. Stent GS. That was the molecular biology that was. *Science* 1968;**160**(3826):390–5.
8. Levinthal C, Crane HR. On the unwinding of DNA. *Proc Natl Acad Sci USA* 1956;**42**(7):436–8.
9. Watson JD, Crick FH. The structure of DNA. *Cold Spring Harbor Symp Quant Biol* 1953;**18**:123–31.
10. Holmes FL. The DNA replication problem, 1953–1958. *Trends Biochem Sci* 1998;**23**(3):117–20.
11. Delbrück M. On the replication of desoxyribonucleic acid (DNA). *Proc Natl Acad Sci USA* 1954;**40**(9):783–8.
12. Delbrück M, Stent GS. On the mechanism of DNA replication. In: McElroy WD, Glass B, editors. *A symposium on the chemical basis of heredity.* Baltimore, MD: The John Hopkins Press; 1957. pp. 699–736.
13. Bloch DP. A possible mechanism for the replication of the helical structure of desoxyribonucleic acid. *Proc Natl Acad Sci USA* 1955;**41**(12):1058–64.
14. Butler JA. The action of ionizing radiations on biological materials; facts and theories. *Radiation Res* 1956;**4**(1):20–32.

15. Stent GS, Jerne NK. The distribution of parental phosphorus atoms among bacteriophage progeny. *Proc Natl Acad Sci USA* 1955;**41**(10):704—9.

16. Fuerst CR, Stent GS. Inactivation of bacteria by decay of incorporated radioactive phosphorus. *J Gen Physiol* 1956;**40**(1):73—90.

17. Levinthal C. The mechanism of DNA replication and genetic recombination in phage. *Proc Natl Acad Sci USA* 1956;**42**(7):394—404.

18. Taylor JH, Woods PS, Hughes WL. The organization and duplication of chromosomes as revealed by autoradiographic studies using tritium-labeled thymidine. *Proc Natl Acad Sci USA* 1957;**43**(1):122—8.

19. Meselson M. Explorations in the land of DNA and beyond. *Nat Med* 2004;**10**(10):1034—7.

20. Meselson M, Stahl FW, Vinograd J. Equilibrium sedimentation of macromolecules in density gradients. *Proc Natl Acad Sci USA* 1957;**43**(7):581—8.

21. Cavalieri LF, Rosenberg BH, Deutsch JF. The subunit of deoxyribonucleic acid. *Biochem Biophys Res Commun* 1959;**1**(3):124—8.

22. Menninger J.R. *A determination of the mass per length of DNA using X-ray diffraction* [Ph. D. thesis]. Harvard University, <http://booksgooglecoil/books?id=9kg8twAACAAJ/>; 1964.

23. Rolfe R. The molecular arrangement of the conserved subunits of DNA. *J Mol Biol* 1962;**4**(1):22—30.

Chapter 7

Defining the Genetic Code

Evolving Ideas on the Nature of the Genetic Code and the Unraveling of Its General Attributes

Chapter Outline

M. Fry: Landmark Experiments in Molecular Biology. DOI: http://dx.doi.org/10.1016/B978-0-12-802074-6.00007-2
© 2016 Elsevier Inc. All rights reserved.

The question of the genetic code was arguably the most difficult among several problems that were raised by the Watson and Crick model of DNA as the genetic material.[1] There were two challenging elements to the riddle of the code. First, it had to be understood how the four nucleotides in DNA encrypt 20 different amino acids in proteins. Second, the general properties of a nucleic acid—based code had to be unraveled. Once these questions were answered, it was necessary to determine the specific code words that encrypt each of the amino acids in protein. Historically, these two tasks were accomplished in two distinct phases within less than 15 years after the introduction in 1953[2] of the double helix model of DNA. In the first period (1953—61), the nature and general properties of the nucleic acid code were largely elucidated. This task was initially pursued through purely theoretical conjectures and later by experimental work. A second exclusively experimental decrypting phase (1961—66) was dedicated to the identification of specific trinucleotide sequences ("codons") that encrypt each of the 20 amino acids. This chapter focuses on the earlier stage of the definition of the principles and the fundamental properties of the code. Chapter 11, "The Surprising Discovery of Split Genes and of RNA Splicing" describes the experimental deciphering of the code. For a broad analytical review of the historical, political, sociological, and scientific aspects of the two phases of the breaking of the genetic code, the reader is directed to Lili Kay's commendable book *Who Wrote the Book of Life? A History of the Genetic Code*.[3] A more recent historical account of the definition of the code, focusing on the contribution of Crick to this endeavor is in Robert Olby's biography of Crick (Ref. 4, pp. 221—288 and 261—288).

7.1 ROLES OF THEORY AND EXPERIMENT IN THE DEFINITION OF THE GENETIC CODE

The double helix model of DNA opened the problem of the mechanics of DNA replication and also raised the much more challenging question of how sequences of nucleotides in DNA can encode the sequences in protein of the chemically distinct amino acids. The complexity of this second problem was the likely reason for the relatively long period (1953–61) that was needed to define the code and to determine its general properties by a combination of theory and experimental work. Whereas the mode of DNA replication was determined in two successive and separate phases of theory first and then experiment (see Chapter 6: "Meselson and Stahl Proved That DNA Is Replicated in a Semiconservative Fashion"), the genetic code was defined by a mixture of linked theoretical deliberations and experiment. Very early attempts to solve the problem of the genetic code involved abstract number-theory approaches to cryptanalysis. It was soon realized, however, that experimental data was vitally needed for the elucidation of the nature of the code. This evolving understanding is best illustrated by thoughts that Sydney Brenner (Box 7.1) had on the subject in the 1950s. He opened an undated talk that was most likely delivered in 1957, by resolutely stating[5]:

The coding problem is essentially a theoretical problem. It concerns the relationship between nucleic acids and proteins and in particular [between] the order of the components in the nucleic acids and the order of the components in the proteins.

BOX 7.1 Sydney Brenner

Sydney Brenner, a founding father of molecular biology, was born in the South Africa town of Germiston to immigrant parents from Lithuania and Latvia. Graduating early from high school and intending to study medicine, he was judged too young to practice medicine, and was thus directed to first complete a BSc degree in anatomy and physiology and an MSc in cytogenetics at the University of Witwatersrand. After receiving in 1951 an MBBCh medical degree, he moved to England to earn a DPhil degree at Oxford University. After a short period back in South Africa, he settled at the Medical Research Council Unit in Cambridge. There he made seminal contributions to the definition of the triplet-based genetic code and to the discovery of mRNA. He was also at the cradle of Crick's adaptor hypothesis and his Central Dogma. At later time, he established the nematode *Caenorhabditis elegans* as model organism for the study of organismal development and particularly of the development of the neural system. For this pioneering work he shared with Robert Horvitz and John Sulston the 2002 Nobel Prize in Physiology and Medicine. During his illustrious career, he received numerous prizes and awards and was associated with several Universities and Institutes in the United States and in Asia. He authored in 2001 a short autobiography and a more comprehensive biography of his was published in 2010. There are also several filmed interviews with him that are available online. A 2007 interview http://downloads.sms.cam.ac.uk/1139457/1139464.mp4 is a part of the Sara Harrison and Alan Macfarlane series *Encounters with*

(Continued)

Sydney Brenner
(1927–)

FIGURE 7.1   Sydney Brenner.

However, while discussing particular features of the code such as whether the code words are nucleotide doublets or triplets, whether each amino acid is encoded by a single codon ("nondegenerate code") or by several codons ("degenerate code"), and the issues of codon overlap, Brenner went on to comment that:

> [...] we cannot make any progres [sic] with it [the coding problem] if we consider it to be completely abstract. [...] you cannot make a theory unless you have some facts, if possible a few facts of an unusual nature.

This realization led Brenner to conclude:

> I started this talk by saying that the coding problem was a theoretical problem, but my conclusion is exactly the opposite. As far as we can see the coding problem is now an experimental problem and the role of theory will be to organize and make sense of the facts as they come out of the laboratory.

It was thus that by applying his analytical prowess Brenner recognized already at an early stage that the coding problem could not be solved by theory alone and that the nature of the code could be construed only from meaningful experimental data.

7.2   CHALLENGES OF THE CODING PROBLEM

The crux of the coding problem was the question of how combinations of just four nucleotides in nucleic acid can be translated into arrangements of 20 different amino acids in protein. This problem was most pointedly posed by Crick and his associates in 1957[6]: "The problem of how, in protein synthesis, a sequence of four things (nucleotides) determines a sequence of many more things (amino acids) is known as the coding problem." To address this principal problem, several subordinate questions had to be answered.

1. The main challenge that investigators of the code faced in the years 1953–1961 was the problem of the organization of the code. One question was how many nucleotides constitute a codon that encrypts a single amino acid. A related problem was whether each amino acid was encoded by just a single unique codon (a state that was dubbed "nondegeneracy") or was each amino acid encoded by several different codons ("degenerate code"). Other issues were whether the code was a string of discrete or of overlapping codons and whether or not codons were separated from one another by "commas." Stepwise theoretical and experimental advances culminated in 1961 with the recognition that degenerate trinucleotide codons encoded amino acids and that the code consisted of a string of commaless nonoverlapping codons.

2. A substantial initial difficulty was posed by the distinctly different chemistries of nucleotides and amino acids. Early ideas on a code that were based on "lock-and-key" fit between amino acids and internucleotide spaces in DNA or RNA were soon rejected for their lack of stereochemical basis. The missing chemical link between the nucleotide and amino acid–based languages was provided by Crick's prophetic adaptor hypothesis[7] and by the eventual biochemical isolation by Zamecnik and Hoagland of the actual adaptor; transfer RNA (tRNA)[8] (see Chapter 8: "The Adaptor Hypothesis and the Discovery of Transfer RNA").

3. There was an initial uncertainty about the identity of the nucleic acid that served as the direct template for proteins. Early on the physicist George Gamow speculated that DNA itself was the immediate template for the synthesis of protein.[9,10] However, the biochemist Alexander Dounce raised at an even earlier time the possibility, which had later been proven correct, that it was not DNA but rather its RNA transcripts that served as templates for proteins.[11,12]

7.3 ROOTS OF THE IDEA OF BIOLOGICAL INFORMATION TRANSFER AND OF A GENETIC CODE

7.3.1 The Beadle and Tatum Experiment and Their "One Gene–One Enzyme" Edict

Two major experimental observations planted the idea that nucleic acids, especially DNA, might encode the sequence of amino acids in proteins. One was the 1944 discovery by Avery, MacLeod, and McCarty that DNA was the determinant of phenotypic properties in *pneumococcus* (see Chapter 3: "Avery, MacLeod, and McCarty Identified DNA as the Genetic Material"). The second observation was the finding by the geneticist George Beadle and the biochemist Edward Tatum (Box 7.2) that mutating discrete genes in the bread mold *Neurospora crassa* inactivated unique proteins.

BOX 7.2 George Beadle and Edward Tatum

Born to a Nebraska farming family, George Beadle received BSc and MSc degrees from the College of Agriculture in Lincoln, Nebraska. While serving as a teaching assistant at Cornell University, he completed in 1931 a PhD thesis on the cytogenetics of corn. For the next several years, he studied at Caltech the genetics of corn and of *Drosophila*. After a short visit to the Paris laboratory of Boris Ephrussi at the Institute of Biology and Physical Chemistry, he turned to the study of the genetics of the fungus *N. crassa*. After becoming in 1937 Professor of Biology at Stanford University, he collaborated with Edward Tatum in studies of *Neurospora* genetics. For this work that brought about the "One Gene–One Enzyme" hypothesis, the two shared with Joshua Lederberg the 1958 Nobel Prize in Physiology and Medicine. In 1946 he moved back to Caltech where he conducted studies on the genetics of corn. In 1961 he became Chancellor and then President of the University of Chicago.

Born to a father who was Professor of Pharmacology at the University of Wisconsin, Edward Tatum studied in the same University and at Chicago University, obtaining BA degree in Chemistry in 1931, an MSc degree in Microbiology in 1932, and a PhD degree in Biochemistry in 1934. After short periods at the University of Wisconsin and at Utrecht University, he worked between 1937 and 1945 at Stanford University first as a Research Associate and then as an Assistant Professor. It was there that he collaborated with Beadle, taking charge of the biochemical aspects of the studies on of *Neurospora*. His research work focused on the biochemistry, nutrition, and genetics of microorganisms and of *Drosophila*. He pursued these interests when he moved in 1945 to Yale University and after his return in 1948 to Stanford as Professor of Biology and then Professor of Biochemistry (Fig. 7.2).

(A) George W. Beadle (1903–1989)

(B) Edward L. Tatum (1909–1975)

FIGURE 7.2 George Beadle and Edward Tatum.

Specifically, these two Stanford scientists irradiated with X-rays haploid strain of the fungus *N. crassa*. Crossing wild type with irradiated fungi produced mutants that lost their ability to synthesize the amino acid arginine and that had, therefore, to be grown in arginine-containing medium. Crosses between mutant fungi identified four distinct mutant strains that were defective in different steps of the arginine biosynthetic pathway.[13] From these results, Beadle and Tatum reasoned that each mutant lacked a specific gene that encoded one of the four enzymes that participated in the production of arginine. Based on this finding, these two scientists set forth their famous dictum: "One gene−One enzyme."[14−16]

7.3.2 Early Ideas on the Storage and Transmission of Genetic Information

Early ideas on the transfer of information in biological systems were far removed from today's understanding of the biochemical basis of a nucleic acids code and its translation into protein. In his famous and influential 1944 book *What Is Life?*,[17] the physicist and 1933 Nobel laureate Erwin Schrödinger (1887−1961) speculated on the mechanism of heredity. Although he recognized the central role of mutations in evolution and attempted to explain the "code script" of life, he was unaware of nucleic acids and at the time could not have any inkling about the actual chemical basis of storage and translation of genetic information. The identification by Avery, MacLeod, and McCarty of DNA as the likely genetic material (see Chapter 3: "Avery, MacLeod, and McCarty Identified DNA as the Genetic Material") and subsequent developments brought nucleic acids to the fore.

An early proposal that nucleic acids could determine the phenotypic properties of cells and organisms was made by the biochemist Kurt Stern (1902−81), an expert on proteins and on biological oxidation who escaped Nazi Germany to work first in Yale University and later at the Brooklyn Polytechnic Institute. In a series of lectures that he presented at Yale, in the New York Mount Sinai Hospital, and at the Hebrew University in Jerusalem, Stern crystallized his thoughts on "nucleoprotein" (actually nucleic acids) as the building blocks for highly diverse genes. In a 1947 paper he summarized his ideas by speculating that variable sequence (or orientation) of bases in nucleic acid might be responsible for the diversity of genes[18]:

If one admits the possibility that the bases may vary in their sequence or orientation with reference to backbone, this would constitute a principle of modulation which does not affect the over-all stoichiometry of composition of the nucleoprotein chain, and which is fully capable of accounting for all possible genotypes.

Incorporating the imprecise and sometime erroneous ideas of the time on the structure of nucleoproteins, Stern constructed model helices that possessed variable grooves. Based on this model and drawing an analogy from the contemporaneous technology of the phonograph, he offered a mechanism for the hereditary transmission of information and its expression[18]:

Just as the sound recorded on a phonograph disc may be reproduced in two ways, viz., by direct impression from master record and by playing it back, the genic modulation of chromosomal nucleoproteins would be transmitted to the daughter cells by reproduction through contact with the newly laid-down nucleoprotein chain, whereas it might exert its specific biological functions on the cell by directing the course of chemical reactions occurring at its surface, much the same as an active surface promotes in a specific manner heterogeneous catalysis.

It should be noted, however, that despite his elegant early idea of replication and although he explicitly raised the point that permutated sequences of bases in nucleic acids could be the basis for the diversity of genes, Stern did not make the linkage between coding nucleic acids and encoded proteins.

The idea that sequences of amino acids in protein may be determined by the base sequences of nucleic acids was made in 1950 by the University of Oxford physical chemist and 1956 Nobel Prize laureate, Cyril Hinshelwood (1897–1967). His work on the kinetics of polymerization chain reactions led to his thinking about bacterial replication. Putting his sights on nucleic acids and proteins, he visualized their coordinated autosynthesis[19]:

[…] In the synthesis of protein, the nucleic acid, by a process analogous to crystallization, guides the order by which the various amino acids are laid down; in the formation of nucleic acids the converse holds, the protein molecule governing the order in which the different nucleotides are arranged […] This suggests some sort of correspondence between the units in the two kinds of polymer. In a protein about 23 different amino acids may occur, whereas in a nucleic acid only 5 basic units are found (including DNA's and RNA's bases) […]Clearly these cannot be one-to-one correspondence between the position of an individual nucleotide in the nucleoprotein and the position of an individual nucleotide in the nucleic acid part.

To explain how five nucleotides encode 23 (his number) amino acids, Hinshelwood assumed that the growing protein chain was lodged between two nucleic acid polymers. The number of possible internucleotide permutations; $5^2 = 25$ wondrously approximated the number of amino acids. However, as Kay pointed out,[3] this view of bidirectional synthesis of nucleic acids to protein and protein to nucleic acids was far from the idea of a code. Rather, it described bidirectional correspondence between the polymerization of physically interacting nucleic acids and protein.

Two year after Hinshelwood made his proposal and one year before the discovery of the structure of DNA by Watson and Crick, the University of Rochester biochemist Alexander Dounce (Box 7.3) applied more biologically inclined terms to suggest how a nucleic acid *template* might dictate the ordered arrangement of amino acids in protein.

BOX 7.3 Alexander Dounce

After concluding his undergraduate studies in Hamilton College, New York, Alexander Dounce did a doctoral thesis in organic chemistry in Cornell University under the guidance of the 1946 Nobel Prize laureate James Sumner who was the first to crystallize an enzyme (urease) and to prove that enzymes are proteins. After obtaining a PhD degree in 1935, he crystallized with Sumner the enzyme catalase. Moving in 1941 to the University of Rochester, he conducted during World War II military research on uranium poisoning. After the war his research focused on the cell nucleus and on mitochondria. Within this framework he developed methods to isolate nuclei, devising the still widely used Dounce tissue homogenizer. He was also the first to use the detergent sodium dodecyl sulfate to isolate DNA. He is particularly remembered for his early insightful ideas about the nature of the genetic code and for his proposition that the direct template for protein synthesis is RNA and not DNA (Fig. 7.3).

Alexander Dounce
(1909–1997)

FIGURE 7.3   Alexander Dounce.

Dounce did his PhD thesis at Cornell University under the direction of the protein crystallization pioneer and 1946 Nobel laureate James Sumner. During his final doctoral examination, Dounce was asked by Sumner how proteins can synthesize other proteins. This question should be understood in the context of the period's belief that the synthesis of polypeptide chains was conducted by the reverse action of proteolytic enzymes (see Chapter 8: "The Adaptor Hypothesis and the Discovery of Transfer RNA"). Thus Sumner was inquiring how is it that proteinaceous enzymes that are made of protein templates, synthesize proteins molecules, and so on ad infinitum. According to Marshall Nirenberg,[20] this question stuck in Dounce's mind since 1935 until he conjectured almost two decades later that nucleic acids serve as templates for protein synthesis. Drawing on Avery's identification of DNA as the genetic material[21] and on Beadle and Tatum's "One Gene−One Enzyme" hypothesis,[13] Dounce suggestion that nucleic acids were templates for proteins thus circumvented the difficulty of explaining how proteins can direct the synthesis of other proteins. Specifically, he speculated that[11]:

> [Avery's work makes it] *increasingly probable that nucleic acids may be the patterns or templates which must govern gene action and hence protein synthesis, since according to Beadle et al a gene appears to act by causing the synthesis of specific enzymes or other proteins.*

As will be discussed later, several months after Dounce published this hypothesis in the journal *Enzymologia*, he offered in an article in *Science* a more specific and prescient prediction that the direct template for protein synthesis should be RNA and not DNA.

7.3.3 George Gamow's Idea of a Code Based on Stereochemical Fit Between Amino Acids and Crevices in DNA

The discovery of the double helical structure of DNA by Watson and Crick[2] and their subsequent statement[1]; "It seems likely that the precise sequence of bases [in DNA] is the code which carries the genetical information" inspired the Russian-American theoretical physicist George Gamow (Box 7.4) to speculate about the nature of the code.

BOX 7.4 George Gamow

The physicist George Gamow (originally named Georgiy Antonovich Gamov) was born in Odessa in the Ukraine. After graduating in 1928 from Leningrad University, he traveled to the University of Göttingen that was then a hub of theoretical physics. There he developed a quantum theory of radioactivity that successfully explained the different half-lives of radioactive isotopes. He next worked at the Copenhagen Institute of Theoretical Physics where he came up with a "liquid drop" model of atomic nuclei that offered a theoretical basis for nuclear fission and fusion. Although he was elected in 1931 at the age of 28 as a corresponding member of the USSR Academy of Sciences, he felt stifled in the Soviet Union and had made several failed attempts to defect. Finally, in 1933 he and his physicist first wife successfully left Russia, first to Europe and then to the United States. During his tenure as Professor of Physics at George Washington University and later at the University of Colorado, he contributed to nuclear physics, to nucleosynthesis, and to the Big Bang theory. Most prominent was the famous 1948

(Continued)

BOX 7.4 (Continued)

paper "The Origin of Chemical Elements" by Gamow, Ralph Alpher, and an invented bogus author, the eminent physicist Hans Bethe who was added because his last name began with a B (beta) (thus the "alpha–beta–gamma paper"). That article explained present day levels of hydrogen and helium by reactions that occurred during the big bang. In the 1950s and 1960s his interest shifted to the problem of the genetic code. Although he did not make any decisive discovery himself, his enthusiastic promotion of ideas and creation of the RNA Tie Club significantly advanced the definition of the code. He was also an exceptionally gifted popularizer of science, publishing more than 15 popular science books on nuclear physics, astronomy, and astrophysics. A series of books had a "Mr Tompkins" as protagonist who explored different fields of science including biology (Fig. 7.4). He was famed in the scientific community for his pranks and practical jokes. In his last years he drank heavily. It is rumored that before he died in 1968 he said "*Finally my liver is presenting the bill.*" By that he summed up "a life of carefree drinking, eating, good humor, sports, and an unusual friendly egoism."

(A)   (B)

George Gamow   Gamow's 1953
(1904–1968)   popular book on biology

FIGURE 7.4   George Gamow and his book on biology.

Shortly after Watson and Crick published their model of DNA, Gamow wrote to them to introduce himself as a physicist who is uninformed about biology though he authored a popular book of biology: *Mr. Tompkins Learns the Facts of Life* (Fig. 7.4B). Having his interest piqued by the assumed coding capacity of DNA, he ventured to speculate on the possible nature of the code. Specifically, Gamow proposed in his letter that coding by the four nucleotide "letters" should be perceived as a number-theory problem (a facsimile of Gamow's Jul. 8, 1953, letter to Watson and Crick is in Ref. 22):

> [. . .] *each organism will be characterized by a long number written in quadrucal [?] system with figures 1, 2, 3, 4, standing for the four bases (or by*

several such numbers, one for each chromosome). It seems to me more logical to assume that different properties (single genes?) of any particular organism are not "located" in definite spots of chromosome, but are rather determined by different mathematical characters of the entire number.

Yet, already at that early stage Gamow recognized that the problem of how four nucleotides in DNA encode the chemically different 20 amino acids could not be treated as a purely numerical question. In Oct. 1953 he wrote to Linus Pauling to present his just hatched idea of a key-and-lock mechanism for the encoding of 20 amino acids by just four nucleotides in DNA. Page 3 of the letter is shown in Fig. 7.5A.

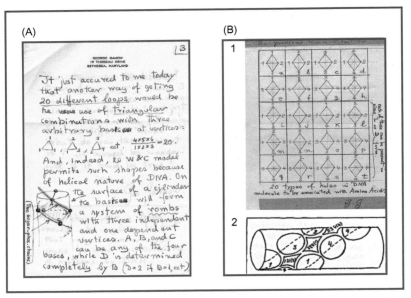

FIGURE 7.5 Gamow's early model of the code proposed that the different amino acids are accommodated in a stereospecific manner within diamond-shaped crevices that are formed between 3 + 1 adjacent bases in the DNA double helix. (A) Gamow's description of his model in an Oct. 1953 letter to Pauling (page 3 of the 5-page letter; a facsimile of the full letter can be found in the Oregon State University Collection: Ava Helen and Linus Pauling Papers 1873–2013: http:// scarc.library.oregonstate.edu/coll/pauling/dna/corr/sci9.001.43-gamow-lp-19531022.html.). Note at the middle of the page triangles with different groupings of 3 bases. Calculation showed that there were a total of 20 different combinations of the bases—a number that fitted the number of different amino acids. At the left bottom of the page Gamow drew a sketch of a DNA double helix with 3 + 1 bases marked at the four corners of a diamond-shaped cavity. (B) (1) Page 5 of Gamow's letter: Scheme of the 20 different diamond-shaped cavities that can be formed by combinations of 3 + 1 bases. (2) Gamow's sketch of the imagined internucleotide cavities in cylinder-shaped DNA double helix. *From Gamow G. Possible mathematical relation between deoxyribonucleic acid and proteins.* Biol Medd Kgl Dan Vidensk Selsk *1954;22(3):1−11.*

Gamow specified in the letter the crux of his idea of the code as follows:

> ...*Ever since I read the article of Watson and Crick last June, I was trying to figure out how a long number written in a four digital system (i.e. nucleic acid molecule) can determine unickly [sic] a correspondingly long word based on 20-letter-alphabet (i.e. an enzyme molecule). Along the lines of key & lock ideas, one could think about different amino acids as fitting into quadrangular loops formed by four bases in the DNA chain. [...] It just accurred [sic] to me today that another way of getting 20 different loops would be the use of triangular combinations with three arbitrary bases at vertices.... (4 × 5 × 6/ 1 × 2 × 3 = 20). And, indeed, the W&C model permits such shapes because of helical nature of DNA. On the surface of a cylinder the bases will form a system of rombs with three independent and one dependent vertices. A, B, and C can be any of the four bases, while D is determined completely by B (D = 2 if B = 1, etc.). Thus, there are 20 different types of such rombs, and I wonder [whether] the 20 amino acids "vital for life" are just those which would fit into these 20 different "locks". The shape of these loops should also explain the fact that only L-amino acids could be used for building of proteins by DNA.*

Shortly thereafter Gamow expounded on his idea of the code in two formal papers that were published in 1954.[9,10] As was already implied in his letter to Linus Pauling, Gamow assumed that DNA served as the direct template for protein synthesis. As to the mechanism of translation, he adhered to the general principle of key-and-lock, envisioning stereospecific accommodation of amino acids within crevices in the DNA. In his model, spaces between two adjacent base pairs in the DNA double helix were cavities in the shape of diamonds. The contours of each cavity were determined by just three nucleotides, whereas the identity of the fourth nucleotide was inevitably dictated by its hydrogen-bond pairing with the third base. This model of different combination of the three independent and one dependent base yielded 20 different cavities—a value that happily matched the number of different amino acids (Fig. 7.5B). This hypothetical scheme was of an *overlapping* triplet code based on a combinatorial scheme in which 4 nucleotides arranged 3-at-a-time would specify 20 amino acids.

A central feature of the Gamow model was that his code was overlapping. In general, a code can be fully overlapping, partially overlapping, or nonoverlapping. These three types of a triplet-based code are schematically depicted in Fig. 7.6. Gamow's scheme of the code was of the overlapping type since any base pair was shared by two adjacent diamond-shaped cavities. This feature of the code was criticized by Crick and was later shown to be in conflict with experimental data (see below).

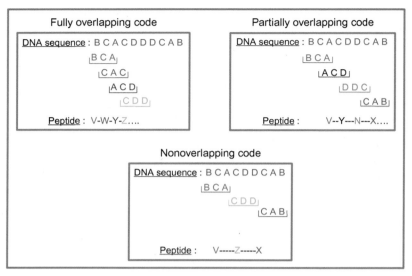

FIGURE 7.6   Three theoretically possible types of three-base code "words." A three nucleotide codon in overlapping or in partially overlapping codes shares, respectively, 2 or 1 bases with a neighboring codon. Thus a single base mutation in fully or partially overlapping code would, respectively, alter in the encoded protein three or two amino acids. By contrast, a single base mutation in a nonoverlapping code should alter no more than a single amino acid. Also, it is evident that the number of combinations of next neighbor amino acids in a protein is constrained in fully or partially overlapping codes but is unrestricted in a nonoverlapping code.

7.3.4   Crick's Astute Criticism of Gamow's Hypothesis

Gamow and Crick met in person for the first time in Dec. of 1953 when Crick was working at the Brooklyn Polytechnic. An animate debate on possible features of the code that took place on that occasion rekindled Crick's interest in the coding problem.[4] Later Crick was recruited by Gamow to the RNA Tie Club (see below) and much of his intellectual energy was then invested in the question of the code. In early 1955 Crick evaluated and incisively criticized cardinal elements of Gamow's model of an overlapping code of stereochemical fit between amino acids and cavities in DNA or RNA.

In a manuscript titled *"On Degenerate Templates and the Adaptor Hypothesis"* that was circulated in early 1955 among members of the RNA Tie Club, Crick presciently hypothesized that adaptor molecules exist (later identified as tRNA that are charged with specific amino acids by specialized enzyme; see Chapter 8: "The Adaptor Hypothesis and the Discovery of Transfer RNA," for a full description of the hypothesis). Apart from this epic prediction and except for a discussion of the question of degeneracy of the code (see below), the first part of Crick's paper was dedicated to critical examination of Gamow's model of a code based on stereochemical match between amino acids and overlapping cavities in the DNA. A principal feature of an overlapping code and to a lesser extent of a partially overlapping code is that they impose restrictions on the number of possible combinations of pairs of adjacent amino acids in protein. For an overlapping triplet-based code, only $4^4 = 256$ overlapping codon combinations were possible. By contrast, with no constraints on the next-neighbor combinations of the 20 different amino acids, all $20^2 = 400$ possible dipeptide permutations should be present in proteins. Crick was well informed about the progress and results of the protein sequencing work that Fred Sanger was accomplishing at the time in Cambridge (Ref. 4, p. 227). Consulting the then already available amino acid sequences of the insulin A and B chains and of β-corticotrophin, he showed that whereas Gamow's scheme imposed restrictions on possible next-neighbor groupings of amino acids, no such constraints appeared to limit the identities of contiguous residues in pairs or triplets of amino acids in the sequenced proteins. The narrow sample of sequenced proteins that was available in early 1955 was broadened later when amino acid sequences of additional proteins were determined. Examining this wider catalog of sequences, Sydney Brenner affirmed in 1957 Crick's early conclusion that the code was nonoverlapping[23] (*vide infra*).

In addition to his argument on the unlikelihood of an overlap, Crick also pointed out that Gamow's code did not distinguish between opposite directions of amino acid sequences. Thus it implausibly permitted equal translation of both A−B−C−D and D−C−B−A type chains. Notably, Crick concluded that simple analysis of extant data was superior to the treatment of the code question as an abstract number-theory problem (citations are from a facsimile of the paper archived in the Francis Crick Papers collection, Profiles in Science, National Library of Medicine http://profiles.nlm.nih.gov/ps/access/SCBBGF.pdf):

> *...I have set out these [analyses of extant amino acid sequences] at length, not to flog a dead horse, but to illustrate some of the simplest ways of testing a code. It is surprising how quickly, with a little thought, a scheme can be rejected. It is better to use one's head for a few minutes than a computing machine for a few days!*

In addressing the structural aspect of Gamow's model, Crick also rejected the idea of possible stereochemical fit between amino acids and cavities in the DNA duplex:

> *...I cannot conceive of <u>any</u> structure [for RNA or DNA] acting as a direct template for amino acids, or at least as a specific template. In other words, if one considers the physical-chemical nature of the amino acid side chains we do not find complimentary features on the nucleic acid. Where are the knobly hydrophobic surface to distinguish valine from leucine and isoleucine? Where*

are the charged groups in specific positions to go with the acidic and basic amino acids?

In rejecting Gamow's scheme of stereochemical matching between amino acids and DNA, Crick raised in the same communication an alternative model of reading of the code. In a farsighted theoretical prediction he hypothesized that on the one hand adaptor molecules form complementary hydrogen bonds with the nucleic acid template and on the other hand they carry an amino acid that eventually forms peptide linkage with the growing polypeptide chain. The adaptor molecules were thus proposed to serve as bridges between the disparate languages of nucleic acid and protein. As is described in Chapter 8, "The Adaptor Hypothesis and the Discovery of Transfer RNA," the actual adaptor molecules—tRNA—were eventually isolated and characterized in independent parallel experiments of Zamecnik and Hoagland.

It should be noted that despite Crick's rather shattering criticism of the Gamow model of steric fit between DNA and amino acids, he also pointed to some important ideas that Gamow introduced. Such were the notion of a degenerate code (encoding of the same amino acid by different nucleotide sequences) and the concept of overlapping code. Crick also generously acknowledged Gamow's valuable contribution to the advancement of the study of the code problem:

[. . .] It is obvious to all of us [members of the RNA Tie Club] *that without our President* [Gamow] *the whole problem would have been neglected and few of us would have tried to do anything about it.*

7.4 RNA ASSUMED CENTER STAGE

7.4.1 Alexander Dounce Proposed That RNA, and Not DNA, Functioned as the Direct Template for Protein Synthesis

Almost a year before Gamow published his model of DNA as the template for amino acid polymerization, Alexander Dounce raised an alternative speculation that the genetic information in DNA is copied into RNA molecules that then serve as the direct templates for protein synthesis. In an early version of what will later become Crick's Central Dogma, he hypothesized in a 1953 *Nature* paper that[12]:

. . .it could conceivably happen that the deoxyribonucleic acid gene molecules would act as templates for ribonucleic acid synthesis, and that the ribonucleic acids synthesized on the gene templates would then in turn become templates for protein synthesis.

More strikingly, in the same theoretical paper Dounce also visualized a genetic code in which nucleotide triplets in RNA encoded each of the 20 amino acids.[12] In his model, amino acids paired during protein synthesis with an individual nucleotide in RNA and the two bordering left and right nucleotides determined the specificity of binding. Although Dounce was in error about the determinants of codon specificity, he correctly predicted that the coding units are nucleotide triplets. Marshall Nirenberg, who eventually deciphered the code (see Chapter 11: "The Surprising Discovery of Split Genes and of RNA Splicing"), admitted to not having been aware of Dounce's speculation until after the code was broken. Retrospectively, however, he expressed admiration for the foresight of Dounce's prediction of a triplet-based RNA code at a time when very little was known about protein synthesis.[20]

7.4.2   Watson Was Also Focusing on RNA as the Bearer of the Genetic Code

After his historical sojourn in Cambridge ended in 1954, Watson embarked on a second postdoctoral period in Caltech. There he ventured to solve the structure of RNA by taking an approach similar to the one that served Crick and him in the elucidation of the structure of DNA. Thus in collaboration with Alexander Rich (see Chapter 9: "The Discovery and Rediscovery of Prokaryotic Messenger RNA") who was then a postdoctoral fellow in the laboratory of Linus Pauling, the two young investigators attempted to determine the structure of RNA fibers by interpreting their X-ray diffraction patterns.[24] Even before they had any fiber diffraction patterns, Watson had the idea of two forms of cytoplasmic RNA; a single-stranded helical structure and a double-stranded form (for a detailed description of this idea, see Ref. 4, pp. 228−231). However, lack of base complementarity in RNA samples from different sources and their X-ray images led Watson and Rich to conclude that RNA was a single-stranded helix.[25] Considering the possibility that RNA rather than DNA may be the direct template for protein synthesis, Watson entertained a Gamow's style idea of stereochemical fit between crevices in the RNA helix and amino acids. Specifically, he speculated that the RNA single-strand helix was endowed with differently shaped cavities into which different amino acids could fit in a stereospecific manner. It should be noted that Watson came up with this scheme some months before Crick circulated his above described critical paper of Gamow's scheme of key-and-lock fit between crevices in DNA and amino acids. Writing to Crick who was then in Cambridge, Watson described structural features of RNA as construed from its X-ray diffraction patterns. Presuming that the RNA helix possessed cavities, he speculated on possible stereochemical correspondence between these nooks in the RNA and amino acids. Notably, although Watson was quite satisfied that the RNA helix had cavities in the approximate size of an amino acid, he acknowledged the difficulty of proving that there were 20 differently shaped cavities in the RNA. In his Dec. 11, 1954, letter to Crick, Watson wrote (a facsimile of the letter is in the Francis Crick Papers, Profiles in Science, National Library of Medicine; http://profiles.nlm.nih.gov/ps/retrieve/ResourceMetadata/SCBBJN#transcript):

The main point is that when we tilt the bases almost 11 [$\overset{\circ}{A}$] to fiber axis we produce very suitable A.A. [amino acids] cavities. In fact we can say that if we were to create a A.A. [he probably meant protein] *making beast we should arrange for a pitch of not greater than 11-12 $\overset{\circ}{A}$ and probably not less than 8 $\overset{\circ}{A}$ (if smaller will be very tricky to fit A.A. backbone [...] So I do not think the x-ray picture is against us. However it will be a nasty job to prove existance* [sic] *of 20 specific equities* [Watson's likely meaning here was 'cavities'].

It appears that Crick gave a serious thought to Watson's idea of replacing DNA by RNA in a Gamow style model of an overlapping code. This is evident from the following underlined remark in his 1955 informal RNA Tie Club communication (http://profiles.nlm.nih.gov/ps/access/SCBBGF.pdf):

Gamow boldly assumed that code would be of the overlapping type. That is, if we denote the sequence of base pairs by 1 2 3 4 5 6 [...] we assumed that the first amino acid was coded by 1 2 3, and the next by 2 3 4, not by 4 5 6. <u>*Watson and I, thinking mainly about coding by hypothetical RNA structures rather than by DNA,*</u> *did not seriously consider this type of coding.*

7.4.3   Gamow Too Changed His Focus to RNA

By May 1954, after a visit to Caltech, Gamow became convinced that rather than DNA, it was single-stranded RNA that had crevices into which amino acids could fit. His revised views were incorporated at a later time into a review article on the coding problem that he authored with Alex Rich and MartynasYčas (1917−2014).[26] Gamow's conversion to RNA entailed his enterprising establishment in 1954 of the RNA Tie Club, a virtual "think tank" dedicated to solving the puzzle of the code. The glue that held together the club was Gamow's infective enthusiasm for the code and his interpersonal and organizational skills. The club had 20 members (for the 20 amino acids) and 4 honorary members (for the 4 bases). The members were from diverse scientific disciplines and from geographically distant institutions (Fig. 7.7).

In addition to biochemists, biophysicists, and biologists such as Francis Crick ("tyrosine," also holding the title of pessimist), James Watson ("proline," the designated Club's optimist), Sydney Brenner, Max Delbrück, Erwin Chargaff, Alexander Dounce, and Alexander Rich, Gamow also recruited to the club leading theoretical physicists and mathematicians such as Richard Feynman, Edward Teller, and Nicholas Metropolis. Although he was not a club member, the mathematician, physicist, and inventor John von Neumann was also attracted by Gamow to the problem of the genetic code.

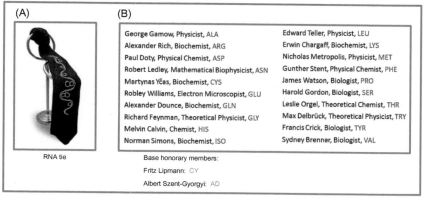

(A)	(B)	
	George Gamow, Physicist, ALA	Edward Teller, Physicist, LEU
	Alexander Rich, Biochemist, ARG	Erwin Chargaff, Biochemist, LYS
	Paul Doty, Physical Chemist, ASP	Nicholas Metropolis, Physicist, MET
	Robert Ledley, Mathematical Biophysicist, ASN	Gunther Stent, Physical Chemist, PHE
	Martynas Yčas, Biochemist, CYS	James Watson, Biologist, PRO
	Robley Williams, Electron Microscopist, GLU	Harold Gordon, Biologist, SER
	Alexander Dounce, Biochemist, GLN	Leslie Orgel, Theoretical Chemist, THR
	Richard Feynman, Theoretical Physicist, GLY	Max Delbrück, Theoretical Physicist, TRY
	Melvin Calvin, Chemist, HIS	Francis Crick, Biologist, TYR
	Norman Simons, Biochemist, ISO	Sydney Brenner, Biologist, VAL
RNA tie	Base honorary members:	
	Fritz Lipmann: CY	
	Albert Szent-Gyorgyi: AD	

FIGURE 7.7   (A) Members of the Club sported a woolen RNA necktie with a green and yellow helix of Gamow's design. (B) List of members of the RNA Tie Club. Each of the 20 members was assigned one of the 20 amino acids. Two (out of four) honorary members were assigned pyrimidine or purine base; Fritz Lipmann got cytosine and Albert Szent-Gyorgyi adenine.

With little regard to the inherent biochemical nature of the code, Gamow and his physicist and mathematician colleagues tried to uncover its features by applying number-theory and statistical approaches (see for instance, Ref. 27). As was comprehensively detailed by Lilly Kay,[3] much of the theoretical work of these scientists was influenced by developments in military cryptanalysis and by the early use of computers. Together with Feynman and with Teller Gamow contemplated, and eventually rejected, different theoretical schemes for the code. Also, his consultations with von Neumann and Metropolis resulted in the testing of some theoretical models of the code in the Maniac computer at the Los Alamos Laboratories.[3] Yet, the treatment of the genetic code as an abstract number-theory problem failed to disclose most of its essential properties.[3] Although Gamow did not succeed in exposing attributes of the code, his contribution as an initiator and a facilitator of the study of the genetic code was substantial.

7.5   THEORY PREDICTED A COMMALESS TRIPLET-BASED BUT NONDEGENERATE CODE

In May of 1956 Crick, John Griffith, and the eminent chemist and molecular biologist Leslie Orgel (1927−2007) circulated among members of the RNA Tie Club a speculative communication on the combinatorial properties of the genetic code (available online at: http://profiles.nlm.nih.gov/ps/access/ SCBBXN.pdf). One year later their theoretical conjectures were presented in a formal publication.[6] This paper was farsighted in predicting that different nucleotide triplets encode each of the amino acids and that nonoverlapping codons are translated continuously without intervening "commas" that separate codons from one another. On the other hand, this article[6] proposed a scheme of *nondegenerate* code under which each amino acid was encoded by only a single codon.

Crick, Griffith, and Orgel proposed two basic tenets for the code: (1) *Code words are comprised of three nucleotides*: A code word could neither be a single nucleotide that suffices for only 4 amino acids, nor could it be 2 nucleotide that would encode only 16 amino acids. A minimum number of nucleotides per code word should thus be three. (2) *The code is probably nonoverlapping*: An overlapping code imposes strict restrictions on the allowed next-neighbor combinations of amino acids in protein chains. Yet, protein sequences that were already available in 1957 revealed no such restrictions. It was thus concluded that the code was most likely nonoverlapping, although a partial overlap could not be completely ruled out at that stage.

With these tenets in place, it had to be explained how only 20 out of the $4^3 = 64$ possible trinucleotide permutations encode amino acids. Theoretically, there were two alternatives; either each amino acid was encoded by several different nucleotide triplets (a so-called *degenerate code*) or each amino acid was encoded by a single triplet and the remaining 44 trinucleotides were nonreadable (a *nondegenerate code*). Crick and associates proposed a scheme of nondegenerate code by setting abstract rules that rendered most of the triplets nonreadable and left the number of meaningful ("sense") triplets at exactly 20. In stating the essence of their speculative scheme, Crick et al. wrote[6]:

> We suppose that there are certain sequences of three nucleotides with which an amino acid can be associated and certain others for which it is not possible.

To explain how only 20 triplets could be meaningful whereas the other 44 could not encode amino acids, the following rules were proposed:

1. Sequences of three identical nucleotides (eg, AAA) are nonsensical. Suppose that such a triplet encoded amino acid α, than a dipeptide $\alpha\alpha$ in a protein should have been be encoded by the sequence AAAAAA: such a combination was assumed to be prohibited since the reading frame was in danger of being subject to shifts and misreading. Thus it was hypothesized that all four repeated single nucleotide triplets (AAA, TTT, CCC, and GGG) must be eliminated from the list of allowable triplets.
2. Removal of the four triplets of identical nucleotides left 60 triplets that Crick et al. grouped into 20 sets of permutated three triplets. In each set only one triplet was presumed to encode an amino acid. Taking one such set ABC, BCA, and CAB as an example, if BCA encoded amino acid β then the dipeptide $\beta\beta$ should have been encoded by the sequence BCABCA. Here again there could be a chance for out-of-frame reading mistakes: BC<u>ABC</u>A (CAB) and BCA<u>BCA</u> (ABC).

Thus CAB and ABC were assumed as nonreadable and only BCA remained as an operational codon. This scheme left, therefore, only a single sense triplet within each set bringing the number of meaningful triplets to exactly 20 (60 − 40 = 20)!

As illustrated in Fig. 7.8A, since only "sense" triplets were read whereas "nonsense" trinucleotides were precluded, the proposed scheme obviated the need for "commas" that separate codons from one another.[6] Crick later found that the code was indeed commaless but not because of unreadable triplets but because reading of the code started at a fixed point (see below).

Notably, in assessing their clever abstract scheme, Crick Griffith and Orgel were well aware of the limitations of applying theory without experimental corroboration. This appreciation was evident by their choice of Francis Bacon's aphorism as a motto to their RNA Tie Club communication: "It cannot be that axioms established by argumentation can suffice for the discovery of new works, since the subtlety of nature is greater many times

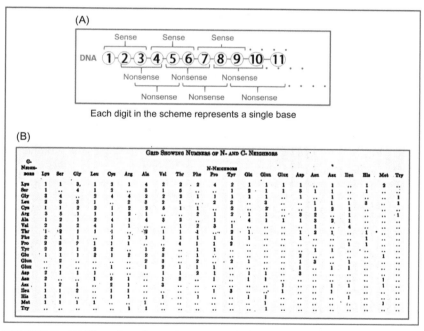

FIGURE 7.8   (A) Illustration of a nondegenerate code in which only 20 out of the possible 64 different triplets encode amino acids. (B) Brenner's tabulation of all the dipeptide combinations in proteins that have been sequenced by 1957. *Panel (A): After Crick FH, Griffith JS, Orgel LE. Codes without commas.* Proc Natl Acad Sci USA *1957;43(5):416−21. Panel (B): From Brenner S. On the impossibility of all overlapping triplet codes in information transfer from nucleic acid to proteins.* Proc Natl Acad Sci USA *1957;43(8):687−94.*

than the subtlety of argument." They also explicitly wrote in their subsequent formal paper[6]:

> *The arguments and assumptions which we have had to employ to deduce this code are too precarious for us to feel much confidence in it on purely theoretical grounds. We put it forward because it gives the magic number -20- in a neat manner and from reasonable physical postulates [...]. Some direct experimental support is therefore required before our idea can be regarded as anything more than a tentative hypothesis.*

7.6 EXPERIMENTAL RESULTS PROVED THE CODE TO BE NONOVERLAPPING

Abstract ideas on the code coalesced in the second half of the 1950s to a degree that permitted design of experiments to test specific properties of the code. The first attribute of the code that was definitively determined by experiment was its nonoverlapping nature. For several years after Watson and Crick introduced their double helix model of DNA, efforts were made to decide by theory alone whether the code was overlapping or not. As described, clever models that were proposed by the best minds of the time could not discern between overlapping or nonoverlapping code. In a 1959 review of the coding problem, Cyrus Levinthal pointed at the shortcomings of a purely theoretical approach to the problem[28]:

> *In the absence of experimental data, it is clear that such* [theoretical] *speculations can go on more or less indefinitely, and that very elaborate codes can be imagined.*

In the end it was experiments that decided between the competing abstract models. Two decisive experimental approaches proved the code to be nonoverlapping. First, there was the amassing by Brenner of amino acid sequences of proteins. Revealing fully random combinations of adjacent amino acids in the proteins, these sequences could not be directed by an overlapping code which put a limit on the number of combinations of neighboring residues.[23] This conclusion was confirmed shortly thereafter as the statistical sample of sequenced proteins was expanded.[29-31] The second proof was attained in the elegant experiments of the groups of Heinz Fraenkel-Conrat in Berkeley (Box 7.5) and of Gerhard Schramm (1910-1969) in the University of Tübingen.

Essentially, these experiments demonstrated that a point mutation in the RNA genome of Tobacco Mosaic Virus (TMV) resulted in the substitution of a single amino acid in the coat protein of the virus and not of two or three residues as would have been respectively expected for a fully or partially overlapping code.[32] It was thus experiment, and not theory, that provided definitive answer to the question whether or not the code was overlapping.

BOX 7.5 Heinz Fraenkel-Conrat

Born in Breslau, Germany, Heinz Fraenkel-Conrat received in 1933 a medical doctorate from the University of Breslau. Like his father, who was Professor of Gynecology, he could not stay on in Germany after the Nazi rise to power and had to move in 1933 to Scotland. He earned a PhD degree in biochemistry at the University of Edinburgh in 1936 and immigrated to the United States in the same year. From 1952 until his retirement in 1981, he served on the faculty of the University of California at Berkeley. Much of his initial work on TMV was done in collaboration with his wife Beatrice (Bea) Singer who also became a professor in Berkeley. For his demonstration that RNA is the genetic material of TMV he shared with Gerhard Schramm and Alfred Hershey the 1958 Albert Lasker Award for Basic Research (Fig. 7.9).

Heinz Fraenkel-Conrat
(1910–1999)

FIGURE 7.9   Heinz Fraenkel-Conrat.

7.6.1 Brenner's Quasi-Experimental Proof

In principle, fully or partially overlapping codes possessed greater density of information than a nonoverlapping code (Fig. 7.6). On the other hand, the sharing of nucleotides among adjacent overlapping codons restricted the possible next-neighbor combinations of amino acids. As noted, a fully overlapping code limits the number of possible codon combinations to $4^4 = 256$, whereas a code of discrete adjacent "sense" triplets allows every possible permutation of next-neighbor amino acids and the 20 amino acids could combine to yield $20^2 = 400$ possible dipeptides.

Starting in the mid-1950s, Sydney Brenner compiled amino acid sequences of an increasing number of proteins. Tabulating data that had been accumulated until 1957, he showed that dipeptides of the various proteins exhibited unrestricted next-neighbor combinations of amino acid (Fig. 7.8B). The observed unconstrained grouping of amino acids excluded, therefore, the possibility of an overlapping code and indicated instead that codons were arrayed in nonoverlapping order.[23]

7.6.2   Point Mutation in the RNA Genome of TMV Resulted in Substitution of a Single Amino Acid in the Encoded Protein

A definitive experimental demonstration of the nonoverlapping nature of the code was attained in 1960 by a series of experiments with TMV that were conducted in the Berkeley laboratory of Heinz Fraenkel-Conrat (Box 7.5) and concurrently by members of a University of Tübingen group that was led by Gerhard Schramm. TMV is a rod-shaped single-stranded RNA virus that infects tobacco plants and other members of the Solanaceae family (Fig. 7.10A).

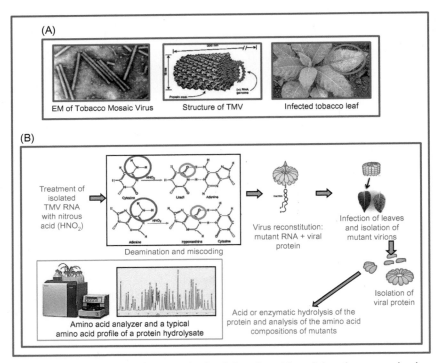

FIGURE 7.10   (A) Electron micrograph of rod-shaped particles of the tobacco mosaic virus (TMV), a scheme of the TMV structure (*red ellipse*), and virus-infected tobacco plant leaf. (B) Scheme of the Tsugita and Fraenkel-Conrat experiment.[32] Isolated TMV RNA was treated with nitrous acid that deaminated cytosine and adenine and converted them into uracil and hypoxanthine, respectively. The transition mutations caused miscoding and production of altered proteins. Wild-type or mutated RNA genomes were mixed with TMV coat protein and the reassembled viruses were used to infect tobacco plants. Progeny viruses that were recovered from the infected tobacco leaves were disassembled and their isolated coat protein was hydrolyzed into its constituent amino acids. Appearance of substituted amino acid in the mutant virus protein was monitored in an automatic amino acid analyzer by quantification of the amino acids in hydrolysates of mutant or wild-type coat protein.

Leaves of virus-infected tobacco plants discolor and assume typical mosaic-like patterns that gave the virus its name. TMV was historically the first virus to be discovered and it became a highly useful research tool. Thus, for instance, it was the first virus to have been crystallized in 1935 by the 1946 Nobel laureate Wendell Meredith Stanley (1904−71) whose studies in the Rockefeller Institute also showed that TMV remained active even after crystallization.[33] Readers interested in the history of TMV research should consult Angela Creager's definitive book[29] and some useful short reviews[34−37] on the subject.

RNA was shown to be the genetic material of TMV: In 1955 Fraenkel-Conrat and the Berkley biophysicist and virologist Robley Williams (1908−95) showed that the protein and RNA components of TMV could be disassembled and then reconstituted. They also demonstrated that whereas each component alone was incapable of infecting tobacco plants, careful reassembly of the 6490 nucleotide-long viral RNA with its coat protein reconstituted virulent viruses.[38] Shortly thereafter, independent studies by Alfred Gierer (1929−) and Gerhard Schramm in the University of Tübingen and by the Berkeley group of Fraenkel-Conrat, Beatrice Singer (1923−2005), and Robley Williams[39] showed that the isolated RNA of the virus, but not its protein component, could productively infect tobacco leaves under defined conditions. That the RNA component was the genetic material of the virus was also revealed in experiments to reconstitute hybrid viruses. Mixing protein of wild-type virus with RNA of an HR mutant strain reconstituted viruses that produced in infected plants symptoms that were characteristic to the HR mutant.[40] It was thus established by 1957 that the hereditary material of the virus was its single-stranded RNA whereas the viral protein served as coat that packaged the RNA genome. Intriguingly, Fraenkel-Conrat admitted at the time that being a protein chemist, he had been initially disappointed that it was the viral RNA and not its protein that carried the genetic information.[41] It should be pointed out here that while he was contemplating the problem of flow of genetic information that culminated in his famous "*Central Dogma*," Francis Crick was imagining an experiment with TMV that would prove that the genetic material was the viral RNA and not its protein. In an Apr. 13, 1955, letter to Watson, Crick described an experiment that, as described below, was eventually conducted two years later by Fraenkel-Conrat. In this letter which again demonstrated Crick's exceptional insight, he wrote (cited in Ref. 4, p. 240):

> [...] *have thought of a beautiful experiment, but don't know who to get to try it. It seems to me possible that if the RNA of TMV is got out very carefully, and if virus-associated protein (i.e. the stuff found in the leaf) is added to this, and if the pH lowered slowly the protein might form up round the RNA and make intact protein again. If (this is a much bigger if) this then proves infective (and Schramm's recent* Nature *letter encourages this hope), then a beautiful*

experiment becomes possible i.e. to use the RNA of one strain and the protein of a different strain, and to see what enzymes alter infection. Do you know anyone who might try this? I can't spark up much enthusiasm here among the virus boys.

Mutagenesis of TMV validated the nonoverlapping nature of the code: After the RNA component of TMV had been recognized as its genetic material, The Tübingen laboratory of Schramm reported that exposure of the viral RNA to nitrous acid (HNO_2) was mutagenic.[42-44] Most of the scored mutations were transitions of cytosine to uracil, of adenine to hypoxanthine, or of guanine to xanthine. Hypoxanthine and xanthine that do not occur naturally in RNA were read as guanine. Using this mutagen, the Tübingen and the Berkeley teams obtained hundreds of mutant strains of TMV. With the coat protein being the only viral protein that could be obtained in workable quantities and purity, both the Tübingen and Berkley groups took a similar approach to identify amino acid substitutions in the coat protein of mutant viruses. Coat proteins that were purified from wild-type or mutant viruses were hydrolyzed and their amino acid compositions were determined by use of the automatic amino acid analyzer (Fig. 7.10B). Heinz Günther Wittmann (1927−90), known for his later studies on ribosomes, was the first to attempt identification of substituted amino acid in the coat protein of mutant viruses.[45] Although he could not detect such substitutions, his failure was almost expected since the coat protein constitutes only ∼8% of the total TMV gene products and mutations could well occur in different proteins of the virus. Fraenkel-Conrat and his associate Tsugita had better luck and in 1960 they detected three single amino acid substitutions in the coat protein of a particular TMV mutant strain that they marked "mutant 171."[32] Specifically, amino acid composition analysis revealed that aspartic acid, threonine, and proline residues of the wild-type protein were, respectively, replaced by alanine, serine, and leucine (Fig. 7.11). The deciphering of the actual code, that is, determination of the identity and order of nucleotides in triplets that encode each amino acid, was in its very early stages at the time of the Tsugita and Fraenkel-Conrat experiment. Also, methods to determine the nucleotide sequences in RNA molecules (or in DNA for that matter) were still nonexistent. Thus the identified amino acid substitutions could not be verified at the RNA level. Now that the code had been deciphered it can be seen that the amino acid replacements that Tsugita and Fraenkel-Conrat recorded were the result of single base substitutions in corresponding codons (Fig. 7.11).

Soon after the publication of the breakthrough paper of Tsugita and Fraenkel-Conrat, Wittmann[46] and the Berkeley researchers[47] identified large numbers of additional single amino acid substitution mutations. Had the code been fully or partially overlapping, then a point mutation in the RNA should have resulted in the respective replacement of three or two adjacent

Eliminated residue	Substituted residue	Mutation (*prospective*)
Aspartic acid	Alanine	GAC→GCC or GAT→GCT
Threonine	Serine	ACT→AGT or ACC→AGC
Proline	Leucine	CCT→CTT or CCC→CTC
		or CCA→CTA or CCG→CTG

FIGURE 7.11   Identified amino acid substitutions in the coat protein of nitrous acid—induced TMV mutant. The left column lists the original amino acids in wild-type virus and the respective substituted residues are listed in the middle column. The right column shows the perfect correspondence between the identified substitutions and point mutations in codons that were deciphered only several years after the identifications of the TMV amino acid substitutions. *Middle column: Tabulated after results in Fraenkel-Conrat H, Singer B. Virus reconstitution and the proof of the existence of genomic RNA*. Philos Trans R Soc B *1999;354(1383):583—6.*

amino acids. However, in all the documented cases of nitrous acid—induced point mutations in the TMV RNA, only a single amino acid had been consistently substituted and three or two substitutions were never observed. Thus results of this experiment elegantly confirmed that codons were discrete entities without partial or full overlap.

In 1960 the same year that point mutation in TMV RNA were shown to result in the substitution of a single amino acid, the Tübingen team,[30] and 5 months later the Berkeley group,[31] respectively, published sequences of 157 and 158 amino acids of the TMV coat protein. The two sequences had high degree of similarity but they were not fully identical. Taking into account the crude techniques of protein sequencing that were available at the time, the achievement of the two competing groups was no mean feat. After reading the report of the Berkley group in the *Proceedings of the National Academy*, Gamow promptly wrote to Wendell Stanley who was a coauthor of the article. Considering the observed randomness of next-neighbor amino acids in the sequences TMV protein, he completely backed away from his original belief in overlapping code (Gamow's Nov. 30, 1960, letter to Stanley, cited in Ref. 29, p. 306):

> *Please accept my congratulations for the beautiful 158-jewel neck-lace. Some six or seven years ago, I would spend some sleepless nights attempting to decode it using an overlapping code. But now I am disillusioned in overlapping codes, and believe that each a.a.* [amino acid] *is determined by an independent triplet of bases in RNA. I added your data to the previously known protein sequences (which enlarged the statistical sample by almost a factor of two), and found that the randomness, as testified by Poisson formula, is even better than it was before.*

7.6.3   Study of the Mutant Sickle Cell Anemia Hemoglobin S Gave Credence to the Conclusion That the Code Was Nonoverlapping

The conclusion that the code was not overlapping gained independent support from the 1959 discovery by Vernon Ingram (1924–2006) that hemoglobin S, the variant hemoglobin that causes inherited sickle cell anemia, differed from normal hemoglobin A by a substitution of glutamic acid with valine at the sixth position of the β-polypeptide chain of the protein.[48] Thus, in that case, a mutation which was later shown to be the result of a GAG→GTG point mutation in the DNA produced a substitution of only a single and not of two or three amino acids in the affected protein.

7.7   THE CODE WAS EXPERIMENTALLY PROVEN TO BE A SEQUENCE OF CONTIGUOUS TRIPLET CODONS

7.7.1   Theory Alone Was Not Sufficient to Reach Definitive Conclusions on the Composition and Organization of Codons

Theoretical considerations, in particular the conjectural analysis of Crick, Griffith, and Orgel,[6] strongly favored a code that was comprised of three nucleotide codons. Therefore, a triplet-based code was consensually accepted in the late 1950s. Yet, because of the lack of empirical evidence, there was still no conclusive decision on the number of bases in a codon and on the organization of the code. Thus, for instance, as late as 1959, the eminent molecular biologist Robert Sinsheimer (1920–) was still able to suggest a clever scheme of two-base codons.[49] Cyrus Levinthal raised in the same year a different theoretically plausible scheme of encoding amino acids by noncontiguous nucleotide sequences.[28] The uncertainties in regard to the constitution of codons and their organization were ultimately removed when it was finally shown by experiment that code words consisted of three nucleotides and that the nucleic acid code was organized in an uninterrupted linear string of contiguous codons.

7.7.2   Laying the Groundwork for Defining the Code: I. Seymour Benzer Constructed a High-Resolution Map of the rII Locus of T4 Phage

The indecision about the number of nucleotide in a codon and about the structure of the code was resolved when Crick and associates at the MRC laboratories in Cambridge provided in 1961 a definitive experimental proof that the code was an uninterrupted string of three nucleotide codons.[50] This classical work stands out as a remarkable example of the power of incisive experiment to dispel uncertainties that theory failed to resolve. Yet, this landmark experiment could not be accomplished without a vital earlier groundwork that was done by Seymour Benzer (Box 7.6). Benzer's experiment was the subject

BOX 7.6 Seymour Benzer

Born in Brooklyn, New York, Benzer graduated from high School at 15 and attended Brooklyn College to major in physics. While he worked at Purdue University on his doctoral thesis in solid state physics, he was involved during World War II in the development of crystal rectifiers for military radar. After receiving a PhD degree in 1947 he was recruited to the Physics Department of Purdue but because of his early interest in biology he chose to study phages in the Caltech laboratory of Max Delbrück. Returning after 2 years to Purdue, he devised the rII recombination assay and used it to construct a map of unprecedented fine resolution of the inner structure of the rII locus. He classified his mapped mutations into point, deletion, insertion, inversion, missense, and nonsense mutations. Demonstrating that mutations could occur at many different sites in the same gene, he deduced that the minimum unit of mutation was a single base pair in DNA. This idea was fundamental to linking the structure of DNA to the reality of genetics. Side-by-side with the discovery of Fred Sanger that proteins were composed of definite sequences of amino acids, his work became a cornerstone of the new science of molecular biology. After having been urged by Delbrück to start something new, he moved to Caltech to initiate studies on the behavioral genetics of *Drosophila*. His belief that single genes determine specific behavioral proclivities of *Drosophila* had been met with the distrust of most of his colleagues in the molecular biology community. Yet, despite doubts and in the face of opposing views, he was able to prove that mutations in single genes affected behavioral attributes of the fly such as circadian rhythms, phototaxis, locomotion, and even learning and memory. His originality and enterprising spirit were revealed again toward the end of his career when he worked on the extension of the lifespan of *Drosophila* and identified in fly and humans cells homologous cancer-related genes (Fig.7.12).

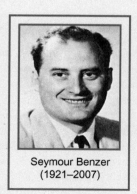

Seymour Benzer
(1921–2007)

FIGURE 7.12   Seymour Benzer.

of a detailed study by the Yale University historian Frederick Holmes (1932–2003) that was edited and published posthumously by his colleague William Summers.[51]

A physicist by training who turned to genetics, Benzer worked at Purdue University to devise assay for scoring recombination events of mutants of bacteriophage T4. This assay was then applied it to construct high-resolution maps of the functionally related cistrons *A* and *B* of the rII locus in the viral genome. An excellent description and analysis of this work can be found in Jeffrey Miller's book.[52] Briefly, Benzer undertook to map at the maximum possible resolution "subgenic" elements within the rII locus of the phage genome. In one of a series of articles that produced maps of ever-higher resolution, Benzer lucidly described the question that he undertook to attack[53]:

> *From the classical researches of Morgan and his school* [Morgan, T.H. "The Theory of the Gene", 1926] *the chromosome is known as a linear arrangement of hereditary elements, the "genes." These elements must have an internal structure of their own. At this finer level, within the "gene" the question rises again: what is the arrangement of* sub-*elements? Specifically, are they linked in a linear order analogous to the higher level of integration of the genes in the chromosomes?*

Benzer went on to explained why microorganisms (or their phages) that produce large number of progeny within a short span of time, were the system needed for fine mapping of intragenic elements[53]:

> *Mapping of a genetic structure is done by observing the recombination of its parts, and recombination involving parts of the structure that are very close together is a rare event. Observation of such rare event requires very many offspring and a selective trick for detecting the few individuals in which the event is recorded. It is for this reason that microorganism are the material of choice for studies of genetic fine structure, and have made it feasible to extend the fineness genetic mapping by orders of magnitude.*

To produce a high-resolution map of the inner structure of a gene, Benzer chose a region in the genome of T4 bacteriophage that he designated rII. He initially isolated mutant strains of the phage that lysed cells more rapidly and thus produced larger, ragged plaques. Phages that displayed such a phenotype were marked "r" (for r̲apid lysis). Mutations responsible for the r phenotype were found to be situated in two distinct loci in the viral DNA; rI and rII. For his mapping project Benzer focused on mutations within the rII locus that comprised of two genes: cistrons *A* and *B*. The general principle of fine intragenic mapping by recombination is illustrated in Fig. 7.13.

To map relative distances between loci within the rII region, Benzer coinfected *Escherichia coli* cells with two rII mutant phages. If the mutations were situated in different loci along the rII region, then crossing events between the genomes of the two mutant strains would have produced some

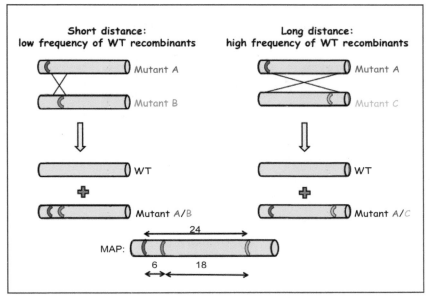

FIGURE 7.13 Gene mapping by recombination frequency. Bacteria are infected by two differ-ent mutant strains of bacteriophage T4. Crossing between the genomes of the two mutants results in progeny that either carry the two mutations or become wild type. The closer the two muta-tions are, the lower are the chances of their recombination and the frequency of production of wild-type phages. Following crossing over and recombination between two mutant viruses, wild-type offspring are selected and their frequency is scored. The higher the frequency of recombina-tion events, the greater is the distance between the two mutations and vice versa. Recombination frequencies are translated into map units that mark the relative distances between pairs of muta-tions. Scoring the frequencies of recombination events between many pairs of mutations yields a map of relative distances between many genetic markers along the mapped genomic locus.

recombinants with wild-type phenotype. The frequency of such wild-type recombinants could then be used as a measure of the relative distance between the two mutations (Fig. 7.13).

In practice, the selection assay for rII recombinants that Benzer devised was based on the different host range of wild-type and rII mutant phages. *E. coli* B allowed growth of both wild-type and rII mutant viruses that produced small and large plaques, respectively. By contrast, strain K of the bacterium (abbreviation for *E. coli* K (λ)) was permissive for the growth of wild-type phages but did not enable replication of rII mutants.

The selection assay was then based on the fact that mutants in the rII region were conditional in having been able to replicate and produce progeny under one set of conditions (growth in *E. coli* B) but not under another (lack of growth in *E. coli* K). The principal features of the selection assay are illustrated in Fig. 7.14.

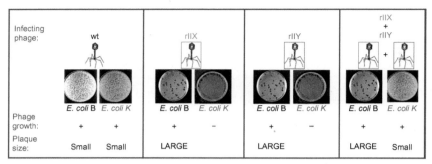

FIGURE 7.14   Scheme of the selection assay for wild-type recombinants in the rII region of bacteriophage T4. Wild-type phages grew in both *E. coli* B and *E. coli* K, producing in both small plaques. Mutants in the rII region could replicate and produce large plaques in *E. coli* B but were unable to grow in *E. coli* K. Double infection of *E. coli* B with a mixture of rIIX and rIIY phages that carried mutations at two different loci within the rII region produced large plaques. At the same time, although each of the mutants alone could not grow in *E. coli* K, crossing between genomes of the two coinfecting mutant viruses generated some wild-type recombinant viruses that could grow and produce small plaques in *E. coli* K cells. The frequency of production of wild-type offspring was in direct proportion to the relative distance between the two mutations.

Pairs of mutations in the rII region could be resolved from one another and their relative distances were determined by coinfecting the same host with the two mutant viruses. Scoring wild-type progeny that could grow in *E. coli* K allowed detection of recombination events at a frequency of 10^{-8}.[54]

Starting in 1955,[54] Benzer gradually screened ever-greater numbers of pairs of rII mutants. Scoring recombination frequencies of the mutant phages, he constructed progressively finer maps of the rII region[53,55] until a map with an estimated resolution of one to two base pairs was completed.[56] Fig. 7.15 shows the progressively more detailed maps of the rII region that Benzer constructed from 1955 to 1959.

Thus, although technologies for the direct sequencing of DNA base were to be introduced only 17 years into the future,[57,58] Benzer attained already in 1960 mapping resolution of the rII region that bordered on the actual nucleotide sequence of the DNA. While it is beyond the scope of this chapter, it should be mentioned that another fundamental outcome of Benzer's mapping project was his classification of the different types of mutations within the rII region into point mutations (missense and nonsense), deletions and insertions.[55]

7.7.3   Laying the Groundwork for Defining the Code: II. Generation of Deletion or Insertion Mutations by Acridine Dyes

Beside Benzer's maps of the rII region, a second essential tool for the determination of the structure of codons and of their organization was the discovery that linear tricyclic acridine dyes such as proflavin or acridine

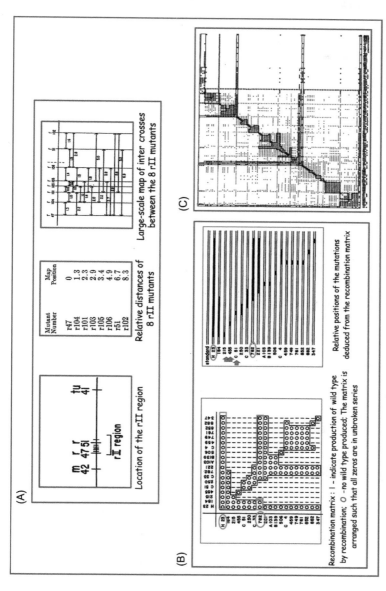

FIGURE 7.15 Succeeding steps in the mapping by Benzer of the rII region of the T4 genome. (A) Proof of principle and an early limited map. Left panel: Linkage map of T4 DNA and the location of the rII region. Center panel: Map positions of 8 rII mutations. Right panel: Map of the relative distances between the 8 mutations. (B) Left panel: Recombination matrix for 19 rII mutants. "T" marks appearance of wild-type recombinants and "O" signifies no such recombinants. The *red-boxed* uppermost and eighth rows correspond to the two *red-circled* rII mutants in the right hand panel. Right panel: Relative positions of the same mutations as deduced from the matrix. (C) Recombination matrix of 145 rII mutants. *Panel (A): Data from Benzer S. Fine structure of a genetic region in bacteriophage.* Proc Natl Acad Sci USA 1955;**41**(6):344–54. *Panels (B): From Benzer S. On the topology of the genetic fine structure.* Proc Natl Acad Sci USA 1959;**45**(11):1607–20. *Panel (C): From Benzer S. On the topology of the genetic fine structure.* Proc Natl Acad Sci USA 1959;**45**(11):1607–20.

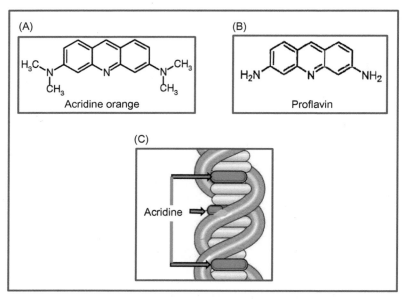

FIGURE 7.16 (A) Structure of acridine orange. (B) Structure of proflavin. (C) Illustration of the intercalation of planar acridine molecules between stacked base pairs in DNA.

orange mutagenize DNA by adding or deleting a base or bases. Most chemical mutagens were found to produce a wide spectrum of mutations in the rII region of the T4 genome. All these mutant viruses lost the ability to grow in *E. coli* K. Yet, mutations generated by almost all the mutagens that were tested tended to revert, resulting in the development of small plaques of revertant wild-type phages in *E. coli* K. Brenner, Benzer, and Barnett noticed, however, that mutants that were generated by acridine orange or proflavin (Fig. 7.16A and B) almost never reverted. Also, unlike many other mutagens, acridines were not integrated into the DNA.[59] Ernst Freese (1925–90), a former physicist and a student of both Werner Heisenberg and Enrico Fermi who later became a leading expert on mutagenesis, proposed that acridines induced transversion mutations.[60] However, based on their analysis of the properties of the acridine-induced mutants, Brenner, his then technician Leslie Barnett (1910–2012), Crick, and the doctoral student and Leslie Orgel's wife Hassia Alice Orgel (1929–) rejected the Freese theory suggesting instead that these dyes acted by producing loss or gain of a nucleotide.[61] Indeed, later biophysical studies revealed that the planar acridines intercalate between neighboring base pairs in an orientation perpendicular to the axis of the DNA helix[62,63] (Fig. 7.16C).

The space occupied by acridine along the DNA is that of one base pair (3.4 Å). It had been proposed, therefore, that a chromosome with intercalated acridine pairs with a homolog chromosome with no acridine, the pairing shifts one base pair out of register.[63] It was later shown that the induction of deletion or insertion mutations by acridine arised in vitro in phage at sites of acridine-stimulated cleavage of T4 DNA by the virus-encoded type II topo-isomerase. It was deduced from experimental results that acridine-induced mutations were dictated by acridine-topoisomerase reaction and not by slipped pairing in repeated sequences as was suggested earlier.[64]

7.7.4   By Employing the rII System and Acridine, Crick, Brenner, and Associates Conclusively Showed That the Genetic Code Consisted of Triplet Codons Arranged in Nonoverlapping and Commaless Sequence

After he was recruited to the MRC laboratory in Cambridge, Sydney Brenner shared an office with Francis Crick. The two scientists were engaged in lengthy and intense discussions on unsettled research problems (for a description of their exceptionally probing dialog, see Ref. 4, pp. 268—270). The most prominent outcome of the dialogue between Crick and Brenner was arguably the experimental definition of the code. As indicated, the right choices of mutagens and of the mutated gene were the two essential components of their ultimately successful collaboration.

Choice of acridine mutagen: Brenner became familiar with acridine dyes early in his student days in South Africa when he conducted cytological studies. This knowledge came into use when he developed in the late 1950s an interest in mutagenesis by agents that could either induce predictable base substitutions or were able to generate specific type of mutations. Pursuing this line of investigation he focused on acridine orange and proflavin.[59,61] At the same time, Crick and Leslie Orgel were contemplating a situation in which a mutation that disrupts normal base pairing in DNA or RNA might be countered by a mutation at a different site, designated suppressor mutation that restores normal base pairing. According to Olby,[65] the idea of looking for suppressor mutations excited Crick so much that he, the great theoretician, ventured to perform experiments with his own hands. Alas, use of any of a number of different mutagens failed to induce suppressor mutations. At that point Crick turned to the acridines that Brenner was working on.

Choice of the rII system: The other essential component of the endeavor to define the code was the choice of gene that was analyzed. Most auspiciously the Cambridge group became closely familiar with the bacteriophage rII work when Benzer spent the 1957—58 year in Cambridge. Thus with this choice of suitable mutagens and gene, a collaboration was forged between Crick, Brenner, Barnett, and Richard J. Watts-Tobin (1934-) to solve the structure of the code. Much had been written about the historical and technical aspects of this classic experiment of the Cambridge group. The interested reader may find additional material on the discovery of the triplet code in Refs. 3,4,52,66.

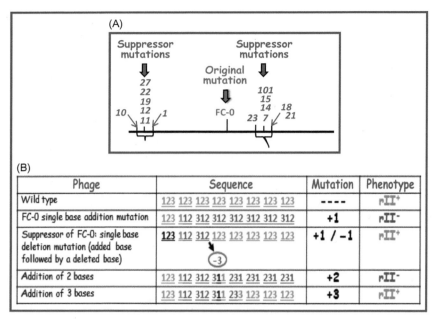

FIGURE 7.17   (A) Locations of the FC-0 mutation in cistron *B* of the rII region and of its upstream and downstream suppressor mutations (the distinct suppressor italicized mutations were numbered by Crick et al.[50]). (B) Scheme of selected major results of the analysis of Crick et al.[50] Bases are designated numbers 1, 2, or 3; wild-type base triplets are in *green*, inserted bases are in *red*; a deleted base " − 3" in the third row is *encircled* and out-of-frame bases are in *purple*.

The actual experiment: Using acridine to mutagenize T4 phage, Crick and his Cambridge associates focused on a specific FC-0 zero mutation in the highly resolved mapped cistron *B* of the rII region of bacteriophage T4. There they identified at loci different from the original mutation, multiple acridine-induced suppressor mutations that displayed wild-type or pseudo wild-type phenotype.[50] It was reasoned that the original zero mutation consisted of a deleted or an added single nucleotide. Suppressor mutations that were located either to the left or to the right of the original FC-0 mutation corrected the original zero mutation by the reciprocal addition or deletion of a base at a different site (Fig. 7.17A and B; third row of the table).

Coinfection and selection experiments proved that the suppressor mutations were located at sites that were distinct from the original FC-0 mutation. Moreover, Crick and his coauthors also isolated suppressors of suppressors that were again located at sites that were distinct from both the original FC-0 mutation and of its suppressor.[50] Similar to the phenotype of phages with a deleted or added single base, deletion or addition of two bases also resulted in rII⁻ mutant phenotype (Fig. 7.17B; fourth row of the Table). By contrast, however, deletions or additions of three bases restored the reading frame, affecting return to the wild-type phenotype (Fig. 7.17B, fifth row of the Table).

This milestone experiment established several fundamental features of the code.

1. *The code is read from a fixed point*: The observed suppression of single base addition or deletion zero mutations by the reciprocal deletion or addition of one base at a different site, indicated that the code was read in frames from a fixed point. This finding obviated a need for commas to separate one codon from another.

2. *Codons are nucleotide triplets*: Whereas deletions or additions of one or two bases resulted in mutant phenotype, deletion or addition of three bases restored wild-type phenotype. Such restoration of wild-type phenotype by elimination or addition of three bases indicated that code words consisted of three nucleotides or their multiples. Since mapping of very close mutations suggested that both acridine and hydrazine added or deleted a single base at a time, it was argued that codons were nucleotide triplets rather than sextets.[50] Crick described in his 1988 book *What a Mad Pursuit* his (rare) venture into bench work and vividly evoked the moment of discovery of the triplet nature of codons (Ref. 67, p. 133):

   One glance at the crucial plate was sufficient. There were plaques in it! The triple mutant was showing the wild type behavior (phenotype). Carefully we double-checked the numbers on the petri dishes to make sure we had looked at the correct plate. Everything was in order. I looked across at Leslie [Orgel]. "Do you realize," I said, "that you and I are the only people in the world who know it's a triple code?

3. *The code is degenerate*: Results of the experiment indicated that each amino acid was encrypted by several different triplets ("degenerate code"). Had only a single triplet encoded any one of the amino acids, then the code should have consisted of 20 sense and 44 nonsense codons. In such a case the suppressor mutations could have occurred only at short distances away from a primary mutation since at greater distances there would be a high probability of targeting a nonsense triplet. In fact, however, most of the suppressor mutations were localized at large distances away from the primary mutation. Thus most or all of the 64 triplets must have encoded amino acids. It was inescapably concluded, therefore, that several different nucleotide triplets specified each amino acid.[50]

4. *There are nonsense codons*: Although most of the combinations of deletion and addition of an original mutation and suppressor yielded wild-type phenotype, a few did not. Crick and his coinvestigators conjectured that in such cases the deletion and suppressing mutations created what they have called a "barrier." The subsequently determined actual dictionary of the genetic code by Marshall Nirenberg at the NIH and by Har Gobind Khorana at the University of Wisconsin identified these presumed barriers as nonsense codons (see Chapter 9: "The Discovery and Rediscovery of Prokaryotic Messenger RNA").

7.8 POSTSCRIPT: PLACE AND MERIT OF THEORY AND EXPERIMENT IN THE ELUCIDATION OF THE NATURE OF THE GENETIC CODE

Defining the genetic code was a significantly more weighty undertaking than the contemporaneous elucidation of the mechanics of replication (see Chapter 6: "Meselson and Stahl Proved That DNA Is Replicated in a Semiconservative Fashion"). Similar to the question of the mode of DNA replication, efforts to define the genetic code also opened with a phase of theorization. Adoption of information theory concepts (see Lilly Kay's book[3] for a multilayered discussion of the evolving perception of information in biology) generated a series of abstract combinatorial models that did not readily succumb to critical testing by experiment. Also, most of the deliberated models paid little or no attention to the largely unknown inherent biochemical aspects of the code. Yet, one attempted explanation for the ability of nucleic acids to encode the chemically distinct proteins employed the paradigm of key-and-lock fit between amino acids and the encoding nucleic acid.[9,10,26] Although Crick handily rejected this idea by arguing that it had no stereochemical grounding (http://profiles.nlm.nih.gov/ps/access/SCBBGF.pdf), he was unable to foretell at that early stage the complex biochemical reality of the code. Thus, though the code problem appeared initially to be amenable to theoretical modeling, it was promptly understood that its complexity defied treatment by theorization alone. This realization was presciently pronounced by Brenner already in 1957[5]: "... *As far as we can see the coding problem is now an experimental problem and the role of theory will be to organize and make sense of the facts as they come out of the laboratory.*" Indeed, rather than by theoretical leaps, the nature of the code was exposed stepwise by painstaking experimental work. The nonoverlapping nature of the code was revealed by cataloging dipeptide combinations in actual sequences of diverse proteins[23] and by demonstrating that point mutations in TMV RNA resulted in single and not in multiple amino acid substitutions.[32] Also, although theoretical arguments favored a triplet-based code, theory by itself failed to solve the question of how 64 different trinucleotides code for just 20 amino acids. It was ultimately an experiment that definitively proved the code to be triplet-based, degenerate, and commaless.[50]

The principally experiment-based definition of the code served as foundation for the subsequent purely experimental deciphering of the code by Marshall Nirenberg and by Har Gobind Khorana (see Chapter 9: "The Discovery and Rediscovery of Prokaryotic Messenger RNA"). Notably though, theory was not completely absent from studies of aspects of the code. One such facet was the problem of wedding the different chemistries of the nucleic acid template and the translated protein. Here history provided an exceptional case of prophetic theoretical prognostication in the form of Crick's adaptor hypothesis.[7] Yet, parallel to, and without cross-pollination by this abstract hypothesis, Zamecnik and Hoagland carried out a series of experiments that culminated in the identification of tRNA as the actual adaptor[8] (see Chapter 8: "The Adaptor Hypothesis and the Discovery of Transfer RNA"). Thus even in this case of insightful theoretical prediction, the actual experimental discovery of tRNA was not guided by Crick's theoretical leap. The limited role of theory in defining the genetic code can be explained in retrospect by the complexity of the problem. The intricate biochemistry of the code and its usage involved unknown variables that theory alone was largely unable to predict and that could only be uncovered by experiment.

REFERENCES

1. Watson JD, Crick FH. Genetical implications of the structure of deoxyribonucleic acid. .Nature 1953;**171**(4361):964−7.
2. Watson JD, Crick FH. Molecular structure of nucleic acids: a structure for deoxyribose nucleic acid. Nature 1953;**171**(4356):737−8.
3. Kay LE. Who wrote the book of life? A history of the genetic code. Stanford, CA: Stanford University Press; 2000.
4. Olby R. Francis Crick: Hunter of life's secrets. New York: Cold Spring Harbor Laboratory Press; 2009.
5. Brenner S. Untitled typescript of a talk Wellcome Library Archives; 1957.
6. Crick FH, Griffith JS, Orgel LE. Codes without commas. Proc Natl Acad Sci USA 1957; **43**(5):416−21.
7. Crick FH. On protein synthesis. Symp Soc Exp Biol 1958;**12**:138−63.
8. Hoagland MB, Stephenson ML, Scott JF, Hecht LI, Zamecnik PC. A soluble ribonucleic acid intermediate in protein synthesis. J Biol Chem 1958;**231**(1):241−57.
9. Gamow G. Possible relation between deoxyribonucleic acid and protein structures. Nature 1954;**173**(4398):318.
10. Gamow G. Possible mathematical relation between deoxyribonucleic acid and proteins. Biol Medd Kgl Dan Vidensk Selsk 1954;**22**(3):1−11.
11. Dounce AL. Duplicating mechanism for peptide chain and nucleic acid synthesis. Enzymologia 1952;**15**(5):251−8.
12. Dounce AL. Nucleic acid template hypotheses. Nature 1953;**172**(4377):541.
13. Beadle GW, Tatum EL. Genetic control of biochemical reactions in Neurospora. Proc Natl Acad Sci USA 1941;**27**(11):499−506.
14. Beadle GW. Genes and biological enigmas. Am Scientist 1948;**36**(1):69−74.
15. Beadle GW. The genetic basis of biological specificity. J Allergy 1957;**28**(5):392−400.
16. Beadle GW. Physiological aspects of genetics. Ann Rev Physiol 1960;**22**:45−74.
17. Schrödinger E. What is life? Cambridge: Cambridge University Press; 1944 [reprinted 1967].
18. Stern KG. Nucleoproteins and gene structure. Yale J Biol Med 1947;**19**(6):937−49.
19. Caldwell PC, Hinshelwood C. Some considerations on autosynthesis in bacteria. J Chem Soc 1950;**4**(0):3156−9.
20. Nirenberg M. A career begins at the National Institutes of Health [Harris R, interviewer]; 1955−1996.
21. Avery OT, Macleod CM, McCarty M. Studies on the chemical nature of the substance inducing transformation of pneumococcal types: induction of transformation by a desoxyribo- nucleic acid fraction isolated from pneumococcus type III. J Exp Med 1944;**79**(2):137−58.

22. Watson JD. *Genes, girls, and Gamow: after the double helix*. New York, NY: Vintage Books; 2001.

23. Brenner S. On the impossibility of all overlapping triplet codes in information transfer from nucleic acid to proteins. *Proc Natl Acad Sci USA* 1957;**43**(8):687−94.

24. Rich A, Watson JD. Physical studies on ribonucleic acid. *Nature* 1954;**173**(4412):995−6.

25. Rich A, Watson JD. Some relations between DNA and RNA. *Proc Natl Acad Sci USA* 1954;**40**(8):759−64.

26. Gamow G, Rich A, Ycas M. The problem of information transfer from the nucleic acids to proteins. *Adv Biol Med Phys* 1956;**4**:23−68.

27. Gamow G, Ycas M. Statistical correlation of protein and ribonucleic acid composition. *Proc Natl Acad Sci USA* 1955;**41**(12):1011−19.

28. Levinthal C. Coding aspects of protein synthesis. *Rev Modern Phys* 1959;**31**(1):249−55.

29. Creager ANH. *The life of a virus: tobacco mosaic virus as an experimental model 1930−1965*. Chicago, IL: The University of Chicago Press; 2001.

30. Anderer FA, Uhlig H, Weber E, Schramm G. Primary structure of the protein of tobacco mosaic virus. *Nature* 1960;**186**(4729):922−5.

31. Tsugita A, Gish DT, Young J, Fraenkel-Conrat H, Knight CA, Stanley WM. The complete amino acid sequence of the protein of tobacco mosaic virus. *Proc Natl Acad Sci USA* 1960;**46**(11):1463−9.

32. Tsugita A, Fraenkel-Conrat H. The amino acid composition and C-terminal sequence of a chemically evoked mutant of TMV. *Proc Natl Acad Sci USA* 1960;**46**(5):636−42.

33. Stanley WM. Isolation of a crystalline protein possessing the properties of tobacco-mosaic virus. *Science* 1935;**81**(2113):644−5.

34. Fraenkel-Conrat H, Singer B. Virus reconstitution and the proof of the existence of genomic RNA. *Philos Trans R Soc B* 1999;**354**(1383):583−6.

35. Creager AN, Scholthof KB, Citovsky V, Scholthof HB. Tobacco mosaic virus. Pioneering research for a century. *Plant Cell* 1999;**11**(3):301−8.

36. Zaitlin M. The discovery of the causal agent of the tobacco mosaic disease. In: Kung SD, Yang SF, editors. *Discoveries in plant biology*. Hong Kong: World Publishing Co.; 1998. pp. 105−10.

37. Fraenkel-Conrat H. Protein chemists encounter viruses. *Ann NY Acad Sci* 1979; **325**(1):309−20.

38. Fraenkel-Conrat H, Williams RC. Reconstruction of active tobacco mosaic virus from its inactive protein and nucleic acid components. *Proc Natl Acad Sci USA* 1955; **41**(10):690−8.

39. Fraenkel-Conrat H, Singer B, Williams RC. Infectivity of viral nucleic acid. *Biochim Biophys Acta* 1957;**25**(1):87−96.

40. Fraenkel-Conrat H, Singer B. Virus reconstitution. II. Combination of protein and nucleic acid from different strains. *Biochim Biophys Acta* 1957;**24**(3):540−8.

41. Williams G. *Virus hunters*. New York, NY: Knopf; 1959.

42. Gierer A, Mundry KW. Production of mutants of tobacco mosaic virus by chemical alteration of its ribonucleic acid in vitro. *Nature* 1958;**182**(4647):1457−8.

43. Schuster H. Die reactionsweise der deoxribonuleicsäure mit salpetriger säure. *Z Naturfor B* 1960;**15**:298−304.

44. Schuster H, Schramm G. Bestimmung der biologisch wirksamen einheit in der ribosenucleinsäure des tabakmosaikvirus auf chemischem wege. *Z Naturfor B* 1958;**13**:697−704.

45. Wittmann HG. Vergleich der proteine des normalstamms und einer nitritmutante des tabakmosaikvirus. *Z Vererbungsl* 1959;**90**(4):463−75.

46. Wittmann HG. Comparison of the tryptic peptides of chemically induced and spontaneous mutants of tobacco mosaic virus. *Virology* 1960;**12**(4):609–12.

47. Tsugita A, Fraenkel-Conrat H. The composition of proteins of chemically evoked mutants of TMV RNA. *J Mol Biol* 1962;**4**(2):73–82.

48. Ingram VM. Abnormal human haemoglobins. III. The chemical difference between normal and sickle cell haemoglobins. *Biochim Biophys Acta* 1959;**36**(2):402–11.

49. Sinsheimer RL. Is the nucleic acid message in a two-symbol code? *J Mol Biol* 1959;**1**(3):218–20.

50. Crick FH, Barnett L, Brenner S, Watts-Tobin RJ. General nature of the genetic code for proteins. *Nature* 1961;**192**(4809):1227–32.

51. Holmes FL. *Reconceiving the gene: Seymour Benzer's adventures in phage genetics*. New Haven, CT: Yale University Press; 2006.

52. Miller JH. *Discovering molecular genetics: a case study course with problems & scenarios*. Cold Spring Labor: Cold Spring Harbor Press; 1995.

53. Benzer S. On the topology of the genetic fine structure. *Proc Natl Acad Sci USA* 1959;**45**(11):1607–20.

54. Benzer S. Fine structure of a genetic region in bacteriophage. *Proc Natl Acad Sci USA* 1955;**41**(6):344–54.

55. Benzer S. Genetic fine structure. *Harvey Lect* 1960–1;1–21.

56. Benzer S. On the topography of the genetic fine structure. *Proc Natl Acad Sci USA* 1961;**47**(3):403–15.

57. Sanger F, Nicklen S, Coulson AR. DNA sequencing with chain-terminating inhibitors. *Proc Natl Acad Sci USA* 1977;**74**(12):5463–7.

58. Maxam AM, Gilbert W. A new method for sequencing DNA. *Proc Natl Acad Sci USA* 1977;**74**(2):560–4.

59. Brenner S, Benzer S, Barnett L. Distribution of proflavin-induced mutations in the genetic fine structure. *Nature* 1958;**182**(4641):983–5.

60. Freese E. The difference between spontaneous and base-analogue induced mutations of phage T4. *Proc Natl Acad Sci USA* 1959;**45**(4):622–33.

61. Brenner S, Barnett L, Crick FHC, Orgel A. The theory of mutagenesis. *J Mol Biol* 1961;**3**(1):121–4.

62. Lerman LS. Structural considerations in the interaction of DNA and acridines. *J Mol Biol* 1961;**3**(1):18–30.

63. Lerman LS. The structure of the DNA-acridine complex. *Proc Natl Acad Sci USA* 1963;**49**(1):94–102.

64. Masurekar M, Kreuzer KN, Ripley LS. The specificity of topoisomerase-mediated DNA cleavage defines acridine-induced frameshift specificity within a hotspot in bacteriophage T4. *Genetics* 1991;**127**(3):453–62.

65. Olby R. Francis Crick, DNA and the central dogma. In: Holton G, editor. *The twentieth century science: studies in the biography of ideas*. New York, NY: W. W. Norton & Co; 1972. p. 227–80.

66. Yanofsky C. Establishing the triplet nature of the genetic code. *Cell* 2007;**128**(5):815–18.

67. Crick FH. *What mad pursuit: a personal view of scientific discovery*. New York, NY: Basic Books; 1988.

The Adaptor Hypothesis and the Discovery of Transfer RNA

Crick's Prescient Hypothesis of an *Adaptor* Molecule and the Independent Identification by Zamecnik and Hoagland of Transfer RNA and Activating Enzymes

Chapter Outline

M. Fry: Landmark Experiments in Molecular Biology. DOI: http://dx.doi.org/10.1016/B978-0-12-802074-6.00008-4
313

Proteins have been known to function as enzymes and were believed for a long time to be the material that genes were made of. Having been considered the key class of biological molecules, proteins were intensely investigated since the late 19th century. Curiously, however, the mode of the synthesis of proteins remained largely obscure. The prevailing belief from the 1930s to the early 1950s was that amino acids were condensed into proteins by reverse action of proteolytic enzymes. It was thus believed that under favorable conditions proteases catalyzed the formation of α-peptide linkages between amino acids rather than their breakage. A view that proteins were synthesized by the concerted action of proteolytic enzymes of different specificities stood in contrast to the later established concept of template-directed synthesis of proteins (a recommended comprehensive historical and philosophical analysis of the study of protein synthesis is Ref. 1). The early belief in enzyme-catalyzed synthesis of proteins was abandoned when it was realized that each protein has a unique sequence of amino acids that must be dictated by a corresponding nucleic acid sequence. This newer insight raised in turn the question of how a 4-letter language of DNA and RNA was translated into the 20 amino acid letters in proteins. This problem was solved in the second half of the 1950s with the discovery of transfer RNA (tRNA) and of the enzymes that activate and link particular amino acids to specialized tRNA molecules. The history of this discovery is unusual in that it entailed two independent lines of investigation, one experimental and the other theoretical.

One track was the experimental work of mainly Paul Zamecnik (1912–2009) and Mahlon Hoagland (1921–2009) at the Huntington

Laboratory of the Massachusetts General Hospital and Harvard University. Their groundbreaking experiments culminated in the isolation of low-molecular-size RNA that carried amino acids to the site of protein synthesis. Hoagland and Zamecnik first termed this new kind of RNA "small RNA" (S-RNA). Shortly thereafter it was renamed "soluble RNA" (sRNA) and it later got its final designation "transfer RNA" (tRNA). This last currently accepted term is also used in this chapter. Specific activating enzymes (later named aminoacyl-tRNA synthetases) were found to activate amino acids and to enable their charging onto cognate tRNA molecules such that covalently linked aminoacyl−tRNA complexes were formed. Hydrogen bonding between particular charged tRNA molecule and a specific coding sequence in the RNA template allowed the formation of α-peptide linkage between the tRNA-carried amino acid and a growing polypeptide chain and the subsequent release of an uncharged tRNA molecule.

In parallel to and totally independent of the Zamecnik and Hoagland work, Francis Crick, then at Cambridge, put forward a prophetic hypothesis of an "adaptor" molecule that acts as a bridge between the different languages of the nucleic acid template and the protein product. Remarkably, neither Crick nor Zamecnik and Hoagland were aware of each other's respective experimental work and theoretical prediction.

Different aspects of the early history and science of protein synthesis, of tRNA, and of amino acid−activating enzymes have been reviewed in several excellent articles.[2−10] The focus of this chapter is limited to the evolution of ideas about protein synthesis and to the discovery of tRNA as the element that links nucleotide sequences in the RNA template to corresponding amino acids that combine to form proteins.

8.1 THE EARLY DAYS: POLYMERIZATION OF AMINO ACIDS INTO PROTEIN WAS BELIEVED TO BE CATALYZED BY SYNTHETIC REVERSE ACTION OF PROTEOLYTIC ENZYMES

The concept of template-directed biosynthesis is an established cornerstone of our present view of the cellular synthesis of nucleic acids and of proteins. However, before and even for a short time after the introduction of the template principle, proteins were thought to be the products of polymerization of amino acids by reverse synthetic action of proteolytic enzymes of different specificities.

Work at the turn of the 20th century in the laboratories of the two great German chemists, Franz Hofmeister (1850−1922) and Emil Fischer (1852−1919), simultaneously established that proteins were linear chains of α-amino acids. In parallel, using organic chemistry synthesis methods, Fischer and his many students created more than a hundred different peptides.[11] The cumbersome organic synthesis methods that Fischer devised were greatly improved by his once assistant, and later a leading protein chemist on his own, Max Bergmann (1886–1944). Together with the Greek organic chemist Leonidas Zervas (1902–80), Bergmann developed a more efficient method to synthesize oligopeptides by using carboxybenzyl groups to protect amines from electrophiles.[12] After the Nazi rise to power in 1933, Bergmann left Germany for the United States. There he joined the Rockefeller Institute at which he established a highly productive protein research program. His Rockefeller group endorsed two lines of thought about proteins and their synthesis that proved in time to be misleading.

8.1.1   First Misconception: Proteins Lack Definitive Sequence of Amino Acids

Based on analyses of the amino acid composition of hydrolyzates of various proteins, Bergmann promoted the idea that proteins do not have defined sequences of amino acids. He thought instead that amino acids were arrayed in periodic arrangements such that each individual amino acid residue recurred at constant intervals in the protein molecule.[13–15] In this model individual molecules of any given single protein had statistically similar composition of amino acids but no common precise sequence. This view, which became in the late 1930s the dominant mode of thinking about the structure of proteins,[16,17] was compatible with the parallel proposal that proteolytic enzymes mediated synthesis of proteins of such structure (see below). The way of thinking about the structure of proteins was transformed only when the eminent British chemist and the 1958 and 1980 twice laureate of Nobel Prizes, Fred Sanger (1918–2013) showed in the early 1950s that proteins have definite sequences of amino acids. In his 1952 presentation of the amino acid sequence of bovine insulin, Sanger pointed at the contrast between the new insight of exact amino acid sequence and the former view of the indefinite structure of proteins[18]:

> It has frequently been suggested that proteins may not be pure chemical entities but may consist of mixtures of closely related substances with no absolute unique structure. The chemical results obtained so far [i.e. the amino acid sequence of insulin] suggest that this is not the case, and that a protein is really a single chemical substance, each molecule of one protein being identical to every other molecule of the same pure protein.

8.1.2   A Second Misconception: Proteins Are Synthesized by Reverse Synthetic Action of Proteolytic Enzymes

The dominant idea from the 1930s to the early 1950s about the mechanism of protein biosynthesis was that it was catalyzed by reverse action of proteolytic enzymes. The University of Toronto biochemists Hardolph Wasteneys (1881−1961) and Henry Borsook (1897−1984) reported in 1930 that concentrated proteolysis products of egg albumin formed protein-like masses that they termed "plasteins."[19] Three years later, Carl Voegtlin (1879−1960) and his associates found that soluble peptides that were released from ruptured cells precipitated out of solution in the presence of sulfhydryl compounds and oxygen.[20] Both groups raised the possibility that the so-called plasteins and the protein precipitates were products of de novo protein synthesis by proteolytic enzymes that were liberated from the cells. Bergmann and his then coinvestigator at the Rockefeller Institute Joseph Fruton (1912−2007) (who later became a leading protein chemist and historian of biochemistry at Yale University) further developed the notion that reverse action of proteases was responsible for amino acid polymerization. Typically, proteases cleave peptide bonds that link specific paired amino acids. For instance, cleavage by trypsin occurs at the carboxyl side of lysine or arginine, except when either is adjacent to proline. It was claimed that these enzymes maintained the same specificity in forming α-peptide bonds as in their breakage. Indeed, results of the Bergmann group indicated that under appropriate conditions different proteolytic enzymes linked together only those amino acids whose peptide bonds they could also break. Heinz Fraenkel-Conrat (see Chapter 7: "Defining the Genetic Code") who was working then in the Bergmann laboratory demonstrated that the protease papain could catalyze formation of peptide bonds between specific amino acids.[21] Shortly thereafter Bergmann and Fruton expanded this observation by showing that chymotrypsin also paired specific amino acids.[22] The same two researchers demonstrated in subsequent experiments that a mixture of catheptic enzymes generated heterotypic peptide chains.[23] It was thus proposed that protein biosynthesis was conducted by the concerted synthetic action of multiple proteolytic enzymes of different specificities. Bergmann stated this model of protein synthesis in 1942[24]:

The proteins generated by living cells [...] must be the product of coupled reactions in which many substrates participate. The synergy of proteolytic enzymes in a sequence of coupled reactions requires a certain harmony between the specificities of the synergistic enzymes

8.1.3   The Problem of the Required Energy for Protein Synthesis

Formation of a peptide bond by extraction of water is an endergonic reaction. After moving from Toronto to Caltech, Borsook conducted with Hugh Huffman[25] (1899−1950) and later with Jacob Dubnoff[26] (1909−72) calorimetric measurements of protein synthesis. Their results showed that the formation of peptide bonds required up to >3000 calories/mole. Thus the equilibrium between free and polymerized amino acids was heavily tilted to the side of free amino acids and against peptide bond formation. The model of catalysis of protein synthesis by proteolytic enzymes had to answer, therefore, the crucial question of the source of energy for the generation of peptide bonds. Remarkably, based on his expertise in bioenergetics, the great biochemist and 1953 Nobel Prize laureate Fritz Lipmann (1899–1986) hypothesized already in 1941 that a phosphorylating enzyme was likely to be involved in the endergonic peptide bond formation in living cells.[27] A similar theoretical proposal was put forward in the very same year by another leading student of bioenergetics, the Danish biochemist Herman Kalckar[28] (1908–91). Although Bergmann and Fruton speculated on possible sources of energy for protease-catalyzed protein synthesis, there was no experimental evidence for any of their conjectures. This quandary appeared to be solved for a while when Fruton found that proteolytic enzymes were capable of transpeptidation between amino acids and peptides.[29,30] In this reaction the enzymes elongated a peptide chain by catalyzing the transfer of an amino acid or of a peptide to a longer acceptor peptide. Importantly, the initial splitting of ester, amide, or peptide links met the energy requirements since there was not much difference in energy between destroyed and formed peptide bonds.[31] Despite these seemingly encouraging results, the idea of enzyme-catalyzed biosynthesis of proteins was invalidated by results of experiments that indicated that protein synthesis was an energy-consuming reaction (see Section 8.2). Also, whereas proteolytic enzymes cleaved synthetic dipeptides that had valine or aminobutyric acid to similar extents, the protein synthesis machinery was found to incorporate into protein only valine but not aminobutyric acid.[31] Most importantly, a template-directed rather than enzyme-catalyzed protein synthesis machinery was discovered after a cell-free system for protein synthesis was developed.

8.2 CONSTRUCTION AND CHARACTERIZATION OF A CELL-FREE PROTEIN SYNTHESIS SYSTEM

The theory of protease-catalyzed synthesis of proteins was abandoned when the general features of the biosynthesis of proteins were revealed by the construction and dissection of a cell-free protein synthesis system. A detailed historical review of the landmark development of in vitro protein synthesis system, which is mostly dedicated to the seminal work of Paul Zamecnik (Box 8.2) is in Ref. 5. The focus of this concise section is on technical developments that were crucial for the assembly and use of cell-free protein synthesis system: one was the use of radioactively labeled amino acids and the other was the employment of differential centrifugation to separate subcellular components of the protein synthesis machinery.

8.2.1 Radioactively Labeled Amino Acids Were an Essential Tool for the Monitoring of Protein Synthesis In Vivo

Emergent availability in the second half of the 1940s of radioactive isotopes, mainly ^{14}C, and their use to prepare radiolabeled amino acids, provided an indispensable tool for the dynamic tracking of protein synthesis. Specifically the measurement of the incorporation of radiolabeled free amino acids into protein served to monitor protein synthesis. The only carbon isotope that was available until the mid-1940s was ^{11}C whose half-life of 20 minutes was too short to allow meaningful biological experimentation. Knowledge and technologies that were accumulated in the course of the development of nuclear bombs under the Manhattan Project opened the way to the production of new radioactive isotopes. So, after the end of World War II the United States Atomic Energy Commission (AEC) authorized mass production of scientifically and medically useful radioisotopes and their delivery to research laboratories and to clinical centers (for a recent comprehensive history of radioisotopes in biological research and medicine, see Ref. 32). Paul Zamecnik of the Huntington Laboratory of the Massachusetts General Hospital attempted without success to radioactively tag amino acids with ^{14}C (half-life 5730 ± 40 years) which became available after 1945. He thus turned to the physical and organic chemist Robert Loftfield (1919−2014) who had then just concluded his PhD thesis as the first graduate student of the eminent organic chemist and 1965 Nobel laureate Robert Woodward (1917−79). Loftfield was working in 1946 in the Radioactivity Center of MIT and Zamecnik challenged him with the task of preparing ^{14}C-labeled amino acids. Using ^{14}C-cyanide as a starting compound, Loftfield successfully prepared ^{14}C-tagged alanine and glycine[33] (for Loftfield's personal take on this historical achievement, see http://hscdm. unm.edu/hslic/oralhist/PDF/LoftfieldROH.pdf). Shortly thereafter Jacklyn Melchior (1919−2009) and Harold Traver from the University of California synthesized ^{35}S-labeled cysteine[34] and methionine.[35] The ^{35}S- and ^{14}C-tagged amino acids were then used in attempts to follow protein synthesis in living animals and in tissue slices and homogenates. After some initial failures to resolve authentic protein synthesis from other metabolic reactions of the radiolabeled amino acids, it became possible to show that ^{14}C-labeled amino acids were incorporated into protein in whole animals and in slices of tissue and in cell homogenates.[36−40]

Some technical aspects of these early experiments should be pointed out. First, early measurements of radioactivity required decarboxylation of ^{14}C-alanine and monitoring of the released isotope in the gaseous phase. These cumbersome and slow measurements which slowed the progress of research[40] were abandoned after instrumentation that was devised in the late 1940s permitted the counting of radioactivity in solid phase.[8] A second difficulty was the interpretation of incorporation of labeled free amino into protein. Technically, this was done by exposing for a period animals or tissues to the radioactive amino acid and then using trichloroacetic acid (TCA) that precipitated proteins but left the free amino acid in the soluble phase. In some cases the TCA precipitates were also heated to destroy nucleic acids and maintain proteins only in the pellet. Yet, the presence of radioactivity in the precipitated phase was open to equivocal interpretations. One was that it represented true incorporation of the radiolabeled amino acid into newly synthesized protein. However, alternative interpretations were also possible. First, it could be that rather than being incorporated into de novo synthesized proteins the labeled amino acid was exchanged with unlabeled residues in preexisting proteins. Second, it was speculated that the incorporated amino acid did not form covalent α-peptide linkage. Finally, there was uncertainty whether the amino acid maintained its chemical integrity during the incorporation experiment.[2] Although incorporation was assumed right from the start to represent de novo protein synthesis, the doubts on its true significance lingered on until it was shown in the mid-1950s that the presence of incorporated labeled amino acids in protein reflected bona fide synthesis of polypeptide chains (see below).

The use of radioactively labeled amino acids bore immediate fruit when it was discovered that protein synthesis was an energy-requiring reaction. First, experiments by Ivan Franz (1916−2009), Loftfield, and Zamecnik showed that protein synthesis in tissue slices necessitated the supply of oxygen and that it did not occur in an atmosphere of nitrogen.[37,40] Also the incorporation of labeled amino acid into protein was inhibited when the production of ATP was blocked by sodium azide.[36] Finally, promptly after Lipmann reported in 1948 that dinitrophenol (DNP) abolished the production of ATP by uncoupling the electron transport system from oxidative phosphorylation,[41] the neighboring Massachusetts General Hospital laboratory of Zamecnik showed that DNP also inhibited the incorporation of labeled alanine into protein in tissue slices.[42] It thus became evident at an early stage that protein synthesis depended on supply of ATP and that incorporation did not depend on respiration en bloc but rather on the production of ATP.

To sum up, utilization of ^{14}C-labled amino acids rapidly proved to be an indispensable tool for the monitoring of incorporation of free amino acids into protein in living cells. Use of radiolabeled amino acids also led to the important observation that protein synthesis was an energy-consuming reaction that required supply of ATP by oxidative phosphorylation.

8.2.2 Differential Centrifugation Was a Second Critical Tool in the Construction of a Cell-Free Protein Synthesis System

Albert Claude (1899−1983) who shared the 1974 Nobel Prize with his Rockefeller Institute colleagues Christian de Duve (1917−13) and George Palade (1912−2008) pioneered in the 1930s and 1940s techniques

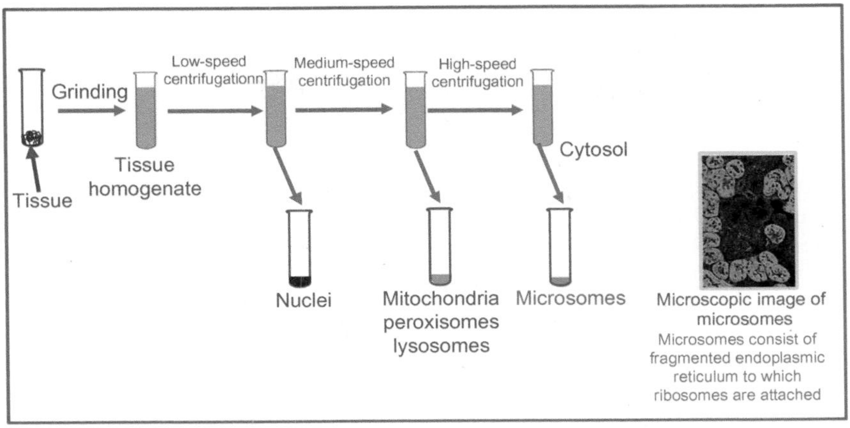

FIGURE 8.1 Scheme of the separation of subcellular components by differential centrifugation. Low-speed centrifugation is usually in the $700-1000 \times g$ range for 10 minutes, medium speed is $15-20,000 \times g$ for $15-20$ minutes, and high-speed is at $80-100,000 \times g$ for 60 minutes. Pelleting of free ribosomes and small polysomes can be achieved by centrifugation of the cytosolic fraction at $300,000 \times g$ for 2 hours or at $\sim 150,000$ for 3 hours.

for cell fractionation by differential centrifugation. Tissue homogenates were subjected under this procedure to repeated cycles of centrifugation at different speeds. As a result, the different centrifugal forces separately sedimented subcellular components according to their particular densities. By combining differential centrifugation with electron microscopy of the resolved subcellular fractions, Claude identified mitochondria, microsomes (see Chapter 9: "The Discovery and Rediscovery of Prokaryotic Messenger RNA"), and the endoplasmic reticulum. After protein synthesis was demonstrated in cell homogenates,[43,44] differential centrifugation became critical for the identification of subcellular fractions that participated in the in vitro incorporation of amino acids and for the combination of isolated inactive fractions into a functioning reconstituted cell-free system. Fig. 8.1 is a scheme of the separation of subcellular components by differential centrifugation as it was used in the 1950s and later for studies on protein synthesis in vitro.

8.2.3 Philip Siekevitz Constructed the First Cell-Free Protein Synthesis System

After serving for almost 4 years in the military during World War II, Philip Siekevitz (Box 8.1) embarked on doctoral studies in the University of California, Berkeley under the guidance of the biochemist David Greenberg (1895−1988).

BOX 8.1 Philip Siekevitz

A child of the Great Depression, Philip Siekevitz worked for 2 years after high school to save money for college. Following his graduation in 1942 from the Philadelphia College of Pharmacy and Science, he aimed at graduate studies in biochemistry but his plans were deferred by 3½ years of army service. After the end of World War II, he used the GI Bill to attend the University of California, Berkeley, where he conducted research on amino acid metabolism under David Greenberg. There he was among the first to use radioactive amino acids to follow protein synthesis in cells and tissue slices. Upon earning a PhD degree in 1949, he joined the laboratory of Paul Zamecnik at the Massachusetts General Hospital where he also studied the biochemistry of mitochondria with Fritz Lipmann. Most prominently, at that time he constructed the first cell-free protein synthesis system. This system was the foundation for the subsequent studies of Zamecnik and Hoagland that were capped by the identification of tRNA molecules and their activating enzymes. After a period in the University of Wisconsin, he was invited in 1954 by George Palade to join the Rockefeller University at which the two scientists collaborated for the next 20 years. By using radioactive amino acids to follow the synthesis and secretion of proteins in the pancreas, they discovered that pancreatic secretory enzymes were synthesized on ribosomes and then transported across the endoplasmic reticulum membranes into its lumen. Ultimately, the enzymes appeared in the zymogen granules that were secreted into the intestinal lumen. The life, work, and personality of Philip Siekevitz were appealingly described after his death by David Sabatini[45] (Fig. 8.2).

Philip Siekevitz
(1918–2009)

FIGURE 8.2   Philip Siekevitz.

In his studies on the synthesis of serine from glycine, Siekevitz became proficient at monitoring the incorporation of ^{14}C-labeled amino acids by liver slices. He also became familiar with work that Greenberg conducted in 1948 with Theodore Winnick (1912−2007) and Felix Friedberg (later at Howard

University) in which radiolabeled amino acids were incorporated into protein in rat liver homogenates that contained adenylic acid or ATP.[43] After earning a PhD degree in 1949 Siekevitz joined the laboratory of Paul Zamecnik (Box 8.2) at the Huntington Laboratories in Boston. There he initiated a project to dissect the rat liver cell-free protein synthesis system into its main components. Prior to Siekevitz's work in Boston, the Caltech laboratory of Borsook demonstrated that all four differentially sedimenting fractions of Guinea pig liver homogenates supported uptake of labeled amino acids.[46] As pointed out elsewhere (Ref. 5, p. 465), these early results were difficult to interpret because not all the measured uptake of the radioactive amino acid was due to true synthesis of protein and also because the measured radioactivity was not much higher than background levels. The experiments that Siekevitz performed in Boston established for the first time a reproducible in vitro system for the synthesis of proteins and identified its principal components and energy requirements. Siekevitz presented his results in a single-author 1952 paper in the *Journal of Biological Chemistry*.[47] Based on his previous experience, Siekevitz chose homogenates of rat liver as a starting material for his cell-free system. Synthesis of protein was monitored by the incorporation of ^{14}C-labeled alanine into acid-insoluble material. To ensure as much as was then possible that the incorporation reflected true integration of the radioactive amino acid into growing polypeptide chains, proteins were precipitated by cold and hot TCA to remove nucleic acids and the pellets were washed with alcohol and with alcohol/ether/chloroform to remove lipids. Siekevitz also used ninhydrin that releases CO_2 from free amino acids but not from protein to show that the acid-insoluble material did not contain free ^{14}C-alanine.

Disassembly and Reconstitution of the In Vitro Protein Synthesis System

A principal novelty of the Siekevitz study was the separation by differential centrifugation of amino acids incorporating cell homogenates into subfractions. He showed that whereas the resolved subfractions lost most of the ability to incorporate amino acids, this capacity was regained when certain fractions were combined.[47] Examination of the incorporation of ^{14}C-alanine into protein by subcellular fractions that were separated by differential centrifugation indicated that the highest specific activity (radioactivity per protein) was found in the microsomes (Fig. 8.3A). Although Siekevitz cautioned that this so-called microsomal fraction was impure and that it contained mitochondria, the enrichment in labeled protein in this fraction implied that the microsomes could be the site of protein synthesis.[47] In a next critical experiment combination of resolved nonactive subfractions of the homogenate reconstituted an active protein-synthesizing system. None of the centrifugally resolved subfraction conducted significant incorporation on its own (Fig. 8.3B). However, putting together the microsomal or supernatant subfractions with mitochondria,

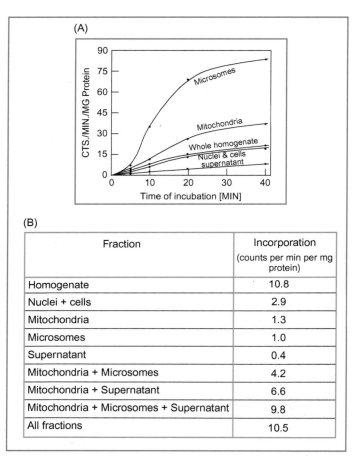

FIGURE 8.3 (A) Incorporation of ^{14}C-alanine by rat liver cell homogenate and by its differentially centrifuged subfractions. The specific activity of incorporated radiolabeled alanine was highest in a microsomal fraction that contained some mitochondria. (B) Results of a reconstitution experiment. The reaction buffer included ^{14}C-alanine, α-ketoglutarate as substrate for oxidation by the mitochondria, phosphate, and adenosine-5-phosphate (AMP) which could be converted to ATP. Whereas the isolated mitochondrial, microsomal, and supernatant subfractions alone had very low activity, their combination together restored almost completely the protein synthesis capacity. *Panel (A): Graph from Siekevitz P. Uptake of radioactive alanine in vitro into the proteins of rat liver fractions.* J Biol Chem *1952;195(2): 549–65. Panel (B): Results derived from Siekevitz P. Uptake of radioactive alanine in vitro into the proteins of rat liver fractions.* J Biol Chem *1952;**195**(2): 549–65.*

restored $\sim$40% or $\sim$65%, respectively, of the activity of the original homogenate. Most significantly, mixing the mitochondria with both the microsomal and the supernatant fractions reestablished the activity of homogenate in full (Fig. 8.3B).

Siekevitz's described initial results were not definitive because the separation technique that he employed yielded subcellular fractions that were cross-contaminated. To obtain a mitochondria-free microsomes he took advantage of the complete precipitation of nucleoproteins at an acidic pH. By precipitating at pH $\sim$5 the supernatant that remained after sedimentation of the mitochondria he obtained, therefore, a purified microsomal fraction. Using this fraction he was able to show that whereas the pH 5 microsomes could not incorporate amino acids on their own, incorporation activity was restored when they were supplemented with the mitochondrial and supernatant subfractions.[47]

Protein Synthesis In Vitro Depended on Supplied ATP

Siekevitz's next step was to define the basis for the requirement for mitochondria for protein synthesis. He established that the extent of ^{14}C-alanine incorporation in the isolated and combined subfractions was directly proportional to the amount of added mitochondria, to its contribution to oxygen consumption (respiration), to the oxidation of α-ketoglutarate, and to PO_4 esterification.[47] Also, whereas the uncoupling agent DNP did not affect oxygen consumption, it did inhibit both ATP production and incorporation of ^{14}C-alanine into protein. Similarly, trapping of the ATP by a glucose-hexokinase system also depressed incorporation. Based on these and other data, Siekevitz concluded that protein synthesis did not depend on respiration per se but rather on ATP that was provided by oxidative phosphorylation.[47]

A First Clue for the Activation of Amino Acids Prior to Their Incorporation

In addition to the reconstitution of an in vitro protein-synthesizing system from differentially centrifuged subcellular components and to the finding that incorporation of free amino acid required ATP, Siekevitz also obtained an early indication for the activation of amino acids prior to their joining into a polypeptide chain. Specifically, he preincubated the supernatant fraction with ^{14}C-alanine and ATP-generating mitochondria. When the preincubated mixture was added to concentrated microsomes, the subsequent incorporation of the labeled alanine into protein took place without a need for ATP. Furthermore, incorporation without ATP also occurred when the preincubated fraction was boiled for several minutes in neutral or acidic pH before it was added to the microsomes. Considering this result Siekevitz presciently stated[47]:

> By incubating the mitochondria with α-ketoglutarate or with succinate and cofactors, a soluble factor was formed which enabled the microsomal fraction obtained by precipitation at pH 5 to incorporate alanine into its proteins. This factor is probably an "activated alanine". It is stable to 100^0 for 7 minutes at

either neutral or acid (1.0 N HCl) pH. Though this factor is probably not ATP, it may be a compound derived from it.

This insightful observation was a first step on the road to the unraveling the biochemistry of the activation of amino acid and of their linking to tRNA.

8.2.4 Characterization of the In Vitro Protein Synthesis System

By effectively using the in vitro protein synthesis system that Siekevitz constructed, Paul Zamecnik and Mahlon Hoagland (Box 8.2) identified and characterized in the 1950s tRNA and the amino acid–activating enzymes.

BOX 8.2 Paul Zamecnik and Mahlon Hoagland

Paul Charles Zamecnik was born in Cleveland, Ohio. After graduating from the Dartmouth Medical School in 1934, he elected research over the practice of medicine. A clinical observation that he had made as an intern initiated his life-long interest in the question of protein synthesis. Because he was an MD and not a PhD in chemistry, Max Bergmann turned down his application to do research in his Rockefeller Institute laboratory. He thus traveled to Denmark to study under the protein chemist Kaj Linderstrøm-Lang at the Carlsberg Laboratory. After gaining this experience, he was accepted by Bergmann but he soon left for an offered position at the Huntington Laboratory at the Massachusetts General Hospital. Working there with Fritz Lipmann, he also established his independent laboratory. Together with Hoagland and others they discovered tRNA and deciphered the biochemistry of amino acid activation. He also prepared an in vitro protein synthesis system from *Escherichia coli* that was later used by Marshall Nirenberg in his work to break the genetic code. Showing in 1978 that oligonucleotides could penetrate cells, he developed antisense RNA that blocked translation of viral mRNA. For his fuller biography, see Ref. 48.

Mahlon Hoagland was born in Boston to Hudson Hoagland, a behavioral biologist whose dedication to science both lured and deterred his son. Deciding on medical career to avoid competition with his father, he attended Harvard Medical School. Just months before his scheduled graduation in 1945, he was diagnosed with tuberculosis and had to temporarily leave school. Only after recuperating, he was able to conclude in 1947 his medical studies. Due to reactivation of the disease, he abandoned his hopes for a career in surgery and instead did research with Joseph Aub, the director of the Huntington Laboratories. There he developed an interest in the work on protein synthesis that was done in the Zamecnik laboratory. Joining that lab in 1953, he conducted experiments that culminated in the discovery of tRNA and amino acid–activating enzymes. In 1958 he worked with Crick at Cambridge in an unsuccessful attempt to decipher the genetic code by using tRNA. Upon his return to the United States, he served in succession on the faculties of the Harvard and Dartmouth Medical

(Continued)

BOX 8.2 (Continued)
Schools. In 1970 he became Director of the Worcester Foundation, which his father founded. Throughout his career, he had an active interest in teaching and in scientific education. His biography can be found in Ref. 49 (Fig. 8.4).

Paul Zamecnik
1912–2009

Mahlon Hoagland
1921–2009

FIGURE 8.4   Paul Zamecnik and Mahlon Hoagland.

These two investigators made their landmark discoveries with several associates; mainly with Mary Louise Stephenson (1921–2009) (see Fig. 8.16A later in this chapter), and with Elizabeth Keller (1918–97) who later collaborated in Cornell University with the 1968 Nobel laureate Robert Holley (1922–93) in the sequencing of alanine tRNA.

ATP Hydrolysis Was Established as the Source of Energy for Amino Acid Incorporation

Taking off where Siekevitz had left, Zamecnik and Keller first defined more precisely the source of energy for protein synthesis. In a 1954 paper[50] they showed that when 3-phosphoglycerate was supplied as a substrate for glycolysis, mitochondria-depleted rat liver homogenate could support under anaerobic conditions incorporation of ^{14}C-leucine into acid-precipitable protein (Fig. 8.5A).

Incorporation of ^{14}C-leucine was stimulated by different high-energy phosphate compounds or precursors thereof, but not by ATP alone. Also, the addition of ATP to the high-energy compounds did not increase the incorporation (Fig. 8.5B). Reasoning that the effect of added ATP was thwarted by adenylic compounds that were present in the extract, Zamecnik

(A)

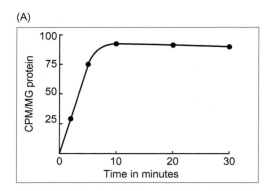

(B)

			C.p.m. per mg. protein	
Experiment No.	Flask No.	Additions, µM	Incomplete homogenate*	Complete homogenate*
1	1		24	10
	2	HDP 10, ATP 1	84	15
	3	3-PGA 10, ATP 1	48	66
	4	PC 20, ATP 1	111	82
2	1		14	
	2	ATP 1	9	
	3	PC 20	121	
	4	" 20, ATP 1	131	
	5	PPA 10	107	
	6	" 10, ATP 1	120	

Table II
Effect of high-energy phosphate compounds and precursors on incorporation of C^{14}-lucine into protein

HDP – Hexose diphosphate; PGA – 3-phosphoglycerate;
PC – phosphocreatine; PPA – phosphoenolpyruvate

FIGURE 8.5 (A) Time course of incorporation of ^{14}C-leucine into acid-precipitable material in the presence of 3-phophoglycerate and under anaerobic conditions. Mitochondria were removed from rat liver homogenate by centrifugation at $15,000 \times g$. The supernatant was incubated at 37°C in an atmosphere of 95% N_2, 5% CO_2, and in the presence of 3-phophoglycerate; ATP and ^{14}C-leucine. (B) Different high-energy phosphate compounds sustain incorporation in vitro of ^{14}C-leucine under anaerobic conditions. *Results in panels (A) and (B) are from Zamecnik PC, Keller EB. Relation between phosphate energy donors and incorporation of labeled amino acids into proteins. J Biol Chem 1954;209(1):337−54.*

and Keller next showed that after adenylic compounds were removed from the cell extracts by dialysis or by adsorption to activated charcoal, ATP plus phosphocreatine or phosphoenolpyruvate, but not ATP alone, stimulated incorporation.[50] Thus ATP whose hydrolysis provided the required energy for amino acids incorporation had to be continually replenished by a regenerating system such as, for instance, phosphocreatine/creatine phosphokinase.

Both Ribosomes and a 100,000 × g Supernatant Fraction Were Necessary for Incorporation

Zamecnik and Keller also made progress in identifying the subcellular fractions that executed the synthesis of protein. They removed mitochondria by differential centrifugation and then applied centrifugation at $105,000 \times g$ for about an hour to separate pelleted microsomes from a supernatant fraction that is commonly designated "100K supernatant" (Fig. 8.6A). Each of these fractions was tested for its ability to support the incorporation of radiolabeled amino acid under anaerobic conditions and in the presence of an ATP regenerating system. The only fraction that was capable of incorporating measurable amounts of the labeled precursor was the supernatant that remained after the removal of mitochondria but which still included ribosomes. By contrast, neither a ribosome-depleted supernatant fraction nor the isolated ribosomes themselves could carry out incorporation. However, recombination of the 100K supernatant with the ribosomal fraction restored 70% of the incorporation activity of the full extract (Fig. 8.6A). Execution of proteins synthesis required, therefore, the combined contributions of components of both the ribosomal and supernatant fractions. The reader's attention should be turned to the fact the maximum number of counts per minute that Zamecnik and Keller recorded with their highly inefficient radiation monitoring equipment was only 80 (!). Yet these low numbers were still significant since the measured background count was zero (Fig. 8.6A).

Incorporation In Vitro Reflected Authentic α-Peptide Bond Formation

As discussed, the unanswered question whether or not incorporation represented true integration of labeled amino acid into polypeptide chains lingered on for several years. Zamecnik and Keller attacked this problem by subjecting to acid hydrolysis proteins that were isolated after the incorporation of ^{14}C-leucine. Measurement of free amino acids that were released from the progressively hydrolyzed proteins revealed overlapping kinetics of release of unlabeled amino acids and of ^{14}C-leucine (Fig. 8.6B). Similar indistinguishable kinetics was monitored for proteins that were doubly labeled with both ^{14}C-leucine and ^{14}C-valine.[50] Hence it appeared that the incorporation of labeled amino acids by the cell-free system could be reliable taken as true reflection of assimilation of the radioactive residue into protein by the formation of α-peptide linkage.

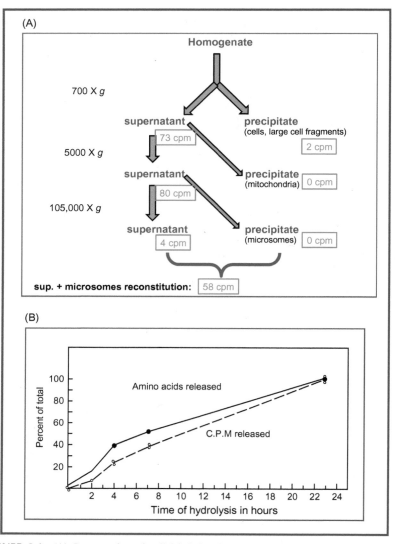

FIGURE 8.6 (A) Incorporation of radiolabeled amino acid required both the ribosomal and the $100,000 \times g$ supernatant subfractions. Subcellular fractions were separated from homogenate of rat liver cells by differential centrifugation, and the isolated subfractions or their combinations were assayed for incorporation in vitro of radioactively labeled amino acid under anaerobic conditions and in the presence of an ATP regenerating system. The applied centrifugal forces are listed at left and the counts per minutes of incorporated radioactivity are boxed in *red*. (B) Release of unlabeled and ^{14}C-labeled amino acids from acid-hydrolyzed protein. Following incorporation in vitro of ^{14}C-leucine, the isolated labeled proteins were dissolved in acid (2 M HCl) and the proteins were hydrolyzed at 100°C. Kinetics of release of free unlabeled amino acids was monitored by the appearance of ninhydrin-reacting material and the release with time of free ^{14}C-leucine was measured by counting acid-soluble radioactivity. The graph is from Ref. [50]. *Panel (A): Adapted from Zamecnik PC, Keller EB. Relation between phosphate energy donors and incorporation of labeled amino acids into proteins.* J Biol Chem *1954;**209**(1):337–54. Panel (B): From Zamecnik PC, Keller EB. Relation between phosphate energy donors and incorporation of labeled amino acids into proteins.* J Biol Chem *1954;**209**(1):337–54.*

8.3   IN THE MEANTIME IN CAMBRIDGE ... CRICK
PRESCIENTLY ENVISAGED AN ADAPTOR MOLECULE

While Zamecnik and his Boston associates were laboriously dissecting the protein synthesis system by conducting "wet" biochemical experiments, Francis Crick on the other end of the Atlantic contemplated the code problem in a purely theoretical manner (which Olby aptly described as Crick's "armchair approach" (Ref. 51, p. 231)). In early 1955 he circulated among the members of the RNA Tie Club a typewritten communication titled "On Degenerate Templates and the Adaptor Hypothesis," which summarized his ideas at the time on the code. This manuscript had never been published and became the most famous unpublished paper in molecular biology. Facsimiles of this historical article (which has already been discussed in Section 7.3.3) including Crick's handwritten modifications is available online both in the Wellcome Trust "Human Genome" website (http://genome.wellcome.ac.uk/assets/wtx030893.pdf) and in the National Library of Medicine "Profiles in Science"; the Francis Crick Papers Collection (http://profiles.nlm.nih.gov/ps/access/SCBBGF.pdf). In it Crick harnessed his critical theoretical thinking to the examination of key features of the code such as its directionally, possible overlap of codons and degeneracy. One hypothesis that he made in this manuscript and which stands out as a rare instance of the power of theoretical thinking in molecular biology was the prediction that nucleic acid adaptor molecules carry amino acids to the site of protein synthesis. Crick hypothesized that different species of the putative adaptor are charged with specific amino acids by specialized enzymes. As importantly, by complementing specific sequences (codons) in the RNA template, the adaptors place the transported amino acids at the site of protein synthesis and mediate their polymerization in an order that the template dictates. This prediction predated the actual physical identification of tRNA by Zamecnik and Hoagland and at the time neither they nor Crick were aware of each other's work or hypothesis. Although the adaptor hypothesis appears at first sight to have arrived from nowhere, in actual fact and as will be described next, it was a logical consequence of Crick's thoughts at the time on the code and protein synthesis.

8.3.1   Crick Precluded the Notion of Stereochemical Fit
Between Amino Acids and Nucleic Acid Template

Crick reached the idea of an adaptor through his critical examination and rejection of Gamow's model of stereochemical fit between amino acids and cavities in nucleic acids. A cardinal argument that he raised against Gamow's idea was that it was bereft of chemophysical basis. Concerning this issue, Crick wrote (this one as well as the next two quotations are from

Crick's unpublished 1955 communication to the RNA Tie Club. Underlined words and terms are his):

> ... *I cannot conceive of <u>any</u> structure [for RNA or DNA] acting as a direct template for amino acids, or at least as a specific template. In other words, if one considers the physical-chemical nature of the amino acid side chains we do not find complimentary features on the nucleic acid. Where are the knobly hydrophobic surface to distinguish valine from leucine and isoleucine? Where are the charged groups in specific positions to go with the acidic and basic amino acids?*

With this argument of lack of steric fit between amino acids and nucleic acids, Crick pointed to hydrogen bonds as the only nucleic acid elements that could determine specificity:

> *What the DNA structure <u>does</u> show (and probably RNA will do the same) is a specific pattern of <u>hydrogen bonds</u>, and very little else. It seems to me, therefore, that we should widen our thinking to embrace this obvious fact.*

This vision of hydrogen bonds as determinants of specificity was the key to Crick's adaptor hypothesis.

8.3.2   The Adaptor Hypothesis Explained How the Nucleotide-Based Language of RNA Could Be Translated Into the Amino Acid–Based Language of Proteins

Crick's own lucid account of his postulate is the best description of the adaptor hypothesis:

> *[...] each amino acid would combine chemically, at a special enzyme, with a small molecule which, having a specific hydrogen-bonding surface, would combine specifically with the nucleic acid template. This combination would also supply the energy necessary for polymerization. In its simplest form there would be 20 different kinds of adaptor molecule, one for each amino acid, and 20 different enzymes to join the amino acid to their adaptors. Sydney Brenner, with whom I have discussed this idea, call this the "adaptor hypothesis", since each amino acid is fitted with an adaptor to go on to the template.*

Hence, Crick visualized the hypothetical adaptor as a small molecule that complemented the nucleic acid template through hydrogen bonds. With an assortment of 20 different amino acids, he presumed that the minimal number of adaptors should be 20. Each of the 20 different adaptors was presumed to be charged with a particular amino acid by one of 20 specialized enzymes. Based on his idea of degeneracy of the code, Crick also declared that: "it is conceivable that there are more than one adaptor molecule for one amino acid." In fact, except for the proposition that the hydrogen bonding of the adaptor to the template provided the required energy for polymerization, subsequent experimental work proved the veracity of every other general facet of the hypothesis (though not of all its imagined details).

8.3.3 Crick Speculated That the Adaptors Were Likely to Be Made of Nucleic Acid and That Specific Enzymes Charged Them With Amino Acids

In Jan 1957 Crick visited the Boston laboratory of Zamecnik and Hoagland and was exposed to their experimental evidence on the charging of amino acid activation onto tRNA which they then named sRNA.[51] At a later time he delivered a lecture at the 12th symposium of the Society of Experimental Biology with the title *On Protein Synthesis* that was published in 1958.[52] In this analytical review Crick covered all the aspects of protein synthesis as they were known at the time and he also revisited the hypothetical adaptor and speculated on its chemical identity.[52] In reflecting on the chemical constitution of the putative adaptor, Crick focused on the need for hydrogen bonding with the template and deduced that the adaptor was likely to contain nucleotides[52]:

What sort of molecules such adaptors might be is anybody's guess.
They might, for example, be proteins [...] though personally I think that proteins, being rather large molecules, would take up too much space.
[...] They might be unsuspected molecules, such as amino sugars.
[...] But there is one possibility which seems inherently more likely than any other - that they might contain nucleotides. This would enable them to join on to the RNA template by the same 'pairing' of bases as is found in DNA, or in polynucleotides.

It is notable that Crick considered proteins to be unlikely candidates for adaptors because of their large size. This concern about the dimensions of the adaptor also drove him to imagine the nucleic acid adaptor as having minute size of probably three bases. Although he discussed at length in his talk the evidence of Zamecnik and Hoagland on amino acid activation and charging of tRNA,[52] Crick was seemingly disinclined to accept tRNA as synonymous with his hypothetical smaller adaptor molecule (Ref. 51, p. 265). In fact, he proposed that tRNA might be a large precursor that broke down into three bases—long adaptors to which the amino acids were linked.[52] In this view, the three-dimensional activating enzyme accommodated at its one end a minimal-length adaptor and carried on its other end the charged amino acid (Fig. 8.7).

Crick also elaborated in the same lecture on the role of specific enzymes in charging particular adaptor molecules with specific amino acids. He pointed out that the specificity of enzyme-catalyzed charging of the adaptors was a (preferred) alternative to the Gamow-style stereochemical direct fit between template and amino acid[52]:

If the adaptors were small molecules one would imagine that a separate enzyme would be required to join each adaptor to its own amino acid and that the specificity required to distinguish between, say, leucine, isoleucine and valine would be provided by these enzyme molecules instead of by cavities in the RNA.

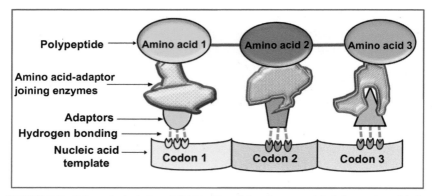

FIGURE 8.7   Schematic depiction of Crick's idea of the adaptors and their interaction with the template and with their activating enzyme. At least 20 different adaptors, of which 3 are sketched in the scheme, had a minimum length of perhaps only 3 bases. Each adaptor was charged with a specific amino acid by one of 20 specialized activating enzymes. The enzymes were supposed to carry at one end the charged amino acid and at the other the adaptor that formed hydrogen bonds with a complementary template sequence ("codon"). The pairing of the adaptors with the template juxtaposed the carried amino acid close enough to the growing polypeptide chain to enable formation of α-peptide bond.

8.3.4   The Adaptor Hypothesis Was Met With Initial Skepticism

With the wisdom of hindsight, a present day observer takes as a given both the logic and the basic veracity of the adaptor hypothesis. Yet, this was not the immediate reaction of Crick's contemporaries. Expecting criticism by experimentalists that there was no empirical evidence for the existence of adaptor molecules, Crick argued in his 1955 communication that the adaptors were hard to detect, possibly because they were "in short supply" relative to the pool of free amino acids. However, despite his caution, the hypothesis was not well received. In a Feb. 10, 1955, letter to Crick Watson expressed his dislike of the adaptor notion and urged continued effort to solve the structure of RNA as a key to the code problem. Beginning the letter with reference to Gamow's drinking problem and ending it by alluding to Crick's and to his own respective scientific fields of viscosity and ornithology prior to their work on nucleic acids, he wrote (see http://profiles.nlm.nih.gov/ps/retrieve/ResourceMetadata/SCBBJL for a facsimile and a transcript of the letter):

> *Gamow was here for 4 days—rather exhausting as I do not live on Whisky. Your TIECLUB note arrived during visit. Am not so pessimistic. Dislike adapters. We must find RNA structure before we give up and return to viscosity and bird watching.*

Watson was not alone in his disapproval of the adaptor idea. As cited in Ref. 53, p. 89, Brenner recalled: "When he circulated it [the Tie Club paper] it was pooh-poohed by everyone." Yet, Brenner himself, who was at the cradle of the adaptor hypothesis and gave it its name, thought that; "It was the only one that made chemical sense."

8.4   BACK IN BOSTON … DISCOVERY OF AMINO ACID–ACTIVATING ENZYMES AND OF THE ACTUAL RNA ADAPTORS

Zamecnik and Hoagland, who were not members of the Tie Club and were thus "out of the loop" and unaware of Crick's adaptor hypothesis, continued in their experiments to elucidate the mechanism of protein synthesis by the use of an in vitro protein synthesis system. Their stepwise efforts first resulted in exposition of the biochemistry of amino acid activation. This feat led in turn to the isolation of a new species of RNA, which was named in time sRNA and was later renamed tRNA. In this work they provided empirical evidence that different tRNA molecules were charged with amino acids by specific enzymes. Based on these data, Zamecnik proposed that the charged tRNA molecules complement specific sequences in the RNA template and enable their borne amino acids to form α-peptide bonds with the nascent polypeptide chain. Acting in this manner, tRNA satisfied the functions that Crick predicted for his hypothetical adaptor.

8.4.1   Discovery of Enzyme-Catalyzed Activation of Amino Acids

The biochemical path of amino acid activation was first outlined by Hoagland in a short 1955 paper.[54] A more detailed paper that Hoagland, Elizabeth Keller, and Paul Zamecnik published several months later validated the earlier report and provided decisive experimental confirmation of the details of the enzyme-catalyzed activation of amino acids.[55]

In his early study Hoagland undertook to determine the fate of amino acids prior to their incorporation into protein. As ATP was found to be essential for protein synthesis,[50] he first constructed a system that included the 100K cytoplasmic fraction of rat liver extract (Fig. 8.6A), ATP, and radiolabeled pyrophosphate, ^{32}PP. The mixture was incubated in the presence of potassium fluoride to inhibit glycolysis and without or with added free amino acids. Hoagland's initial results indicated that amino acids stimulated "incorporation" of ^{32}P into ATP and that the extent of stimulation was proportional to the number of different amino acids that were present in the reaction mixture. Suspecting that the amino acids enhanced the appearance of radioactive label in the ATP by stimulating an ATP $\leftrightarrow$ ^{32}PP + AMP exchange reaction, Hoagland dissected the reaction by setting up an amino acid, ATP, and hydroxylamine exchange reaction. In principle, hydroxylamine disassociated the AMP residue from its complex with an amino acid (AMP–amino acid) forming aminoacyl hydroxamate (Fig. 8.8A).

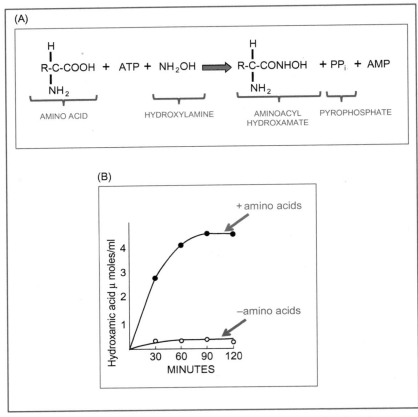

FIGURE 8.8 (A) Accumulation of aminoacyl hydroxamate is proportional to the extent of amino acid activation. Amino acids (AA) are activated in the presence of ATP by forming AMP−AA linkage (see text). Hydroxylamine disassociates AMP from the AMP−AA complexes to generate aminoacyl hydroxamates (AA-hydroxamate) that are quantified by a colorimetric reaction. The amount of AA-hydroxamates is equivalent to the amount of AMP−AA and thus to the amount of activated amino acid. (B) Amino acids stimulated the accumulation of hydroxamate. The reaction mixtures contained a concentrated pH 5 fraction of the 100K supernatant of rat liver extract (Fig. 8.9), ATP, and hydroxylamine without or with added 12 amino acids. *Modified from Hoagland MB, Keller EB, Zamecnik PC. Enzymatic carboxyl activation of amino acids. J Biol Chem 1956;218(1):345−58.*

Levels of amino acid activation could thus be gauged by colorimetric assay of the generated hydroxamates that were equivalent to the amounts of the AMP−amino acid complexes. In practice Hoagland incubated mixtures that contained six different amino acids and after terminating the activation reaction by heat denaturation of the enzymes, he determined in parallel the levels of hydroxamates, inorganic phosphate, and ATP.[54] Based on his quantitative results, Hoagland tentatively proposed the stages of the

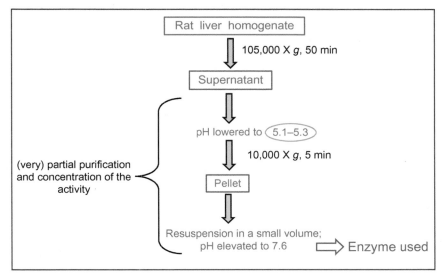

FIGURE 8.9 Preparation of a mixture of concentrated amino acid–activating enzymes. See text for details. *After Hoagland MB, Keller EB, Zamecnik PC. Enzymatic carboxyl activation of amino acids. J Biol Chem 1956;218(1):345–58.*

activation of amino acids by enzymes that were supposedly contained in the 100K cytoplasmic fraction. In his model an activating enzyme first labilized the AMP pyrophosphate (AMP–PP_i) linkage in ATP. In a second stage the amino acid displaced the pyrophosphate and formed through its carboxyl end an activated AMP–amino acid complex while releasing pyrophosphate.[54]

Within few months after the publication of his first short report, Hoagland substantiated its findings in a paper that he coauthored with Elizabeth Keller and Paul Zamecnik.[55] The first step in this second work was the preparation of a concentrated mixture of amino acid activation enzymes. Acidification to pH ~5 of the 100K supernatant fraction of rat liver extract precipitated proteins that included the activating enzymes. The precipitate was pelleted by centrifugation and resuspended in a small volume of a pH 7.6 buffer (Fig. 8.9). This solution served as source for a mixture of active (though impure) amino acid activation enzymes. By using the pH 5 enzyme fraction and assaying levels of formed hydroxamate, Hoagland, Keller, and Zamecnik next demonstrated that a mixture of 12 amino acids greatly stimulated the accumulation of aminoacyl hydroxamates (Fig. 8.8B).

Parallel measurements of the stimulation of ^{32}PP-ATP exchange and of hydroxamate formation by each one of five different amino acids and by a mixture of all the five indicated that the amino acids enhanced hydroxamate accumulation in an additive manner (Fig. 8.10A). This suggested that each one of the amino acids was activated by distinct specialized enzyme that catalyzed formation of respective AMP–amino acid complexes.

(A)

TABLE I
Hydroxamic Acid Formation and PP-ATP Exchange with Amino Acids

Amino acid, 5 µmoles each	Per cent exchange	Hydroxamic acid formed
		µmoles
Tryptophan..............................	2.4	1.1
Leucine.................................	14.7	1.0
Alanine.................................	1.2	0.7
Lysine..................................	0.2	0.3
Valine..................................	5.1	0.1
All 5...................................	17.3	3.0

FIGURE 8.10 (A) Five different amino acids additively stimulate ^{32}PP-ATP exchange and hydroxamate formation. Each of the listed five amino acids or their mixture were added to pH 5 fraction of rat liver cytoplasm and assayed for their stimulation of ^{32}PP-ATP exchange and of hydroxamate formation. The individual amino acids exhibited limited stimulatory activities whereas their mixture had greater enhancing activities. Adding up the stimulation by each one of the amino acids approximated the enhancement that was exerted by their mixture. *Table from Hoagland MB, Keller EB, Zamecnik PC. Enzymatic carboxyl activation of amino acids. J Biol Chem 1956;218(1):345−58.* (B) Isoelectrically resolved fractions of rat liver cytoplasm have different contents of amino acid−specific activating enzymes. Proteins of the rat liver supernatant fraction were precipitated at the three indicated pH values, pelleted by centrifugation, and resuspended in pH 7.6 buffer. Stimulation of hydroxamate accumulation was assayed by adding to each subfraction the designated amino acids: leu—leucine; gly—glycine; val—valine; try—tryptophan; ala—alanine. Results showed that the activation of different amino acids was distributed differently among the three isoelectrically resolved fractions.

The next step that Hoagland and his associates took was, therefore, to demonstrate directly the existence of separate amino acid−specific activating enzymes. Proteins were resolved isoelectrically into three fractions by stepwise addition of acid to the rat liver cytoplasm and collection by centrifugation of proteins that were precipitated at three different ranges of pH. The three iso-electrically separated subfractions of the cytoplasm were then supplemented with each one of four different amino acids and the stimulation of hydroxamate accumulation was assayed. Results that are reproduced in Fig.8.10B revealed that capacities to activate different amino acids were differently distributed among the three subfractions. Thus for instance, leucine greatly stimulated hydroxamate generation when it was added to the pH 5.8−6.2 subfraction whereas the same amino acid exerted minimal stimulation in the pH 6.2−7.0 or pH 5.2−5.8 subfractions. By contrast, whereas tryptophan yielded maximum stimulation when added to the pH 5.2−5.8 subfraction, it was less potent with the pH 5.8−6.2 subfraction and had practically no effect when it was added to the pH 6.2−7.0 subfraction (Fig. 8.10B). The other three amino acids failed to exert significant stimulation in any of the subfractions of the liver extract. These findings strongly indicated, therefore, that leucine and trypto-phan were activated by distinct and separable enzymes.

Importantly, Hoagland, Keller, and Zamecnik also demonstrated that the activation of amino acids was essential for protein synthesis. Specifically, they showed that hydroxylamine which disassociated the amino acid residue from the AMP−amino acid complex, also inhibited the incorporation of ^{14}C-leucine into protein in in vitro reaction mixtures that contained three different ATP regenerating systems.[55]

Based on their findings, Hoagland, Keller, and Zamecnik proposed a general mechanism of enzyme-catalyzed activation of amino as illustrated and described in the legend of Fig. 8.11.

A few months after Hoagland, Keller, and Zamecnik published their results and their model of amino activation, several other groups produced confirmatory data. G. David Novelli who migrated from Lipmann's laboratory to Case Western Reserve University (and later to Oak Ridge) showed with DeMoss in the same year of 1956 that amino acids stimulated ^{32}PP-ATP exchange in extracts of E. coli cells.[56] Also in 1956, the future (1980) Nobel Prize laureate Paul Berg (1926−), who was then at Washington University, documented activation of methionine by extracts of yeast cells.[57−59] The Lipmann laboratory purified in the very same year a tryptophan-specific activating enzyme from beef pancreas.[60] Using different systems, other investigators also described in 1957−58 activation of amino acids by the formation of AMP−amino acid intermediates.[61−64] Thus in less than 2 years the general mechanism and key steps of the activation of amino acids became established.

8.4.2 Isolation and Characterization of tRNA and Its Identification as the Acceptor for Activated Amino Acid

Hoagland, Zamecnik, and Mary Louise Stephenson submitted in mid-January 1957 a two-page preliminary note to the journal Biochimica et Biophysica

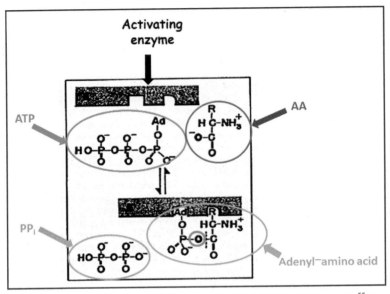

FIGURE 8.11   Amino acid activation according to Hoagland, Keller, and Zamecnik.[55] The activating enzyme was proposed to bind both ATP at its adenyl (Ad) moiety and the amino acid through its R moiety. While bound to the enzyme, the oxygen (*red circle*) in the carboxyl group of the amino acid (AA) attacks the bond between the pyrophosphate (PP$_i$) and AMP in the ATP causing the release of pyrophosphate and formation of an AMP−amino acid compound. *Scheme modified from Hoagland MB, Keller EB, Zamecnik PC. Enzymatic carboxyl activation of amino acids. J Biol Chem 1956;218(1):345−58.*

Acta. In it they announced the identification of an intermediate reaction between the activation of amino acids and their actual incorporation into protein.[65] In a subsequent article that Hoagland, Zamecnik, and associates submitted in Sep. of the same year to the *Journal of Biological Chemistry*, and that was published in early 1958, the key molecule in this intermediate reaction was identified as S-RNA that carried amino acids to the microsomal site of protein synthesis.[66]

In their preliminary communication Hoagland, Zamecnik, and Stephenson reported that in addition to amino acid−activating enzyme the supernatant fraction of rat liver extracts also contained about 5% RNA on a weight basis. They also found that after incubation of the supernatant fraction with ^{14}C-leucine and ATP, the radioactive label was transferred to a low-molecular-size RNA that they initially designated S-RNA (for small RNA).[65] Shortly thereafter they renamed this new RNA species sRNA until it finally got its present designation of tRNA. Results showed that the linkage between the radioactive leucine and the S-RNA resisted acid hydrolysis but that enzymatic digestion of the RNA moiety or its hydrolysis by alkali released free ^{14}C-leucine. When incubated with microsomes, ^{14}C-leucine-S-RNA lost about two-thirds of the linked radiolabeled amino acid which was reciprocally relocalized to newly made protein. As was shown in a separate paper by Keller and Zamecnik,[67] Hoagland, Zamecnik, and Stephenson also affirmed that guanosine triphosphate (GTP) and its regenerating system were necessary for the incorporation of labeled amino acids into protein.

After first emphasizing that the activation of amino acids was an essential first step in the path to their incorporation into protein, Hoagland, Zamecnik, and Stephenson ended their brief report with the statement[65]:

> It is now further postulated that this initial activation of amino acids is followed by transfer of activated amino acids to S-RNA. This latter reaction is ribonuclease sensitive, while the former is not. GTP mediates the transfer of this activated amino acid to peptide linkage via the microsomes by a mechanism as yet unknown.

These conclusions were substantiated and expanded in the next paper by Hoagland and his associates.[66] In it they first identified the subcellular fraction that bound activated amino acids before their incorporation into protein.

Activated Amino Acids Bind to RNA in the Cytoplasm

Rat liver cell homogenate was fractionated by differential centrifugation and each of the separated subcellular fractions was incubated under anaerobic conditions at 37°C with ^{14}C-leucine and with ATP and its regenerating system. At the end of the incubation, RNA was isolated by phenol extraction and its specific activity (counts of radioactivity per minute per milligram of RNA) was measured. Removal of nuclei and mitochondria and isolation of the remaining cytoplasmic fraction increased the specific activity of the labeled RNA by up to fourfold (Fig. 8.12A). By contrast, RNA that was isolated from nuclei, mitochondria, or microsomes had no or low level of radioactivity. Finally, enrichment of the RNA-associated ^{14}C-leucine was attained by precipitating the supernatant fraction at pH 5 such that the pelleted RNA had nearly eightfold higher specific activity than the homogenate (Fig. 8.12A).

The Activated Amino Acid Becomes Bound to Cytoplasmic RNA

In their next experiment, Hoagland and his associates showed directly that the cytoplasm contained RNA that bound the radiolabeled leucine. First, RNA that was extracted from the pH 5-precipitated supernatant fraction bound ^{14}C-leucine four- to fivefold better than either total RNA of ascites cells or microsomal RNA (Fig. 8.12B). With the perspective of hindsight, this result is easily explained by the fact that whereas tRNA was the exclusive type of RNA in the pH 5 precipitate of supernatant fraction (cytoplasm), the dominant RNA component in both the total cell extract and in the microsomal fraction was ribosomal RNA which does not bind activated amino acids.

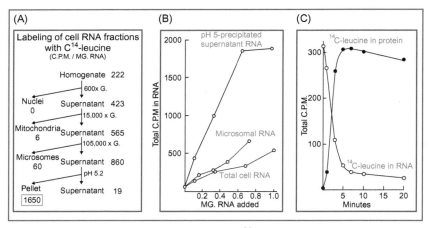

FIGURE 8.12 (A) Specific activity of RNA-linked ^{14}C-leucine in total rat liver homogenate and in subcellular fractions. The homogenate and each of its subfractions were incubated under anaerobic conditions at 37°C with ^{14}C-leucine, ATP, and its regenerating system but without added microsomes. After terminating the reaction, RNA was extracted from each fraction and its specific activity (counts per mg of RNA) was determined. The highest measured specific activity was recorded in the pH 5-precipitated supernatant fraction (*red frame*). (B) Different RNA preparations had different capacity to be charged with ^{14}C-leucine. RNA was extracted from the pH 5 precipitate of the rat liver cytoplasmic fraction, from the microsomes of the same cells, and from whole ascites cells. The various RNA fractions were incubated with ^{14}C-leucine together with pH 5-precipitated activating enzymes, ATP, and regenerating system. Shown are specific activities of RNA samples that were extracted from the different fractions. (C) Kinetics of the transfer of ^{14}C-leucine from prelabeled pH 5 RNA to newly made, microsome-associated protein. RNA that was labeled with ^{14}C-leucine by the pH 5 fraction of the supernatant of rat liver extract was incubated with microsomes under protein synthesis conditions. Shown are levels of radioactivity that were measured at different times in microsome-associated RNA and in microsome-associated newly made protein. *All the three panels of this figure were derived from Hoagland MB, Stephenson ML, Scott JF, Hecht LI, Zamecnik PC. A soluble ribonucleic acid intermediate in protein synthesis.* J Biol Chem *1958;231(1):241−57.*

The tRNA Charged Amino Acid Was Transferred to Newly Made Protein

A key finding was obtained when Hoagland et al. incubated microsomes with ^{14}C-leucine-charged RNA under conditions of protein synthesis. Monitoring the time course of the changing location of the radiolabeled amino acid revealed that it was progressively depleted from the RNA and that this loss was accompanied by reciprocal enrichment of radioactivity in the microsome-associated newly synthesized proteins (Fig. 8.12C). Thus, the gathered evidence showed that following their activation, amino acids were charged onto low-molecular-size cytoplasmic RNA (tRNA). Discharge of the amino acids from the RNA and their polymerization into polypeptide chains happened in the presence of microsomes and under conditions of protein synthesis.

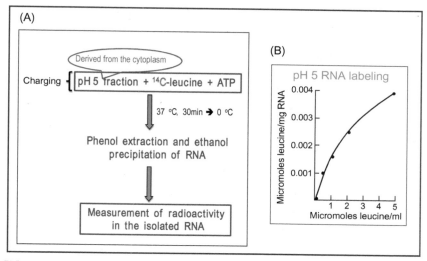

FIGURE 8.13   Charging of pH 5 cytoplasmic RNA with ^{14}C-leucine. (A) Procedure of isolation of cytoplasmic RNA charged with ^{14}C-leucine. (B) Time course of the labeling of the cytoplasmic RNA with ^{14}C-leucine. *Derived from Hoagland MB, Stephenson ML, Scott JF, Hecht LI, Zamecnik PC. A soluble ribonucleic acid intermediate in protein synthesis. J Biol Chem 1958;231(1):241−57.*

The Binding of Activated Amino Acid to the Cytoplasmic RNA Was Demonstrated

In another experiment Hoagland and his associates showed that the enrichment of the cytoplasmic pH 5 fraction with RNA-associated ^{14}C-leucine represented true binding of the radiolabeled amino acid to RNA. After incubating ^{14}C-leucine in a mixture that contained ATP and the cytoplasmic pH 5 fraction that included both the activating enzyme and RNA, proteins were extracted by phenol and the RNA was precipitated with ethanol (Fig. 8.13A). Determination of the radioactivity in RNA samples that were isolated at different incubation times indicated that the RNA became progressively enriched in bound ^{14}C-leucine[66] (Fig. 8.13B).

8.4.3   The Mode of Charging of Activated Amino Acids Onto tRNA Was Uncovered

The details of the loading of activated amino acids onto tRNA were exposed in the second half of the 1950s.

Identification of the tRNA Amino Acid CCA Acceptor End

Following early reports that labeled nucleotides could be incorporated into RNA,[57,68−72] several laboratories made the more definitive observation that the 3′ and 2′-hydroxyl termini of tRNA served as acceptors for cytosine monophosphate (CMP) and adenosine monophosphate (AMP), and to a lesser extent for uracil monophosphate.[57,73] Specifically, the pH 5 cytoplasmic fraction ("100K supernatant") hydrolyzed supplemented radiolabeled cytosine triphosphate (^{14}C-CTP) and adenosine triphosphate (^{14}C-ATP) and the product nucleotide monophosphates; CMP and AMP formed in sequence phosphodiester linkages with the free 3′ or 2′-hydroxyl group of tRNA.[74,75] Although unknown in the 1950s, it was later found that a specific enzyme; *CCA tRNA nucleotidyl transferase* catalyzed the addition of CMP and AMP to the 3′ end of tRNA. Most importantly, Liselotte (Lisa) Hecht, who was at the time a postdoctoral fellow in the Zamecnik laboratory, showed with Mary Stephenson and Zamecnik that tRNA molecules invariably had at their 3′-terminus the three nucleotides CCA[75] and that this trinucleotide end-unit was essential for the charging of amino acids onto the tRNA.[74]

Demonstration of the Linking of Amino Acids to the 3′ or 2′ Hydroxyl Groups of the Terminal Adenosine in the CCA End of tRNA

The laboratories of Lipmann and of Zamecnik, respectively, published in 1958 and 1959 experimental results that provided direct evidence for the loading of activated amino acid onto tRNA by the formation of ester bond with the 3′ or 2′ hydroxyl groups of the terminal adenosine of the 3′ CCA end-unit in tRNA.

Hans Georg Zachau (1930−) and George Acs (1923−2013) working then in the Rockefeller Institute laboratory of Fritz Lipmann charged tRNA with ^{14}C-leucine or ^{14}C-valine and then digested the RNA with ribonuclease (RNase). Electrophoretic resolution of the RNase digestion products identified a small basic fragment that had nearly all the initial radioactivity. Chemical analyses revealed that this fragment is comprised of the radiolabeled amino acid in linkage to adenosine.[76] Zachau, Acs, and Lipmann concluded that breakage of the bond in the AMP−amino acid complex allowed the liberated amino acid to form an ester linkage with the adenosine at the 3′-end of its cognate tRNA molecule (Fig. 8.14).

Hecht, Stephenson, and Zamecnik drew a similar conclusion from results of their independent experiments. They first showed that tRNA that was isolated from fresh pH 5 fraction of the cytoplasm contained the CCA end-unit as well as charged amino acids. However, incubation of the pH 5 fraction without CTP, ATP, and amino acids and in the presence of AMP and phosphate resulted in loss of the CCA end-group and of the charged amino acids.[77] Starting with the acquired stripped tRNA molecules, Hecht, Stephenson, and Zamecnik next showed that the charging of amino acids onto tRNA occurred in three steps. The first demonstrated step was hydrolysis of CTP and the addition of two CMP residues to the end of the tRNA

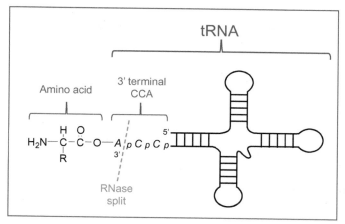

FIGURE 8.14   Linkage of an amino acid to the 3′-terminal adenosine residue in tRNA. In the actual experiment of Zachau et al.[76] the amino acid was radiolabeled and after RNase cleaved the aminoacyl-tRNA at the marked site the radioactive amino acid was found to be linked to adenine. *Illustration after Zachau HG, Acs G, Lipmann F. Isolation of adenosine amino acid esters from a ribonuclease digest of soluble, liver ribonucleic acid. Proc Natl Acad Sci USA 1958;44(9):885—9.*

to form tRNA-pCpC molecules ("p" denotes a 3′−5′ phosphodiester bond). The second step was hydrolysis of ATP and the addition of a terminal AMP residue to the tRNA-pCpC. The third demonstrated step was the linking of an activated amino acid to the terminal adenine nucleotide in the tRNA-pCpCpA. The site of amino acid linkage was identified as either the 3′ or the 2′ hydroxyl group of the terminal AMP residue in tRNA.[77]

8.4.4   Zamecnik's 1960 Model of the Path From Free Amino Acids to Their Incorporation Into Protein

All in all, results gathered by the Zamecnik and Hoagland group and by others provided strong evidence that following the activation of amino acids they were charged onto the low-molecular-size cytoplasmic tRNA. In the presence of microsomes and under conditions of protein synthesis the amino acids were discharged in turn from the tRNA and polymerized into protein. Based on the work that was mainly accomplished in his laboratory, Zamecnik described in a 1960 Harvey lecture the steps that free amino acids take on their way to be incorporated into protein.[78] An illustration from that lecture (Fig. 8.15) encapsulates Zamecnik's ideas in 1960 on protein synthesis which have been proven in time to be mostly correct.

One misconception, however, was that Zamecnik, similar to almost everyone else at the time, thought that ribosomal RNA, in its role as the cytoplasmic bearer of genetic information, served as the template for protein

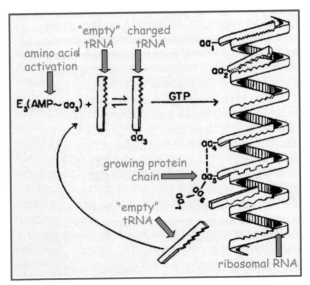

FIGURE 8.15   Zamecnik's 1960 scheme of the path from free amino acids to their polymerization into protein. *Modified from Zamecnik PC. Historical and current aspects of the problem of protein synthesis.* Harvey Lect *1960;54:256—81.*

synthesis. This misconstrued function of ribosomal RNA was abandoned a year or two after Zamecnik's 1960 Harvey lecture when the concept of mRNA emerged and after mRNA molecules were detected in phage-infected and in uninfected *E. coli* cells (see Chapter 9: "The Discovery and Rediscovery of Prokaryotic Messenger RNA").

8.4.5   Zamecnik and Hoagland, Discoverers of Activating Enzymes and of tRNA, and Crick, the Originator of the Adaptor Hypothesis, Finally Learned of Each Other's Attainments

The story of tRNA is a unique case in the annals of molecular biology of autonomous theoretical and experimental lines of investigations that, without cross-pollinating one another, exposed the same biological reality. Hypothetical adaptor molecules and their general mode of operation were theoretically predicted by Crick in an attempt to explain how the 4-letter language of nucleic acids may be translated into the chemically distinct 20 letters language of proteins. At the very same time, the methodical dissection of in vitro protein synthesis system led Zamecnik and Hoagland to the experimental identification of amino acid activation by specialized enzymes, of tRNA, and of ribosome-requiring polypeptide synthesis. It was only after their discoveries of amino acid activation and of tRNA

(A)

Zamecnik, Hoagland, and Mary Stephenson at the time of their discoveries

(B)

...and 35 years later

The winged theoretician Crick and the grappling experimenters Hoagland and Zamecnik

FIGURE 8.16   (A) Zamecnik (left), Hoagland (center), and Mary Stephenson at different times. (B) Artist's rendition of Hoagland's 2004 portrayal of the struggling experimentalists (Zamecnik and he) and of Crick, the winged theoretician. *Panel (A): Modified from Zamecnik PC. From proteins synthesis to genetic insertion.* Annu Rev Biochem *2005;74:1−28. Panel (B): Modified from Hoagland M. Enter transfer RNA.* Nature *2004;431(7006):249.*

were accomplished that Hoagland and Zamecnik (Fig. 8.16A) learned of the adaptor hypothesis.

In 2004 Hoagland described in a charming short memoirs how Zamecnik and he became belatedly aware of the adaptor hypothesis[79]:

> *In late 1956 Jim Watson, recently appointed to the Department of Biology at Harvard, learned of our findings through the efficient Boston-Cambridge (Massachusetts) grapevine and paid us a visit. After hearing our account of the discovery of sRNA, he asked if we knew of Francis Crick's adaptor hypothesis. Acknowledging our ignorance and somewhat miffed that a molecular biologist had foretold the existence of the intermediate we had discovered, we couldn't help but admire Francis's prescience. An image arose before me: we explorers, slashing and sweating our way through a dense jungle, rewarded at last by a vision of a beautiful temple − looking up to see Francis, on gossamer wings of theory, gleefully pointing it to us!" (Fig.8.16B)*

Crick also learned of the Zamecnik and Hoagland results only after the event. As mentioned, he visited in Jan. of 1957 the Boston laboratory of Hoagland

and Zamecnik and learned of their discoveries of the activation of amino acids, their loading onto tRNA and subsequent microsome-requiring polymerization into polypeptide chains. Having been greatly impressed he wrote to a friend: "I visited Hoagland on my way back to England and now bursting with ideas about protein synthesis" (cited in Ref. 51, p. 247). As noted (Section 8.3.3), Crick originally imagined his hypothetical adaptor to be small (probably three nucleotide long). Thus, he first tended to think that the larger tRNA was a precursor that was processed into the smaller adaptor. Yet, recognizing the importance of tRNA, he invited Hoagland to spend a year with him in Cambridge. There they attempted, without success, to decipher the genetic code by isolating amino acid–specific tRNA molecules and identifying their coding trinucleotides (anticodons).[7]

8.5 THE CODING PROPERTIES OF tRNA WERE SHOWN TO DETERMINE THE IDENTITY OF INCORPORATED AMINO ACIDS

The work of Zamecnik and Hoagland had left one unturned stone in the study of the function of tRNA in the translation of RNA into protein. Crick's hypothesis assumed that the adaptor acted as a bifunctional agent. While an adaptor molecule was purported to carry at one of its ends a specific amino acid, another end was hypothesized to pair by hydrogen bonds with a codon in the RNA template. If this hypothesis was true, then only the complementarity between the nucleotide sequences of the tRNA and the RNA template, and not the nature of the carried amino acid, should determine the final sequence of amino acids in the protein product. Although Zamecnik and Hoagland uncovered most of the steps in the process of activation of free amino acids, their loading onto tRNA and ultimate polymerization into polypeptide chain, this last proof of the bifunctional mode of operation of tRNA was still missing.

The French-Polish scientist François Chapeville (born Franciszek Chrapkiewicz in 1924) worked in the early 1960s with Seymour Benzer and in collaboration with Fritz Lipmann and others. These investigators undertook to determine whether the specificity of incorporation of amino acids into protein resided with the carried amino acid or with the tRNA itself. Experimental results that these investigators presented in a 1962 article in the *Proceedings of the National Academy of Sciences*[80] provided unambiguous evidence that the specificity of incorporation of amino acids into protein was determined by the tRNA molecule and not by the amino acid that it carried. The principle of their elegant experiment was as follows: An amino acid that was already linked to tRNA was chemically modified. If specificity resided with the tRNA, the *modified* amino acid would have been incorporated into protein. If however, the amino acid itself determined the specificity, the modified residue *would not* be present in the synthesized protein.

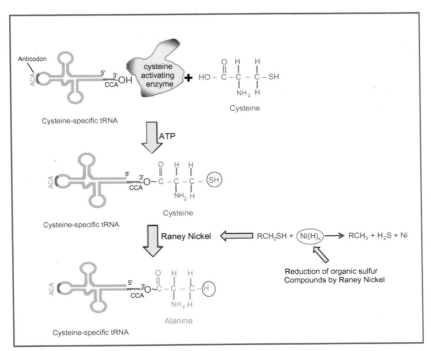

FIGURE 8.17 Scheme of the strategy to modify cysteinyl-tRNA to alanyl-tRNA. See text for details. *Modified from Chapeville F, Lipmann F, Von Ehrenstein G, Weisblum B, Ray Jr WJ, Benzer S. On the role of soluble ribonucleic acid in coding for amino acids. Proc Natl Acad Sci USA 1962;48(6):1086–92.*

Under this general strategy, Chapville and his associates charged tRNA with the sulfur-containing labeled amino acid ^{14}C-cysteine. They next exposed the cysteinyl-tRNA to the strong reducing agent of sulfur-containing organic molecules Raney Nickel that converted tRNA-linked cysteine residues to alanine (Fig. 8.17). In practice, this procedure yielded a population that comprised of 60% ^{14}C-alanyl-tRNA and 40% ^{14}C-cysteinyl-tRNA.[80]

Next the extent of incorporation into protein of untreated ^{14}C-cysteinyl-tRNA was compared to the incorporation into protein of the Raney Nickel–treated ^{14}C-cysteinyl-tRNA of which 60% were modified into ^{14}C-alanyl-tRNA. The in vitro protein synthesis system contained ribosomes, ATP and GTP, and a regenerating system. An added template was the synthetic polyribonucleotide poly(UG) that had been shown to serve as template for cysteine but not for alanine.[81–84] Results revealed similar extents of incorporation of both types of ^{14}C-labeled aminoacyl-tRNA molecules. They also showed that not only ^{14}C-cysteine but also ^{14}C-alanine must have been incorporated into protein. Modified ^{14}C-cysteine was incorporated at a level of 447 counts per minute (cpm) (30% of an input of 1490 cpm). Of those,

60% were alanine and thus 268 cpm of ^{14}C-alanine appeared to have been incorporated into protein (Fig. 8.18A(2)). It seemed, therefore, that although the poly(UG) did not encode alanine, it was still incorporated into protein after having been generated by the modification of cysteinyl-tRNA.

Next, Chapville and his associates ascertained that the modified cysteinyl-tRNA indeed delivered alanine to the poly(UG)-dictated protein. They compared incorporation into protein of ^{14}C-labeled unmodified cysteine, of alanine that were linked to their respective cognate tRNA species, or of Raney Nickel−treated cysteine that was carried by the cysteine-specific tRNA. Results indicated that the poly(UG) template dictated incorporation of unmodified ^{14}C-cysteine (346 cpm above background) but not of alanine (only 11 cpm above background) (Fig. 8.18B). By contrast, after the conversion by Raney Nickel of 60% of the cysteine in cysteinyl-tRNA to alanine, 460 cpm were detected in the synthesized protein. This indicated that a significant amount of alanine that was carried by the cysteine-specific tRNA was incorporated into protein (Fig. 8.18B).

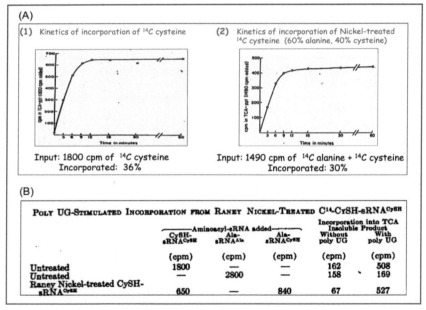

FIGURE 8.18   (A) Time course of poly(UG)-directed incorporation of unmodified (1) and Raney Nickel−modified (2) tRNA-linked ^{14}C-cysteine. The modified ^{14}C-cysteinyl-tRNA comprised of 60% ^{14}C-alannyl-tRNA and 40% ^{14}C-cysteinyl-tRNA. (B) Poly(UG), a template encoding cysteine polymerization, dictated incorporation into protein of ^{14}C-alanine that was generated by Raney Nickel modification of cysteinyl-tRNA. See text for details. *Panels (A) and (B): From Chapeville F, Lipmann F, Von Ehrenstein G, Weisblum B, Ray Jr WJ, Benzer S. On the role of soluble ribonucleic acid in coding for amino acids.* Proc Natl Acad Sci USA 1962; 48(6):1086−92.

Chapeville and his associates concluded their experiments by a direct experimental proof of the incorporation into protein of alanine residues which were generated by the reduction of the cysteine in cysteinyl-tRNA. Nickel-treated ^{14}C-cysteinyl-tRNA was added to an in vitro protein synthesis system and the synthesized proteins were acid-precipitated and then subjected to oxidative hydrolysis. Paper electrophoresis of the products of hydrolysis resolved spots not only of ^{14}C-cysteic acid but also of ^{14}C-alanine. This result confirmed that a part of the cysteine residues in molecules of cysteinyl-tRNA was converted to ^{14}C-alanine and that the latter was incorporated into protein despite the fact that it was carried by cysteine-specific tRNA.[80]

The classic experiments of Chapeville and his associates established, therefore, that the specificity of amino acids incorporation into protein resided with their tRNA carriers and not with the amino acids themselves. This conclusion was best put by the authors themselves[80]:

Since an amino acid, once attached, no longer participates in coding, it follows that the code (i.e., the correspondence between nucleotide sequence in template RNA and amino acid sequence in protein) is embodies in the precise structures and interrelationships of the set of sRNA adaptors and activating enzymes.

8.6 POSTSCRIPT: BACK TO CRICK'S VISIONARY VIEW OF THE ADAPTOR AS AN ESSENTIAL NEXUS IN THE FLOW OF GENETIC INFORMATION

Present day readers with their full knowledge of the experimental identification of amino acid activation and of tRNA are best positioned to evermore appreciate Crick's prescient hypothesis of the adaptor.

In his 1958 Society of Experimental Biology talk[52] Crick underlined the critical role of the adaptor (which in time proved to be synonymous with tRNA) in the flow of genetic information.[52] He used in this seminal lecture extant experimental data to prophetically set forth the general tenets of the path of genetic information from DNA down to its translation into proteins. Crick's basic precept was his "*sequence hypothesis*" (this and the next quotations are all from Cricks published lecture[52]):

In its simplest form it assumes that the specificity of a piece of nucleic acid is expressed solely by the sequence of its bases, and that this sequence is a (simple) code for the amino acid sequence of a particular protein.

As an inference from this basic principle Crick put then forward his famous "*Central Dogma*" (which had later been reduced to the simplistic scheme DNA → RNA → protein):

This states that once 'information' has passed into protein it cannot get out again. In more detail, the transfer of information from nucleic acid to nucleic acid, or from nucleic acid to protein. In more detail, the transfer of information from nucleic acid to nucleic acid, or from protein to nucleic acid is impossible. Information means here the precise *determination of sequence, either of bases in the nucleic acid or of amino acid residues in the protein.*

Much had since been written about of the historical circumstances of the introduction of the Dogma and of the reactions that it had raised. No less attention had been given to the substance of the Dogma and to its reevaluation in light of later developments (ie, reverse transcription, prions). Here we point to only one derivative aspect of the Dogma. This is Crick's vision of the role of the adaptor in the flow of genetic information. Rejecting any Gamow's-style idea of a possible stereochemical fit between crevices in nucleic acids and amino acids, he raised the alternative hypothesis of an adaptor as the vehicle that enables the crossing of the border between nucleic acid and protein in the forward flow of genetic information. Considering the chemistry of the RNA template, Crick wrote:

... [Template] *RNA presents mainly a sequence of sites where hydrogen bonding could occur. One would expect, therefore, that whatever went on to the template in a* specific *way did so by forming hydrogen bonds. It is therefore a natural hypothesis that the amino acid is carried to the template by an 'adaptor' molecule, and that the adaptor is part which actually fits on to the RNA. In its simplest form one would require twenty adaptors, one for each amino acid.*

And:

If the adaptors were small molecules one would imagine that a separate enzyme would be required to join each adaptor to its own amino acid and that the specificity required to distinguish between, say, leucine, isoleucine, and valine would be provided by these enzyme molecules [...]. Enzymes, being made of proteins, can probably make such distinctions more easily than can nucleic acids.

Crick next put the amino acid—carrying adaptor in the context of protein synthesis:

Each adaptor molecule containing, say, a di- or trinucleotide would each be joined to its own amino acid by a special enzyme. These molecules would then diffuse to the microsomal particles and attach to the proper place on the bases on the RNA by base-pairing, so that they would then be in a position for polymerization to take place.

Finally, he underlined the role of the adaptor as a nucleic acid intermediate that negotiates the language of nucleic acid with that of amino acids and protein:

It will be seen that we have arrived at the idea of common intermediates without using the direct experimental evidence in their favour; but there is one important qualification, namely that the nucleotide part of the intermediates must be specific for each amino acid, at least to some extent [...].

Because successful theoretical predictions are very rare in the life sciences in general and in molecular biology, the remarkable insight of Crick in the case of the adaptor/tRNA is evermore admirable.

REFERENCES

1. Bartels D. The multi-enzyme programme of protein synthesis—its neglect in the history of biochemistry and its current role in biotechnology. *Hist Philos Life Sci* 1983;**5**(2):187–219.
2. Zamecnik PC. Historical aspects of protein synthesis. *Ann NY Acad Sci* 1979; **325**(1):269–302.
3. Siekevitz P, Zamecnik PC. Ribosomes and protein synthesis. *J Cell Biol* 1981;**91**(3, Pt 2): 53s–65s.
4. Burian RM. Technique, task definition, and the transition from genetics to molecular genetics: aspects of the work on protein synthesis in the laboratories of J. Monod and P. Zamecnik. *J Hist Biol* 1993;**26**(3):387–407.
5. Rheinberger H-J. Experiment and orientation: early systems of in vitro protein synthesis. *J Hist Biol* 1993;**26**(3):443–71.
6. Darden L, Craver C. Strategies in the interfield discovery of the mechanism of protein synthesis. *Stud Hist Philos Sci Part C* 2002;**33**(1):1–28.
7. Rheinberger H-J. A history of protein biosynthesis and ribosome research. In: Nierhaus KH, Wilson DN, editors. *Protein synthesis and ribosome structure: translating the genome.* Weinheim: Wiley VCH; 2004.
8. Zamecnik PC. From proteins synthesis to genetic insertion. *Annu Rev Biochem* 2005;**74**:1–28.
9. Pederson T. 50 years ago protein synthesis met molecular biology: the discoveries of amino acid activation and transfer RNA. *FASEB J* 2005;**19**(12):1583–4.
10. Giege R. The early history of tRNA recognition by aminoacyl-tRNA synthetases. *J Biosci* 2006;**31**(4):477–88.
11. Fruton JS. *Contrasts in scientific style: research groups in the chemical and biochemical sciences.* Philadelphia, PA: Memoirs of the American Philosophical Society; 1990.
12. Bergmann M, Zervas L. Über ein allgemeines verfahren der peptid-synthese. *Ber Dtsch Chem Ges* 1932;**65**(7):1192–201.
13. Bergmann M, Niemann C. On blood fibrin: a contribution to the problem of protein structure. *J Biol Chem* 1936;**115**(1):77–85.
14. Bergmann M, Niemann C. On the structure of proteins: cattle hemoglobin, egg albumin, cattle fibrin, and gelatin. *J Biol Chem* 1937;**118**(1):301–14.
15. Bergmann M, Niemann C. Newer biological aspects of protein chemistry. *Science* 1937;**86** (2226):187–90.
16. Astbury WT, Bell FO. Some recent developments in the X-ray study of proteins and related structures. *Cold Spring Harbor Symp Quant Biol* 1938;**6**:109–21.
17. Ogston AG. On the numerical consequences of certain hypotheses of protein structure. *Trans Faraday Soc* 1945;**41**:670–6.
18. Sanger F. The arrangement of amino acids in proteins. *Adv Protein Chem* 1952;**7**:1–66.
19. Wasteneys H, Borsook H. The enzymatic synthesis of protein. *Physiol Rev* 1930;**10**(1):110–45.
20. Voegtlin C, Maver ME, Johnson JM. The influence of the oxygen tension on the reversal of proteolysis (protein synthesis) in certain malignant tumors and normal tissues. *J Pharm Exp Ther* 1933;**48**(2):241–65.

21. Bergmann M, Fraenkel-Conrat H. The role of specificity in the enzymatic synthesis of proteins: synthesis with intracellular enzymes. *J Biol Chem* 1937;**119**(2):707−20.

22. Bergmann M, Fruton JS. Some synthetic and hydrolytic experiments with chymotrypsin. *J Biol Chem* 1938;**124**(1):321−9.

23. Bergmann M, Fruton JS. The specificity of proteinases. *Adv Enzymol* 1941;**1**:63−98.

24. Bergmann M. A classification of proteolytic enzymes. *Adv Enzymol* 1942;**2**:49−68.

25. Borsook H, Huffman HH. Some thermodynamical considerations of amino acids, peptides and related substances. In: Schmidt CLA, editor. *Chemistry of the amino acids and proteins*. Springfield, IL: H. M. Thomas Publishing, Co.; 1938. p. 865.

26. Borsook H, Dubnoff JW. The biological synthesis of hippuric acid in vitro. *J Biol Chem* 1940;**132**(1):307−24.

27. Lipmann F. Metabolic generation and utilization of phosphate bond energy. *Adv Enzymol* 1941;**1**:99−162.

28. Kalckar HM. The nature of energetic coupling in biological synthesis. *Chem Rev* 1941;**28**:71−178.

29. Fruton JS. The role of proteolytic enzymes in the biosynthesis of peptide bonds. *Yale J Biol Med* 1950;**22**(3):263−71.

30. Fruton JS, Johnston RB, Fried M. Elongation of peptide chains in enzyme-catalyzed transamidation reactions. *J Biol Chem* 1951;**190**(1):39−53.

31. Loftfield RB, Grover JW, Stephenson ML. Possible role of proteolytic enzymes in protein synthesis. *Nature* 1953;**171**(4362):1024−5.

32. Creager AN. *Life atomic: a history of radioisotopes in science and medicine*. Chicago, IL: Chicago University Press; 2013.

33. Loftfield RB. Preparation of ^{14}C-labeled hydrogen cyanide, alanine and glycine. *Nucleonics* 1947;**1**(3):54−7.

34. Melchior JB, Tarver H. Studies in protein synthesis in vitro; on the synthesis of labeled cystine (S35) and its attempted use as a tool in the study of protein synthesis. *Arch Biochem* 1947;**12**(2):301−8.

35. Melchior J, Tarver H. Studies on protein synthesis in vitro; on the uptake of labeled sulfur by the proteins of liver slices incubated with labeled methionine (S35). *Arch Biochem* 1947;**12**(2):309−15.

36. Winnick T, Friedberg F, Greenberg DM. Incorporation of C14-labeled glycine into intestinal tissue and its inhibition by azide. *Arch Biochem* 1947;**15**(1):160.

37. Frantz Jr ID, Loftfield RB, Miller WW. Incorporation of C14 from carboxyl-labeled DL-alanine into the proteins of liver slices. *Science* 1947;**106**(2762):544−5.

38. Anfinsen CB, Beloff A, Hastings AB, Solomon AK. The in vitro turnover of dicarboxylic amino acids in liver slice proteins. *J Biol Chem* 1947;**168**(2):771.

39. Borsook H, Deasy CL, Haagen-Smit AJ, Keighley G, Lowy PH. Isolation of a peptide in guinea pig liver homogenate and its turnover of leucine. *J Biol Chem* 1948;**174**(3):1041.

40. Zamecnik PC, Frantz Jr ID, Loftfield RB, Stephenson ML. Incorporation in vitro of radioactive carbon from carboxyl-labeled dl-alanine and glycine into proteins of normal and malignant rat livers. *J Biol Chem* 1948;**175**(1):299−314.

41. Loomis WF, Lipmann F. Reversible inhibition of the coupling between phosphorylation and oxidation. *J Biol Chem* 1948;**173**(2):807−8.

42. Frantz ID, Zamecnik PC, Reese JW, Stephenson ML. The effect of dinitrophenol on the incorporation of alanine labeled with radioactive carbon into the proteins of slices of normal and malignant rat liver. *J Biol Chem* 1948;**174**(2):773−4.

43. Winnick T, Friedberg F, Greenberg DM. The utilization of labeled glycine in the process of amino acid incorporation by the protein of liver homogenate. *J Biol Chem* 1948;**175**(1):117–26.

44. Siekevitz P, Zamecnik PC. In vitro incorporation of 1-C14-DL-alanine into proteins of rat liver granular fractions. *Fed Proc* 1951;**10**:246.

45. Sabatini DD. Philip Siekevitz: bridging biochemistry and cell biology. *J Cell Biol* 2010;**189**(1):3–5.

46. Borsook H, Deasy CL, Haagen-Smit AJ, Keighley G, Lowy PH. The uptake in vitro of C¹⁴-labeled glycine, L-leucine, and L-lysine by different components of guinea pig liver homogenate. *J Biol Chem* 1950;**184**(2):529–44.

47. Siekevitz P. Uptake of radioactive alanine in vitro into the proteins of rat liver fractions. *J Biol Chem* 1952;**195**(2):549–65.

48. Pederson T. Paul C. Zamecnik, 1912–2009. <http://www.nasonline.org/publications/biographical-memoirs/memoir-pdfs/zamecnik-paul.pdf>; 2011.

49. Pederson T. Mahlon Hoagland. <http://www.nasonline.org/publications/biographical-memoirs/memoir-pdfs/hoagland-mahlon.pdf>; 2011.

50. Zamecnik PC, Keller EB. Relation between phosphate energy donors and incorporation of labeled amino acids into proteins. *J Biol Chem* 1954;**209**(1):337–54.

51. Olby R. *Francis Crick: hunter of life's secrets.* New York, NY: Cold Spring Harbor Laboratory Press; 2009.

52. Crick FH. On protein synthesis. *Symp Soc Exp Biol* 1958;**12**:138–63.

53. McElheny VK. *Watson & DNA: making a scientific revolution.* New York, NY: Basic Books; 2004.

54. Hoagland MB. An enzymic mechanism for amino acid activation in animal tissues. *Biochim Biophys Acta* 1955;**16**(2):288–9.

55. Hoagland MB, Keller EB, Zamecnik PC. Enzymatic carboxyl activation of amino acids. *J Biol Chem* 1956;**218**(1):345–58.

56. Demoss JA, Novelli GD. An amino acid dependent exchange between ³²P labeled inorganic pyrophosphate and ATP in microbial extracts. *Biochim Biophys Acta* 1956;**22**(1):49–61.

57. Heidelberger C, Harbers E, Leibman KC, Takagi Y, Potter VR. Specific incorporation of adenosine-5′phosphate-³²P into ribonucleic acid in rat liver homogenates. *Biochim Biophys Acta* 1956;**20**(2):445–6.

58. Berg P. Acyl adenylate: the synthesis and properties of adenyl acetate. *J Biol Chem* 1956;**222**(2):1015–23.

59. Berg P. Acyl adenylate: the interaction of adenosine triphosphate and L-methionine. *J Biol Chem* 1956;**222**(2):1025–34.

60. Davie EW, Koningsberger VV, Lipmann F. The isolation of a tryptophan-activating enzyme from pancreas. *Arch Biochem Biophys* 1956;**65**(1):21–38.

61. Ogata K, Nohara H. The possible role of the ribonucleic acid (RNA) of the pH 5 enzyme in amino acid activation. *Biochim Biophys Acta* 1957;**25**(3):659–60.

62. Koningsberger VV, Van Der Grinten CO, Overbeek JT. Possible intermediates in the biosynthesis of proteins. I. Evidence for the presence of nucleotide-bound carboxyl-activated peptides in baker's yeast. *Biochim Biophys Acta* 1957;**26**(3):483–90.

63. Schweet RS, Bovard FC, Allen E, Glassman E. The incorporation of amino acids into ribonucleic acid. *Proc Natl Acad Sci USA* 1958;**44**(2):173–7.

64. Weiss SB, Acs G, Lipmann F. Amino acid incorporation in pigeon pancreas fractions. *Proc Natl Acad Sci USA* 1958;**44**(2):189–96.

65. Hoagland MB, Zamecnik PC, Stephenson ML. Intermediate reactions in protein biosynthesis. *Biochim Biophys Acta* 1957;**24**(1):215−16.

66. Hoagland MB, Stephenson ML, Scott JF, Hecht LI, Zamecnik PC. A soluble ribonucleic acid intermediate in protein synthesis. *J Biol Chem* 1958;**231**(1):241−57.

67. Keller EB, Zamecnik PC. The effect of guanosine diphosphate and triphosphate on the incorporation of labeled amino acids into proteins. *J Biol Chem* 1956;**221**(1):45−60.

68. Friedkin M, Lehninger AL. Oxidation-coupled incorporation of inorganic radiophosphate into phospholipide and nucleic acid in a cell-free system. *J Biol Chem* 1949;**177**(2):775−88.

69. Goldwasser E. Incorporation of adenosine-5'-phosphate into ribonucleic acid. *J Am Chem Soc* 1955;**77**(22):6083−4.

70.· Herbert E, Potter VR, Hecht LI. Nucleotide metabolism. VII. The incorporation of radioactivity from orotic acid-6-C14 into ribonucleic acid in cell-free systems from rat liver. *J Biol Chem* 1957;**225**(2):659−74.

71. Canellakis ES. Incorporation of radioactive uridine-5'-monophosphate into ribonucleic acid by soluble mammalian enzymes. *Biochim Biophys Acta* 1957;**23**(1):217−18.

72. Logan R. The incorporation of 8-^{14}C-adenine into calf thymus nuclei in vitro. *Biochim Biophys Acta* 1957;**26**(1):227−8.

73. Canellakis ES. On the mechanism of incorporation of adenylic acid form adenosine triphosphate into ribonucleic acid by soluble mammalian enzyme systems. *Biochim Biophys Acta* 1957;**25**(1):217−18.

74. Hecht LI, Stephenson ML, Zamecnik PC. Dependence of amino acid binding to soluble ribonucleic acid on cytidine triphosphate. *Biochim Biophys Acta* 1958;**29**(2):460−1.

75. Hecht LI, Zamecnik PC, Stephenson ML, Scott JF. Nucleoside tri-phosphates as precursors of ribonucleic acid end groups in a mammalian system. *J Biol Chem* 1958;**233**(4):954−63.

76. Zachau HG, Acs G, Lipmann F. Isolation of adenosine amino acid esters from a ribonuclease digest of soluble, liver ribonucleic acid. *Proc Natl Acad Sci USA* 1958;**44**(9):885−9.

77. Hecht LI, Stephenson ML, Zamecnik PC. Binding of amino acids to the end group of a soluble ribonucleic acid. *Proc Natl Acad Sci USA* 1959;**45**(4):505−18.

78. Zamecnik PC. Historical and current aspects of the problem of protein synthesis. *Harvey Lect* 1960;**54**:256−81.

79. Hoagland M. Enter transfer RNA. *Nature* 2004;**431**(7006):249.

80. Chapeville F, Lipmann F, Von Ehrenstein G, Weisblum B, Ray Jr WJ, Benzer S. On the role of soluble ribonucleic acid in coding for amino acids. *Proc Natl Acad Sci USA* 1962;**48**(6):1086−92.

81. Nirenberg MW, Matthaei JH. The dependence of cell-free protein synthesis in *E. coli* upon naturally occurring or synthetic polyribonucleotides. *Proc Natl Acad Sci USA* 1961;**47**(10):1588−602.

82. Martin RG, Matthaei JH, Jones OW, Nirenberg MW. Ribonucleotide composition of the genetic code. *Biochem Biophys Res Commun* 1962;**6**(6):410−14.

83. Lengyel P, Speyer JF, Basilio C, Ochoa S. Synthetic polynucleotides and the amino acid code. III. *Proc Natl Acad Sci USA* 1962;**48**(2):282−4.

84. Speyer JF, Lengyel P, Basilio C, Ochoa S. Synthetic polynucleotides and the amino acid code. IV. *Proc Natl Acad Sci USA* 1962;**48**(3):441−8.

Chapter 9

The Discovery and Rediscovery of Prokaryotic Messenger RNA

Initial Detection of Messenger RNA Was Overlooked Until It Was Discovered Again

Chapter Outline

M. Fry: Landmark Experiments in Molecular Biology. DOI: http://dx.doi.org/10.1016/B978-0-12-802074-6.00009-6

The model of the double helical structure of DNA that Watson and Crick proposed[1] and the predicted genetic implications of this structure[2] brought to the fore three fundamental unanswered questions. The first problem to be solved was the question of the mechanics of duplication of the complementary chains in double-stranded DNA. In addressing the question of the mode of DNA replication, three competing abstract models of conservative, semiconservative, and dispersive duplication were first formulated. Only then were these models put to experimental test and to the elegant demonstration by Meselson and Stahl that DNA was replicated in semiconservative fashion[3] (see Chapter 6: "Meselson and Stahl Proved That DNA Is Replicated in a Semiconservative Fashion"). The more complex problem of the nature of the code that enables translation of sequences of four bases in DNA into corresponding sequences of 20 different amino acids was solved more slowly. Following a decade-long stage (early 1950s to 1961s) of theoretical considerations that defined its principal properties (see Chapter 7: "Defining the Genetic Code"), the actual code was deciphered experimentally in the first half of the 1960s (see Chapter 10: "The Deciphering of the Genetic Code").

A third question, which is the subject of this chapter, was the matter of the mode of transmission of genetic information from DNA to protein. As was detailed in Chapter 7, "Defining the Genetic Code," George Gamow offered the idea that DNA serves as the direct template for protein synthesis.[4,5] This idea was eventually replaced by an even earlier hypothesis that the templates for protein synthesis were RNA transcripts of the DNA.[6-8] Experimental results that were amassed during the 1940s and early 1950s indicated that cytoplasmic protein synthesis takes place on ribosomes. It was also established that ribosomal RNA comprised the bulk of the cytoplasmic RNA. Based on these two observations, a consensus was formed that the genetic information was copied from DNA to ribosomal RNA transcripts that then directed protein synthesis. This *"One Gene—One Ribosome—One Protein"* postulate of ribosomal RNA functioning as the carrier of genetic information from DNA to the proteins synthesizing machinery remained a universally accepted paradigm for a relatively long period (1955—60). Moreover, without a competing theory in existence until rather late in the game, experimental evidence that contradicted the idea of ribosomal RNA as template for proteins as well as emerging indications for the existence of messenger RNA (mRNA) was ignored, misinterpreted or was left unexplained. Only when the accumulated weight of experimental evidence reached a critical level, direct experiments could be designed and executed to identify mRNA. Thus, in contrast to the important roles that theory played in the elucidation of the mode of DNA replication and of the nature of the genetic code, the identification of mRNA as the actual template for protein synthesis was achieved primarily by experiment. This chapter describes the circumstances of the erroneous identification of ribosomal RNA as the bearer of genetic information, the gradual accumulation of empirical evidence that contested this notion, the realization that mRNA should exist, and ultimately its eventual detection by experiment.

Much of the historical background to the discovery of mRNA and of subsequent developments were extensively covered in a 2014 Cold Spring Harbor meeting titled, *mRNA: From discovery to synthesis and regulation in bacteria and eukaryotes.* Most of the material that was presented in this meeting is available online: http://library.cshl.edu/Meetings/mRNA/.

9.1 EARLY EXPERIMENTS WITH GIANT *ACETABULARIA* CELLS DEMONSTRATED THAT NUCLEUS-DICTATED GENETIC INFORMATION WAS TRANSMITTED TO AND EXPRESSED IN THE CYTOPLASM

Decades before the discovery of the structure of DNA and the discovery of the roles of DNA and RNA in the storage and transmission of genetic information, the Danish-German marine biologist Joachim Hämmerling (1901—80)[9] showed experimentally that the cell nucleus contained hereditary information that was transmitted to and expressed in the cytoplasm. This insight evolved from Hämmerling's lifelong study of the marine algae *Acetabularia*.[9] During its vegetative stage, this unicellular organism reaches the amazing size of 3—5 cm. The giant cell consists of a base, a long stalk, and a cap whose shape is characteristic for each *Acetabularia* species (Fig. 9.1).

Starting his studies in the late 1920s, Hämmerling showed that *Acetabularia* contained a single nucleus that resided in the base of the cell. He then discovered that nuclear transplantation could be achieved with relative ease by cutting the nucleated base from one cell and grafting it onto a different cell whose base was removed beforehand. In an original experiment that he published in 1934,[11] Hämmerling transplanted the nucleus of one species, *Acetabularia crenulata* onto the stem of another species, *Acetabularia mediterranea* whose cap had a different morphology. The critical result of this experiment was that the morphology of the cap that developed in the grafted cell was that of *A. crenulata* (Fig. 9.1). Essentially, therefore, the development of a cap that matched the species of the base rather than that of the stalk showed that the nucleus dictated the morphological information that was expressed in the cap cytoplasm. Hämmerling proposed, therefore, that the nucleus cell transmitted factors (or "substances" in his words) to the cytoplasm that determined the morphology of the cap.[11]

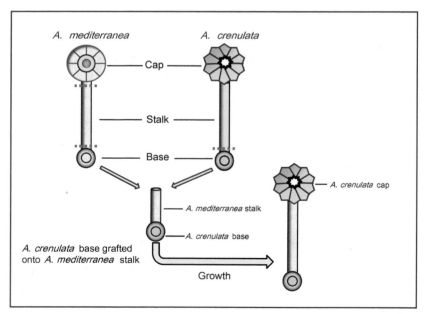

FIGURE 9.1 Scheme of Hämmerling's *Acetabularia* nuclear transplantation experiment.[10] The *dotted red lines* mark the cuts that were made in the two cells in order to generate their grafted parts.

A series of additional experiments yielded more detailed information. Species-specific cap could still be formed weeks after removal of the nucleus. Since development of the cell body did not require attendant presence of the nucleus, it appeared that the needed genetic information was transmitted to the cytoplasm long before it was expressed. Moreover, the cytoplasmic genetic information was found to be highly stable. Enucleated cells that were grown for weeks under conditions that prevented cap formation were able to develop a typical cap upon the restoration of normal growth conditions. It is today's commonplace knowledge that the genetic information for cell morphology is stored in the nucleus and is transmitted and expressed in the cytoplasm. However, as nothing was known in the 1930s on the chemical constitution of the hereditary material and on the way that genetic information is transported to and translated in the cytoplasm, Hämmerling's claim that the nucleus dispatches morphogenetic substances to the cytoplasm was largely mistrusted.[9] Also, because most of Hämmerling's papers were in German, it was not until the publication of his 1953 article in the *International Review of Cytology* that a larger body of the English-speaking scientific community learned of his findings.[12] However, doubt about Hämmerling's discoveries persisted well into the 1960s. Jacob and Monod posited in their classical 1961 model of the operon that bacterial mRNA was short-lived[13] and their prediction was experimentally verified soon thereafter in both bacteria and bacteriophages.[14,15] These theoretical developments and experimental discoveries led in the 1960s to a consensual conviction that mRNA molecules in every species were labile. The stability and delayed expression of the nucleus-dictated *Acetabularia* cytoplasmic morphogenetic "substances" contrasted, therefore, this conception. As a result, Hämmerling's findings were regarded as an unexplained oddity and continued to be ignored almost throughout his lifetime.[9] Validation and acceptance of the *Acetabularia* results came only after messengers in higher cells were found to be long-lived and to be often preserved in cells for extended periods of time before being expressed.

9.2   RNA WAS LINKED TO PROTEIN SYNTHESIS

Work that was done in the 1930s and 1940s exposed tight association between RNA and protein synthesis. Subsequently, microsomes, the particulate fraction that consists of ribosomes and fragmented endoplasmic reticulum (ER), were discovered. It was also established during that early phase that most of the cytoplasmic RNA was made of the RNA portion of the ribosomes (ribosomal RNA).

9.2.1   The Initial Phase: RNA Was Implicated in Protein Synthesis

Starting in the late 1930s, experiments revealed linkage between protein synthesis and the accumulation of RNA in the cytoplasm. In the late 1930s, the

Swedish cytologist and geneticist Torbjörn Caspersson (1910–97)[16] of the Karolinska Institute in Sweden and the American Drosophila geneticist Jack Schultz (1904–71)[17] used microspectrography to detect RNA in living cells. Their studies revealed close correlation between vigorous protein synthesis in dividing cells and increased levels of RNA in their cytoplasm.[10,18] Subsequent experiments with embryonic tissue, cancer cells, protein secreting glands, growing yeast cells, and other cell types confirmed the universality of the correlation between rising protein synthesis and elevated levels of RNA in the cytoplasm.[19] In parallel studies, the Belgian cytologist Jean Brachet (1909–88) of the Université Libre de Bruxelle used differential staining and in situ RNase digestion of tissues to independently show that RNA was localized in the cytoplasm and that variations in protein synthesis were accompanied by corresponding changes in the concentration of the cytoplasmic RNA.[20–22] Parallel to Caspersson, Brachet too suggested that RNA plays a role in protein synthesis. However, Brachet disputed Caspersson's theory that identified the nucleus as the primary site of protein synthesis in the cell.[23] Indeed, in subsequent experiments, Brachet enucleated Amoeba proteus cells and showed that their RNA-rich cytoplasm could maintain protein synthesis for days despite the absence of a nucleus.[24] These results reinforced the linkage between cytoplasmic RNA and protein synthesis and were in conflict with the idea that the nucleus (and DNA) was directly involved in protein synthesis. Historically, Brachet's work on cytoplasmic RNA and protein synthesis had wide impact on his era's thinking about cytoplasmic inheritance, embryology, and development. These aspects of his work, which are beyond the scope of this chapter, were extensively discussed elsewhere.[25–28]

9.2.2 The Discovery of Microsomes (and Ribosomes)

The French microscopist Charles Garnier (1875–1958) was the first to report in 1899 that the cytoplasm of glandular cells contained basophilic component that he named ergastoplasm ("work plasm" in Greek).[29,30] Similar basophilic areas of the cytoplasm were observed in the early 1900s in other cell types and ergastoplasm became the accepted designation for a basophilic domain in the cell cytoplasm. A major development was the discovery that the basophilia was owed to RNA. Working at the Rockefeller Institute, the Belgian cytologist and 1974 Nobel Prize laureate Albert Claude (1899–1983) developed in the 1930s the groundbreaking technique of cell fractionation which involved rupturing cells followed by differential separation of their constituents by centrifugation. Claude was also the first to use the electron microscope, which until then was used only in physical research, to study cells. Claude identified already in 1938 a component of Rous sarcoma virus as "ribose nucleoprotein" (later named RNA)[31] that he also detected in normal uninfected chicken cells.[32] Applying a combination of cell fractionation, cytochemical analyses, and electron microscopy, Claude subsequently showed that the cytoplasmic RNA was located in subcellular organelles that he originally named "small granules" and later "microsomes."[33] Subsequent analysis revealed that the microsomes were vesicle-like particles composed of fragments of the ER and of ribosomes that were comprised of protein and RNA at about 1:1 ratio. Later more accurate measurements revealed these proportions to be 60% RNA to 40% protein by weight. Fifteen years after the initial description of microsomes, their ribonucleoprotein components were named for the first time "ribosomes" by Richard Brooke Roberts in a 1958 meeting of the Biophysical Society (for a detailed history of the discoveries of microsomes and ribosomes, see Ref. 34). It should be pointed out that because of the limited sensitivity of the analytical methods that were used in the 1940s the microsomal/ribosomal RNA that comprises 80% of the cellular RNA content was regarded for a significant length of time as the major or even the only form of cytoplasmic RNA.

The link between RNA, microsomes/ribosomes, and protein synthesis was substantiated in the early 1950s by Tore Hultin (1919—) of the Wenner-Gren Institute in Stockholm[35] and by Elizabeth Keller (1918—97) and Phillip Siekevitz (1918—2009) in the Harvard University and Massachusetts General Hospital Laboratory of Paul Zamecnik (1912—2009).[36—38] In these experiments, cells were exposed to radioactively labeled amino acids for a short period of time and then extracted. Fractionation of the cells extracts revealed that most of the newly made labeled protein was associated with the microsomes. Subsequent experiments demonstrated that the microsomes were the site at which energy consuming protein synthesis took place.[39]

In sum, data that were gathered until the first half of the 1950s linked RNA and protein synthesis, identified ribosomal RNA as the bulk of the cytoplasmic RNA and identified microsomes/ribosomes as the cellular site of protein synthesis.

9.2.3   Ribosomes Were Conceived to Be Analogous to Small RNA Viruses

At the time that microsomes/ribosomes and their linkage to protein were discovered, a parallel line of studies explored the structure of small RNA viruses. Eventually, investigators drew parallel lines between these viruses that carry their genetic information within their RNA, and the similarly sized ribosomes. Starting in the late 1930s, the arrangement of the protein subunits in the protein coat of the rod-shaped Tobacco Mosaic Virus (TMV) was investigated by X-ray diffraction analysis.[40—44] This technique was also used to investigate the organization of the viral RNA genome.[45] The studies of TMV were later extended to X-ray diffraction analyses of crystals of other

RNA viruses such as the rod-shaped Turnip Yellow Mosaic Virus or the spherical Bushy Stunt Virus[46] (for a recent review of the history of the subject, see Ref. 47). By the mid-1950s, infective viruses were successfully reconstituted from their protein and RNA elements[48] and the viral RNA was shown to be the self-replicating genetic material of the virus.[49]

Among those who were actively interested in the structure of RNA viruses was James Watson. Beginning during his initial sojourn in Cambridge, and especially after he was instructed by Lawrence Bragg to cease his work with Crick on the structure of DNA (Refs. 50,51, see also Chapter 5: "Discovery of the Structure of DNA"), he engaged in X-ray diffraction analysis of the structure of TMV.[51] Although he and Crick later returned to their pursuit of the structure of DNA, Watson completed and published his findings on the helical arrangement of the protein subunits of the TMV capsid.[43] In his later work in Caltech, Watson continued his studies of the structure of small RNA viruses, collaborating with Donald Caspar (1927−) and exchanging data and ideas with Rosalind Franklin who was then studying TMV at Birkbeck College (see Ref. 47 and also Watson's May 27, 1955, letter to Crick that summarized his and Caspar's results on the structure of TMV: http://profiles.nlm.nih.gov/ps/retrieve/ResourceMetadata/SCBBJJ). Upon his second visit to Cambridge University in 1955, Watson developed with Crick a general theory of the packaging of the RNA genome of rod-shaped or spherical small viruses within a shell of regularly arranged protein subunits.[52] As Watson later wrote, these studies coalesced with his and Crick's interest in ribosomes[53]:

> At that time (1955) there was no good evidence for RNA existing free from protein; All RNA was thought to exist either as viral component or to be combined with protein in ribonucleoprotein particles. It thus seemed logical to turn our attention to study of ribonucleoprotein particles (ribosomes) since upon their surface protein was synthesized. [...] Then we were struck by the morphological similarity between ribosomes and small RNA-containing viruses [...]

In their theory on the structure of RNA viruses, Crick and Watson argued that because of their small size, the viral RNA genomes could encode only one or a few proteins. They speculated that in analogy, the RNA component of the ribonucleoprotein microsomal particles should also encode protein of limited size[54]:

> For this reason [the arrangement of RNA within the protein coat] it is quite possible that microsomes, which also appear spherical and are of similar size to the smaller viruses [...] may also have cubic symmetry, and perhaps possess protein subunits. In any case, since microsomes contain only a limited amount of RNA, we would predict, by an extension of our first argument, that no very large protein molecule will be found that is not an aggregate.

Two years later, Crick restated in a programmatic article[55] the idea of structural (and implicitly, functional) analogy between microsomes and small RNA viruses whose RNA component had been shown to dictate synthesis of type-specific viral proteins.[56]

9.3   THE RIBOSOME AS CARRIER OF GENETIC INFORMATION: EMERGENCE OF A "ONE GENE–ONE RIBOSOME–ONE PROTEIN" HYPOTHESIS

In contrast to the spirited theoretical activity that was part of the search for the mode of DNA replication and of the studies on the nature of the genetic code, the problem of the identity of the RNA template for protein synthesis entailed much lesser theoretical thinking. In effect, the field became dominated between the mid-1950s and 1960s by the single idea that the population of ribosomal RNA molecules represented a collection of transcripts of different individual genes. Thus, the RNA component of each ribosome was considered to be a template for the synthesis of a specific protein (Fig. 9.2).

In an influential 1958 theoretical article, Crick listed several arguments in support of this proposition[55]: (1) Experimental evidence revealed strong positive correlation between levels of cytoplasmic RNA and protein synthesis activity in the cytoplasm. (2) Cytoplasmic RNA was equated with

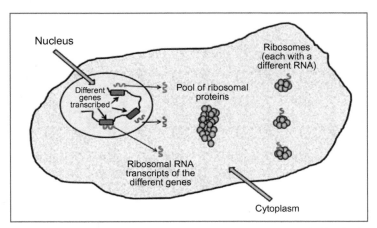

FIGURE 9.2   Illustration of the "One Gene–One Ribosome–One Protein" idea. Each nuclear gene was thought to be transcribed into a unique molecule of ribosomal RNA. Upon their migration to the cytoplasm, the RNA transcripts associated with nonspecific ribosomal proteins to form ribosomes. Each ribosome was thus believed to dictate the synthesis of a specific protein.

ribosomal RNA. (3) Ribosomes were identified as the site of protein synthesis. These points as well as analogies drawn between ribosomes and RNA viruses—whose RNA genome was shown to encode the viral proteins—led to the proposition that each ribosome contained an RNA transcript of a different gene and that this RNA served as template for the synthesis of a specific protein. This premise, that was coined the "*One Gene—One RNA—One Ribosome—One Protein*" hypothesis, was persuasively stated by Crick[55]:

> *What can we guess about the structure of the microsomal particle? On our assumptions the protein component of the particles can have no significant role in determining the amino acid sequence of the proteins which the particles are producing. We therefore assume that their main function is a structural one, though the possibility of some enzyme activity is not excluded. The simplest model then becomes one in which each particle is made of the same protein, or proteins, as every other one in the cell, and has the same basic arrangement of the RNA, but that different particles have, in general, different base-sequences in their RNA, and therefore produce different proteins. This is exactly the type of structure found in tobacco mosaic virus, where the interaction between RNA and protein does not depend upon the sequence of bases of the RNA.*

In discussing the ribosomal RNA as template for protein synthesis, Crick emphasized its stability in distinction from what he named "metabolic RNA"[55]:

> *The first, [of two proposed types of RNA, the second being "metabolic RNA"] which we may call "template RNA" is located inside the microsomal particles. It is probably synthesized in the nucleus [...] under the direction of DNA, and carries the information for sequentialization. It is metabolically inert during protein synthesis, though naturally it may show turnover whenever microsomal particles are being synthesized (as in growing cells), or breaking down (as in certain starved cells).*

9.4   ACCUMULATING EXPERIMENTAL EVIDENCE CONTRADICTED THE "ONE GENE—ONE RNA—ONE RIBOSOME—ONE PROTEIN" HYPOTHESIS

The "*One Gene—One RNA—One Ribosome—One Protein*" percept was in the mid-1950s the consensually accepted account of information transfer from DNA to protein. Yet, experimental data that were accumulating during the very same years (Table 9.1) put this notion under question.

Three main lines of evidence stood in conflict with the idea that ribosomal RNA was the cytoplasmic carrier of genetic information and template for protein.

TABLE 9.1 Experimental Evidence That Stood in Conflict With the "One Gene—One Ribosome—One Protein" Percept

Observation	References
Base compositions of cytoplasmic (ribosomal) RNA and of DNA were different	57
Unlike the heterogeneously sized proteins, ribosomal RNA had only two fixed sizes	60,61,62
Whereas evidence indicated that the carrier of genetic information was labile, ribosomal RNA was found to be stable	64

9.4.1 Different Base Compositions of Cytoplasmic RNA and DNA

A first inconsistency emerged from a study by the Soviet researchers Andrey Belozersky (1905—72), the founder in 1965 and namesake of the A.N. Belozersky Institute of Physico-Chemical Biology of Moscow State University, and Alexander Spirin (1931—) who later became a widely recognized authority on ribosomes. Belozersky and Spirin performed comparative analyses of the base ratios of the DNA of 19 bacterial species and of their corresponding RNA, which was predominantly ribosomal. Belozersky and Spirin originally published their findings in Russian and later in English.[57] Their data revealed that although the (G + C) to (A + T) ratios in the DNA of the different species extended over a sixfold range, the ratios in cellular RNA covered a much narrower range of only 1.2-fold (Fig. 9.3).

It appeared, therefore, that the base compositions of the (mostly ribosomal) bacterial RNA did not correspond to the base composition of the respective genomic DNA. Notably, however, regression analysis that Belozersky and Spirin conducted revealed a statistically significant correlation between the GC ratios of RNA and DNA. This observation raised the possibility that a minor fraction of the RNA might have possessed DNA-like GC content.[57]

This prediction was later proven to be correct when messenger RNA (mRNA) was identified and its base composition was found to be DNA-like (see below). Nevertheless, at the time, the distinctly different base constitutions of the largely ribosomal cytoplasmic RNA and of DNA were incongruent with the concept of ribosomal RNA molecules as a population of copies of different DNA genes. Seeking to preserve the One gene—One RNA—One ribosome—One protein hypothesis, Crick offered explanations that could potentially settle the contradiction between the experimental observations of Belozersky and Spirin and that postulate.[58] Of six scenarios that Crick raised to perhaps resolve the conflict between experiment and hypothesis, only the

DNA

Species	Bases (moles per cent) G	A	C	T	Pu/Py	(G+T)/(A+C)	(G+C)/(A+T)
Clostridium perfringens	15·8	34·1	15·1	35·0	1·00	1·03	0·45
Staphylococcus pyogenes aureus	17·3	32·3	17·4	33·0	0·98	1·01	0·53
Pasteurella tularensis	17·6	32·4	17·1	32·9	1·00	1·02	0·53
Proteus vulgaris	19·8	30·1	20·7	29·4	1·00	0·97	0·68
Escherichia coli	26·0	23·9	26·2	23·9	1·00	1·00	1·09
Proteus morganii	26·3	23·7	26·7	23·3	1·00	0·98	1·13
Shigella dysenteriae	26·7	23·5	26·7	23·1	1·01	0·99	1·15
Salmonella typhosa	26·7	23·5	26·4	23·4	1·01	1·00	1·13
Salmonella typhimurium	27·1	22·9	27·0	23·0	1·00	1·00	1·18
Erwinia carotovora	27·1	23·3	26·9	22·7	1·02	0·99	1·17
Corynebacterium diphtheriae	27·2	22·5	27·3	23·0	0·99	1·01	1·20
Aerobacter aerogenes	28·8	21·3	28·0	21·9	1·00	1·03	1·31
Mycobacterium vadosum Kras.	29·2	20·7	28·5	21·6	1·00	1·03	1·37
Brucella abortus	29·0	21·0	28·9	21·1	1·00	1·00	1·37
Alcaligenes faecalis	33·9	16·5	32·8	16·8	0·98	1·03	2·00
Pseudomonas aeruginosa	33·0	16·8	34·0	16·2	0·99	0·97	2·03
Mycobacterium tuberculosis BCG	34·2	16·5	33·3	16·0	1·03	1·01	2·08
Sarcina lutea	36·4	13·6	35·6	14·4	1·00	1·03	2·57
Streptomyces griseus	36·1	13·4	37·1	13·4	0·98	0·98	2·73

G+C/A+T = 0.45→2.73 (x6-fold)

RNA

Species	Bases (moles per cent) G	A	C	U	Pu/Py	(G+U)/(A+C)	(G+C)/(A+U)
Clostridium perfringens	29·5	28·1	22·0	20·4	1·36	1·00	1·06
Staphylococcus pyogenes aureus	28·7	26·9	22·4	22·0	1·25	1·03	1·05
Pasteurella tularensis	29·8	27·3	21·0	21·9	1·33	1·07	1·03
Proteus vulgaris	31·0	26·3	24·0	18·7	1·34	0·99	1·22
Escherichia coli	30·7	26·0	24·1	19·2	1·31	1·00	1·21
Proteus morganii	31·1	26·0	23·7	19·2	1·31	1·01	1·21
Shigella dysenteriae	30·4	25·9	24·4	19·3	1·29	0·99	1·21
Salmonella typhosa	30·8	26·1	24·0	19·1	1·32	1·00	1·21
Salmonella typhimurium	31·0	26·1	23·8	19·1	1·33	1·00	1·21
Erwinia carotovora	29·5	26·5	23·7	20·3	1·27	0·99	1·14
Corynebacterium diphtheriae	31·6	23·1	23·8	21·5	1·21	1·13	1·24
Aerobacter aerogenes	30·3	26·0	24·1	19·6	1·29	1·00	1·19
Mycobacterium vadosum Kras.	31·7	23·8	23·5	21·0	1·25	1·12	1·23
Brucella abortus	30·2	25·4	24·9	19·5	1·26	0·99	1·23
Alcaligenes faecalis	30·9	25·7	24·1	19·3	1·31	1·01	1·22
Pseudomonas aeruginosa	31·6	25·1	23·8	19·5	1·31	1·05	1·24
Mycobacterium tuberculosis BCG	33·0	22·6	26·1	18·3	1·25	1·05	1·45
Sarcina lutea	32·7	23·2	24·2	19·9	1·27	1·11	1·32
Streptomyces griseus	31·1	23·8	25·2	19·9	1·22	1·04	1·29

G+C/A+T = 1.06→1.29 (x1.2-fold)

FIGURE 9.3 Belozersky and Spirin's comparison of the base ratios in the DNA and RNA of diverse bacterial species. Note that in contrast to the base ratios of DNA that extended over a sixfold range, the base ratios of RNA of the same species extended over a much narrower range of only 1.2-fold. It should be kept in mind that the analyzed RNA was overwhelmingly ribosomal. The Tables are from Belozersky AN, Spirin AS. A correlation between the compositions of deoxyribonucleic and ribonucleic acids. *Nature* 1958;**182** (4628):111–12.

three main ones are cited here. One possibility was that only a part of the DNA-encoded proteins and thus the base composition of the RNA should not correspond to the base composition of the entire DNA. Alternatively, Crick speculated that most amino acids could have different representations in DNA and RNA, and the comparison of base ratios in the two nucleic acids was, therefore, irrelevant to their coding capacities. A third speculation was that although the RNA-to-protein code was uniform, the DNA-to-RNA code could have been variable. Remarkably, however, before listing his six different possible explanations, Crick admitted that "... in my view they all, at the moment, appear unattractive."[58]

9.4.2   Set Sizes of Ribosomal RNA

A second characteristic of ribosomal RNA that was inconsistent with its proposed function as informational RNA was its fixed sizes. Determination of the size of ribosomes and their RNA was initiated in 1958 when *Escherichia coli* 70S ribosomes were dissociated into two respective 50S and 30S large and small subunits.[59] Shortly thereafter the Caltech biophysicist Paul O.P. Ts'o (1929–2009) isolated two RNA species of 28S and 18S from the 74S ribosomes of pea epicotyls.[60] Corresponding RNA molecules of 23S and 16S were isolated from bacterial ribosomes in Harvard by Charles Kurland (Box 9.6)[61] and independently by Alexander Spirin in Moscow.[62] Considering the highly variable sizes of cellular proteins, the uniform sizes of ribosomal RNA in both prokaryotes and eukaryotic appeared to be incompatible with their proposed role as encoders of proteins.

9.4.3   Stability of Ribosomal RNA

Observed stability of ribosomal RNA constituted a third facet that was in conflict with its suggested role as template for protein synthesis. As is detailed in the following section, studies in the Pasteur Institute and elsewhere demonstrated that the activity of a specific carbohydrate-hydrolyzing enzyme was rapidly induced or depressed in response to a change in the type of sugar that served as the carbon source in the growth medium of *E. coli*. The henceforth-described PaJaMo experiment documented fast kinetics of the induction of β-galactosidase protein by exposing *E. coli* cells to an analog of its substrate lactose. A favored interpretation to the rapid induction of the synthesis of β-galactosidase and its fast shut down was that the enzyme was encoded by a rapidly turning-over "messenger" molecule.[63] Thus, if ribosomal RNA was indeed template for proteins such as β-galactosidase it should have been labile.

To assess the stability of ribosomal RNA, Matthew Meselson together with Rick Davern employed the strategy that he and Stahl used successfully to demonstrate the semiconservative mode of DNA replication (see Chapter 6: "Meselson and Stahl Proved That DNA Is Replicated in a Semiconservative Fashion"). After labeling *E. coli* RNA with a heavy ^{15}N isotope, the cells were transferred to a medium that contained the lighter isotope ^{14}N. Ribosomal RNA that was then isolated during succeeding cell divisions was resolved by density-equilibrium centrifugation. Results revealed that unlike the semiconservatively replicating DNA, the ^{15}N-labeled ribosomal RNA did not generate replicated ^{15}N/^{14}N hybrid progeny molecules and remained stable in its native ^{15}N constitution for several generations (Ref. 64; Fig. 9.4). This documented stability of ribosomal RNA stood, therefore, in clear contrast to the proposed instability of at least some messenger molecules.

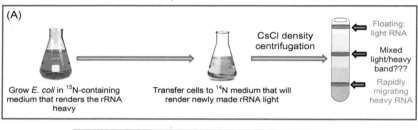

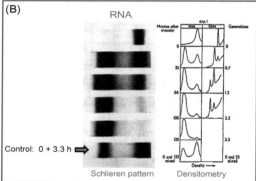

FIGURE 9.4 Ribosomal RNA is stable. (A) Scheme of the experiment: *Escherichia coli* cells that were grown in medium that contained a heavy isotope of nitrogen ^{15}N were transferred to a medium containing the lighter isotope ^{14}N and cell aliquots were drawn at different times after the transfer. RNA samples that were isolated from the cells were resolved by CsCl density-equilibrium centrifugation. (B) Results of the density-gradient resolution of ribosomal RNA. The heavy ribosomes were fully conserved throughout more than three generations and ribosomes that were generated after the shift to ^{14}N-containing medium, were exclusively light. Parallel controls of semiconservatively replicating DNA (right hand of the densitometry patterns) showed the formation of heavy–light hybrid molecules. *Panel (B): Adapted from Davern CI, Meselson M. The molecular conservation of ribonucleic acid during bacterial growth.* J Mol Biol *1960;2 (3):153–60.*

9.5   EXPERIMENT BRED A NEW THEORY: THE PAJAMO EXPERIMENT AND THE HYPOTHESIS OF AN UNSTABLE MESSENGER

An alternative to the theory of ribosomal RNA as the bearer of genetic information was engendered by results of an experiment whose primary aim was to identify the mechanism of enzyme induction. Starting in 1940, Jacque Monod (Box 9.1) at the Pasteur Institute strove to elucidate the underlying mechanisms of enzyme adaptation.

BOX 9.1 Jacques Monod

Jacques Monod was born in Paris to an American mother and to French Huguenot father. While already an active scientist, he was during World War II Chief of Staff of Operations for the Home Resistance Forces. In preparation for the Allied landings, he commanded parachute drops of weapons and railroad bombings. His scientific contributions encompassed studies on enzyme induction, on the *lac* operon and, together with Jacob, of the operon model of gene regulation.[13] Together with Jeffries Wyman (1901–1995) and Jean-Pierre Changeux (1936–), he also formulated the theory of protein allostery. In recognition of his scientific achievements, he shared the 1965 Nobel Prize in Physiology and Medicine with Jacob and André Lewoff. From 1971 and until his death in 1976, he served as the Director of the Pasteur Institute.

He was not only a biologist but also a fine musician (cellist and conductor) and a highly regarded writer on the philosophy of science. His most famous philosophical tract is "Chance and Necessity"[65] whose title was inspired by a line attributed to Democritus, *Everything existing in the universe is the fruit of chance and necessity*. In this book, he argued that life on earth emerged by a rare (and a likely irreproducible) chemical accident and was maintained by the processes of selection among randomly generated mutations and ensuing evolution. Retaining this view, he was a devout atheist who famously wrote: "The first scientific postulate is the objectivity of nature: nature does not have any intention or goal." Accounts of his scientific work and personal qualities are in a volume edited by Agnes Ullman.[66] The fascinating story of his life and personal and intellectual friendship with Albert Camus has been attractively told by Sean Carrol[67] (Fig. 9.5)

| Jacques Monod (1910–76) | Monod's "Chance and Necessity" |

FIGURE 9.5   Jacques Monod and his "Chance and Necessity."

As succinctly described by Monod himself in his 1965 Nobel lecture,[68] the basis to his studies was the so-called diauxy phenomenon that was first described at the turn of the 20th century. This condition was observed when bacterial cells were grown in a medium that contained readily consumable glucose as carbon source and a second, less easily useable carbohydrate. After the cells exhausted the glucose and following a lag period, activity of a specific hydrolytic enzyme was induced such that the second carbon source could be utilized (Fig. 9.6A).

Focusing on the consumption of lactose as carbon source for *E. coli*, Monod and associates used antibodies against its hydrolyzing enzyme β-galactosidase (Fig. 9.6B) to demonstrate that induction of this enzyme was due to its de novo synthesis.[70] In practice, β-galactosidase activity was routinely induced by the introduction into the growth medium of the nonhydrolyzable lactose analog iso-propyl IPTG (Fig. 9.6C). Although the amount (and corresponding activity) of β-galactosidase increased progressively in the presence of the inducer, enzyme induction ceased promptly when the inducer was removed (Fig. 9.6D). Genetic analysis identified a gene marked "Z" that encoded β-galactosidase and two additional genes that encoded the lactose catabolism enzymes galactoside permease and transacetylase. Induction of the three enzymes was found to be governed by a distinct controlling gene designated "i." The Pasteur researchers were also able to identify a null "i⁻" mutant that expressed β-galactosidase in a constitutive fashion without a need for inducer (Fig. 9.8A).

As described in detail elsewhere (Ref. 50, pp. 388−411; and Ref. 71), the mode of action of the "i" regulatory gene was vigorously debated in the 1950s leading to the formulation of two contrasting models of its manner of operation.

Monod believed that the regulatory gene acted positively by producing a molecule that combined and cooperated with the lactose inducer to initiate the expression of β-galactosidase (Fig. 9.8B1). In contrast, drawing analogy from their studies on the bacteriophage λ repressor, François Jacob (Box 9.2) and the Pasteur Institute microbial geneticist Èlie Wollman (1917−2008) maintained that the "i" regulatory gene produced a repressor and that its removal by the lactose inducer enabled the expression of β-galactosidase (Fig. 9.8B2).

The enigma of the mode of regulation of β-galactosidase induction was solved when Arthur Pardee (Box 9.3), an American biochemist on leave from UC Berkeley, joined the Monod laboratory in 1957.

There he conducted an elegant experiment that established the "i" gene product as a negative rather than positive regulator of β-galactosidase synthesis. This experiment that was later dubbed "PaJaMo" or more playfully "PaJaMa" for its authors Pardee, Jacob and Monod was based on Jacob and Wollman's work on bacterial conjugation. In mating Hfr⁺ "male" donor *E. coli* cells with F⁻ "female" bacteria, the male cells inserted their chromosome into the female bacteria in a polar and orderly manner without transferring any cytoplasm.

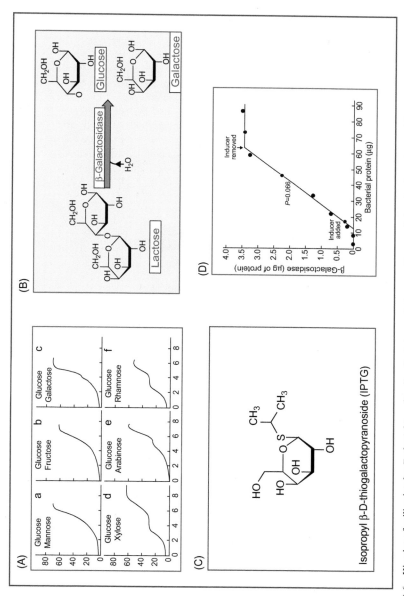

FIGURE 9.6 Kinetics of utilization by *Escherichia coli* cells of mixtures of glucose with one of six other less consumable carbon sources. Following utilization and exhaustion of the glucose, the cells go through lag periods of variable length until synthesis of a specific hydrolyzing enzyme is induced and utilization of the second carbon source commences. The curves are from Ref. 68. (B) β-Galactosidase hydrolyzes the disaccharide sugar lactose into its constituent monosaccharides: glucose and galactose. (C) β-D-1-thiogalactopyranoside (IPTG): a nonhydrolyzable inducer of β-galactosidase. (D) Kinetics of induction of β-galactosidase. Note that the progressive accumulation of the enzyme protein promptly ceased when the inducer was removed. *Panels (A) and (D): From Volkin E, Astrachan L. Phosphorus incorporation in Escherichia coli ribonucleic acid after infection with bacteriophage T2. Virology 1956;2(2):149–61.*

BOX 9.2 François Jacob

François Jacob was born to an old and respected Jewish-French family (his grand-father was a General of the artillery). Intending to follow the steps of an uncle and become a surgeon, his medical studies and career were interrupted by World War II. He served in the African Corps of de Gaule's Free French movement and was severely wounded during the invasion to France. Although he completed his studies of medicine after recuperation, he chose to engage in scientific research rather than practice medicine. Collaborating with Èlie Wollman at the Pasteur Institute, he made significant contributions to the genetics of bacteria and phage. His work with Monod on the regulation of the lac operon was crowned by their celebrated operon model of gene regulation.[13] He also collaborated with Brenner and Meselson to provide the first physical evidence for the existence of mRNA.[14] In this autobiography titled *"The Statute Within,"*[72] he wonderfully described his childhood, education, military service, and introduction to science. In recognition of his scientific achievements, he shared the 1965 Nobel Prize in Physiology and Medicine with Monod and André Lewoff (Fig. 9.7).

François Jacob
(1920–2013)

Jacob's "The
Statute Within"

FIGURE 9.7 François Jacob and his autobiography.

In the PaJaMo experiment, the chromosome of wild-type Hfr$^+$ male bacteria that carried a functional "i" gene and an inducible β-galactosidase "Z" gene was injected into F$^-$ bacteria that lacked both genes. Due to their genetic makeup, expression of β-galactosidase by the male bacteria was entirely dependent on the addition of an inducer, whereas the female cells were incapable of expressing the enzyme without or with inducer (Fig. 9.10).

Monitoring of the kinetics of β-galactosidase synthesis in the hetero-merozygotic bacteria revealed that upon introduction of the donated Z$^+$ allele, enzyme activity was expressed instantly *in the absence of inducer* (Ref. 63; Fig. 9.11A). After a period, however, the induced expression of β-galactosidase tapered off and could be maintained only if inducer was added to the medium (Fig. 9.11B).

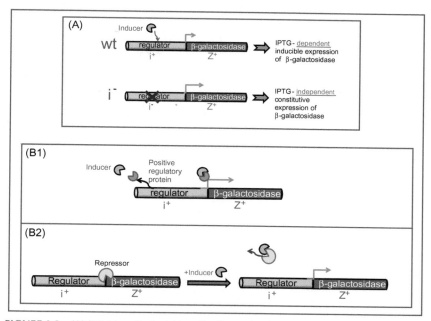

FIGURE 9.8   (A) Wild-type *Escherichia coli* cells require the presence of an inducer that inter-acts with the regulatory gene "i" in order to induce the structural gene "Z" to express β-galactosidase. Null "i" mutant of the bacteria expresses β-galactosidase in a constitutive man-ner without need for inducer molecule. (B) Two contrasting models of the mode of operation of the "i" regulatory gene. (B1) *Positive regulation model*: The inducer molecule binds to a protein product of the "i" gene and it is the formed bipartite complex that acts to induce the synthesis of β-galactosidase. (B2) *Negative regulation model*: The inducer molecule acts to induce the syn-thesis of β-galactosidase by removing a repressor molecule that the "i" gene produces.

Historically, it was the famed Hungarian-American physicist-turned-biologist Leo Szilard (1898−1964), who during a visit to the Pasteur Institute, interpreted the results of the PaJaMo experiment as reflecting negative regulation of β-galactosidase synthesis (Ref. 50, pp. 395−398). According to this interpretation, the "i" gene produced a repressor rather than a co-inducing molecule. The immediate synthesis of β-galactosidase occurred when the male-contributed active "Z^+" gene was introduced into a female cytoplasm that had no "i" gene product which could block the expression of the "Z" gene. After some time, however, the male-contributed active "i" gene produced a repressor that silenced the "Z" gene. Under these conditions of renewed repression, the expression of β-galactosidase became again dependent on the presence of added inducer. This interpretation of the results of the PaJaMo experiment is illustrated in Fig. 9.12.

BOX 9.3 Arthur Pardee

Arthur Pardee received his BSc and MSc degrees from the University of California at Berkeley and a PhD degree from Caltech under the mentorship of Linus Pauling. After joining the Berkeley faculty in 1949, he took in the late 1950s a sabbatical leave at the Pasteur Institute in Paris to work with François Jacob and Jacques Monod. On that occasion he performed the famous PaJaMo experiment that predicted the existence of mRNA. In 1961 he became Professor in Biochemical Sciences at Princeton and in 1975 he moved on to become Professor of Biological Chemistry and Molecular Pharmacology at the Harvard Medical School and the Division of Cell Growth and Regulation at the Dana-Farber Cancer Institute (Fig. 9.9).

Arthur Pardee
(1921–)

FIGURE 9.9   Arthur Pardee.

Importantly, in addition to providing evidence for the negative regulation of gene expression by a putative repressor molecule, the PaJaMo experiment was also the source of the equally important hypothesis of a short-lived "messenger" that carries information from DNA to the protein synthesis machinery. The origin of this idea was the observed rapid kinetics of β-galactosidase synthesis immediately after the introduction of the active "Z" gene into the female cytoplasm and its prompt switching off following the accumulation of repressor molecules (Figs. 9.11 and 9.12).

The rapid kinetics of induction as observed in the PaJaMo experiment was reconfirmed in a follow-up study that Pardee conducted with his then graduate student Monica Riley (1926–2013). In this experiment, they closely monitored the kinetics of inducer-independent β-galactosidase expression after the introduction of an active "Z" gene into recipient female cells.[73] Results of the experiment affirmed that the rate of accumulation of β-galactosidase in the heteromerozygote was maximal right from the start.

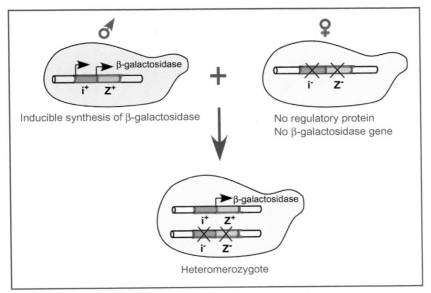

FIGURE 9.10   Scheme of the PaJaMo experiment. Wild-type Hfr$^+$ male *Escherichia coli* cells were conjugated with F$^-$ female bacteria whose regulatory "i" and the β-galactosidase encoding "Z" genes were missing. The male cells injected their DNA bearing functional "i" and "Z" genes into the female cells without transferring any cytoplasm. Kinetics of β-galactosidase synthesis was monitored in the presence of streptomycin that repressed the donor male cells but not the female bacteria or the heteromerozygote. Thus, measured enzyme synthesis was exclusively that of the heterozygotes.

Also, β-galactosidase expression was diminished proportionally to the extent of radioactive inactivation of the "Z" gene by incorporated ^{32}P, suggesting that maintenance of enzyme synthesis depended on the integrity of the genetic material.[73]

Two alternative interpretations were offered for the observed kinetics of β-galactosidase induction and repression in the heteromerozygotes. One hypothesis was that the male-contributed "Z" gene itself served as template for the β-galactosidase protein. According to this model, the subsequent shutoff of protein synthesis was due to the blocking of translation by the repressor.[63] An alternative idea was that the introduced Z gene was copied into a rapidly made messenger molecules that served as template for β-galactosidase synthesis. In this model, the prompt shutoff of β-galactosidase synthesis by the accumulated repressor reflected rapid degradation of the messenger molecules.[63,73] This hypothesis of a short-lived messenger was later fully formulated in Jacob and Monod's model of the *lac* operon.[13]

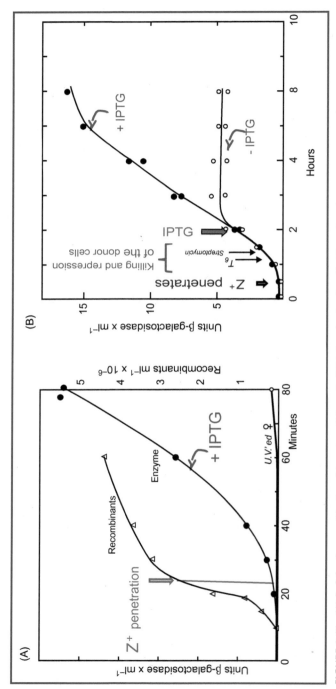

FIGURE 9.11 (A) Left hand curve: Kinetics of insertion of the male DNA into the female cells. The point of penetration of the Z^+ gene is marked. Right hand curve: Kinetics of β-galactosidase expression in the heteromerozygotes in the presence of IPTG inducer. (B) Kinetics of β-galactosidase expression in the heteromerozygotes with or in the absence of IPTG. Note that the enzyme was expressed without inducer for about 2 hours and then its synthesis tapered off. Addition of IPTG allowed continuous induction of enzyme. *Panel (A): From Monod J, Pappenheimer Jr AM, Cohen-Bazire G. The kinetics of the biosynthesis of beta-galactosidase in Escherichia coli as a function of growth. Biochim Biophys Acta 1952;9(6):648–60. Panel (B): Curves adapted from Pardee AB, Jacob F, Monod J. The genetic control and cytoplasmic expression of "Inducibility" in the synthesis of β-galactosidase by E. coli. J Mol Biol 1959;1(2):165–78.*

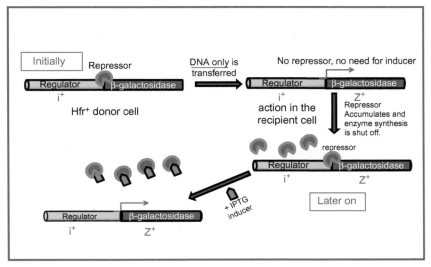

FIGURE 9.12 Scheme illustrating the Pasteur group's interpretation of the results of PaJaMo experiment (Fig. 9.11). The Hfr^+ male cells injected their active "i" and "Z" genes into F^- female cells without transferring their cytoplasmic repressor molecules. Because of the initial lack and then insufficient levels of repressor molecules in the heteromerozygotic cells, the introduced "Z" gene was expressed and β-galactosidase was synthesized without a need for IPTG inducer. However, expression of the inserted "i" gene led in time to the accumulation of critical levels of repressor which then blocked the expression of the "Z" gene such that β-galactosidase synthesis tapered off. Introduction of IPTG defeated the negative action of the repressor and β-galactosidase synthesis could continue in the heteromerozygotes.

In sum, based on their indirect evidence, Pardee, Riley, Jacob, and Monod excluded the possibility that the carrier of information for β-galactosidase synthesis could be a stable intermediate. Rather, either protein was translated directly from the transduced DNA or an unstable messenger served as template for protein synthesis[73]:

> *The main conclusion to be drawn from the kinetic study of enzyme synthesis following gene transfer may be stated as follows: the z^+ gene which determines the structure of β-galactosidase in E. coli functions without delay, at maximal rate, when it is transferred by sexual recombination into the cytoplasm of a cell which possessed only an inactive allele of the gene.*
>
> *This result evidently agrees with the model involving no stable intermediates, whether the gene itself acts directly as a template or via an unstable intermediate. The result is not compatible with the model involving the formation of stable intermediate templates, since these templates would accumulate in the cytoplasm and the rate of enzyme synthesis should start at zero and increase gradually.*

As is described below, 3 years after the original PaJaMo experiment, two groups provided direct evidence for the existence of a minor unstable RNA fraction in phage-infected[14] and in uninfected[15] *E. coli* cells. In addition, these experiments also excluded the possibility that DNA served as the direct template for protein synthesis.

Despite the elegance of the PaJaMo experiment and the appeal of the new theory of an unstable messenger, its significance and that of the subsequent work of Riley and Pardee were not immediately grasped. In Sep. 1959, leading researchers assembled in Copenhagen at a meeting that was organized by the Danish microbiologist Ole Maaløe (1914−88). Among the attendees were most of the scientists who later provided physical evidence for the existence of mRNA: Sydney Brenner, François Gros, Jacob, Meselson and Watson. In his talk at that earlier meeting, Jacob presented the PaJaMo results and the emergent theories of the repressor and of the messenger (which he then named "the tape" or "X"). In that presentation, Jacob directly addressed the paradox between the observed very rapid synthesis of β-galactosidase and the much longer time that was needed to assemble specific ribosomes that carry the needed information. As Jacob later attested in an interview with Horace Judson (Ref. 50, p. 410), the audience was unresponsive:

> *Everybody was there, [...] I remember I gave a talk, and discussed this paradox, and at that time nobody reacted to that. Just nobody.*

When he listened to Jacob in Copenhagen, Crick was already primed to the design and results of the PaJaMo experiment that were described to him beforehand by Monod when they met in London in the fall of 1958. Recalling at a later time, his reaction to this earlier disclosure Crick said: *"[...] we were very puzzled with the experiment; it didn't fit in with the ideas we had. We messed up by the ribosomes. And we tried various ways to get out of it"* (Ref. 50, p. 407). After listening to Jacob's presentation in Copenhagen, Crick had a similar reaction. In an interview with Judson, he admitted that (Ref. 50, p. 411):

> *Well, we were unsympathetic, in the sense that we recognized they'd got an interesting experiment, but we weren't clear that there wasn't some snag, you see. It wasn't until we were converted on the road to Damascus that we saw it all, you see!*

In mentioning "the road to Damascus," Crick was referring to an Apr. 15, 1960 (Good Friday) meeting at which different lines of evidence coalesced to bring forth a working hypothesis of a putative mRNA (see Section 9.9).

To summarize the state of affairs up to 1960, the search for the cytoplasmic informational RNA was a case in point of resistance to contradictory experimental results and a lack of openness to new theory. The hypothesis of ribosomal RNA as template for protein synthesis persisted, therefore, for several years despite the accumulation of conflicting experimental evidence (Table 9.1; Figs. 9.3 and 9.4), and it was not reassessed when the alternative idea of an unstable messenger was introduced.

9.6 A FIRST SIGHTING OF RAPIDLY TURNING-OVER RNA IN PHAGE-INFECTED CELLS

Although the PaJaMo experiment[63] and especially its follow-up experiment[73] advanced the idea of an unstable cytoplasmic messenger, the chemical nature of this entity was left unspecified. Early insight that the messenger was RNA emerged from studies on the metabolism of RNA in bacteriophage-infected *E. coli*. Following his groundbreaking demonstration that DNA is the genetic material of T bacteriophage (Ref. 74, see Chapter 4: "Hershey and Chase Clinched the Role of DNA as the Genetic Material"), Alfred Hershey monitored the kinetics of incorporation of ^{32}P radioactive phosphorous into the DNA of progeny T2 bacteriophage. In the course of these experiments, he noticed that at a very early time after the infection, ^{32}P was also incorporated into RNA and that the labeled RNA disappeared rapidly at a later time.[75] Three years passed until this finding was affirmed and extended by two Oak Ridge researchers, Elliot "Ken" Volkin and Lazarus Astrachan (Box 9.4).

Infecting *E. coli* with T2 phage in ^{32}PO$_4$-containing medium, they observed that although the radioactive label was accumulated continuously in the phage DNA, it was incorporated into RNA at an early time and then the radioactivity in RNA declined.[69] Volkin and Astrachan presented their data on the instability of rapidly labeled T2 phage RNA in a 1956 McCullum Pratt Conference in Baltimore on the chemical basis of heredity. They also showed there that the base composition of the labeled RNA was similar to that of the phage DNA. However, in discussing these results, they speculated that the observed labile RNA could be a precursor for the phage DNA.[76] Oddly enough, the description by Volkin and Astrachan of DNA-like rapidly turning-over RNA in T2 phage-infected cells did not gain its due measure of recognition. Remarkably, Crick, Jacob, and Watson who were all directly responsible for the 1961 "rediscovery" of labile mRNA in phage-infected and uninfected *E. coli* cells (*vide infra*) attended the 1956 McCullum Pratt meeting at which the T2 RNA results were reported.[76] However, according to Judson (Ref. 50, p. 323), Crick did not attend Volkin's talk and read it only a year later in its published form.[76] Yet, Judson also noted that both Crick and Brenner were aware of the Oak Ridge results. Asked at a later time whether he was aware of those findings Jacob was cited as saying: "The question is what you mean, to be aware? To be

BOX 9.4 Elliot "Ken" Volkin and Lazarus Astrachan

After earning a BSc degree at Penn State University, Volkin got a master's and doctoral degrees in biochemistry at Duke University. In 1948, he moved to the Oak Ridge National Laboratory serving there as scientific director of the biochemistry section in the Biology Division from 1969 to 1980 and working until the end of his scientific career in 1984. His discovery, with Astrachan of what they called "DNA-like-RNA" in phage-infected *E. coli* was the first, and unduly ignored, direct observation of mRNA.

Lazarus Astrachan was 32 when with a newly minted doctorate he joined Volkin at Oak Ridge to conduct their landmark discovery of phage "DNA-like-RNA." In 1961 he moved to Case Western Reserve University at which he spent his entire career until his retirement in 1990 (Fig. 9.13).

Elliot "Ken" Volkin   Lazarus Astrachan
(1927–2011)   (1925–2003)

FIGURE 9.13   Elliot Volkin and Lazarus Astrachan.

aware is to have it on the front of consciousness, and clearly we were not, at this level, even if we knew" (Ref. 50, p. 414). Years later, Watson provided a likely explanation for the lack of reaction to the Volkin and Astrachen's 1956 presentation[77]:

> There were [...] two potential bombshells. [the other mentioned "bombshell" that Watson was referring to had been Arthur Kornberg's report of DNA synthesis by cell-free extracts of E. coli]. One was the report by Elliot Volkin and Larry Astrachan from Oak Ridge that the unstable RNA made after phage T2 infection had a base composition very similar to that of T2 DNA. Years later, we realized what they were studying was the very RNA made off DNA templates involved in protein synthesis. Their talk pushed the alternative idea that this RNA was a precursor on the way to being transformed enzymatically into DNA.

In 1958, the Oak Ridge group reported that exposure of *E. coli* bacteria to radioactive ^{32}P isotope at the time of their infection by T7 bacteriophage resulted in rapid labeling of a minor fraction of RNA.[78] After the short exposure to the radioactive isotope ("pulse"), it was diluted by the addition of an excess of unlabeled ^{31}P phosphorous ("chase"). Radioactivity was monitored in the phage RNA and in DNA at different time points both the pulse and chase periods. Results showed that although the labeled RNA progressively dematerialized during the chase period, radioactivity in the phage DNA increased progressively (Fig. 9.14A). Hence, unlike the phage DNA, the pulse-labeled RNA fraction appeared to be short-lived. Hydrolysis of this rapidly turning-over RNA fraction and quantification of its separated four bases revealed close similarity between the base compositions of this RNA and of the DNA of T7 bacteriophage (Fig. 9.14B).

Naming this fraction "DNA-like-RNA," Volkin and Astrachan now argued that the different kinetics of label accumulation in DNA and RNA discounted the option that this RNA was a precursor for DNA. Rather, they now ventured that it could be template for the viral proteins[78]:

> It is possible that the specific kind of RNA synthesized by the host under the influence of the infecting phage may serve as the proper functional unit (template?) for the synthesis of phage-specific protein.

9.7   RAPIDLY SYNTHESIZED DNA-LIKE RNA WAS ALSO IDENTIFIED IN YEAST CELLS

A somewhat neglected 1960 paper by the Lithuanian-American biochemist Martynas Yčas (1917−2014) and Walter Vincent of the New York Medical Center in Syracuse suggested that DNA-like-RNA may not be restricted to phage-infected bacterial cells.[79]

As illustrated in Fig. 9.15, these investigators labeled with ^{32}P yeast cell RNA for a brief period of time and then compared the base composition of the labeled RNA to the compositions of the yeast DNA and of its cellular (mostly ribosomal) RNA. The results of this analysis showed that the nucleotide constitution of the pulse-labeled RNA was similar to that of the cell DNA and that it was clearly different from the nucleotide composition of the ribosomal RNA.[79] These findings provided, therefore, evidence for the existence in yeast cells of a rapidly synthesized DNA-like RNA fraction that was distinct from the bulk ribosomal RNA.

9.8   BELIEF IN RIBOSOMAL RNA AS THE CARRIER OF GENETIC INFORMATION PERSISTED IN SPITE OF CONTRADICTORY EVIDENCE AND THE IDENTIFICATION OF DNA-LIKE RNA

By 1960 the proposed function of ribosomal RNA as conveyer of genetic information was contradicted by several lines of evidence (Table 9.1). Also, other data pointed to the existence of a minor fraction of labile DNA-like RNA.[57] In parallel, indirect evidence from the PaJaMo experiment raised the

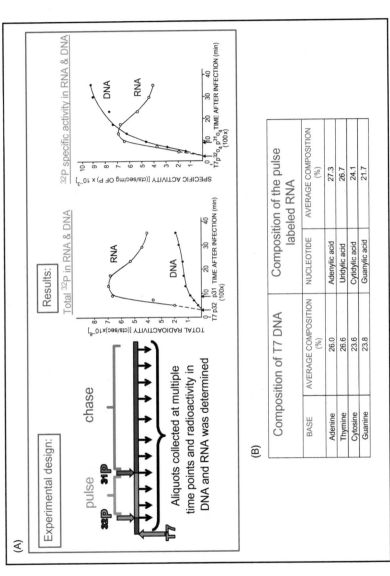

FIGURE 9.14 (A) Experimental design and results of Volkin et al.[78] Shortly after infecting *Escherichia coli* with T7 bacteriophage, the cells were exposed for a brief period of time to ^{32}P ("pulse") that was next "chased" by high excess of unlabeled ^{31}P phosphorous. Aliquots were drawn at different time points during the pulse and chase periods and total and specific radioactivity in the phage RNA and DNA was determined. Kinetics of the labeling of the phage RNA revealed that although both its total and specific radioactivity increased progressively for about 10 minutes, they declined thereafter. By contrast, the total and specific radioactivity of the phage DNA increased continuously. The graphs are from Ref. 78. (B) Base composition of the rapidly turning-over phage RNA was similar to the base composition of the phage DNA. Data are from Ref. 78. *Panel (A): Graphs adapted from Volkin E, Astrachan L, Countryman JL. Metabolism of RNA phosphorus in Escherichia coli infected with bacteriophage T7. Virology 1958;**6**(2):545–55. Panel (B): Data from Volkin E, Astrachan L, Countryman JL. Metabolism of RNA phosphorus in Escherichia coli infected with bacteriophage T7. Virology 1958;**6**(2):545–55.*

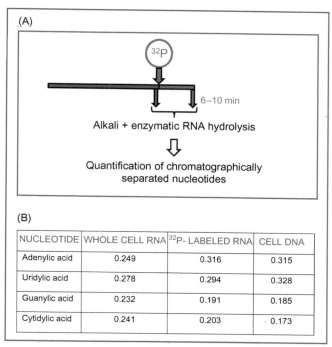

NUCLEOTIDE	WHOLE CELL RNA	^{32}P- LABELED RNA	CELL DNA
Adenylic acid	0.249	0.316	0.315
Uridylic acid	0.278	0.294	0.328
Guanylic acid	0.232	0.191	0.185
Cytidylic acid	0.241	0.203	0.173

FIGURE 9.15 (A) Experimental design of the Yčas and Vincent experiment.[79] Yeast cells were pulse labeled with ^{32}P, the radioactive RNA was hydrolyzed and its nucleotides were chromatographically separated and quantified. (B) Comparison of the base compositions of the yeast cell (mostly ribosomal) RNA, its pulse-labeled RNA, and DNA. *Panel (B): Data from Yčas M, Vincent WS. A ribonucleic acid fraction from yeast related in composition to desoxyribonucleic acid.* Proc Natl Acad Sci USA *1960;46(6):804—11.*

alternative theory of an unstable messenger whose chemical nature was unspecified.[63,80] Yet, the strongly held belief that ribosomal RNA is the carrier of genetic information was hard to abandon. This was manifested in a 1960 work of Masayasu Nomura, Ben Hall, and Sol Spiegelman of the University of Illinois (Box 9.5).

9.8.1 Sucrose-Gradient Centrifugation

To understand the subsequently described experiments, the commonly employed and enormously useful technique of separating macromolecules by sucrose-gradient centrifugation (Fig. 9.17) must be explained. A continuous linear gradient of sucrose is prepared from solutions that contain low (usually 3—5%) and high (20—25%) sucrose concentrations (Fig. 9.17). After a sample that contains a mixture of macromolecules (nucleic acids, proteins, ribosomes, etc.) of different sizes is loaded onto the gradient, it is then

After earning a BA degree in mathematics at Columbia University, Spiegelman taught physics and applied mathematics and did graduate work in Washington University, St Louis. Switching to biology, he became a pioneer molecular biologist. Working first in the University of Illinois and later in Columbia University, he advanced techniques for the detection by hybridization of specific RNA and DNA molecules. In later work, he studied the replication and evolution in vitro of Qβ bacteriophage.

Ben Hall received a PhD degree in physical chemistry at Harvard University. After a time at the University of Illinois, he served for most of his career on the faculty of the University of Washington. In his earlier work, he did pioneering work on DNA and RNA hybridization. Later, he studied yeast genetics and was among the first to apply recombinant techniques in yeast. Using this approach, he prepared a vaccine against hepatitis B that saved innumerable lives around the globe.

Born in Japan, Masayasu Nomura received bachelor's and doctoral degrees from the University of Tokyo. In 1957, he moved to the United States working as postdoctoral fellow in the laboratories of Spiegelman, Watson, and Benzer. Becoming an independent researcher in the University of Wisconsin, his laboratory became a leading center of ribosome research. Among his many achievements were the total in vitro reconstitution of functional ribosomes from ribosomal RNA and proteins, identification of the stepwise process of initiation of protein synthesis and analysis of the translation-level regulation of coordinate synthesis of ribosomal RNA and proteins. After his move to UC Irvine, he studied yeast ribosomes, discovering that they differed significantly from bacterial ribosomes (Fig. 9.16).

| Sol Spiegelman | Benjamin (Ben) Hall | Masayasu Nomura |
| (1914–1983) | (1933–) | (1927–2011) |

FIGURE 9.16   Sol Spiegelman, Ben Hall, and Masayasu Nomura.

centrifuged in a swinging bucket rotor in which the macromolecules migrate horizontally and are separated by size along a gradient of increasing viscosity. Unlike in CsCl density-equilibrium centrifugation, the macromolecules do not reach equilibrium with the relatively low density of the sucrose solutions. Centrifugation is stopped, therefore, after several hours and the

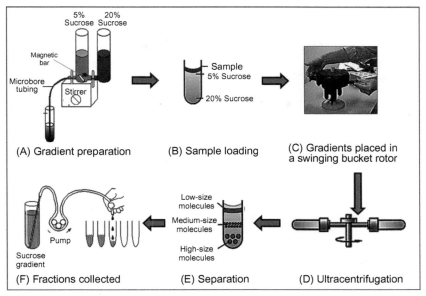

FIGURE 9.17 Separation of macromolecules by sucrose-gradient centrifugation. (A) A continuous linear gradient of low to high sucrose concentrations is prepared as sketched. The concentration of sucrose in the formed gradient is highest at the bottom of the tube and lowest at the top. (B) Sample containing a mixture of macromolecules of different sizes is loaded onto the gradient. (C) The gradients are placed in a swinging bucket ultracentrifuge rotor. (D) During ultracentrifugation, the tubes are swung at 90 degrees such that the different macromolecules are migrating horizontally and resolved from one another. (E) At the end of the centrifugation, the different macromolecules are separated by size. (F) The gradient is pumped from the bottom and fractions that contain separated macromolecules are collected.

macromolecules that were resolved by size are collected into separate fractions (Fig. 9.17).

9.8.2 The Nomura, Hall, and Spiegelman Experiment

This Illinois group monitored the synthesis and intracellular location of RNA that was labeled with ^{32}P at 5 to 7 minutes after the infection of E. coli with T2 phage. Uninfected control cells were exposed to ^{32}P for a similar brief period of 2 minutes.[81] The bulk of the labeled RNA of both phage-infected and uninfected cells was found to cosediment in the ultracentrifugation together with the ribosomes and to co-migrate in starch gel electrophoresis with ribosomal RNA. However, electrophoresis of ribosomes in starch column showed that although pulse-labeled RNA from uninfected cells co-migrated with the ribosomal RNA, the mobility of the ^{32}P-labeled RNA in the T2-infected cells was higher than that of the ribosomal RNA (Fig. 9.18A).

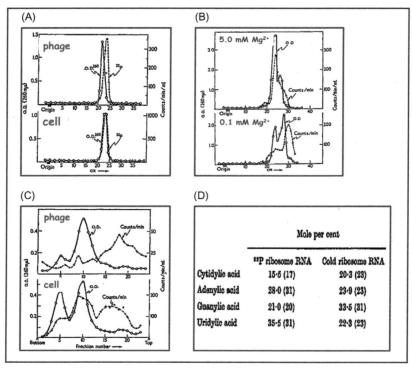

FIGURE 9.18 Nomura, Hall, and Spiegelman's analyses of pulse-labeled RNA from bacteriophage T2-infected and uninfected *Escherichia coli* cells. (A) Starch column electrophoresis of the pulse labeled (*dotted line peaks*) and ribosomal RNA (*continuous line peaks*). After electrophoresis RNA was extracted from slices of the gel and radioactivity and UV adsorption at 260 mµ were measured to identify the respective positions of the pulse-labeled and of ribosomal RNA. Polarity of migration was from left (origin) to right. Note the greater mobility of the phage RNA relative to ribosomal RNA. (B) Starch column electrophoresis of mixtures of unlabeled whole ribosomes and ribosomal RNA and of ^{32}P-pulse-labeled phage RNA was conducted as described in A in the presence of 5.0 or 0.1 mM Mg^{2+}. In the presence of 5.0 mM Mg^{2+}, most of the phage ^{32}P-labeled RNA (*dotted line peaks*) co-migrated with the whole ribosomes (*slowest continuous line peak*) and some with the ribosomal RNA (*faster continuous line peak*). By contrast, at 0.1 mM Mg^{2+}, the mobility of the T2 RNA was higher than the mobility of whole ribosomes or ribosomal RNA. (C) Sucrose-gradient centrifugation of mixtures of phage or cellular pulse-labeled RNA with unlabeled ribosomal RNA. Note that the radioactively pulse-labeled phage or cellular RNA migrated as heterogeneous lighter peak than either the 16S or 23S peaks of ribosomal RNA. (D) Base compositions of phage ^{32}P-pulse-labeled RNA and of cellular ribosomal RNA. Note that despite the significant differences between the two RNA species, the authors named the phage RNA "^{32}P ribosome RNA." *All the shown results are from Nomura M, Hall BD, Spiegelman S. Characterization of RNA synthesized in* Escherichia coli *after bacteriophage T2 infection.* J Mol Biol *1960;2(5):306−26.*

When the ^{32}P-labeled phage RNA was electrophoresed together with unlabeled ribosomes and ribosomal RNA at a relatively high concentration of magnesium (5.0 mM), it migrated mostly together with the whole ribosomes. However, at low concentration of magnesium (0.1 mM), the labeled phage RNA had significantly higher electrophoretic mobility than either the ribosomes or the ribosomal RNA (Fig. 9.18B). Most significantly, sucrose-gradient centrifugation resolved similar <16S heterogeneous peaks of both phage and cell pulse-labeled RNA. Thus, these radioactive rapidly labeled RNA species had distinctly different size than either the 16S or 23S RNA subunits of ribosomal RNA (Fig. 9.18C). Last, the base compositions of the ^{32}P-labeled T2 RNA and of the cellular ribosomal RNA were determined. As seen in Fig. 9.18D, the base ratios of bacteriophage RNA and ribosomal RNA were significantly different. Despite the several distinct differences between the phage RNA and ribosomal RNA, Nomura, Hall, and Spiegelman held on to the "One Gene—One RNA—One Ribosome—One Protein" percept that maintained that the pulse-labeled T2 RNA represented phage-dictated ribosomal RNA. According to this interpretation, upon infection the phage shut off synthesis of cellular ribosomes and initiated synthesis of phage-specific ribosomal RNA. This supposedly virus ribosomal RNA then associated with proteins to form phage-specific ribosomes that were alleged to encode the viral proteins. The different electrophoretic mobility of the phage RNA relative to ribosomes was interpreted as reflection of the failure of the T2 ribosomes to assume their ultimate conformation at low magnesium concentration. The observed similar mobility of phage RNA and ribosomes at higher concentration of the ion was taken as an indication for the formation under those conditions of mature viral ribosomes.[81] The Illinois group also constructed a mathematical model to show that the amount of RNA that was made during the 2 minutes long pulse period sufficed to generate enough phage-specific ribosomes that were necessary to provide for the amount of viral proteins which were made during that interval.[81]

Remarkably, therefore, despite the accumulated lines of evidence that stood in contrast to the "One Gene—One RNA—One Ribosome—One Protein" concept (Table 9.1), and although their own results were incongruent with this view, Nomura, Hall, and Spiegelman interpreted their results to still fit with this consensually held idea.

9.9 "THE PENNY DROPS"—mRNA REPLACED RIBOSOMAL RNA AS THE CARRIER OF GENETIC INFORMATION

The consensually held belief that ribosomal RNA was the information-bearing macromolecule came to an end at a specific point in time. The PaJaMo results and Volkin and Astrachan's observations were finally connected in the spring of 1960 during an auspicious meeting of few leading molecular biologists. Consequently, the theory of ribosomal RNA as the conveyer of genetic information was discarded in one dramatic instance and was replaced by the hypothesis that a minor fraction of unstable mRNA carries the genetic information. This sudden shift in theory occurred when a small group of participants in Microbiological Society Conference in London convened after the meeting in Sydney Brenner's rooms at Cambridge King's College on the afternoon of Good Friday, Apr. 15, 1960. Among those attending were Jacob, Crick, Leslie Orgel, Alan Garen, possibly Ole Maaløe, and Brenner himself. Based on interviews that he conducted in the 1970s with Jacob, Brenner, and Crick, Judson reconstructed this historic gathering (Ref. 50, pp. 415−421). According to his rendition, prior to the gathering, Brenner was already mindful of the conflict between the presumed role of ribosomes as carriers of genetic information and the reported dissimilarity between the base compositions of ribosomal RNA and DNA and the fixed sizes and the stability of ribosomal RNA. The discussion was opened by Jacob's detailed description of the PaJaMo experiment[63] and of then still unpublished kinetic analysis of the β-galactosidase gene expression.[73] The results of both sets of experiments implied either direct translation of protein off DNA or the existence of a labile messenger. Both Brenner and Crick realized that Riley and Pardee's observation that radioactive decay of the Z gene abolished the ability of cells to synthesize β-galactosidase[73] indicated a need for continued activity of the gene. Inescapably, this observation excluded the possibility of a stable cytoplasmic intermediate.[50] As recalled by Crick in his interview with Judson: "That is when the penny dropped and we realized what it was all about." It is worthwhile to reproduce here Judson's vivid description of the instance at which the PaJaMo experiment and Volkin and Astrachan's results were pieced together and the idea of mRNA was crystallized (Ref. 50, p. 418):

> *"Sydney and Francis put the thing together with something which was known which has to do with the phage DNA synthesis." Jacob said. "That there was some kind of very bizarre RNA made after phage infection by T2." ["What Hershey had first found", Judson suggested] "Yes, Hershey and two others, I forget their names -" [obviously, Jacob forgot the names of Volkin and Astrachan].*
>
> *"The one thing that had been kind of stuck was the Volkin and Astrachan RNA," Brenner said. "They had shown that you made an RNA whose apparent base composition [. . .] was the mimic of phage DNA. [. . .] and of course Volkin and Astrachan themselves interpreted it that this was precursor to DNA [. . .] From my point of view what happened at that day was that I clicked together the Volkin and Astrachan RNA with these possibilities".*
>
> *"Well, then what was clear suddenly was that the Volkin-Astrachan RNA was the message," Crick said. "Which they hadn't realized in Paris. In other*

words, that people had already discovered *the messenger RNA and hadn't* realized *it."[...] "And then of course it immediately followed [...] that the ribosome was an inert—was a* reading head, *and had got nothing to do with the message. And with that it just went on. You see. The false idea was just removed in one blow".*

It should be noted that in mentioning Volkin and Astrachan, Brenner was referring to their original notion that the unstable RNA was precursor to phage DNA (Ref. 76, pp. 686−695). Remarkably, however, he seemed not to recall their later observation of the different kinetics of accumulation of radioactivity in phage RNA and DNA that excluded precursor−product relationship between the two. Their conclusion that the rapidly labeled phage RNA could be template for proteins remained, therefore, unmentioned.[78]

9.10 EUREKA MOMENT: CRICK DREW A THEORY OF mRNA WHILE BRENNER AND JACOB DESIGNED EXPERIMENTS FOR ITS DETECTION

The "dropped penny" at the Good Friday gathering had immediate consequences. Crick proceeded straightaway to formulate a theory of transfer of genetic information via rapidly turning-over mRNA molecules, whereas Brenner drafted a plan for experiments aimed to detect the elusive mRNA. Crick drafted a note (never to be published) to the RNA Tie Club on the predicted properties of mRNA (a facsimile of Crick's May 1960 handwritten communication is posted in the NIH Profiles in Science Web site: http://profiles.nlm.nih.gov/ps/retrieve/ResourceMetadata/SCBBFZ). This note had Brenner and Crick as authors and was titled "What are the Properties of Genetic RNA?" Indicating frustration at the intangibility of the hypothetical mRNA, Crick chose as a telling motto to the paper the Baroness Emmuska Orczy's description of her fictional master of disguise Scarlet Pimpernel: "Is he in heaven? / Is he in hell? / That demmed elusive Pimpernel." Referring to the kinetic analysis of the induction of β-galactosidase on the one hand, and to Volkin and Astrachan's results on the other hand, Crick delineated the predicted attributes of the purported "genetic RNA":

On the basis of these two sets of experiments, we propose the following hypothesis:

1. The genetic RNA makes up only a small part (say 10 to 20%) of the RNA of the ribosome.

2. It has the same base ratios as the DNA which controls it.

3. It is unstable at least under some circumstances.

The excitement of the moment and yet hesitations about the new idea were reflected in a letter that Crick wrote shortly thereafter to Arthur Kornberg (Facsimile of this May 10, 1960, letter is posted in: http://profiles. nlm.nih.gov/ps/retrieve/ResourceMetadata/SCBBGZ). After listing the above forecasted features of the messenger, Crick commented: *"This hypothesis explains a number of experiments, but contradicts others, so we are still in two minds about it."*

While Crick was outlining the theoretical requirements from mRNA, Brenner was already designing experiments to demonstrate its existence. In a conversation with Judson, he later described his state of mind at the time (Ref. 50, p. 419):

> *[…] it suddenly hit me between the eyes, that of all things there had to be an RNA which was added to the ribosomes […] it suddenly dawned on me that there was a complete interpretation of the phage business and not just enzyme induction […]. The most important outcome of that was that* here was a way to actually proving it. *I just followed straight on to design the experiment we later did.*

In a fortunate turn of events, Delbrück invited Jacob for a summer visit in Caltech and Meselson (through Beadle) invited Brenner for a visit to the same institution at the same time. Leaving together the Good Friday gathering, Jacob and Brenner instantly started to plan experiments that they intended to conduct in Caltech with the objective of detecting mRNA. Promptly after returning to Paris, Jacob wrote to Meselson to alert him about Brenner's and his planned experiment for Pasadena (a facsimile of Jacob's Apr. 19, 1960, letter to Meselson is posted in the Cold Spring Harbor Laboratory DNA Learning Center Web site: http://www.dnalc.org/view/16480-Gallery-21-Letter-from-Fran-ois-Jacob-s-to-Matt-Meselson.html). Jacob's writing reflected a sense of urgency to do the critical experiment: "I met Sydney in Cambridge last week and we have plan(ned) a few experiments to do together with you if you agree …." He then proclaimed as dead the idea of stable ribosomes as carriers of genetic information and briefly outlined the new hypothesis of an unstable message. Shortly thereafter Brenner too wrote to Meselson (a facsimile of Brenner's May 7, 1960, letter to Meselson is also posted in the Cold Spring Harbor Laboratory DNA Learning Center Web site: http://www.dnalc.org/view/16481-Gallery-21-Letter-from-Sydney-Brenner-to-Matt-Meselson-3-.html). Opening his letter with the enthusiastic statement: "This letter is to tell you about exciting developments here and also to discuss an experiment that we should do in Pasadena …." Brenner went on to chart experiments designed to identify phage-dictated labile RNA and its association with ribosomes.

9.11   BRENNER, JACOB, AND MESELSON DETECTED LABILE RNA IN PHAGE-INFECTED CELLS

The quest for "genetic RNA" started in Caltech in early Jun. 1960. The chosen experimental system was the T4 phage-subverted genetic apparatus of *E. coli*. The experiment was designed to critically distinguish between three competing theories of the mode of information transfer that are illustrated in Ref. 14 and are reproduced in Fig. 9.19.

One model was that upon invasion by the phage, the cell ribosomes become inhibited and synthesis of phage-specific proteins is directed by newly generated phage-specific ribosomes. An alternative model, based on one possible interpretation of the PaJaMo results, was that phage proteins are copied directly from phage DNA. This model also alleged that the function

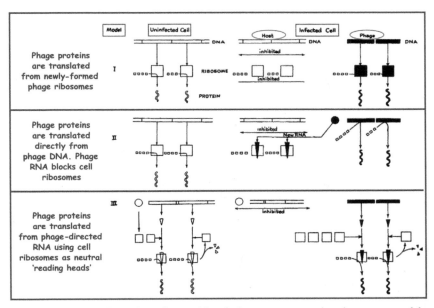

FIGURE 9.19   The three alternative models of phage gene expression that were tested by Brenner, Jacob, and Meselson.[14] In *Model I*, T4 phage inhibited the ability of the host cell ribosomes to serve as template for cellular proteins and dictated synthesis of new phage-specific ribosomes that were used as templates for the synthesis of viral proteins. In *Model II*, the phage DNA itself was the direct template for the synthesis of viral proteins while phage-induced RNA inhibited the host ribosomes. In *Model III*, cell ribosomes that acted as nonspecialized "reading heads" served as a platform for the translation of the viral protein from RNA template that was copied from the phage DNA. *Adapted from Brenner S, Jacob F, Meselson M. An unstable intermediate carrying information from genes to ribosomes for protein synthesis.* Nature 1961;**190**(4776):576–81.

of the phage-induced RNA was to repress the host ribosomes. The third model claimed that a new type of RNA: "mRNA" that is copied off phage DNA, transfers information from DNA to nonspecialized ribosomes that serve as passive "reading heads."[14] Experiments were thus conducted to decide which of the three models corresponded to the actual pattern of expression of T4 phage genes.

To examine the validity of the first model, Brenner, Jacob, and Meselson prepared extracts of T4-infected cells either after pulse labeling the phage RNA with ^{14}C-uridine or after an additional period of chasing the label by an excess of nonradioactive ^{12}C-uridine (Fig. 9.20A). Density-gradient centrifugation of extracts of the pulse-labeled cells revealed that unlike true

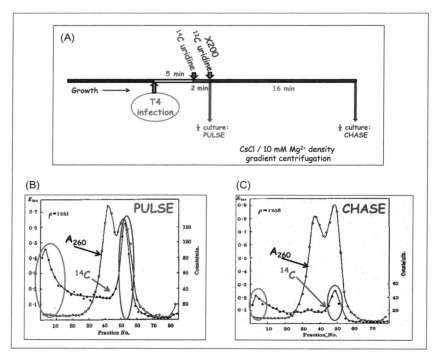

FIGURE 9.20   Design and results of an experiment to test whether new virus-specific ribosomes were formed in phage-infected cells. (A) Time course of T4 phage infection, pulse labeling of the viral RNA, chase of the label and CsCl density-equilibrium resolution of the host ribosomes and of the phage RNA. (B) Density-gradient distribution of the ^{14}C radioactively labeled viral RNA and of the host ribosomes as measured by their absorption at 260 mμ. Most of the radioactively pulse-labeled RNA (*red ellipse*) had higher density than the ribosomes and some of it (*purple ellipse*) was lighter than ribosomes. (C) Both peaks of the phage RNA largely disappeared after the chase period indicating that it was labile. *The density-gradient profiles in B and C were derived from Brenner S, Jacob F, Meselson M. An unstable intermediate carrying information from genes to ribosomes for protein synthesis. Nature 1961;**190**(4776):576–81.*

ribosomes, the density of most of the T4 RNA was higher than that of the ribosomes and that some of the radioactively labeled viral RNA displayed a density that was lower than that of the ribosomes (Fig. 9.20B). Furthermore, similarly to RNA of T2 and T7 phages,[69,78] the activity of the pulse-labeled T4 RNA decreased after the chase period indicating that this RNA was unstable (Fig. 9.20C).[14]

It was also found that the viral RNA fully dissociated from the ribosomes at low magnesium concentration. The different hydrodynamic migration of ribosomes and T4 RNA was inconsistent with the first model of phage RNA being a constituent of newly synthesized phage-specific ribosomes.[14]

Having precluded the first model Brenner, Jacob, and Meselson next put to test the second and third models. Experiments were executed to determine whether it was the phage DNA itself or preexisting cell ribosomes that served as platform for viral protein synthesis. Here came into use the technique that Meselson originally devised with Stahl of using heavy and light isotopes of nitrogen to distinguish between old and newly replicated DNA.[3] To decide between the second and third models, bacterial ribosomes were prelabeled with a heavy ^{15}N isotope and then the cells were infected with T4 phage and transferred to a medium that contained a lighter ^{14}N isotope. Newly synthesized proteins were labeled by radioactive ^{35}S isotope at the time of infection and transfer of the infected bacteria to the ^{14}N-containing medium. Completed proteins were then chased away from the ribosomes by diluting the ^{35}S isotope with an excess of nonradioactive ^{32}S (Fig. 9.21A). Density-gradient centrifugation of cell extracts revealed that ^{35}S-labeled newly made viral proteins were transiently associated with "old" ^{15}N-containing cell ribosomes. Also, there was no hint of formation of newly made ^{14}N-containing phage-specific ribosomes (Fig. 9.21B).[14]

The transient association of the phage proteins with the host ribosomes indicated that the bacterial ribosomes and not the viral DNA was the stage upon which phage protein synthesis took place. These results were consistent, therefore, with the third model in which phage proteins were translated from a labile T4 mRNA template that bound to the cell ribosomes in magnesium-dependent manner.

9.11.1 Facing Lackluster Reception of the Work, Brenner Added Experiments to Substantiate the Existence of Labile RNA in Phage-Infected Bacteria

Due to technical snags, the struggle to obtain the above-depicted results was hard. As Jacob vibrantly described in his autobiography, just as he and Brenner were ready to leave California without reaching conclusive results, Brenner realized that adjusting the concentration of magnesium was critical to their analysis. Thus, the decisive experiment that demonstrated magnesium-dependent association of the pulse-labeled RNA with ribosomes

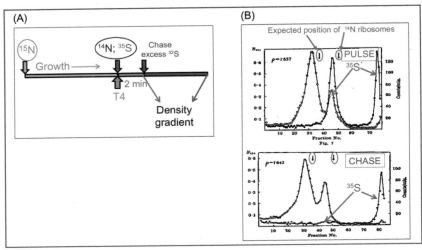

FIGURE 9.21 Design and results of an experiment to identify the site of synthesis of newly made viral proteins. (A) Scheme of the experimental design: cells were grown for extended period of time in a medium containing the ^{15}N heavy isotope. Upon cell infection with T4 phage, the cells were transferred to ^{14}N-containing medium and newly made viral proteins were labeled with ^{35}S isotope. The infected cell culture was then divided into two portions. In one, pulse labeling was stopped after 2 minutes whereas the remaining cells were provided with an excess of nonradioactive ^{32}S that diluted the radioactive isotope and thus could expose dissociation of the pulse-labeled viral proteins from their site of synthesis. Pulse-labeled and chased phage proteins and ribosomes were then resolved by CsCl density-equilibrium centrifugation. (B) Results of the experiment: Top profile: Newly made ^{35}S-labled proteins associated with a peak of cell ribosomes. No trace of phage-directed new ribosomes was detectable at the expected position of ^{14}N-containing lighter ribosomes. Bottom profile: Newly made ^{35}S-labled phage proteins after a period of incubation in the presence of excess nonradioactive ^{32}S (chase). *The density-gradient profiles in (B) were derived from Meselson M, Stahl FW. The replication of DNA in* Escherichia coli. Proc Natl Acad Sci USA *1958;**44**(7): 671−82.*

was performed practically at the last moment with some important controls still missing. Indeed, when the unpolished results were publicly presented by Jacob in Caltech and by Brenner in Stanford they were met with silence or with skepticism (Ref. 50, p. 427). It was thus that upon his return to Cambridge Brenner had to work for four additional months in order to reproduce the Caltech results and to complete the necessary controls.

9.11.2 Brenner, Jacob, and Meselson Were Unaware of a Parallel Study That Was Being Performed in Watson's Harvard Laboratory

While Brenner and Jacob were still toiling away in Caltech, Meselson left for a Gordon Conference in the east coast. According to Judson, Brenner and Jacob telephoned Meselson to tell him of the results after their experiments were finished. When Meselson told Watson, who also attended the Gordon Conference, about the experiments and their outcome, Watson disclosed that his laboratory was also doing experiments along the same lines (Ref. 50, p. 427; see also Meselson's videotaped testament about the parallel efforts at Caltech and Harvard: http://www.dnalc.org/view/15330-Messenger-RNA-Matthew-Meselson.html). Indeed in a remarkable happenstance, while Brenner, Jacob, and Meselson were discovering phage mRNA, Watson's Harvard Laboratory was independently attempting to detect unstable mRNA in uninfected *E. coli* cells. What were the sources of the Harvard Laboratory experiments? As mentioned, Watson developed an interest in RNA viruses while still at Cambridge (this chapter) and moving later to Caltech, he became actively engaged in studies on the structure of RNA (see Chapter 7: "Defining the Genetic Code"). This line of research was expanded in his Harvard Laboratory to studies on the structure of bacterial ribosomes.[53] Yet, Watson was not part of the Good Friday gathering and was uninvolved, therefore, in the theoretical and experimental developments that ensued. As an aside, it should be remarked, however, that Judson speculated that Watson might have heard what transpired on that occasion from Alan Garen, one of the attendants of the Good Friday meeting (Ref. 50, p. 428). More importantly, Watson's interest in RNA and ribosomes coalesced with those of Françoise Gros of the Pasteur Institute (Box 9.6) who joined his Harvard Laboratory in May 1960.

BOX 9.6 Members of Watson's Harvard Group

James Watson (see Chapter 5: "Discovery of the Structure of DNA") headed the group at the Harvard University Biological Laboratories that detected a minor fraction of labile RNA in *E. coli* whose properties corresponded to mRNA.

After completing his PhD work at the Pasteur Institute, Françoise Gros joined its department of biochemistry to study the effects of antibiotics on bacterial metabolism. He was introduced to microbial genetics when he worked in the early 1950s with Spiegelman at the University of Illinois and then with Hotchkiss at the Rockefeller Institute. Upon his return to Paris he joined Monod's department where he studied RNA. This work culminated in the discovery in 1961 of mRNA. During his ensuing long career, he studied mechanisms of transcription and translation and later of development. He also inherited Monod as Head of the Pasteur Institute. For his short autobiographical note, see Ref. 82.

After graduating from Harvard College, Howard Hiatt received in 1948 an MD degree from the Harvard Medical School. Having been also trained in biochemistry, he joined the Watson group in their search of *E. coli* mRNA. He later did independent work on mRNA in mammalian cells. His lifelong career was at Harvard University where he served as Professor of Medicine, Physician-in-Chief at Beth Israel Hospital, and Dean of the School of Public Health where he promoted integration of basic biological sciences into medicine.

(Continued)

BOX 9.6 (Continued)

Walter Gilbert obtained BA and MSc degrees in physics at Harvard University and a PhD in theoretical physics at the University of Cambridge. He returned to Harvard in 1956 to serve there as an Assistant Professor of Physics. Gilbert and Watson became friends during their time together in Cambridge. As described in his illuminating recollections,[83] when they met again in Harvard, Watson invited him to join his laboratory in the search for mRNA. This was the start of his celebrated career in molecular biology that involved the isolation of the *lac* repressor, invention of a method of DNA sequencing that won him the 1980 Nobel Prize in Chemistry and the advancement of the RNA world idea.

Charles Kurland participated in the discovery of *E. coli* mRNA while he was a PhD candidate at Harvard. After postdoctoral work in the Microbiology Institute of the University of Copenhagen, he became Professor of Molecular Biology at Uppsala University, and Professor at Lund University, both in Sweden. His main subjects of research are bacterial ribosomes and the evolution of endoparasitic bacteria and cellular organelles (Fig. 9.22).

| James Watson (1928–) | François Gros (1925–) | Howard H. Hiatt (1925–) | Walter Gilbert (1932–) | Charles Kurland (1936–) |

FIGURE 9.22 Members of the Harvard group that identified rapidly turning-over RNA in uninfected *Escherichia coli* cells.

Prior to that visit Gros conducted in Paris experiments that showed that protein synthesis ceased a few minutes after the addition of the nucleotide analog 5-fluorouracil (5FU) to *E. coli* cells. This observation contradicted the idea that the template for proteins was stable.[84] That Gros was alert early on to the potential existence of a presumed labile RNA fraction was reflected in his May 1, 1960 (shortly before his arrival in Harvard) letter to Watson (Handwritten letter from François Gros to James D. Watson; *CSHL Archives Repository*, Reference JDW/2/2/746/12; http://libgallery.cshl.edu/items/show/38664). There he wrote: "We presently have, I think interesting case of RNA turnover, with bacteria in which uracil RNA is replaced by 5FU. It [illegible word] partly the sRNA but also a more polymerized." He then commented that this fraction perhaps resembled what he dubbed as the "*RNA Volkin.*" (For an additional retrospective description by Gros of the discovery of mRNA, see Ref. 85.)

9.12 THE HARVARD GROUP IDENTIFIED A MINOR FRACTION OF RAPIDLY TURNING OVER RNA IN UNINFECTED BACTERIA

Starting his work in Harvard, Gros first followed the incorporation of ^{14}C-uracil into incomplete ribosomes under conditions of inhibited protein synthesis. These experiments exposed a fraction of unstable labeled RNA that differed from both ribosomal and soluble (transfer) RNA. Both Gros and Watson felt at that point that this rapidly turning-over RNA could be the sought-after unstable informational RNA. At that point, several additional members of the laboratory, Howard Hiatt, Walter Gilbert, and Charles Kurland (Box 9.6), joined the effort to identify mRNA in uninfected *E. coli* cells. In their experiments, the Harvard investigators prepared pulse-labeled RNA by exposing *E. coli* cells to ^{32}P or ^{14}C-uracil for only 10−20 seconds (for a captivating description of how this was actually done, see Gilbert's recollections in Ref. 83). Resolution by sucrose-gradient centrifugation of ribosomes and of the radioactive RNA revealed that at a low concentration of magnesium (0.1 mM), the 70S ribosomes dissociated into their 50S and 30S subunits and the rapidly labeled RNA migrated as a wide peak of <30S that was lighter and separate from both subunits of the ribosome (Fig. 9.23A). Control experiments indicated that short radioactive pulse that was given to T2 phage-infected *E. coli* cells similarly labeled a wide <30S RNA peak that had a distinctly lower molecular size than the ribosome subunits (Fig. 9.23B).

To assess whether or not the pulse-labeled RNA of uninfected cells was able to associate with ribosomes, they were resolved by sucrose-gradient centrifugation in the presence of a high (10 mM) concentration of magnesium (Fig. 9.24A). The ribosomal subunits associated under these conditions to form the heterodimeric 70S ribosome (with some heavier 100S ribosomes). Importantly, while most of the pulse-labeled ^{14}C-RNA still migrated in the gradient as a peak lighter than the 70S ribosomes, a portion of this fraction of radioactive RNA was found to co-migrate with the 70S ribosomes—suggesting that the labile RNA associated with ribosomes (Fig. 9.24B, Ref. 15).

Following deproteinization of the cell extract, the labeled RNA migrated in sucrose gradients as a wide peak of 8S that was distinct from the 16S and 23S ribosomal RNA subunits and from the 4S peak of transfer RNA (tRNA). Similarly to the cellular pulse-labeled RNA, rapidly labeled RNA of phage T2 also migrated in sucrose gradients distinctly from ribosomes and ribosomal RNA (Fig. 9.24C, Ref. 15).

Once the Harvard group and Brenner, Jacob, and Meselson learned about one another's results, Watson asked that their two papers be published at the same time. As documented in a Feb. 22, 1961, letter from

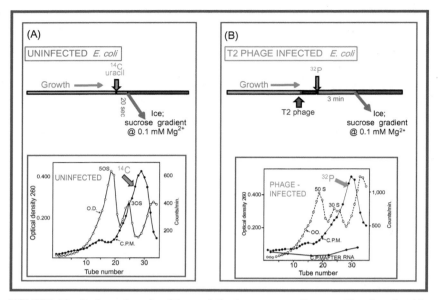

FIGURE 9.23   Design and results of the resolution by sucrose-gradient centrifugation of rapidly labeled RNA from uninfected and T2 phage infected *Escherichia coli* cells. (A) Isolation and resolution of pulse-labeled RNA from uninfected bacteria. Upper panel: Time course of the experiment. Growing *E. coli* cells were exposed to ^{14}C uracil for 20 seconds and rapid cooling of the cell culture terminated the incorporation of the radioactive RNA precursor. The bacteria were broken in the cold and their extract was subjected to sucrose-gradient centrifugation. Lower panel: Sucrose-gradient profile of the labeled RNA and of the ribosomal subunits in cell extracts. Centrifugation of cell extracts was conducted in the presence of a low 0.1 mM concentration of Mg^{2+} at which the 70S ribosomes were dissociated into their 30S and 50S subunits. Radioactivity and UV adsorption in fractions of the gradient identified the respective positions of the pulse-labeled RNA and of the ribosomal subunits. (B) Isolation and resolution of pulse-labeled RNA from T2 bacteriophage-infected bacteria. Upper panel: Time course of the experiment. Following infection by T2 phage, the bacteria were exposed to radioactive ^{32}P isotope for 3 minutes, cooled rapidly, extracted and subjected to sucrose-gradient centrifugation in the presence of 0.1 mM Mg^{2+}. Lower panel: Sucrose-gradient profile of the radioactive pulse-labeled RNA and the ribosomal subunits. The radioactive peak was verified as RNA by showing its complete elimination following digestion with RNase (bottom profile). *Panels (A) and (B): From Gros F, Hiatt H, Gilbert W, Kurland CG, Riseborough RW, Watson JD. Unstable ribonucleic acid revealed by pulse labelling of* Escherichia coli. Nature *1961;**190** (4776):581−85.*

Jacob to Meselson, he consented to the delay in publishing their paper until the Gros et al. paper was completed (Facsimile of the letter in the J.D. Watson CSHL Collection: http://libgallery.cshl.edu/items/show/40970). Ultimately, the two papers were published back-to-back in the same May 13, 1961, issue of *Nature*.[14,15]

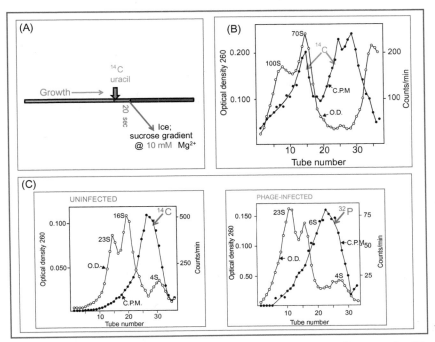

FIGURE 9.24 Design and results of the examination of association with ribosomes of rapidly labeled RNA from uninfected *Escherichia coli* cells in the presence of high concentration of magnesium (A,B), and different hydrodynamic behavior of ribosomal RNA and of pulse-labeled RNA from uninfected or T2 phage infected cells (C). (A) Design of an experiment to examine the sedimentation behavior of pulse-labeled RNA from uninfected *E. coli* cells. The procedure was identical to the protocol in Fig. 9.23A except that sucrose-gradient sedimentation of the RNA and ribosomes was performed in the presence of 10 mM Mg^{2+}. (B) Sedimentation profile of pulse labeled and ribosomes in the presence of high concentration of magnesium. In contrast to the fully resolved peaks of ribosomal subunits and rapidly labeled RNA in the presence of low magnesium (Fig. 9.23A), a portion of the pulse-labeled RNA associated with the 70S ribosomes at higher concentration of magnesium. (C) Sucrose-gradient profiles of ribosomal RNA and pulse-labeled RNA from uninfected and phage-infected cells. The rapidly labeled RNA migrated in both instances as a wide peak that was lighter than the 16S and 23S subunits of ribosomal RNA. *Panels (B) and (C): Derived from Gros F, Hiatt H, Gilbert W, Kurland CG, Risebrough RW, Watson JD. Unstable ribonucleic acid revealed by pulse labelling of* Escherichia coli. Nature 1961;**190**(4776):581−85.

9.13 DESPITE THE DISCOVERY OF RIBOSOME-ASSOCIATED LABILE RNA, IT STILL REMAINED TO BE SHOWN THAT THIS RNA WAS THE CARRIER OF INFORMATION FROM DNA

The fraction of labile RNA that was identified in uninfected and bacteriophage-infected cells satisfied some but not all of the attributes that Jacob and Monod predicted for mRNA.[13] The minor fraction of unstable RNA[14,15] did fulfill two of the forecasted properties: (1) (mRNA should be) ... "very heterogeneous with respect to molecular weight." (2) (mRNA should be) ... "found associated with ribosomes." However, a third requirement remained unsatisfied: (3) "(mRNA) ... should have a base composition reflecting the base composition of DNA."[13] In the absence of DNA sequencing methods that became available only 16 years later, the determination of nucleotide sequences of genes was yet unfeasible. The technically harder task of sequencing RNA molecules was even less approachable in the early 1960s. Yet, very shortly after the identification of rapidly turning-over bacterial and phage RNA, the technique of DNA–RNA hybridization was successfully applied to demonstrate that this fraction of RNA was indeed copied from the respective DNA templates. This experimental achievement satisfied the last missing requirement of mRNA, that is, that its base composition should reflect the base composition of DNA.

9.13.1 Early History of DNA–DNA and DNA–RNA Hybridization

The technique of nucleic acids hybridization was launched in 1956 by Alexander Rich (Box 9.7) and David Davies (1927–), both then at the NIH. In an initial report, they showed that RNA strands of synthetic polyriboadenylic [poly(rA)] and polyribouridylic acid [poly(rU)] could complementarily anneal to form a double-stranded poly(rA) · poly(rU) structure.[86]

BOX 9.7 Alexander Rich, Paul Doty, and Julius Marmur

Alex Rich earned both his Bachelor's and MD degrees from Harvard University. Choosing to pursue science rather than medicine, he did postdoctoral work at Caltech under Linus Pauling. After several years at the NIH he joined in 1958 the MIT faculty where he was Professor of Biophysics. Among his many scientific contributions were the pioneering experiments on nucleic acid hybridization, the first description of polyribosomes and the discovery of left-handed Z-DNA.

After graduating from Penn State University, Paul Doty did his PhD in Columbia University. Following a short period at the Polytechnic Institute of Brooklyn, he joined in 1948 the Chemistry Department of Harvard University where he was among the founders and the first chairman of Harvard's Department of Biochemistry and Molecular Biology. His scientific interest was in the application of physical methods to characterize DNA and proteins. Having worked in the Manhattan project as a graduate student, he had lifelong involvement in efforts to prevent nuclear war. In this capacity, he served as member of the President's Science and Arms Control Advisory Committees.

(Continued)

BOX 9.7 (Continued)

Julius Marmur obtained Bachelor's and Master's degrees from McGill University and a PhD in Bacterial Physiology from Iowa State University. Following periods of research at the NIH, the Rockefeller Institute and Pasteur Institute, he did his pioneering work on DNA hybridization while serving as research associate in Doty's Harvard Laboratory. After 3 years on the faculty of Brandeis University, he joined in 1963 the Yeshiva University Albert Einstein College of Medicine where he served as Professor of Biochemistry and Molecular Genetics for the rest of his career (Fig. 9.25).

Alexander Rich   Paul M. Doty   Julius Marmur
(1924–2015)   (1920–2011)   (1926–1996)

FIGURE 9.25   Alexander Rich, Paul Doty, and Julius Marmur.

Although this result is evident to today's reader, it met with skepticism at the time as it was deemed impossible for a double strand to be generated without the aid of an enzyme such as the recently isolated DNA polymerase I of *E. coli*. Also, it was considered unlikely that long polynucleotides could untangle to properly bond with one another.[87] In 1959, Rich raised the idea that genetic information is copied from nuclear DNA into RNA. If so, the base sequence of the transcribed RNA should be complementary to the sequence of the DNA template.[88] Guided by this notion, he next provided experimental evidence that synthetic poly(rA) strand was capable of annealing with the complementary synthetic DNA strand polythymidylic acid [poly(dT)] to form a poly(rA) · poly(dT) double-stranded RNA · DNA hybrid.[89]

Technically, the DNA and RNA strands were mixed together at 1:1 ratio, heated to allow their full unfolding and then the mixture was cooled slowly. Several lines of evidence indicated that the product of this procedure was indeed a RNA · DNA double-stranded hybrid. First, examination of the UV spectra of each of the single strands and of the product of their annealing revealed that the respective maxima of absorption by poly(rA) and poly(dT) were at 259 and 266 mμ, whereas the maximum absorption of their annealed product was at 262 mμ. In addition, complex formation entailed a drop in the UV adsorption (hypochromic effect) as was expected for transit from disorganized single-strand to a double-stranded nucleic acid whose bases are stacked (Fig. 9.26A). The formation of a duplex structure was also confirmed

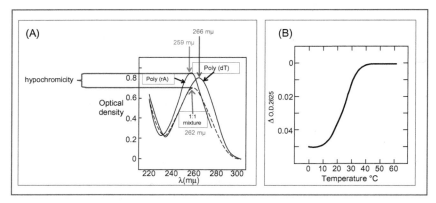

FIGURE 9.26 Formation and melting of poly(rA)·poly(dT) double-stranded RNA·DNA hybrid. (A) UV spectra of single-strands of poly(rA) and poly(dT) and of the product of the annealing of their mixture. Note the different adsorption maxima of each type of nucleic acid and the hypochromicity of the hybridized duplex. (B) Melting curve of the poly(rA)·poly(dT) duplex. Denaturation of the hybrid as function of increasing temperature was monitored by the rising UV adsorption of the nucleic acid (hyperchromicity). In the shown case, the hybrid became completely denatured at 40°C. *Panel (A): From Rich A. A hybrid helix containing both deoxyribose and ribose polynucleotides and its relation to the transfer of information between the nucleic acids.* Proc Natl Acad Sci USA *1960;46(8):1044−53. Panel (B): From Doty P, Marmur J, Eigner J, Schildkraut C. Strand separation and specific recombination in deoxyribonucleic acids: physical chemical studies.* Proc Natl Acad Sci USA *1960;46(4):461−76.*

by changed hydrodynamic properties of the nucleic acid and by X-ray verification of the helical structure of the hybrid.[89] An increase in the UV adsorption of the RNA·DNA duplex after heating (hyperchromic effect) served to indicate that the double-strand was denatured back to its constituent DNA and RNA strands (Fig. 9.26B).

Subsequent work of the Harvard Laboratory of Paul Doty (Box 9.7) provided essential evidence that hybrid formation was dependent on sequence homology. Doty together with Julius Marmur (Box 9.7) and associates[90] and Marmur and Lane[91] showed that biologically active (transforming) double-strand hybrids could be formed by annealing complementary strands under proper conditions of slow cooling of boiled DNA strands in the presence of a cation to counter the negative charge of the interacting strands. Most importantly, hybrids were formed only between homologous strands of denatured DNA, whereas strands from heterologous species were incapable of forming double-stranded hybrids. Thus, sequence homology between the nucleic acid strands was demonstrated to be a mandatory requirement for hybrid formation. It was, therefore, that by 1960 hybridization was recognized as a method to detect complementarity either between two strands of DNA or between RNA and DNA strands. These methodological developments came to fruition when DNA—RNA hybridization was employed to satisfy Jacob and Monod's third requirement from mRNA, that is (that it) "should have base composition reflecting the base composition of DNA."[13]

9.13.2 Hybridization Was Employed to Demonstrate Sequence Complementarity Between Phage RNA and Its DNA

In Feb. 1961, 3 months before the identification of labile RNA in uninfected and phage-infected cells was announced[14,15] Ben Hall and Sol Spiegelman of the University of Illinois reported that RNA—DNA hybridization revealed sequence homology between the T2 phage-dictated RNA and T2 DNA. It is enlightening to cite here the rational to this work[92]:

> *The fact that "T2-RNA" possesses a base ratio analogous to that of T2-DNA is of interest because it suggests that the similarity may go further and extend to a detailed correspondence of base sequence. [...] A direct attack on this problem by complete sequence determination is technically not feasible at the moment. However, some recent findings of Marmur, Doty, et al., suggest the possibility for an illuminating experiment. [...] Presumably, the specificity requirement for a successful union of two strands reflects the need for a perfect, or near perfect, complementarity of their nucleotide sequences. The formation of a double-stranded hybrid during a slow cooling of a mixture of two types of poly-nucleotide strands can be accepted as evidence for complementarity of the input strands.*

To show that rapidly labeled RNA from bacteriophage T2-infected cells was sequence-homologous to T2 DNA, Hall and Spiegelman annealed denatured strands of ^{3}H-labeled phage DNA with pulse-labeled ^{32}P T2 RNA. The formed ^{32}P-RNA $\cdot$ ^{3}H-DNA double-stranded was then resolved by density-equilibrium centrifugation (Fig. 9.27).

In practice, the phage ^{32}P-labeled RNA was fully resolved from denatured ^{3}H-labeled T2 DNA by CsCl density-gradient centrifugation. When the two single strands were merely mixed together under conditions that were not conducive for hybridization, no additional nucleic acid species was detectable. However, under proper hybridization conditions, the peaks of ^{32}P RNA and ^{3}H DNA overlapped indicating that an RNA $\cdot$ DNA hybrid was formed[92] (Fig. 9.28).

Hall and Spiegelman confirmed the sequence specificity of hybrid formation by also showing that ^{32}P-labeled T2 RNA failed to form hybrids with heterologous ^{3}H DNA from *E. coli* or *Pseudomonas* cells or from T5 bacteriophage.

Summarizing their results, Hall and Spiegelman wrote[92]:

> *It is concluded that T2-DNA and T2-specific RNA form hybrids because they possess complementary nucleotide sequences. [...] The demonstration of*

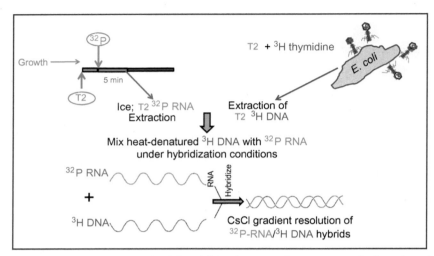

FIGURE 9.27   Illustration of the Hall and Spiegelman experiment to detect the formation of specific hybrid between denatured T2 phage DNA and phage-induced RNA.[92] Following infection of *Escherichia coli* cells by T2 bacteriophage, the viral RNA was pulse labeled with ^{32}P and extracted. Bacteriophage DNA was separately labeled by feeding ^{3}H-thymidine to T2 phage-infected cells and the labeled DNA was isolated from released virus particles. The ^{32}P-labeled RNA and the viral ^{3}H-DNA were mixed and hybridized and formed ^{32}P-RNA · ^{3}H-DNA duplex that was resolved from DNA and RNA single strands by CsCl density-equilibrium centrifugation.

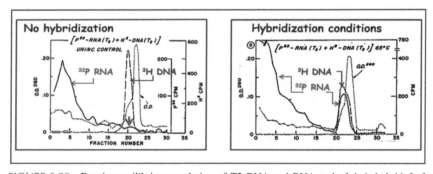

FIGURE 9.28   Density-equilibrium resolution of T2 DNA and RNA and of their hybrid. Left hand panel: Under conditions not conducive for hybridization, a denser ^{32}P-labeled phage RNA was well separated from the viral ^{3}H-labeled single-stranded DNA. The *red arrow* marks the position at which an (undetected) RNA · DNA hybrid should have been. The *dotted peak* is of adsorption at 260 mμ of double-stranded DNA. Right hand panel: When the ^{32}P-labeled viral RNA was mixed and hybridized with the denatured ^{3}H-labeled phage DNA, the appearance of a distinct peak of overlapping ^{32}P and ^{3}H radioactivity signified the formation of an RNA · DNA hybrid. *Results derived from Hall BD, Spiegelman S. Sequence complementarity of T2-DNA and T2-specific RNA.* Proc Natl Acad Sci USA *1961;47(2):137−46.*

sequence complementarity between homologous DNA and RNA is happily con-
sistent with an attractively simple mechanism of informational RNA synthesis
in which a single strand of DNA acts as a template for polymerization of a
complementary RNA strand.

Within months of the publication of the Hall and Spiegelman's paper, Samuel Weiss (1925−) and Tokumasa Nakamoto (1928−) at the University of Chicago reported that *Micrococcus luteus* RNA polymerase transcribed in vitro T2 phage DNA into RNA whose base composition[93] and nearest neighbor base arrangement[94] were similar to those of the DNA template. Most importantly, collaborating with Peter Geiduschek (1928−), these authors used RNA−DNA hybridization to demonstrate homology between the T2 phage DNA template and its in vitro transcribed RNA.[95] Thus, by 1961, it was satisfactorily established both in vivo and in vitro that phage-dictated RNA was copied from the viral DNA.

9.13.3   Hayashi and Spiegelman Identified DNA-Complementing mRNA in Uninfected Bacteria

Just a few months after the May 1961 announcement of the Harvard team that uninfected *E. coli* cells synthesized a minor fraction of rapidly turning-over nonribosomal RNA,[15] Masaki Hayashi and Spiegelman employed RNA−DNA hybridization to show that this RNA was complementary to the bacterial DNA.[96] To obtain workable amounts of the short-lived RNA, it was necessary to selectively repress synthesis of ribosomal RNA that was the major RNA species that the cell produced. Fortunately, it was known at the time that the rate of ribosomal RNA and ribosome synthesis was depressed when protein synthesis was lowered in nutritionally depleted medium. Thus, nutritional step-down suppressed the synthesis of ribosomal RNA and Hayashi and Spiegelman attained enrichment of rapidly synthe-sized RNA that they named "informational RNA" (Fig. 9.29A). The effec-tiveness of the suppression of ribosomal RNA synthesis was validated in three bacterial species by that the purine/pyrimidine ratios in their step-down RNA were similar to the ratios of their respective DNA but dissimilar from the ratios in their ribosomal RNA (Fig. 9.29B). Most importantly, density-gradient centrifugation of the RNA, the DNA, and their hybrids revealed that the nonribosomal pulse-labeled RNA of *E. coli* formed hybrids with homolo-gous *E. coli* DNA but not with heterologous *Pseudomonas* DNA (Fig. 9.29C).

Based on its DNA-like base composition and homologous hybridization, the pulse-labeled RNA was thus demonstrated to consist of transcripts of

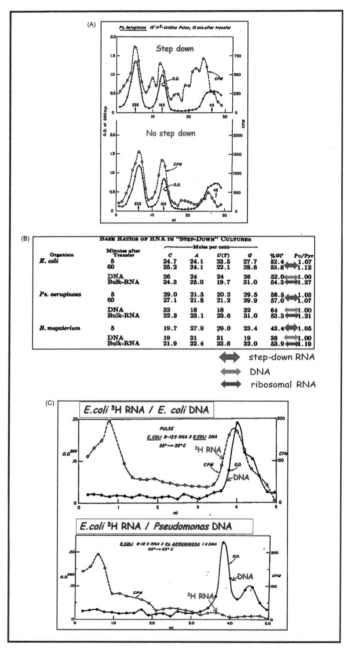

FIGURE 9.29 Detection of DNA-complementing RNA in nutritionally stepped-down *Escherichia coli* cells. (A) Enrichment of "informational RNA" in stepped-down bacteria. Cells that grew in poor and rich media were pulse labeled with radioactive ³H-uridine and their

(*Continued*)

the DNA. Naming these transcripts "informational RNA," Hayashi and Spiegelman speculated about its meaning[96]:

> It is important to emphasize that the word "informational" is not suggested as a substitute for the term "messenger" introduced in the elegant theorization of Jacob and Monod. It seems likely that both terms will be useful. Thus a given messenger RNA is presumed to constitute the structural program for the synthesis of a particular protein. It obviously must, therefore, be informational. However, not all informational RNA need operate as messengers. It is conceivable that complementary RNA molecules will be found which serve regulatory rather than programming functions.

This 1961 proposition that not all the RNA transcripts were protein-coding messenger molecules remained practically unnoticed for a long time. However, the discovery in later decades of regulatory microRNA and long noncoding RNA is affording now regained legitimacy to this prescient speculation of Hayashi and Spiegelman.

9.14 PHAGE AND BACTERIAL PROTEINS, AND IMPLICITLY THEIR mRNA WERE SHOWN TO BE COLLINEAR WITH THEIR DNA GENES

A few years after the identification of mRNA in uninfected and in phage-infected bacterial cells, two groups demonstrated that proteins in *E. coli* and bacteriophage T4 were collinear with their encoding genes and by implication, with their respective mRNA transcripts. Several groups used in the early 1960s bacterial or phage systems to look into the question of collinearity. Among the studied systems were *E. coli* alkaline phosphatase,[97–100] T4 bacteriophage lysozyme,[101] and the rII locus of this phage (see Chapter 7: "Defining the Genetic Code"). There were, however, two groups that provided definitive answers to the question of collinearity. One was a Stanford University team led by the microbial geneticist and biochemist Charles Yanofsky (Box 9.8) and the other was a University of Cambridge group led by Sydney Brenner.

◀ extracted RNA was resolved by sucrose-gradient centrifugation. Although practically all the ^{3}H-RNA was ribosomal in cells that were grown in rich medium, the stepped-down RNA was enriched in <16S RNA that was presumed to represent nonribosomal "informational RNA." (B) Base ratios of stepped-down RNA from three bacterial species. The base ratios of the stepped-down RNA were similar in all three species to the base ratios of their respective DNA and were dissimilar to the ratios in their ribosomal RNA. (C) The stepped-down pulse-labeled RNA was complementary to its homologous DNA. Equilibrium CsCl density centrifugation revealed presence of a peak of *E. coli* ^{3}H-RNA · DNA hybrid that migrated under a peak of unlabeled double-stranded DNA marker that was detected by its UV adsorption. No such hybrid was generated when the *E. coli* ^{3}H-RNA was hybridized to *Pseudomonas* DNA. *Panels (A)–(C): Adapted from Hayashi M, Spiegelman S. The selective synthesis of informational RNA in bacteria. Proc Natl Acad Sci USA 1961;47(10):1564–80.*

BOX 9.8 Charles Yanofsky

Charles Yanofsky's undergraduate studies at the City College of New York were interrupted by his military service in World War II. Returning back from the war, he obtained a Bachelor's degree in 1948, and earned a PhD in microbiology in 1951 from Yale University. Following a period on the faculty of Western Reserve University School of Medicine, he moved in 1958 to the Department of Biochemistry at Stanford University where he remained for the length of his career. The main focus of his research work was on molecular regulatory mechanisms of bacterial transcription. Among his contributions to molecular biology is the demonstration of collinearity between a bacterial gene and its protein product. He also showed with Stuart Brody that suppression of a missense mutation resulted in wild-type protein. He predicted that the cause for this phenomenon was an altered (suppressor) tRNA molecule. Expanding his studies on the *trp* operon, he and his associates discovered the phenomenon of transcriptional attenuation by which expression of the *trp* operon is regulated by a complex multistep process that involves secondary structures of the RNA transcript, the status of tRNA charging and the coupled process of translation (Fig. 9.30).

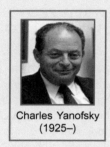

Charles Yanofsky
(1925–)

FIGURE 9.30   Charles Yanofsky.

Yanofsky's interest in gene−protein relationship was initiated during his graduate studies at Yale University. Mentored there by David Mahlon Bonner (1916−64), an associate of Edward Tatum, he aimed at elucidating gene−enzyme relationship under the Beadle-Tatum framework of "One Gene−One Enzyme."[102] Yanofsky began his investigations by isolating niacin-requiring mutants of *Neurospora* to identify respective intermediates in the niacin pathway. At a later time, he switched to tryptophan-requiring mutants of *Neurospora*, again with the aim of identifying intermediate steps in the tryptophan synthesis pathways.[103]

As an independent investigator at Stanford, and for a number of years, Yanofsky undertook to verify or disprove gene−protein collinearity for a selected *E. coli* gene and its protein product. This he did by mapping mutations in a gene that encoded a subunit of the *E. coli* enzyme tryptophan synthase. In parallel, he identified and determined the positions of substituted amino acid in the synthase of each mutant. Ultimately, positions of mutations in the gene were aligned with the mapped substituted amino acid to determine whether or not the two were collinear. Yanofsky and his postdoctoral fellow Irving Crawford launched the tested experimental systems in 1958 when they showed that tryptophan synthase of *E. coli* was comprised of two dissimilar polypeptide chains. One subunit, the A protein (TrpA), hydrolyzed indole glycerol phosphate to indole, and the second subunit, the B protein (TrpB) covalently joined indole and L-serine, forming L-tryptophan.[104] The subsequent discovery that the two subunits mutually activated one another enabled the development of an assay for the detection of altered A proteins by measuring their changed ability to activate the B protein. Employment of this assay resulted in the identification and purification of a number of mutated A proteins. To examine the possibility of structural collinearity between gene and protein, Yanofsky and his associates examined 16 mutant strains whose mutations were mapped to one segment of the tryptophan synthase A gene and protein. The order and relative distances of mutated loci in the A gene were charted first by deletion mapping and were then pinpointed by the determination of recombination frequencies and by three-point genetic tests. In parallel, the location of specific amino acid substitutions in each of the 16 mutants were determined by sequencing tryptic peptides of the A protein segment (Fig. 9.31A). Results of this analysis ultimately revealed that the positions of mutations in the A gene fully corresponded to the respective sites of substituted amino acids in the A protein. Specifically, the relative distances between mutations in the DNA were perfectly aligned with the distances between the corresponding substituted amino acids in the protein product (Ref. 103; illustrated in Fig. 9.31B).

Recognizing that the documented collinearity was restricted for only a limited section of the A gene and protein, Yanofsky and his coworkers next undertook to respectively map mutations and determine the amino acid sequence throughout the whole A gene and A protein. This was a weighty undertaking since the technologies for amino acid sequencing were cumbersome and slow and no method to sequence DNA had yet been invented. Nonetheless, by 1967, the analysis was completed for the entire 268 amino acid-long A protein and its collinearity with the encoding A gene was confirmed.[105]

At the time that the Stanford team was carrying on their heroic effort to demonstrate collinearity between the tryptophan synthase A protein and its encoding gene, the University of Cambridge group used T4 phage to similarly test the collinearity proposition. These investigators mapped a series of translation termination mutations in the gene that encodes the head protein of T4 bacteriophage. The studied mutations were of the *amber* type and they

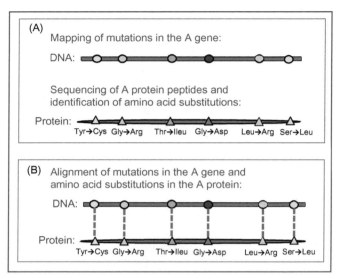

FIGURE 9.31 Scheme of Yanofsky's procedure to determine whether the tryptophan synthase A protein was collinear with its encoding gene. (A) Mutants of *Escherichia coli* with defective A subunit of tryptophan synthase were isolated. Parallel to the mapping of the mutations in the A gene, the positions of corresponding substituted amino acids were mapped in the A protein. (B) Collinearity was validated by identifying full correspondence between the relative distances that separated individual mutations in the A gene and the distances between the matching substituted amino acids in the A protein product.

were so designated after the Caltech graduate student Harris Bernstein who was the first to identify them and whose last name means *amber* in German. These are TAG (UAG in mRNA) stop codons that are known to cause termination of translation by binding release factors that trigger dissociation of the ribosomal subunits and the ensuing release the translated polypeptide chain. Thus, when an amino acid encoding nucleotide triplet is mutated into an *amber* codon, synthesis of the protein is prematurely terminated at the site of the mutation. The Cambridge group figured that the length of the prematurely terminated polypeptide chains should be determined by the location of the in-phase amber mutation. The closer the mutation was to the N-terminus of the T4 head protein, the shorter should the polypeptide chain be (Fig. 9.32).

This indeed proved to be the case as the lengths of a number of prematurely terminated polypeptides were found to correspond to the mapped positions of the respective *amber* stop codons.[106] Similarly to the collinearity between A gene and the tryptophan synthase A protein, the head protein of T4 phage was, therefore, also concluded to be collinear with its encoding gene.[106]

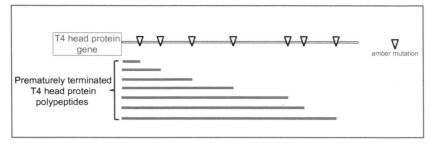

FIGURE 9.32   Scheme of the experimental strategy to test whether the head protein of T4 bacteriophage was collinear with its encoding gene.[106] Mutations that created *amber* stop codons were mapped at different positions along the T4 phage gene that encodes its head protein. In parallel, lengths of prematurely terminated head polypeptides that were produced by each mutant were determined. Documented full correlation between the relative positions of the *amber* mutations and the length of the corresponding polypeptide chains was in line with the proposition that the T4 head protein encoding gene and its protein product were collinear.

9.15   THE DEMONSTRATED COLLINEARITY BETWEEN PROKARYOTIC GENE AND PROTEIN BRED A MISLEADING PRECONCEPTION THAT COLLINEARITY WAS UNIVERSAL IN ALL THE LIVING ORGANISMS

The elegant and convincing demonstrations that bacterial and phage proteins were collinear with their encoding genes had a longstanding, and not entirely positive, impact on the thinking of molecular biologists until well into the 1970s. The long chain of triumphs of molecular biology from Avery's 1944 identification of DNA as the genetic material and through the fundamental discoveries of the 1950s and 1960s were based mostly on results that were obtained with *E. coli* and a few other bacterial species and their bacteriophages. The enormity of what was revealed in studies of prokaryotes led to the belief that basic mechanisms of gene action were universal to all branches of life. Paraphrasing the 1926 proclamation of the Dutch microbiologist Albert Jan Kluyver (1888−1956): *"From the elephant to butyric acid bacterium—it is all the same,"* Jacques Monod wrote more authoritatively and famously in 1954: *"Anything found to be true of* E. coli *must also be true of elephants."* To large extent Monod's edict guided molecular biologists in the 1960s to extrapolate from principles that were discovered in prokaryotes to the whole tree of life. The collinearity of proteins (and mRNA) with their encoding genes was a case in point. Taking collinearity as a universal rule, scientists were misled for years to think that it was valid for eukaryotes. As is shown in Chapter 11: "The Surprising Discovery of Split Genes and of RNA Splicing" despite accumulating experimental evidence that suggested otherwise, investigators were blinded by the collinearity preconception and misinterpreted or ignored data that indicated that proteins of eukaryotes were not collinear with their encoding genes. The noncollinearity in eukaryotes was ultimately recognized when it experiments convincingly showed in 1977 that pre-mRNA molecules that were collinear with the transcribed gene were processed in the nucleus to produce a spliced, noncollinear mRNA that was then translated in the cytoplasm into noncollinear proteins (see Chapter 11: "The Surprising Discovery of Split Genes and of RNA Splicing").

9.16   POSTSCRIPT: THE DISCOVERY OF mRNA WAS DRIVEN BY EXPERIMENT AND NOT BY THEORETICAL PROPOSITIONS

It becomes evident in hindsight that the discovery of mRNA was impeded by protracted adherence to the untested and misleading hypothesis that the ribosomes were the bearers of genetic information (the "One Gene—One RNA—One Ribosome—One Protein" hypothesis). This theory resisted the progressive accumulation of conflicting experimental evidence. Moreover, this preconception was not revaluated in the face of indirect and direct evidence that suggested the existence of an unstable informational RNA fraction in prokaryotes. The emerging competing theoretical concept of a rapidly turning-over messenger molecule was not raised as an abstract speculation but as one explanation for the results of the PaJaMo experiment. It should lastly be stressed that it was experiment alone that was responsible for the physical detection of mRNA and for the demonstration of its complementarity to the DNA. Thus, pure theory had no role in the discovery of mRNA. Rather, it was gradually detected and characterized in bacteria and phage exclusively by experiment. A likely explanation to the failure of theory in this case is the high complexity of the question of the transmission of genetic information from DNA to the protein synthesis machinery. The intricacy of the problem most likely precluded the possibility of a theoretical prediction of the existence of mRNA (*vide infra*).

Unlike the significant contribution of theory to the solution of the problems of the mechanics of DNA replication (see Chapter 6: "Meselson and Stahl Proved That DNA Is Replicated in a Semiconservative Fashion") and of the nature of the genetic code (see Chapter 7: "Defining the Genetic Code"), the search for the mode of transfer of information from the DNA to the protein synthesis apparatus, which was crowned by the discovery of mRNA, involved very little pure theory. The exceptional theoretical brilliance of the adaptor hypothesis (see Chapter 8: "The Adaptor Hypothesis and the Discovery of transfer RNA") notwithstanding, the case of mRNA offers an example to the ineptness of theory in predicting the existence of a yet unknown biological pathway or component. A string of findings between the late 1930s and early 1950s revealed that RNA was the likely direct template for protein synthesis. Initial observations of positive correlation between RNA and protein syntheses[10,18−21] were followed by the realization that the bulk of the cellular RNA was contained within ribosomes and that the ribosomes were the site of protein synthesis.[35−38] In addition, a (false) analogy was drawn between ribosomes and ribosome-sized viruses whose proteins are encoded by their own RNA genomes.[53] Put together, these observations led to the formulation of the "One Gene−One Ribosome−One Protein" hypothesis.[55] This notion claimed that the RNA component of each ribosome is transcript of a specific gene that serves as template for the synthesis of a particular protein. Without being challenged by alternative theories, this model was consensually adopted throughout most of the 1950s. Interestingly, the hypothesis of ribosomes as carriers of genetic information resisted the progressive accumulation of contradicting experimental observations (Table 9.1) and was not questioned even after a phage-directed unstable RNA fraction was detected.[69,76,78,107] The first theoretical challenge to the *One Gene−One Ribosome−One Protein* dogma was based on experiment. Claiming that the template for proteins may be a labile molecule (later identified as mRNA) this alternative model was formulated on the basis of the indirect experimental results of the PaJaMo experiment.[63,73] This hypothesis was independently supported by the direct observations of Volkin and Astrachan of labile RNA in phage-infected bacteria.[69,76,78,107] These two lines of evidence led eventually to the direct experimental detection of an unstable phage and bacterial RNA fraction and to the rejection of the idea of ribosomes as bearers of genetic information.[14,15] Subsequent demonstration of complementarity between this RNA and DNA indicated that mRNA molecules were transcripts of DNA.[92,96] In retrospect, it was implausible that theory alone could have predicted the existence of mRNA because of the very fragmentary nature of the understanding in the 1950s of the complexities of RNA and protein syntheses. Thus, unawareness of essential variables defied construction of a productive theoretical model. The case of the discovery of mRNA offers, therefore, an excellent example for the weakness of theory in the face of a complex biological problem.

REFERENCES

1. Watson JD, Crick FH. Molecular structure of nucleic acids: a structure for deoxyribose nucleic acid. *Nature* 1953;**171**(4356):737−8.
2. Watson JD, Crick FH. Genetical implications of the structure of deoxyribonucleic acid. *Nature* 1953;**171**(4361):964−7.
3. Meselson M, Stahl FW. The replication of DNA in *Escherichia coli*. *Proc Natl Acad Sci USA* 1958;**44**(7):671−82.
4. Gamow G. Possible relation between deoxyribonucleic acid and protein structures. *Nature* 1954;**173**(4398):318.
5. Gamow G. Possible mathematical relation between deoxyribonucleic acid and proteins. *Biolo MeddKgl Dan Vidensk Selsk* 1954;**22**(3):1−11.
6. Boivin A. Directed mutation in colon bacilli, by inducing principle of desoxynucleic acid nature: its meaning for the general biochemistry of heredity. *Cold Spring Harbor Symp Quant Biol* 1947;**12**:7−17.
7. Dounce AL. Duplicating mechanism for peptide chain and nucleic acid synthesis. *Enzymologia* 1952;**15**(5):251−8.

8. Dounce AL. Nucleic acid template hypotheses. *Nature* 1953;**172**(4377):541.

9. Harris H. Joachim Hammerling. 9 March 1901–5 August 1980. *Biogr Mem Fellows R Soc* 1982;**28**:110–26.

10. Caspersson T, Schultz J. Nucleic acid metabolism of the chromosomes in relation to gene reproduction. *Nature* 1938;**142**(3589):294–5.

11. Hämmerling J. Über formbildende substanzen bei *Acetabularia* mediterranea, ihre räumliche und zeitliche verteilung und ihre herkunft. *Wilhelm Roux' Arch Entwicklungsmech Organ* 1934;**131**:1–81.

12. Hämmerling J. Nucleo–cytoplasmic relationships in the development of *Acetabularia*. *Intl Rev Cytol* 1953;**2**:475–98.

13. Jacob F, Monod J. Genetic regulatory mechanisms in the synthesis of proteins. *J Mol Biol* 1961;**3**(3):318–56.

14. Brenner S, Jacob F, Meselson M. An unstable intermediate carrying information from genes to ribosomes for protein synthesis. *Nature* 1961;**190**(4776):576–81.

15. Gros F, Hiatt H, Gilbert W, Kurland CG, Risebrough RW, Watson JD. Unstable ribonucleic acid revealed by pulse labelling of *Escherichia coli*. *Nature* 1961;**190** (4776):581–5.

16. Klein G. Torbjörn Caspersson, 15 October 1910–7 December 1997. *Proc Am Phil Soc* 2003;**147**(1):73–5.

17. Anderson TF. Jack Schultz, May 7 1904–April 29, 1971. *Natl Acad Sci Biogr Mem* 1975;**47**:392–423.

18. Caspersson T, Schultz J. Pentose nucleotides in the cytoplasm of growing tissues. *Nature* 1939;**143**(3623):602–3.

19. Caspersson T. The relations between nucleic acid and protein synthesis. *Symp Soc Exp Biol* 1947;**1**:127–51.

20. Brachet J. The metabolism of nucleic acids during embryonic development. *Cold Spring Harbor Symp Quant Biol* 1947;**12**:18–27.

21. Brachet J. The localization and the role of ribonucleic acid in the cell. *Ann NY Acad Sci* 1950;**50**(8):861–9.

22. Brachet J. La localisation des acides pentosenucléiques dans les tissus animaux et les oeufs d'Amphibiens en voie de développement. *Arch Biol* 1941;**53**:207–57.

23. Thieffry D, Burian RM. Jean Brachet's alternative scheme for protein synthesis. *Trends Biochem Sci* 1996;**21**(3):114–17.

24. Brachet J. Recherches sur les interactions biochimiques entre le noyau et le cytoplasme chez les organismes unicellulaires. I. *Amoeba proteus*. *Biochim Biophys Acta* 1955;**18** (2):247–68.

25. Sapp J. Jean Brachet, L'Hérédité Générale and the Origins of Molecular Embryology. *Hist Philos Life Sci* 1997;**19**(1):69–87.

26. Alexandre H. Contribution of the Belgian school of embryology to the concept of neural induction by the organizer. *Int J Dev Biol* 2001;**45**(1):67–72.

27. Sapp J. Interview of Jean Brachet by Jan Sapp. Arco Felice, Italy, December 10, 1980. *Hist Philos Life Sci* 1997;**19**(1):113–40.

28. Sapp J. *Beyond the gene: cytoplasmic inheritance and the struggle for authority in genetics*. New York, NY: Oxford University Press; 1987.

29. Haguenau F. The ergastoplasm: its history, ultrastructure, and biochemistry. In: Bourne GH, Danielli JF, editors. *International review of cytology*. New York, NY: Academic Press; 1958. pp. 425–83.

30. Haguenau F. Electron microscopy in France: early findings in the life sciences. In: Tom M, editor. *Advances in imaging electron microscopy*. New York, NY: Elsevier; 1996. pp. 93–100.
31. Claude A. Concentration and purification of chicken tumor I agent. *Science* 1938;**87**(2264): 467–8.
32. Claude A. A fraction from normal chick embryo similar to the tumor producing fraction of chicken tumor I. *Exp Biol Med* 1938;**39**(2):398–403.
33. Claude A. The constitution of protoplasm. *Science* 1943;**97**(2525):451–6.
34. Rheinberger HJ. From microsomes to ribosomes: "strategies" of "representation". *J Hist Biol* 1995;**28**(1):49–89.
35. Hultin T. Incorporation in vivo of ^{15}N-labeled glycine into liver fractions of newly hatched chicks. *Exptl Cell Res* 1950;**1**(3):376–81.
36. Keller EG. Turnover of proteins of cell fractions of adult rat liver in vivo. *Fed Proc* 1951;**10**:206.
37. Siekevitz P, Zamecnik PC. In vitro incorporation of 1-C14-DL-alanine into proteins of rat liver granular fractions. *Fed Proc* 1951;**10**:246.
38. Siekevitz P. Uptake of radioactive alanine in vitro into the proteins of rat liver fractions. *J Biol Chem* 1952;**195**(2):549–65.
39. Zamecnik PC, Keller EB. Relation between phosphate energy donors and incorporation of labeled amino acids into proteins. *J Biol Chem* 1954;**209**(1):337–54.
40. Bernal JD, Fankuchen I. Structure types of protein crystals from virus-infected plants. *Nature* 1937;**139**(3526):923–4.
41. Bernal JD, Fankuchen I. X-ray and crystallographic studies of plant virus preparations: I. Introduction and preparation of specimens, II. Modes of aggregation of the virus particles. *J Gen Physiol* 1941;**25**(1):111–46.
42. Bernal JD, Fankuchen I. X-ray and crystallographic studies of plant virus preparations. III. *J Gen Physiol* 1941;**25**(1):147–65.
43. Watson JD. The structure of tobacco mosaic virus. I. X-ray evidence of a helical arrangement of sub-units around the longitudinal axis. *Biochim Biophys Acta* 1954;**13**(1):10–19.
44. Caspar DLD. Structure of tobacco mosaic virus: radial density distribution in the tobacco mosaic virus particle. *Nature* 1956;**177**(4516):928.
45. Franklin RE. Structure of tobacco mosaic virus: location of the ribonucleic acid in the tobacco mosaic virus particle. *Nature* 1956;**177**(4516):929.
46. Caspar DL. Structure of bushy stunt virus. *Nature* 1956;**177**(4506):475–6.
47. Creager AN, Morgan GJ. After the double helix: Rosalind Franklin's research on Tobacco Mosaic Virus. *Isis* 2008;**99**(2):239–72.
48. Fraenkel-Conrat H, Williams RC. Reconstitution of active Tobacco Mosaic Virus from its inactive protein and nucleic acid components. *Proc Natl Acad Sci USA* 1955;**41** (10):690–8.
49. Gierer A, Schramm G. Infectivity of ribonucleic acid from tobacco mosaic virus. *Nature* 1956;**177**(4511):702–3.
50. Judson HF. *The eighth day of creation*. Expanded ed. New York, NY: Cold Spring Harbor Laboratory Press; 1996.
51. Watson JD. *The annotated and illustrated double Helix*. New York, NY: Simon & Schuster; 2012.
52. Crick FH, Watson JD. Structure of small viruses. *Nature* 1956;**177**(4506):473–5.
53. Watson JD. *The involvement of RNA in the synthesis of proteins. Nobel lectures, physiology or medicine 1942–1962*. Amsterdam: Elsevier Publishing Company; 1964. pp 785–808.

54. Crick FH, Watson JD. Virus structure: general principles. In: Wolstenholme GEW, Millar ECP, editors. *Ciba foundation symposium—the nature of viruses*. Chichester: John Wiley & Sons; 1956.

55. Crick FH. On protein synthesis. *Symp Soc Exp Biol* 1958;**12**:138−63.

56. Fraenkel-Conrat H. The role of the nucleic acid in the reconstitution of active tobacco mosaiv virus1. *J Am Chem Soc* 1956;**78**(4):882−3.

57. Belozersky AN, Spirin AS. A correlation between the compositions of deoxyribonucleic and ribonucleic acids. *Nature* 1958;**182**(4628):111−12.

58. Crick FH. Biochemical activities of nucleic acids. The present position of the coding problem. *Brookhaven Symp Biol* 1959;**12**:35−9.

59. Tissieres A, Watson JD. Ribonucleoprotein particles from *Escherichia coli*. *Nature* 1958;**182**(4638):778−80.

60. Ts'o POP, Squires R. Quatitative isolation of intact RNA from microsomal particles of pea seedlings and rabbit reticulocytes. *Fed Proc* 1959;**18**:341.

61. Kurland CG. Molecular characterization of ribonucleic acid from *Escherichia coli* ribosomes: I. Isolation and molecular weights. *J Mol Biol* 1960;**2**(2):83−91.

62. Spirin AS. The "temperature effect" and macromolecular structure of high-polymer ribonucleic acids of various origin. *Biokhimiia* 1961;**26**:454−63.

63. Pardee AB, Jacob F, Monod J. The genetic control and cytoplasmic expression of "Inducibility" in the synthesis of β-galactosidase by *E. coli*. *J Mol Biol* 1959;**1**(2):165−78.

64. Davern CI, Meselson M. The molecular conservation of ribonucleic acid during bacterial growth. *J Mol Biol* 1960;**2**(3):153−60.

65. Monod J. *Chance and necessity*. New York, NY: Vintage Books; 1972.

66. Ullmann A, editor. *Origins of molecular biology. A tribute to Jacques Monod*. Revised ed. Washington, DC: ASM Press; 2003.

67. Carrol SB. *Brave genius: a scientist, a philosopher, and their daring adventures from the french resistance to the nobel prize*. New York, NY: Crown; 2013.

68. Monod J. *From enzymatic adaptation to ellosteric eransitions. Nobel lectures, physiology or medicine 1963−1970*. Amsterdam: Elsevier Publishing Company; 1965. pp. 188−209.

69. Volkin E, Astrachan L. Phosphorus incorporation in *Escherichia coli* ribo-nucleic acid after infection with bacteriophage T2. *Virology* 1956;**2**(2):149−61.

70. Monod J, Pappenheimer Jr AM, Cohen-Bazire G. The kinetics of the biosynthesis of beta-galactosidase in *Escherichia coli* as a function of growth. *Biochim Biophys Acta* 1952;**9**(6):648−60.

71. Pardee AB. PaJaMas in Paris. *Trends Genet* 2002;**18**(11):585−7.

72. Jacob F. *The statue within: an autobiography*. Cold Spring Harbor, NY: Cold Spring Harbor Laboratory Press; 1995.

73. Riley M, Pardee AB, Jacob F, Monod J. On the expression of a structural gene. *J Mol Biol* 1960;**2**(4):216−25.

74. Hershey AD, Chase M. Independent functions of viral protein and nucleic acid in growth of bacteriophage. *J Gen Physiol* 1952;**36**(1):39−56.

75. Hershey AD. Nucleic acid economy in bacteria infected with bacteriophage T2. *J Gen Physiol* 1953;**37**(1):1−23.

76. Volkin E, Astrachan L. RNA metabolism in T2-infected *Escherichia coli*. In: McElroy WD, Glass B, editors. *A symposium on the chemical basis of heredity*. Baltimore, MD: The Johns Hopkins Press; 1957. pp. 686−95.

77. Watson JD. *Genes, girls, and Gamow: after the double helix*. New York, NY: Vintage Books; 2001.

78. Volkin E, Astrachan L, Countryman JL. Metabolism of RNA phosphorus in *Escherichia coli* infected with bacteriophage T7. *Virology* 1958;**6**(2):545−55.
79. Yčas M, Vincent WS. A ribonucleic acid fraction from yeast related in composition to desoxyribonucleic acid. *Proc Natl Acad Sci USA* 1960;**46**(6):804−11.
80. Riley M, Pardee AB. Beta-galactosidase formation following decay of 32P in *Escherichia coli* zygotes. *J Mol Biol* 1962;**5**(1):63−75.
81. Nomura M, Hall BD, Spiegelman S. Characterization of RNA synthesized in *Escherichia coli* after bacteriophage T2 infection. *J Mol Biol* 1960;**2**(5):306−26.
82. Gros F. The mobility principle: how I became a molecular biologist. *J Biosci* 2006;**31**(3):303−8.
83. Gilbert W. Life after the helix. *Nature* 2003;**421**(6921):315−16.
84. Naono S, Gros F. Synthese par *E. coli* d'une phosphatase modifiee en presence d'une analogue pyrimidique. *Compt Rend Acad Sci (Paris)* 1960;**250**(6):3889−91.
85. Gros F. The messenger. In: Ullmann A, editor. *Origins of molecular biology: a tribute to Jacques Monod*. Revised ed. Washington, DC: ASM Press; 2003. pp. 143−51.
86. Rich A, Davies DR. A new two stranded helical structure: polyadenylic acid and polyuridylic acid. *J Am Chem Soc* 1956;**78**(14):3548−9.
87. Rich A. The excitement of discovery. *Annu Rev Biochem* 2004;**73**:1−37.
88. Rich A. An analysis of the relation between DNA and RNA. *Ann NY Acad Sci* 1959;**81**(3):709−22.
89. Rich A. A hybrid helix containing both deoxyribose and ribose polynucleotides and its relation to the transfer of information between the nucleic acids. *Proc Natl Acad Sci USA* 1960;**46**(8):1044−53.
90. Doty P, Marmur J, Eigner J, Schildkraut C. Strand separation and specific recombination in deoxyribonucleic acids: physical chemical studies. *Proc Natl Acad Sci USA* 1960;**46**(4):461−76.
91. Marmur J, Lane D. Strand separation and specific recombination in deoxyribonucleic acids: biological studies. *Proc Natl Acad Sci USA* 1960;**46**(4):453−61.
92. Hall BD, Spiegelman S. Sequence complementarity of T2-DNA and T2-specific RNA. *Proc Natl Acad Sci USA* 1961;**47**(2):137−46.
93. Weiss SB, Nakamoto T. On the participation of DNA in RNA biosynthesis. *Proc Natl Acad Sci USA* 1961;**47**(5):694−7.
94. Weiss SB, Nakamoto T. The enzymatic synthesis of RNA: nearest-neighbor base frequencies. *Proc Natl Acad Sci USA* 1961;**47**(9):1400−5.
95. Geiduschek EP, Nakamoto T, Weiss SB. The enzymatic synthesis of RNA: complementary interaction with DNA. *Proc Natl Acad Sci USA* 1961;**47**(9):1405−15.
96. Hayashi M, Spiegelman S. The selective synthesis of informational RNA in bacteria. *Proc Natl Acad Sci USA* 1961;**47**(10):1564−80.
97. Garen A, Levinthal C. A fine-structure genetic and chemical study of the enzyme alkaline phosphatase of *E. coli*. I. Purification and characterization of alkaline phosphatase. *Biochim Biophys Acta* 1960;**38**(0):470−83.
98. Echols H, Garen A, Garen S, Torriani A. Genetic control of repression of alkaline phosphatase in *E. coli*. *J Mol Biol* 1961;**3**(4):425−38.
99. Garen A, Echols H. Genetic control of induction of alkaline phosphatase synthesis in *E. coli*. *Proc Natl Acad Sci USA* 1962;**48**(8):1398−402.
100. Garen A, Echols H. Properties of two regulating genes for alkaline phosphatase. *J Bacteriol* 1962;**83**(2):297−300.

101. Streisinger G, Mukai F, Dreyer WJ, Miller B, Horiuchi S. Mutations affecting the lysozyme of phage T4. *Cold Spring Harb Symp Quant Biol* 1961;**26**:25−30.

102. Beadle GW, Tatum EL. Genetic control of biochemical reactions in Neurospora. *Proc Natl Acad Sci USA* 1941;**27**(11):499−506.

103. Yanofsky C, Carlton BC, Guest JR, Helinski DR, Henning U. On the colinearity of gene structure and protein structure. *Proc Natl Acad Sci USA* 1964;**51**(2):266−72.

104. Crawford IP, Yanofsky C. On the separation of the tryptophan synthetase of *Escherichia coli* into two protein components. *Proc Natl Acad Sci USA* 1958;**44**(12):1161−70.

105. Yanofsky C, Drapeau GR, Guest JR, Carlton BC. The complete amino acid sequence of the tryptophan synthase A protein (alpha subunit) and its colinear relationship with the genetic map of the A gene. *Proc Natl Acad Sci USA* 1967;**57**(2):296−8.

106. Sarabhai AS, Stretton AO, Brenner S, Bolle A. Co-linearity of the gene with the polypeptide chain. *Nature* 1964;**201**(4914):13−17.

107. Volkin E, Astrachan L. Intracellular distribution of labeled ribonucleic acid after phage infection of *Escherichia coli*. *Virology* 1956;**2**(4):433−7.

Chapter 10

The Deciphering of the Genetic Code

Nirenberg and Khorana Decrypt the Genetic Code

Chapter Outline

M. Fry: Landmark Experiments in Molecular Biology. DOI: http://dx.doi.org/10.1016/B978-0-12-802074-6.00010-2
421

The principal attributes of the genetic code were defined between 1953 and 1961 by a combination of theoretical and experimental approaches. As was detailed in Chapter 7, "Defining the Genetic Code," by the end of that period it became evident that the code was a commaless, nonoverlapping string of 64 code words of 3 nucleotides each. These code words received from Sydney Brenner their current common designation "codons." It was also realized that the code was most likely degenerate, that is, that each amino acid was encoded by several different codons. Yet, one major task that still remained by the early 1960s was the deciphering of the bilingual "dictionary" of the code. In other words, it had to be discovered which specific 3-nucleotide sequences encoded each one of the 20 amino acids. This undertaking was masterfully accomplished in the first half of the 1960s by the American research teams of Marshall Nirenberg (1927—2013) at the National Institute of Health (NIH) and of Har Gobind Khorana (1922—2011) at the Institute for Enzyme Research in the University of Wisconsin. The decrypting of the code by these two groups was built upon several prior or contemporaneous developments. First, the earlier definition of the basic characteristics of the code that established that it is comprised of nonoverlapping, degenerate triplet codons afforded a theoretical foundation for the deciphering work. Second the development of bacterial cell-free protein synthesis system provided a vital experimental tool for the breaking of the code. Third, tRNA and amino acid—activating enzymes which were discovered in the late 1950s (see Chapter 8: "The Adaptor Hypothesis and the Discovery of Transfer RNA") offered both a conceptual framework and an essential experimental handle for the decoding work. Finally the work to decipher the code was critically linked to the contemporaneous realization that messenger RNA (mRNA) was the direct template for protein translation (see Chapter 9: "The Discovery and Rediscovery of Prokaryotic Messenger RNA").

Although the definition of the principal features of the code involved both theoretical considerations and empirical findings, its actual decrypting was accomplished exclusively by experiment. This chapter is dedicated to the historical and scientific circumstances of the experimental deciphering of the genetic code. Here we first survey the work of Marshall Nirenberg and then describe the independent experiments and results of Gobind Khorana.

10.1 A CRUCIAL FIRST STEP: CONSTRUCTION OF CELL-FREE PROTEIN SYNTHESIS SYSTEM FROM *ESCHERICHIA COLI*

Cell-free protein synthesis systems from animal cells were established already in the early 1950s.[1] The use of these in vitro systems gainfully defined the energy requirements for protein synthesis, identified ribosomes as the site of protein synthesis, and afforded the isolation of tRNA and of its amino acid—charging enzymes (see Chapter 8: "The Adaptor Hypothesis and the Discovery of Transfer RNA"). Yet, attempts to prepare similar in vitro systems from bacterial cells were met with difficulties and it was not until 1960 that reliable cell-free protein synthesis systems were prepared from *Escherichia coli* bacteria. As was established in animal cell-free systems, protein synthesis in bacterial cell extracts was also measured by the incorporation of radiolabeled amino acid into acid-insoluble protein. Although the activation of amino acids and their charging onto tRNA were documented in extracts of bacterial cells,[2] detection of bona fide incorporation of radioactive amino acids into protein proved to be more elusive. Attempts to monitor true cell-free incorporation of radiolabeled amino acids in bacterial extracts were frustrated by the presence of surviving intact cells. Active incorporation of labeled amino acid by these living bacteria masked the relatively low levels of authentic cell-free protein synthesis. To overcome this difficulty, instead of measuring the incorporation of radiolabeled amino acids, it was attempted to monitor cell-free protein synthesis in bacterial extracts by assaying increase in the activities of in vitro—synthesized specific bacterial or phage proteins such as β galactosidase from *Bacillus megaterium*,[3] bacteriophage T2 antigen,[4] or *Bacillus subtilis* amylase.[5] Unfortunately, these experiments failed to yield conclusive results.[5]

Marvin Lamborg (1927—) and Paul Zamecnik (see Chapter 8: "The Adaptor Hypothesis and the Discovery of Transfer RNA") were the first to successfully prepare an in vitro protein synthesis system from *E. coli* cells. In a paper in the Jan. 1960 issue of *Biochimica et Biophysica Acta*,[6] they described what appeared to be true in vitro synthesis of protein by a bacterial extract. Their technical innovation was the use of alumina (granular aluminum oxide, Al_2O_3) to grind and completely rapture the bacterial cells. The obtained cell-free extract sustained the incorporation of ^{14}C-leucine into acid-precipitable protein in reaction mixtures that contained a nonradioactive mixture of all the amino acids, magnesium and potassium ions, ATP, GTP, and ATP-/GTP-regenerating system. Removal of any single component or depletion of most of the amino acids significantly diminished or completely abolished the capacity of the bacterial extract to carry out incorporation. Also, addition to the extract of ribonuclease or of the protein synthesis inhibiting antibiotic chloramphenicol completely obliterated incorporation. Significantly, incorporation was stimulated by the addition of the $100,000 \times g$ cytoplasmic supernatant fraction that had already been known to contain tRNA and amino acid–activating enzymes (see Chapter 8: "The Adaptor Hypothesis and the Discovery of Transfer RNA"). Several lines of evidence indicated that it was the extract and not residual intact cells that carried out the observed synthesis of protein. First, unlike the extract, incorporation by whole cells did not require the addition of magnesium, ATP and GTP, or their regenerating system and it was not affected by exogenously added RNase.[6] Second, measurements of the ability of serially diluted bacterial suspensions to incorporate radiolabeled amino acid revealed that incorporation became undetectable at a concentration of 10^5 cells/mL. In parallel, counting of viable cells that remained in the extracts indicated that their number rarely exceeded 10^5 cells/mL.[6] Thus by grounding the bacteria with alumina, Lamborg and Zamecnik achieved full or nearly complete rupture of the cells and the observed incorporation of radioactively labeled amino acid represented bona fide cell-free synthesis of protein.

Successful preparation of a cell-free protein synthesis system from *E. coli* was also reported later in 1960 by three members of the Harvard group of James Watson: the Swiss molecular biologist Alferd Tissières (1917–2003), David Schlessinger who was then a PhD student, and Françoise Gros who was visiting from the Pasteur Institute (see Chapter 9: "The Discovery and Rediscovery of Prokaryotic messenger RNA").[7] Similar to Lamborg and Zamecnik, the Harvard team ruptured *E. coli* cells with alumina. They also confirmed that the incorporation of ^{14}C-alanine by the resulting cell extract required ATP and GTP and their regenerating system as well as magnesium ions. In addition to these confirmatory results, Tissières, Schlessinger, and Gros fractionated the crude extract by differential centrifugation and showed that protein synthesis required the combination of the ribosomes with the 100K cytoplasmic supernatant fraction.[7] Sucrose gradient centrifugation of the ribosomes revealed that the newly made radiolabeled proteins were mostly associated with a minor fraction of heteromeric 70S ribosomes, whereas the 30S and 50S ribosomal subunits carried 15- to 40-fold less of the nascent protein. Finally, effective chasing of the newly synthesized labeled proteins from the ribosomes indicated that their association with the ribosomes was transient.[7]

As we shall see, the bacterial cell-free system was an essential tool in the work to decipher the genetic code. This was stressed by Nirenberg himself in one of several interviews that he held with the historian Ruth Harris in late 1995 and early 1996 (http://history.nih.gov/archives/downloads/Nirenberg% 20oral%20history%20Chap%203-%20%20RNA%20and%20PolyU.pdf):

> **Ruth Harris:** *Could you please give the background for your use or development of the cell-free system that you used to study protein synthesis?*

Marshall Nirenberg: Yes. This was largely based on the work of Paul Zamecnik and his colleagues. Zamecnik was the leader in protein synthesis, so I used his and his colleagues' techniques predominantly, their cell-free systems. We had all the goodies that were needed for cell-free protein synthesis: an ATP generating system with phosphoenolpyruvate (PEP) that catalyzed the phosphorylation—that combination was the substrate from which phosphate was used to synthesize ATP, and cell extracts—magnesium.

10.2 MARSHALL NIRENBERG AND HIS SUPPORTIVE SCIENTIFIC ENVIRONMENT AT THE NIH

After receiving a PhD degree from the University of Michigan, Marshall Nirenberg (Box 10.1) came in 1957 to the NIH as an American Cancer Society postdoctoral fellow.

BOX 10.1 Marshall W. Nirenberg

Nirenberg was born in New York City and as a boy he developed rheumatic fever. Because of his health problems and the beneficial effect of a subtropical climate, the family moved to Orlando, Florida. In 1948 he received a Bachelor degree and in 1952 a Master's degree in zoology from the University of Florida at Gainesville. In 1957 he was awarded a PhD degree in biochemistry from the University of Michigan, Ann Arbor. He then joined the National Institutes of Health (NIH) in 1957 as a postdoctoral fellow. In time he became a research biochemist and then Laboratory Chief, spending his whole career in the same institution. Between 1961 and 1965, he, together with a growing group of associates, deciphered the genetic code. For this triumphant achievement, he was bestowed with many awards and honors including the 1968 Nobel Prize in Medicine and Physiology that he shared with Har Gobind Khorana and with Robert Holley. A valuable source on his life and work is his recently published biography by Franklin Portugal[8] (Fig. 10.1).

Marshall Nirenberg (1927–2010)

FIGURE 10.1   Marshall W. Nirenberg.

In another one of his interviews with Ruth Harris titled "A Career Begins at the National Institute of Health," Nirenberg described the human and scientific environment that he found at the NIH. There he recapped his early period as postdoctoral fellow with DeWitt Stetten (1909–90), when he was still undecided what area of research to choose (http://history.nih.gov/archives/downloads/Nirenberg%20oral%20history%20Chap%201-%20Childhood.pdf). At the end of that period, when Nirenberg was recruited in 1959–60 as an independent researcher, he developed an active interest in protein synthesis. The person who gave him the opportunity to pursue his independent interests was Gordon Tomkins (Box 10.2), who was then the head of the Molecular Biology Section in the NIH Institute of Arthritis, Metabolism, and Digestive Diseases (NIAMD; later renamed NIADDK). In his interview with Harris, Nirenberg described admiringly Tomkins' scientific brilliance and gifts that marked him as a veritable Renaissance man.

BOX 10.2 Gordon Tomkins and Leon Heppel

A native of Chicago, Gordon Tomkins received a bachelor's degree in philosophy from the University of California, Los Angeles, an MD degree from Harvard Medical School, and a PhD in biochemistry from Berkeley. In 1953 he was designated head of the Section of Molecular Biology at the NIH NIAMD. There he conducted research on the molecular basis of hormone action. Working in Tomkins' NIAMD section, Nirenberg benefited from discussions that they held during the early phases of his decoding work. In 1968 he was recruited to the Department of Biochemistry at the University of California San Francisco. There he was a highly influential researcher and educator until his untimely death 6 years later at the age of 49.

Born in Utah to a poor Mormon family, Leon Heppel received in 1937 a PhD degree in biochemistry from the University of California, Berkeley and earned in 1941 an MD degree from the University of Rochester. Together with his Rochester classmate, the 1959 Nobel laureate Arthur Kornberg, he was recruited during World War II to the NIH at which he conducted toxicological studies. Staying on at the NIH after the war, he focused on ribonucleases and on the separation and identification of their RNA hydrolysis products. At a later time he studied polynucleotide phosphorylase (PNP) and in collaboration with Severo Ochoa (Box 10.4) he determined the structure of its polyribonucleotide products thus becoming a leading expert in the analysis and determination of the structure of oligoribonucleotides. Having been the Laboratory Chief at the NIAMD at the time that Nirenberg initiated his work on the genetic code, his expertise in nucleic acids was of great value in the early stages of the deciphering of the code (Fig. 10.2).

(Continued)

BOX 10.2 (Continued)

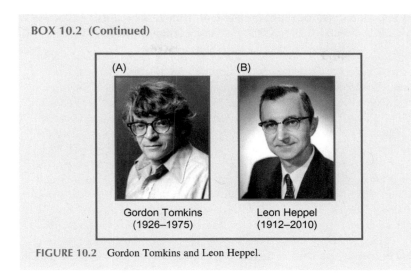

FIGURE 10.2   Gordon Tomkins and Leon Heppel.

Although Tomkins himself did not play an active part in the actual experimental work that led to the breaking of the genetic code, Nirenberg gained
much from the stimulatory discussions that they had at the early stages of
the endeavor to decipher the code. The following citation from his aforementioned interview reflects Nirenberg's admiration for Tomkins:

*Gordon was a live wire. He knew everything that was happening. He was prob-
ably the most articulate person I have ever met in my life. His detailed knowledge of basic science was tremendous. He had a remarkable ability to
synthesize information from different fields, be it physical chemistry or genetics. What he basically did best was to talk to people. I found it wonderful to be
able to talk to him because, even though he wasn't working on what I was, he
followed every experiment, and he knew the details of what I was doing and
what other people had done, as well.*

A second important figure in Nirenberg's immediate NIH environment
was Leon Heppel (Box 10.2) who was at the time the Laboratory Chief.
Heppel was a prominent expert on nucleic acids and nucleic acids processing
enzymes. One of the enzymes that Heppel studied was PNP that catalyzed
template-free synthesis of polyribonucleotides (*vide infra*). This enzyme
proved to be of great value for Nirenberg's decoding work. Heppel's technical knowhow and polynucleotides that he prepared were indispensable
resources during the early stages of the breaking of the code. In an interview with Harris, Nirenberg described Heppel's important contribution

to his work (http://history.nih.gov/archives/downloads/Nirenberg%20oral%20 history%20Chap%203-%20%20RNA%20and%20PolyU.pdf):

> *At the time, there were maybe only five or six people in the world who were experts in nucleic acid biochemistry, Leon Heppel was one of them, and he was the real McCoy. He had a postdoctoral fellow, Maxine Singer. She was an expert on polynucleotide phosphorylase, the enzyme that catalyzes the synthesis of these randomly ordered polynucleotides. [...] Heppel was such an interesting person, a marvelous person. [...] My problem was that he worked all the time, and I didn't want to interrupt him when he was working. [...] When could I get a chance to talk to him? He was an enormous resource for me. He knew everything that he had done, and he had kept most of his experimental products. They were by products of previous experiments that he had done' and he was extremely organized. He had a number of deep freezers and he would store these samples in these freezers. He would say something like, "Yes. I did an experiment that made some of this product three years ago." He would be able to find it, and he would give it to me. His lab was very important to me. I got poly-A, poly-C, poly-U and some other from him.*

Beside the fertile interactions that Nirenberg had with Tomkins and Heppel, the NIH environment also offered exposure to research at the forefront of biochemistry. Thus when he was still a postdoctoral fellow, Nirenberg expanded his scientific horizons by taking several courses at the NIH and in Cold Spring Harbor. One course that the leading enzymologist Earl Stadman (1919–2008) taught was dedicated to bacterial metabolism. Another course on the regulation of gene expression in bacteria educated Nirenberg on much of what was known in the late 1950s on molecular biology. This was complemented by a Cold Spring Harbor course on bacterial genetics that taught Nirenberg methods of DNA isolation and the use of DNA as a template for RNA synthesis. Equally important to his scientific development were daily noontime seminars that at the NIH. It was in one of those seminars that he heard Fritz Lipmann (1899–1986) reporting on the then just-discovered tRNA and amino acid–activating enzymes. This discovery and perhaps other stimuli drove Nirenberg to choose protein synthesis as his field of research. His first venture in this area was an attempt to synthesize in vitro the enzyme penicillinase. Working unaided for about 1½ years, he learned to construct and use bacterial cell-free protein synthesis system and to detect increased activity of the in vitro–synthesized penicillinase.

10.3 THE EARLY PHASE: DEFINED POLYNUCLEOTIDE SEQUENCE DICTATED INCORPORATION OF A SPECIFIC AMINO ACID INTO PROTEIN IN A CELL-FREE SYSTEM

Although Nirenberg was aware that cell-free protein synthesis was one of the hottest areas in biochemistry, which was pursued by large and experienced groups as those of Paul Zamecnik or Fritz Lipmann, he daringly chose it as his topic of study. With hindsight, Nirenberg regarded this choice as crucial to his subsequent decrypting of the genetic code (http://history.nih.gov/archives/downloads/Nirenberg%20oral%20history%20Chap%202%20-NIH%201957_1959.pdf):

> *In retrospect, that was the biggest factor—that was the key decision. For everything else that flowed from the system I had to have guts to go ahead and do it with the knowledge that I was just starting from scratch.*

10.3.1 Matthaei and Nirenberg Established Their First Assay System of RNA-Dependent Cell-Free Protein Synthesis

The endeavor to decipher the code was initiated by Nirenberg in a series of historical experiments that he performed with his 31-year-old postdoctoral fellow Heinrich Matthaei (Box 10.3).

BOX 10.3 Heinrich Matthaei

Matthaei, a native German, received a PhD degree from the University of Bonn in plant physiology. Securing a 2 years NATO research fellowship, he joined the NIH in 1961 as Nirenberg's first postdoctoral trainee. A gifted and meticulous experimentalist, he conducted early critical experiments that launched the project to decrypt the genetic code. His series of experiments were crowned with the breakthrough discovery that the polyribonucleotide poly(U) dictated the synthesis in vitro of polyphenylalanine and thus that the triplet UUU encoded phenylalanine. After his grant expired, he returned to Germany to become for most of his career a member of the Max Planck Society in Göttingen. In an interview on the occasion of the 50th anniversary of his and Nirenberg's discovery, he expressed bitterness for not having been awarded a Nobel Prize together with Nirenberg[9] (Fig. 10.3).

Heinrich Matthaei
(1929–)

FIGURE 10.3 Heinrich Matthaei.

Arriving at the NIH on a NATO research grant, Matthaei who was trained in Germany in plant physiology sought a laboratory that worked on protein synthesis. Since Nirenberg was at the time the only investigator at the NIH who was working on cell-free protein synthesis, Matthaei ended up as his first postdoctoral trainee. Prior to his arrival Nirenberg had already set up a system for cell-free synthesis of penicillinase. When Matthaei joined him they modified the protocols to measure protein synthesis by monitoring the incorporation of radiolabeled amino acid into acid-insoluble material (see Chapter 8: "The Adaptor Hypothesis and the Discovery of Transfer RNA"). The first piece of work that Matthaei and Nirenberg produced was a short paper in which they demonstrated that added RNA "message" was necessary for the synthesis of protein in *E. coli*-derived cell-free system. Their results were published in Apr. 1961 in a sparsely typewritten 4.5-page-long paper with only five cited references, that was titled "The Dependence of Cell-Free Protein Synthesis in *E. coli* Upon RNA Prepared from Ribosomes."[10] In this work they described the fractionation of bacterial extract (Fig. 10.4A) and the subsequent identification of components that were essential for the incorporation in vitro of ^{14}C-valine into acid-precipitable protein (Fig. 10.4B).

The key reported finding was that ribosomal RNA (rRNA) stimulated protein synthesis in the cell-free system. Results indicated that the extent of

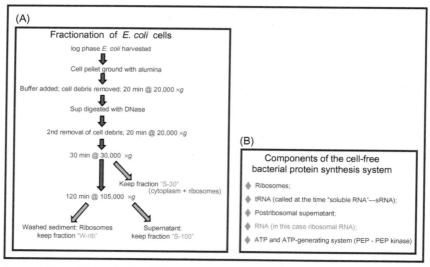

FIGURE 10.4   (A) Matthaei and Nirenberg's protocol of preparation and fractionation by differential centrifugation of *Escherichia coli* cell extract. (B) Components required for cell-free protein synthesis: (1) ribosomes; (2) tRNA (named at that time "soluble RNA" or sRNA); (3) "S-100" supernatant fraction after sedimentation of the ribosomes at $105,000 \times g$ for 2 hours; (4) template RNA (rRNA in the hands of Matthaei and Nirenberg[10]); (5) ATP and ATP-generating system (phosphoenolpyruvate (PEP) and PEP kinase).

incorporation of the radiolabeled valine was proportional to the amount of rRNA that was added to the cell-free system (Fig. 10.5A).

The kinetics of incorporation indicated that the rRNA acted in stoichiometric rather than catalytic fashion. It was retrospectively realized that the template for protein synthesis was actually not rRNA but rather a small contaminating fraction of mRNA. Indeed, significant incorporation was documented even without added rRNA (Fig. 10.5A) and as Matthaei and Nirenberg showed some months later,[11] this synthesis was due to endogenous mRNA template that was present in the cell-free system (*vide infra*).

In well-controlled experiments Matthaei and Nirenberg also defined conditions and components that were required for the RNA-directed synthesis of protein. Although the levels of counted radioactivity were low (~100 counts per minute above background), the obtained results (Fig. 10.5B) were unambiguous. Strong or complete inhibition of incorporation by the inhibitors of protein synthesis chloramphenicol and puromycin or by depletion of all 20 amino acids showed that incorporation reflected *bona fide* protein synthesis. Eradication of incorporation by the addition of RNase to the system and lack of an effect of added DNase indicated that although an exogenously added RNA template was essential for protein synthesis, there was no role for DNA in the system (Fig. 10.5B).

Matthaei and Nirenberg realized that some RNA must have dictated protein synthesis. Commenting on their results they wrote: "It is possible that part or all of the rRNA used in our study corresponds to template or messenger RNA."[10] It is noteworthy that at that very early time (Apr. 1961), Nirenberg

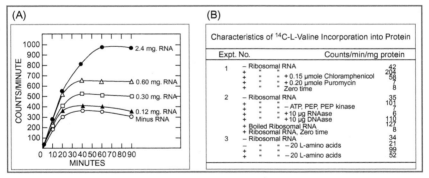

FIGURE 10.5 (A) The extent of cell-free incorporation of ^{14}C-valine into acid-insoluble protein was proportional to the amount of added ribosomal RNA. The cell-free system contained S-30 fraction (Fig. 10.4A); buffer; magnesium ions; ATP and ATP-generating system; ^{14}C-valine and 19 unlabeled amino acids; saturating concentration of tRNA, and the indicated amounts of added ribosomal RNA. (B) Identification of the components and conditions that were required for protein synthesis. *Panels (A) and (B): Reproduced from Matthaei H, Nirenberg MW. The dependence of cell-free protein synthesis in* E. coli *upon RNA prepared from ribosomes.* Biochem Biophys Res Commun 1961;4(6):404−8.

had already used the term "messenger RNA." The likely source for this idiom was Gordon Tomkins who, after receiving from Jacob and Monod a preprint of their historical 1961 paper of the operon model,[12] had told Nirenberg (and most likely many others) about their emergent concept of mRNA (http://history.nih.gov/archives/downloads/Nirenberg%20oral%20history%20Chap%203-%20%20RNA%20and%20PolyU.pdf).

Several months after their first report,[10] Matthaei and Nirenberg published a paper in the Oct. 15, 1961, issue of the *Proceedings of the National Academy*[11] in which they substantiated and expanded their initial findings. In this second work they showed that the incorporation of ^{14}C-valine into protein in the bacterial cell-free system also occurred without the addition of external RNA and that the conditions and requirements for this incorporation were the same as for exogenous RNA-directed protein synthesis.[11] Complete inhibition of incorporation by added RNase was taken as indication that endogenous RNA, which they now termed "mRNA," served as template for protein synthesis.[11] In agreement with previous reports[7,13] they found that DNase also inhibited endogenous incorporation, albeit to a more limited degree than RNase.[11] This proved to be a technically useful finding and in subsequent works Nirenberg preincubated the system with DNase to lower endogenous incorporation to background levels and make exogenous RNA-directed incorporation more prominent.[14]

10.3.2 Synthetic Polyuridylic Acid Dictated Synthesis In Vitro of Polyphenylalanine

Following the observation that *E. coli* rRNA (or rather traces of mRNA in it) stimulated cell-free protein synthesis,[10] Nirenberg prepared a list of potential RNA messengers to be tested in the cell-free system. He thus collected from neighboring laboratories both natural (viral and ribosomal) RNA and synthetic polyribonucleotides and together with Matthaei examined their capacity to direct protein synthesis in vitro. The outcome of this enterprise was a breakthrough discovery that Nirenberg and Matthaei announced in a paper which was also published in the Oct. 15, 1961, issue of the *Proceedings of the National Academy*.[15] Much of the historical of the personal background to this landmark experiment was described in vibrant detail by Horace Judson (Ref. 16, pp. 452−465).

This landmark paper opened with rather conventional biochemistry. In a case that is not likely to happen today, Figure 3 and Table 1 of this report[15] were the exact same as Figure 1 and Table 1 of their previous paper[10] (reproduced in Fig. 10.5). By presenting these data, Nirenberg and Matthaei probably intended to stress their claim that exogenously added RNA stimulated protein synthesis in a cell-free system. They then expanded this observation by showing that the added rRNA stimulated to similar extent incorporation of seven different ^{14}C-labeled amino acids. Also, the specific requirement for the RNA template was underlined by the finding that the incorporation of ^{14}C-valine was stimulated by rRNA but not by salmon sperm DNA, by synthetic poly(A), or by the polyanion polyglucose carboxylic acid.[15] In contrast, several types of natural RNA stimulated incorporation equally well or better than *E. coli* rRNA. Notably the RNA of Tobacco Mosaic Virus (TMV) was the most efficient template[15] (see below).

After this first part of the paper, Nirenberg and Matthaei introduced a series of celebrated experiments that proved to be the first crucial step on the road to break the genetic code. Essentially, they showed that synthetic polyuridylic acid (poly(U)) specifically dictated the synthesis of polyphenylalanine. With this result they identified, therefore, the first code word UUU that encoded the amino acid phenylalanine. Before analyzing in detail these experiments, their historical background is introduced.

Historical Background to the Poly(U) Experiment

The observation that externally added *E. coli* rRNA (or rather traces of mRNA that it contained) served as template for cell-free protein synthesis prompted Nirenberg to look for other RNA species that could also act as messengers. In addition to the aforementioned TMV RNA and yeast and mammalian rRNA, Nirenberg also obtained from neighboring laboratories synthetic polyribonucleotides such as poly(U) and poly(A). He presumed that because these homopolymers were comprised of a repeating single nucleotide they would encode polymers of a single amino acid. Thus to substantiate this assumption it had to be shown that these polyribonucleotides indeed dictated polymerization of only one and not of other amino acids. To that end Nirenberg and Matthaei had to have each of the 20 amino acids radioactively labeled. Although Nirenberg wanted to purchase these reagents (that were excessively expensive at the time), Matthaei insisted on preparing them with his own hands. Relying on his previous experience in the University of Bonn, Matthaei fed algae with ^{14}C-bicarbonate and radiolabeled proteins that were synthesized in vivo were extracted and hydrolyzed, the 20 radioactive amino acids were fractionated and each acid was isolated.

While Matthaei was busy preparing the essential reagents, Nirenberg had his mind on another outcome of their work. Having been impressed by the capacity of TMV RNA to greatly stimulate protein synthesis in vitro,[15] he reasoned that the viral RNA served as messenger for the coat protein of the virus. Excited by this prospect he contacted the premier expert on TMV, Heinz Fraenkel-Conrat (see Chapter 7: "Defining the Genetic Code") at the University of California, Berkeley, offering to collaborate in testing the hypothesis that mutated TMV RNA would encode substituted amino acids in the coat protein. Having been invited to pursue this project in Berkeley, before leaving for California Nirenberg left with Matthaei an experimental plan to test the coding specificity of poly(U). It was thus Matthaei alone who conducted the actual poly(U) experiment.

Before the reader becomes immersed in the poly(U) historical experiment, a word is in place about the experiments that Nirenberg conducted in the Berkeley laboratory of Fraenkel-Conrat and his postdoctoral fellow Akira Tsugita (see Chapter 7: "Defining the Genetic Code"). Although the idea that TMV RNA encoded the virus coat protein was sound, Nirenberg and Fraenkel-Conrat did not realize at the time that the coding RNA strand was not the viral RNA itself, which they used as template in the cell-free system, but rather its complementary strand. Thus Tsugita misidentified peptides of the in vitro—synthesized protein. In the end the incorrect results were published[17] but then retracted.

Analysis of the Poly(U) Experiment

Matthaei undertook to find out which amino acid was polymerized in a poly (U) supplemented cell-free system and to ascertain that no other amino was incorporated into protein under the direction of this template. The experiment was conducted in two stages. First Matthaei prepared separate cell-free assay systems, each with a poly(U) template and a pool of few radiolabeled amino acids. Results revealed that the incorporation of radioactivity into protein took place with only one group of pooled amino acids that included labeled phenylalanine, tyrosine, and lysine. In the next stage incorporation of each of the three radiolabeled amino acids was tested separately. This experiment showed that it was the radiolabeled phenylalanine that was incorporated under the direction of the poly(U) template, whereas neither tyrosine nor lysine was incorporated. In the morning of Saturday, May 27, 1961, Matthaei meticulously marked this breakthrough result in his notebook under the identifier "27-Q." With Nirenberg out of town, he showed his results to Gordon Tomkins. According to Matthaei,[9] when Tomkins grasped that the experiment had just deciphered the first code; UUU for.phenylalanine, he said: "*the Nobel is Henry's (Heinrich).*" Matthaei called Nirenberg to tell him about the "fantastic" result that he obtained. The lore is that Nirenberg responded, "*Heinrich, it is going to be one of the most exciting times in both of our lives.*" Matthaei and Nirenberg (back from California) spent then the next few months in work to complete the experiments that were eventually incorporated into the formal paper of Oct. 15, 1961.[15] The principal result in this section of the manuscript is shown in Fig. 10.6.

Technically, incorporation was conducted by incubating *E. coli*-derived cell-free protein synthesis system without or with a chosen polyribonucleotide and with ^{14}C-phenylalanine. Following termination of the reaction, the synthesized proteins were precipitated with trichloroacetic acid (TCA), centrifuged, washed again with TCA, and the pellets were counted for radioactivity. Results showed that without added template there was no incorporation of ^{14}C-phenylalanine into acid-insoluble material. By contrast, when

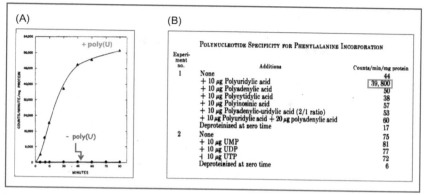

FIGURE 10.6   (A) Kinetics of incorporation of ^{14}C-phenylalanine in a cell-free system without or with added poly(U). (B) Specificity of poly(U)-directed phenylalanine incorporation. Note the strong incorporation of labeled phenylalanine in the presence of poly(U) (*boxed in red*) versus background level radioactivity in the presence of five other polyribonucleotide templates. *Panels (A) and (B): Reproduced with modifications from Nirenberg MW, Matthaei JH. The dependence of cell-free protein synthesis in* E. coli *upon naturally occurring or synthetic polyribonucleotides.* Proc Natl Acad Sci USA *1961;47(10):1588–602.*

poly(U) was present, there was robust incorporation of the radiolabeled phenylalanine (Fig. 10.6A). The specificity of incorporation was demonstrated by showing that the incorporation of ^{14}C-phenylalanine was dictated by poly(U) but not by poly(A), poly(C), poly(I), poly(AU), or a mixture of poly(A) and poly(U) (Fig. 10.6B).

In a converse experiment Nirenberg and Matthaei demonstrated the amino acid specificity of the coding by poly(U). Although poly(U) directed massive incorporation of ^{14}C-phenylalanine, there was no measurable incorporation of mixtures of 16 other ^{14}C-labeled amino acids or of ^{35}S-cysteine (Fig. 10.7A).

Notably, one element of good fortune that Nirenberg and Matthaei had was their choice of a nonphysiological high concentration of magnesium (10 mM) in their cell-free reaction mixtures. This concentration permitted the translation of poly(U) (and of other synthetic polyribonucleotides) without a need for the initiation codons that are essential for protein synthesis at lower physiological magnesium concentrations.

In determining that the product of phenylalanine incorporation was indeed polyphenylalanine, Nirenberg benefitted from the advice of the Israeli immunologist Michael Sela (1924−) who was working then in the NIH. Sela educated Nirenberg about the unique properties of polyphenylalanine and also provided him with an unlabeled polypeptide.[14] As a consequence, Nirenberg and Matthaei were able to show that the solubility in various inorganic or organic solvents of the product of poly(U)-directed incorporation of ^{14}C-phenylalanine was the same as that of unlabeled polyphenylalanine (Fig. 10.7B). Another

(A)

SPECIFICITY OF AMINO ACID INCORPORATION STIMULATED BY POLYURIDYLIC ACID

Experiment no.	C14-amino acids present	Additions	Counts/min/mg protein
1	Phenylalanine	Deproteinized at zero time	25
		None	68
		+ 10 µg polyuridylic acid	38,300
2	Glycine, alanine, serine, aspartic acid, glutamic acid	Deproteinized at zero time	17
		None	20
		+ 10 µg polyuridylic acid	33
3	Leucine, isoleucine, threonine, methionine, arginine, histidine, lysine, tyrosine, tryptophan, proline, valine	Deproteinized at zero time	73
		None	276
		+ 10 µg polyuridylic acid	899
4	S35-cysteine	Deproteinized at zero time	6
		None	95
		+ 10 µg polyuridylic acid	113

(B)

COMPARISON OF CHARACTERISTICS OF PRODUCT OF REACTION AND POLY-L-PHENYLALANINE

Treatment	Product of reaction	Poly-L-phenylalanine
6 N HCl for 8 hours at 100°	Partially hydrolyzed	Partially hydrolyzed
12 N HCl for 48 hours at 120–130°	Completely hydrolyzed	Completely hydrolyzed
Extraction with 33% HBr in glacial acetic acid	Soluble	Soluble
Extraction* with the following solvents: H₂O, benzene, nitrobenzene, chloroform, N,N-dimethylformamide, ethanol, petroleum ether, concentrated phosphoric acid, glacial acetic acid, dioxane, phenol, acetone, ethyl acetate, pyridine, acetophenone, formic acid	Insoluble	Insoluble

FIGURE 10.7   (A) Poly(U) encoded incorporation of ^{14}C-phenylalanine but not of 17 other labeled amino acids. (B) Chemical properties of the product of ^{14}C-phenylalanine incorporation correspond to polyphenylalanine. *Panels (A) and (B): Reproduced from Nirenberg MW, Matthaei JH. The dependence of cell-free protein synthesis in* E. coli *upon naturally occurring or synthetic polyribonucleotides.* Proc Natl Acad Sci USA 1961;47(10):1588–602.

indication that the in vitro–incorporated ^{14}C-phenylalanine was polymerized into polyphenylalanine was that acid hydrolysis of the labeled product of poly(U)-directed incorporation released stoichiometric amounts of ^{14}C-phenylalanine.

10.3.3   The News Quickly Spread That a Defined Nucleotide Sequence Dictated Polymerization of a Specific Amino Acid

After completing the experiments, Nirenberg and Matthaei submitted their paper to the *Proceedings* on Aug. 3, 1961. Shortly thereafter, Nirenberg left for the 5th International Congress of Biochemistry that took place in Moscow between Aug. 10 and 16. There he was scheduled to present a 15-minute talk on his and Matthaei's results.[18] The dramatic circumstances of Nirenberg's the public announcement in Moscow that UUU encodes phenylalanine were told and retold by many. Briefly, having been a young researcher at the early phases of his career, Nirenberg was not known at the time to most of the more established investigators. Thus, as Meselson told Judson (Ref. 16, pp. 463–464), because he was not a "member of the club," Nirenberg's talk attracted only approximately 35 listeners.[14] Fortunately, among the few who did attend the talk were three Harvard University students of ribosomes and of mRNA, Alfred Tissières, Walter Gilbert, and Matthew Meselson. Crick who was told by Gilbert about the poly(U) result immediately arranged for Nirenberg to present his work to a much wider audience. According to Crick, Nirenberg's talk in the last session of the congress which he chaired "electrified the audience" (though Meselson had a photograph of some attendees dozing off).[16]

The news of the identification of the first code word by Nirenberg and Matthaei spread rapidly. First in scientific circles and almost instantly thereafter in the popular press that brought it to the attention of the general public. Most of the reactions were highly positive—putting the deciphering of the code on a scale similar to the revolution in nuclear physics. Typical to this enthusiastic brand of reactions was a Jan. 14, 1962, *New York Times* report by William J. Lawrence titled "STRUCTURE OF LIFE: 'Genetic Code' Discoveries Bring New Understanding of Heredity." This piece (facsimile in http://timesmachine.nytimes.com/timesmachine/1962/01/14/issue.html) opened with the following bold declaration:

> *In its quest for understanding the chemistry of life and the basic mechanism of heredity, whereby all things living reproduce themselves in their own image, the science of biology has reached a new frontier said to be leading to "a revolution far greater in its potential than the atomic or hydrogen bomb.*

Other similarly fervent commentators drew parallels between Nirenberg's work and the working of computers and of high-tech automatic assembly lines (for a multifaceted discussion of the analogies that have been made at the time between the genetic code and information theory, see Ref. 19). Mindful of discoveries in physics that led to the development of nuclear weaponry and to a grave threat to the very existence of humanity, some, however, regarded the decrypting of the code to be dangerous. For instance, the Case Western University human geneticist Arthur G. Steinberg (1912–2006)[20] opined at a meeting of the American Association for the Advancement of Science in Dec. 1961 that "[knowledge acquired from the genetic code] *might well lead in the foreseeable future to a means of directing mutations and changing genes at will."* Also in 1961, on the occasion of the 60th anniversary of the Nobel Prize, Arne Tiselius (1902–71), the 1948

Nobel Laureate and the sitting President of the Nobel Foundation, spoke about the moral and ethical aspects of the recent discovery. He stated that the new knowledge could "lead to methods of tampering with life, of creating new diseases, of controlling minds, of influencing heredity, even perhaps in certain desired directions."[21]

Interestingly, Nirenberg appeared to have met this mixture of contrasting enthusiastic and negative opinions with equanimity. In a Jan. 15, 1962, letter to Crick (a facsimile of the letter in http://profiles.nlm.nih.gov/ps/access/ JJBBFJ.pdf), he commented dryly:

I haven't seen the English newspapers but the American press has been saying that this type of work may result in (1) the cure of cancer and allied diseases (2) the cause of cancer the and end of mankind, and (3) a better knowledge of the molecular structure of God. Well, it's all in a day's work.

10.3.4 Poly(U)-Directed Synthesis of Polyphenylalanine Involved Charging of Phenylalanine Onto Its Specific tRNA and Its Transfer to a Growing Polypeptide

Upon his return from Moscow, Nirenberg completed a work that he did with Matthaei and a new postdoctoral fellow, Oliver Jones, to prove that the poly (U)-directed incorporation of phenylalanine represented bona fide protein synthesis. The results of this set of experiments, that were submitted for publication in late Nov. 1961,[22] demonstrated that translation in vitro of poly(U) to polyphenylalanine required charging of phenylalanine onto its specific tRNA. Thus, poly(U)-directed incorporation of phenylalanine represented authentic synthesis of protein. Results showed that phenylalanyl-tRNA was the direct intermediate in the path to the synthesis of polyphenylalanine (Fig. 10.8A).

Addition of excess of free unlabeled (^{12}C) phenylalanine to the cell-free protein synthesis system diluted out the ^{14}C-labeled phenylalanine and prevented its incorporation into protein. By contrast, excess of unlabeled phenylalanine did not diminish incorporation when ^{14}C-phenylalanyl-tRNA replaced free ^{14}C-phenylalanine. Lack of competition by the unlabeled amino acid indicated that unlike free ^{14}C-phenylalanine, tRNA-linked ^{14}C-phenylalanine was preferentially and directly transferred into protein. Kinetic analysis also showed that poly(U) was obligatory for the incorporation of the tRNA-linked phenylalanine (Fig. 10.8B).

Being a meticulous experimenter, Nirenberg made an extra effort to demonstrate that the charging of phenylalanine onto intact tRNA was a requisite for the in vitro synthesis of polyphenylalanine. He thus first showed that tRNA hydrolysis by RNase or by alkali defeated incorporation (Fig. 10.9(1)). Second, the S-100 fraction was essential for the charging of tRNA and for incorporation (Fig. 10.9(2)). Third, S-100 could be replaced by partially purified phenylalanyl-tRNA synthetase (activating enzyme) (Fig. 10.9(3)).

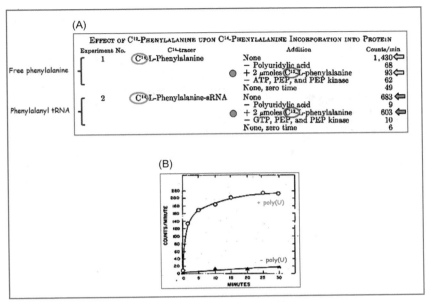

FIGURE 10.8 (A) Poly(U)-dependent incorporation of free ¹⁴C-phenylalanine (*yellow arrows*) but not of tRNA-linked ¹⁴C-phenylalanine (*green arrows*) was inhibited by dilution with excess of unlabeled ¹²C-phenylalanine (*circled in blue*). Red dots mark reaction mixtures that contained the competing nonlabeled phenylalanine. (B) In vitro incorporation of tRNA-linked ¹⁴C-phenylalanine into polyphenylalanine depended on poly(U). *Panels (A) and (B): Modified from Nirenberg MW, Matthaei JH, Jones OW. An intermediate in the biosynthesis of polyphenylalanine directed by synthetic template RNA. Proc Natl Acad Sci USA 1962;48(1):104−9.*

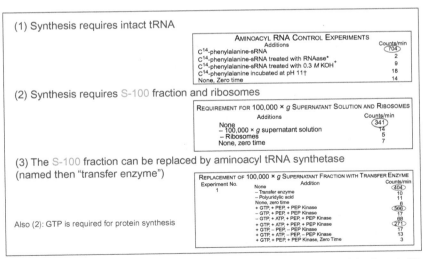

FIGURE 10.9 Intact tRNA and its charging with phenylalanine were essential for the poly(U)-directed in vitro incorporation of phenylalanine into acid-insoluble protein. *Modified from Nirenberg MW, Matthaei JH, Jones OW. An intermediate in the biosynthesis of polyphenylalanine directed by synthetic template RNA. Proc Natl Acad Sci USA 1962;48(1):104−9.*

10.3.5 A Race Was Opened to Complete the Decryption of the Code

While Nirenberg was still in Moscow, Matthaei identified a second code word. He found that poly(C) specifically directed the incorporation of proline into acid-insoluble material and thus that the CCC triplet encoded proline. A note of this finding was added in proof to the *Proceedings* report on the encoding of phenylalanine by UUU.[15] Yet, with only two code words discovered, there still loomed the task of identifying the remaining 62 codons. Nirenberg and his associates accomplished this undertaking in the next 3 years and Khorana and his coworkers independently reached the same goal. Because of its great significance, other laboratories also joined the race to break the code. Nirenberg's ultimately successful pursuit faced, therefore, fierce competition—primarily from the New York University laboratory of Severo Ochoa (Box 10.4).

BOX 10.4 Severo Ochoa and Marianne Grunberg-Manago

Born in Spain, Ochoa earned a baccalaureate degree in 1921. Having been inspired by the Nobel laureate of 1906 Santiago Ramón y Cajal (1852–1934) to pursue experimental research in biology, he attended medical school and conducted research under the motivating teacher Juan Negrín. After concluding his medical studies in 1928, he did research work in Berlin under the Nobel laureate of 1923 Otto Meyerhoff (1884–1951) and later in Heidelberg and in London. His subsequent service as Professor of Physiology at the University of Madrid was cut short in the wake of the Spanish civil war and he left Spain to wander between laboratories in Heidelberg, Plymouth, and Oxford. After reaching the United States in 1940, he first worked in the St Louis laboratory of the Nobel laureates of 1947 Carl and Gerty Cori and since 1942 he was in New York University that remained his base for most of the rest of his career. His voluminous contributions to the study of intermediary metabolism and to enzymology were crowned by his discovery with Marianne Grunberg-Manago of PNP. For this achievement, he shared with Arthur Kornberg the 1959 Nobel Prize in Physiology and Medicine.

Born in St Petersburg to a family of artists, the 9-month-old Grunberg-Manago immigrated to France with her parents. Following studies of both comparative literature and biology at the University of Paris, she received a PhD degree in 1947. After first working on intermediary metabolism in bacteria at the Paris Institut de Biologie Physico Chimique (IBPC), she traveled in 1953 to the University of Illinois at Urbana-Champaign and then to the New York University Laboratory of Severo Ochoa at which she discovered PNP. Upon her return in 1956 to the IBPC she studied RNA degradation by PNP and later protein translation. She was the first female scientist to be elected President of the International Union of Biochemistry and Molecular Biology (1985–88) and the only woman President of the French Academy of Sciences (1995–96) (Fig. 10.10).

(Continued)

BOX 10.4 (Continued)

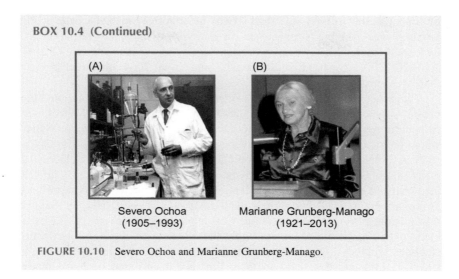

FIGURE 10.10   Severo Ochoa and Marianne Grunberg-Manago.

The endeavor to break the code had two phases. In the first stage experiments were conducted to identify individual codons by matching base compositions of synthetic polyribonucleotide templates with the amino acid compositions of in vitro translated proteins. The three groups of Nirenberg, Ochoa, and Khorana took variations on this approach. In the next phase the 64 RNA triplets were synthesized by different techniques and their coding specificities were determined in a filter-binding assay that the Nirenberg laboratory developed. At that point the Ochoa group left the field and it remained for Nirenberg and Khorana to complete the mission of the deciphering of the code. Here we first describe the work of Nirenberg and Ochoa in the first phase and then of Nirenberg's alone in the second phase. The parallel approach and achievements of Khorana are considered in the last part of this chapter.

10.3.6 Translation of Synthetic Polyribonucleotide Templates Revealed the Composition but Not the Sequence of Code Words

Historical Background

The rapid spreading of the news that the UUU triplet encoded phenylalanine excited other investigators who either speculated on the base composition of codons[23] or did experiments to determine which amino acids were incorporated in vitro under the direction of other exogenously introduced synthetic polyribonucleotides.[24] However, it rapidly became obvious that the serious competition to crack the code was between the small NIH group of Nirenberg and the larger team of the distinguished biochemist and 1959 Nobel laureate Severo Ochoa at New York University. The story of the parallel efforts of these two laboratories to solve the code problem was told by Nirenberg in an interview with Ruth Harris (http://history.nih.gov/archives/downloads/Nirenberg%20oral%20history%20Chap%204%20-%20Race%20to%20decode.pdf) and in a 2004 historical review that he wrote.[14]

This affair was later revisited by Peter Lengyel (1929−) who initiated the pursuit of the code in the Ochoa laboratory.[25] In the wake of the failed Hungarian uprising of 1956, Lengyel escaped Hungary for the United States where he joined the New York University laboratory of Ochoa as a graduate student. After he heard Sydney Brenner's description of the isolation of mRNA at a Cold Spring Harbor Symposium,[26] Lengyel came up with the idea of using synthetic polyribonucleotides of known nucleotide composition as messengers in a cell-free protein synthesis system. Specifically, he thought that specific codons might be defined by identifying amino acids that were incorporated in vitro into protein under the direction of specific polyribonucleotides templates. With Ochoa out of the country, Lengyel outlined in a letter his idea and described some preliminary results that he had already obtained[25] (for a facsimile of his Aug. 19, 1961, letter to Ochoa, see http://www.annualreviews.org/doi/suppl/10.1146/annurev-micro-010312-100615/suppl_file/mi.66.lengyel.supmat.pdf):

> *On the very day when you left for the summer and when I attended the lecture on messenger RNA at the Cold Spring Harbor Symposium, the idea occurred to me that homoribopolynucleotides should be tested as messengers which, if they would act at all, should induce the ribosomes to form homopelypeptides. (E.g. poly(C) could induce the formation of polyleucine.)*

An instructor in the department, Joseph Speyer (1926−98) who had already set up a cell-free protein synthesis system, promptly joined Lengyel in the work on the code. After Ochoa returned to New York, he and other members of the laboratory also participated in the project. Nirenberg became aware of the competing venture of the New York laboratory only in early 1962. At the end of a seminar that he presented at the MIT Lengyel came to the podium to declare that the Ochoa group succeeded in incorporating specific amino acids under the direction of randomly ordered polyribonucleotide templates.[14,25] Shaken by this revelation, Nirenberg soon traveled to New York in an attempt to forge collaboration with the Ochoa group. Although he found Ochoa a gracious and gentlemanly host, it became clear that a collaboration was not to be (for Nirenberg's description of this incidence, see his relevant interview with Ruth Harris: http://history.nih.gov/archives/downloads/Nirenberg%20oral%20history%20Chap%204%20-%20Race%20to%20decode.pdf). Thus, a race was launched to decipher the code. Between 1961 and 1963 the two groups published multiple full-length papers that in the most part reported similar results. At the same time, Har Gobind Khorana confirmed and expanded much of this work by chemically synthesizing polydeoxyribonucleotides of known base compositions and determining their coding specificities in cell-free protein synthesis system (see below).

PNP Was Used to Synthesize Polyribonucleotide Templates

A common key to the parallel work of Nirenberg and of Ochoa was their similar employment of the enzyme PNP to synthesize homopolyribonucleotides from a single nucleotide or to produce statistically defined heteropolyribonucleotides from two, three, or four different nucleotides. Bacterial PNP and DNA polymerase I from *E. coli*, which were discovered at about the same time, were, respectively, the first identified RNA- and DNA-synthesizing enzymes. The discoverers of these nucleic acid—polymerizing enzymes, Severo Ochoa and Arthur Kornberg (1918—2007) shared the 1959 Nobel Prize "for their discovery of the mechanisms in the biological synthesis of ribonucleic acid and deoxyribonucleic acid."

Brief History of the Discovery of PNP

While working as a postdoctoral fellow in Ochoa's laboratory, the French biochemist Marianne Grunberg-Manago (Box 10.4) detected PNP activity in extracts of *Azobacter vinelandii*. She found that partially purified PNP catalyzed exchange between $^{32}PO_4$ and impure amorphous ATP.[27] Based on the observation that the reaction did not occur when crystalline rather than impure ATP was used, she found that the actual active compound was ADP that contaminated the amorphous ATP. Although Ochoa was initially suspicious of the result, he soon became convinced of its veracity and Grunberg-Manago promptly showed that other nucleoside diphosphates: GDP, IDP, UDP, and CDP, but not their corresponding nucleoside triphosphates, could also serve as substrates for the reaction: $NDP \rightleftarrows NMP + PO_4$. Further, liberation of $^{32}PO_4$ was accompanied by disappearance of the nucleotide into nondialyzable acid-insoluble material.[27] Additional experiments demonstrated that this high-molecular-weight product was RNA.[28] Hoping that the enzyme-catalyzed RNA synthesis, Ochoa wanted to name it "RNA synthetase." Grunberg-Manago thought, however, that it was an RNA-degrading enzyme and she prevailed in naming it "PNP."[29] Indeed, although PNP was the first enzyme that was found to catalyze the in vitro synthesis of RNA, subsequent studies by Grunberg-Manago and by others showed that the primary role was that of a phosphorolytic $3' \rightarrow 5'$ exoribonuclease. Thus despite the fact that the enzyme ran in vitro the reverse reaction of RNA synthesis, its native function was to degrade RNA (Fig. 10.11).

Notably, RNA synthesis by PNP did not require a template, and the composition and nearest-neighbor distribution of bases in the product polyribonucleotides were governed solely by the identities and molar ratios of the ribonucleotide diphosphate (rNDP) substrates.

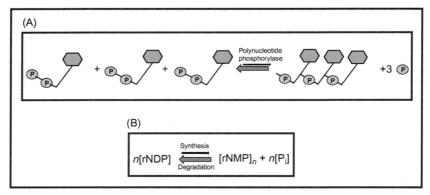

FIGURE 10.11   (A) Catalysis by PNP of RNA degradation and synthesis. The major function of the enzyme (*wide red arrow*) is $3' \rightarrow 5'$ phosphorolytic breakdown of RNA in which phosphate is used to disrupt phosphodiester bonds in RNA and release ribonucleoside diphosphates. In the reverse reaction (*narrow black arrow*) PNP uses nucleoside diphosphates to polymerize RNA chains and liberate phosphate. (B) Scheme of RNA degradation and synthesis by PNP. The designations rNDP and rNMP correspond, respectively, to the ribonucleotide diphosphate and monophosphate forms of adenine, uracil, guanine, and cytosine.

Use of PNP-Generated Polynucleotides to Decipher the Code

PNP polymerized each of the four rNDP substrates into defined homopolymers: poly(A); poly(U); poly(G); and poly(C). Also the enzyme utilized two, three, or more different rNDPs to synthesize heteropolymers whose base compositions could be modulated by changing the molar ratios between the different rNDP substrates. Both the laboratories of Nirenberg and Ochoa used PNP to synthesize homo- or heteropolyribonucleotides that then served as templates in cell-free protein synthesis systems. It was anticipated that triplet sequences that encoded each of the 20 amino acids could be deduced from the amino acid compositions of polypeptides that were translated from the polynucleotide templates.

By Translating In Vitro Synthetic Polyribonucleotide Templates Nirenberg and Ochoa Could Determine the Base Compositions but Not the Sequences of Codons

Nirenberg and Lengyel in the Ochoa laboratory independently came up with the idea of identifying which amino acids would be incorporated in vitro under the direction of defined homopolyribonucleotide templates. Nirenberg and Matthaei were the first to report that poly(U) encoded polyphenylalanine and that poly(C) dictated incorporation of proline.[15] Yet, their attempt to similarly find which amino acid was encoded by poly(A) had failed. It was only in 1962 that Robert Gardner, a medical student working in the Ochoa laboratory, found that this template encoded polylysine.[30] Nirenberg and Matthaei missed this polypeptide because it was soluble in trichloroacetic acid (TCA), the standardly used protein-precipitating acid. After finding that it could be precipitated by a TCA—tungstate mixture, Gardner and his associates identified the product of poly(A) translation to have been polyly-sine.[30] Because the fourth homopolyribonucleotide poly(G) formed secondary structures, its coding specificity could not be determined.

To increase the catalogue of codons, both Nirenberg and Ochoa chose a similar approach of creating heteropolyribonucleotide templates by providing PNP with mixtures of different rNDP substrates at varying molar ratios. In late 1961 and early 1962 the two laboratories published first reports on the amino acid compositions of polypeptides that were translated from such templates. At the NIH, Robert Martin (1935—) had left for a time his independent research project in order to join Matthaei, Oliver Jones, and Nirenberg in these experiments. Using purified PNP from *Micrococcus lysodeikticus* and mixtures of two or three different rNDPs, they synthesized poly(UC), poly(UA), and poly(UG), as well as poly(UCG) and poly(UGA). Each of these polyribonucleotides proved to be an efficient template for cell-free incorporation of specific radiolabeled amino acids.[31] Tabulation of the 15 amino acids that were incorporated under the direction of the various templates revealed the base composition of the code words but not their sequence (Fig. 10.12A).

Thus, for instance, although it was evident that poly(UG) encoded four different amino acids: valine, leucine, cysteine, and tryptophan, it remained unknown which of the possible sequences of the two bases such as UGU, GUG, UUG, GGU encoded each acid. Taking an essentially similar approach, the Ochoa group reported at about the same time comparable results.[34,35] Thus for instance similar to Martin et al.,[31] they also found that poly(UG) encoded cysteine and valine and like their NIH competitors they also defined the base compositions but not the sequences of codons for these two amino acids.[35]

In an attempt to obtain more definitive idea on the sequence of specific codons the laboratories of Nirenberg[32,36] and of Ochoa[37,38] independently took a similar strategy of using different molar ratios of the rNDP substrates to create polyribonucleotides with varying relative content of bases. With a given molar ratios between the different rNDP substrates, the probabilities of appearance of different triplets that the bases could form became calculable. Thus, for instance, having poly(UG) with a U to G ratio of 0.76—0.24, the calculated probabilities of appearance of possible triplets relative to a frequency of 100.0 for the UUU trinucleotide were 32.0 for UUG, 10.6 for UGG, and 3.4 for GGG (Fig. 10.12B). In an attempt to pinpoint specific codons, such heteropolyribonucleotides with varying base ratios were used as templates for cell-free protein synthesis and the incorporated amino acids were identified. However, this approach also failed to reveal the base sequences of codons for specific amino acids and again only the base compositions of codons but not their sequences could be determined (Fig. 10.12C). Encouragingly, however, both Nirenberg[32] and Ochoa[38] independently showed that some of the possible triplet sequences that their

(A)

Amino Acid	Nucleotide Composition of Coding Unit *
Phenylalanine	UUU...
Valine	UG (U > G)
Leucine	UG, UC, (U > G) (U > C)
Cysteine	UG
Tryptophan	UG (U ≶ G)
Glutamic Acid	UGC
Methionine	UG
Glycine	UG
Arginine	UGC
Alanine	UGC
Serine	UC, UGC, (U > C) (U > G or C)
Proline	UC (U < C)
Tyrosine	UA (U > A)
Isoleucine	UA
Lysine	UA (U < A)

(B)

Amino Acid Incorporation into Protein Stimulated by Randomly Mixed Polynucleotides

Polynucleotide	UA	UC	UG	UAC	UGC	UGA	Composition of coding units*
Base-ratio	U = 0.87 A = 0.13	U = 0.39 C = 0.61	U = 0.76 G = 0.24	U = 0.834 A = 0.050 C = 0.116	U = 0.341 G = 0.152 C = 0.503	U = 0.675 G = 0.291 A = 0.034	
Probability of triplet relative to phenylalanine (UUU) = 100%	UUU—100 UUA— 13 UAA— 2.2 AAA— 0.3	UUU—100 UUC—157 UCC—244 CCC—382	UUU—100 UUG— 32 UGG— 10.6 GGG— 3.4	UUU—100 UUA— 6.0 UAA— 0.4 AAA— 0.02 UUC— 13.9 UCC— 1.9 CCC— 0.3 UAC— 0.8 AAC— 0.05 ACC— 0.12	UUU—100 UUG— 46.2 UGG— 21.0 GGG— 1.0 UUC—147 UCC—218 CCC—322 UGC— 68.1 GGC— 31.7 GCC—101	UUU—100 UUG— 43 UGG— 19 GGG— 8.1 UUA— 5.1 UAA— 0.26 AAA— 0.01 UGA— 2.2 GGA— 0.1 GAA— 0.01	
Amino Acid							
Phenylalanine	100	100	100	100	100	100	UUU...
Arginine	0	0	1.1	0	49.3	2.9	UCG...
Alanine	1.9	0	0	1.0	40.4	0.9	UCG...
Serine	0.4	160	3.2	3.6	170	2.3	UUC.. + UCG...
Proline	0	285	0	0	188	0	UCC...
Tyrosine	13	0	0	8	1.0	8.6	UUA...
Isoleucine	12	1.0	1.0	4.8	5.4	8.4	UUA...
Valine	0.6	0	37	0.4	29.8	75	UUG...
Leucine	4.9	79	36	5.1	157	44	UUC.. + UUG...
Cysteine	4.9	0	35	0	5.4	46	UUG... or UGG
Tryptophan	1.1	0	14	0	1.6	23	UGG...
Glycine	4.7	0	12	0.5	9.7	15	UGG...
Methionine	0.6	0	0	0.6	1.5	8	UGA...
Glutamic acid	1.5	0	0	1.2	0.44	6.2	UGA...
Lysine							UAA... (?)

(C)

Polynucleotides	Amino acids			
U	PHE			
C	PRO			
A	LYS			
G	—			
UC	LEU	SER		
UA	LEU	TYR	ILE	ASN
UG	LEU	VAL	CYS	TRP
CA	HIS	THR	GLN	ASN
CG	ARG	ALA		
AG	ARG	GLU		
UAG	ASP	MET		
CAG	ASP	SER		

FIGURE 10.12 (A) Base compositions of different heteropolyribonucleotide templates and amino acids that were incorporated under their direction in a cell-free protein synthesis system. (B) In vitro incorporation of amino acids under the direction of heteropolyribonucleotides with different base ratios and calculated frequencies of the possible triplets. (C) Base compositions (but not base sequences) of code words for amino acids that as determined between 1961 and 1963 by identifying amino acids that were incorporated in vitro under the direction of polyribonucleotide templates. *Panel (A): From Martin RG, Matthaei JH, Jones OW, Nirenberg MW. Ribonucleotide composition of the genetic code.* Biochem Biophys Res Commun *1962;**6**(6):410–14.*[31] *Panel (B): From Matthaei JH, Jones OW, Martin RG, Nirenberg MW. Characteristics and composition of RNA coding units.* Proc Natl Acad Sci USA *1962;**48** (4):666–77.*[32] *Panel (C): From Nirenberg M.* The genetic code. Nobel lecture December 12, 1968. *Amsterdam: Elsevier Publishing Company; 1972.*[33]

experiments defined conformed with tentative code words that were deduced by Tsugita and Fraenkel-Konrat[39] and by Wittmann[40] from amino acid substitutions in TMV mutants (see also Chapter 7: "Defining the Genetic Code").

Despite the advances that Ochoa and Nirenberg have made, the nucleotide sequences of codons remained unknown. By the end of 1962 Crick published a review article titled "The Recent Excitement in the Coding Problem" in which he cuttingly criticized the work that was done up to that time as ill-conceived and sloppy.[41] Opening the article with the statement: *"There are so many criticisms to be brought against this type of experiment that one hardly knows where to begin,"* he went on to count the many faults of the experiments and concluded by stressing *"...the necessity for rigorous experimental proof of each codon."*[41] As history proved, Nirenberg and independently Khorana shared Crick's outlook that it was indeed high time for a change of strategy and that defined RNA triplets should be employed to directly expose their coding specificities.

10.4 THE SECOND PHASE: THE NIRENBERG TEAM DETERMINED THE BASE SEQUENCES OF ALL 64 CODONS

By early 1963 it became evident that codon sequences could not be definitely determined by analysis of the amino acid composition of products of translation of polyribonucleotide templates. It was thus realized that breaking the code necessitated the determination of the coding specificities of individual RNA triplets. At that point in time, Ochoa left the race to decipher the code and it remained for Nirenberg at the NIH and for Khorana in the University of Wisconsin to independently accomplish this enterprise. This section is dedicated to the work of Nirenberg and his associates. A subsequent part of this chapter describes Khorana's achievements.

Nirenberg's successful completion of the decryption of the code had two components: (1) He and his associates developed methods to synthesize every one of the 64 RNA triplets. (2) Nirenberg and Philip Leder devised a filter-binding assay that conclusively identified the amino acid—coding specificities of each of the RNA triplets.

10.4.1 Synthesis of RNA Triplets of Defined Base Sequence

The Nirenberg laboratory devised two methods to produce the 64 different RNA triplet sequences. Philip Leder (Box 10.5) and Richard Brimacombe from the Nirenberg group collaborated with Maxine Singer (1931—) who was then a postdoctoral fellow in the laboratory of Leon Heppel to develop the first synthesis method (for relevant primary material concerning Singer's contribution, see the Maxine Singer Papers in the National Library of Medicine "Profiles in Science": http://profiles.nlm.nih.gov/ps/retrieve/Narrative/DJ/p-nid/206).

BOX 10.5 Philip Leder

Born in Washington, DC, Philip (Phil) Leder graduated from Harvard College in 1956 and received an MD degree from Harvard Medical School in 1960. After a medical residency at the University of Minnesota, he joined in 1962 the laboratory of Marshall Nirenberg as a Research Associate. There he made a major contribution to the development and employment of the methodologies for the determination of the coding specificities of all the 64 RNA triplets. Staying on at the NIH as an independent researcher and moving later to Harvard University, he had made major contributions to molecular biology and immunology. Among his achievements were the sequencing of the first mammalian gene (the β globin gene), discovery of the genetic basis for antibody synthesis and diversity and later studies on the genetic basis of certain malignancies. He won the Lasker Award in 1987 in recognition of his scientific accomplishments and was presented with the National Medal of Science in 1989 (Fig. 10.13).

Philip Leder
(1934–)

FIGURE 10.13   Philip Leder.

As told by Nirenberg,[14] development of this method began when Leder noticed an advertisement in a scientific journal for all the 16 diribonucleotides. These reagents, which were then sold for a stupendous, sum ($1500 per doublet at a time that Nirenberg's annual income was $3000[14]) were purchased and used as nuclei for the addition by PNP of a third ribonucleotide. With the helpful advice of Marianne Grunberg-Manago who was visiting the laboratory, Leder, Singer, and Brimacombe used PNP to extend the dinucleotide primer. Oligomeric products of the reaction were separated by paper chromatography, visualized by UV light, and bands of RNA triplets of designed sequence were excised, extracted, and purified.[42]

A second method to synthesize RNA triplets of defined sequence originated in an observation that Leon Heppel had made already in 1955.[43] He found that pancreatic RNase A used in the presence of high methanol concentrations pyrimidine $2'-3'$ cyclic phosphates to catalyzed addition of the ribonucleotide to oligoribonucleotide primers. The Nirenberg laboratory adapted this method for the synthesis of RNA triplets by using RNase A to add single ribonucleotides to the diribonucleotide cores.

10.4.2   A Filter-Binding Assay Enabled the Determination of the Amino Acid—Coding Specificities of All the 64 Codons

Side-by-side with their development and deployment of methods for the synthesis of RNA triplets, Leder and Nirenberg also devised a filter-binding assay for the identification of amino acids that each of the RNA trinucleotides encoded. The basis for this assay was laid in a short 1963 report by Akira and Hideko Kaji of the University of Pennsylvania.[44] In it they showed that ^{14}C-phenylalanyl-tRNA bound to polysomes that formed along a poly (U) template. Similarly,[14]C-lysyl-tRNA bound to polysomes that organized along a poly(A) template. These binding interactions were specific such that ^{14}C-phenylalanyl-tRNA did not associate with polysomes that formed along poly(A) and ^{14}C-lysyl-tRNA did not form a complex with poly(U) polysomes.[44] Prompted by this observation, Leder and Nirenberg developed a clever assay to identify the triplet sequences that encoded the 20 amino acids. Their methodology was established in a series of experiments that were described in late 1964 in a seminal paper in the journal *Science*.[45] In this first in a line of works that emanated from the Nirenberg laboratory in the next year and a half, the assay system was set up using homopolyribonucleotide or homooligoribonucleotide templates whose coding specificities had already been established (see Section 10.3). In a first step, a reaction was set up to form specific ternary complex of aminoacyl tRNA—ribosome—template. After preincubation of ribosomes with a defined homopolyribonucleotide template such as poly(U), a radiolabeled amino charged onto cognate tRNA molecule was added (Fig. 10.14A).

In the next step reactants and products of the reaction were passed through a nitrocellulose filter. The washed filter was found to have retained only the tripartite complex of ^{14}C-aminoacyl tRNA, ribosome, and the template, whereas free molecules of the poly(U) template, of uncharged tRNA, or of aminoacyl tRNA, did not bind to the filter (Fig. 10.14B). In a final step radioactivity that was retained by the dried filters was counted to quantify the ribosome-bound—labeled aminoacyl tRNA that was in complex with the tested template.

Control experiments confirmed that the retention of the tripartite complex by the filter reflected true complementarity between the ribosome-bound aminoacyl tRNA and the template. First, binding to the filter required that all of the three reactants (ribosomes, aminoacyl tRNA, and a matching template) had to be present. Very little binding was documented when any one of the three components or magnesium ions were omitted. Also, the addition of excessive

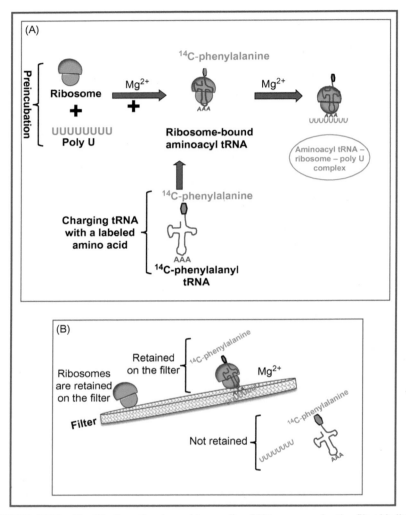

FIGURE 10.14   Identification of amino acid–encoding RNA sequence by the filter-binding assay. (A) Formation of specific aminoacyl-tRNA–ribosome–poly(U) ternary complex. Ribosomes were preincubated in the presence of magnesium ions with poly(U) and then mixed with ^{14}C-phenylalanyl-tRNA. Incubation of the mixture produced specific complexes of ^{14}C-phenylalanyl-tRNA–ribosome–poly(U). (B) Selective retention on a cellulose nitrate filter of the complex of ^{14}C-phenylalanyl-tRNA–ribosome–poly(U). Reactants and products of the magnesium-containing reaction in (A) were passed through a cellulose nitrate filter. The filter retained only ribosomes and tripartite complexes of ^{14}C-phenylalanyl-tRNA that specifically bound to ribosomes in the presence of its matching poly(U) template. Free molecules of ^{14}C-phenylalanine, of poly(U), of uncharged tRNA, and of aminoacyl tRNA were not retained by the filter.

(A)

Modifications	C^{14}-Phe-sRNA bound to ribosomes ($\mu\mu$ mole)
Complete	5.99
− PolyU	0.12
− Ribosomes	0
− Mg^{++}	0.09
+sRNA (deacylated) at 50 min	
0.500 A^{200} units	5.69
2.500 A^{200} units	5.36
+sRNA (deacylated) at zero time	
0.500 A^{200} units	4.49
2.500 A^{200} units	2.08

(B)

Polynucleotide ($m\mu$ mole base residues)	C^{14}-Aminoacyl-sRNA bound to ribosomes ($\mu\mu$ mole)		
	C^{14}-Phe-sRNA	C^{14}-Lys-sRNA	C^{14}-Pro-sRNA
None	0.19	0.99	0.25
PolyU, 25	(6.00)	0.67	0.15
PolyU, 16	0.22	(4.35)	0.17
PolyU, 19	0.21	0.72	(0.80)

FIGURE 10.15 Specific binding to filters of ternary complexes of ribosomes with aminoacyl tRNA and matching RNA template. (A) Requirements for binding. The filters retained in the presence of magnesium ions tripartite complexes of ribosomes with ^{14}C-phenylalanyl-tRNA and poly(U) template. However, there was no significant binding to the filters when poly(U), ribosomes, or magnesium ions were omitted. By contrast, excessive amounts of deacylated tRNA that were added in the course of the reaction or prior its initiation did not compete with aminoacyl tRNA on complex formation or binding to filters. (B) Specificity of the formation and retention by filters of ternary complexes. Ribosome-bound aminoacyl tRNA molecules formed filter-retained three-party complexes only with matching encoding templates (*circled in red*). Because ribosome-bound aminoacyl tRNA did not form complexes with a mismatched template, radioactivity on the filters was at the background level. *Data in panels (A) and (B) are from Nirenberg M, Leder P. RNA codewords and protein synthesis. The effect of trinucleotides upon the binding of sRNA to ribosomes.* Science *1964;**145**(3639):1399−407.*

amounts of free, deacylated tRNA either before or late into the reaction did not compete with the binding of the ternary complex to the filter (Fig. 10.15A).

Most importantly, the formation of the three-party, filter-retained complex was shown to reflect specific interaction between particular aminoacyl tRNA and its complementary template. Ribosome-bound phenylalanyl-tRNA, lysyl tRNA, or prolyl tRNA formed filter-bound ternary complexes only with their respective templates: poly(U), poly(A), or poly(C). Conversely, when ribosome-bound aminoacyl tRNA molecules were reacted with mismatched templates, binding to the filters remained at the background level (Fig. 10.15B).

Nirenberg and Leder next showed that oligoribonucleotides could replace polyribonucleotide in specific filter-retained three-party complexes. Most importantly, they demonstrated that ternary filter-retained specific complexes could be formed with oligomers as short as RNA triplets but that dinucleotides lost the ability to generate complexes (Fig. 10.16).

It should be pointed out, however, that Nirenberg and Leder found that complex formation and filter binding were consistently true for RNA oligomers (down to triplets) that had a terminal phosphate at their 5′ ends. By contrast, oligomers that had phosphate at their 2′ or 3′ end entered into filter-retained complexes at a much lower efficiency or did not form filter-retained complexes at all.[45]

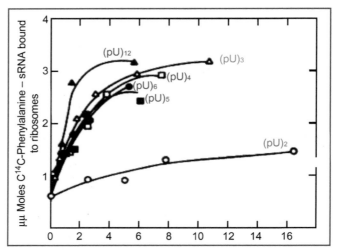

FIGURE 10.16 RNA oligomers down to triplets but not dinucleotides formed filter-retained tripartite complexes with ribosomes and a matching aminoacyl tRNA. Shown are the extents of binding to nitrocellulose filters of complexes of ribosomes, ^{14}C-phenylalanyl-tRNA, and 5′-phosphate oligo(U) chains of different lengths. Oligomers of 12, 6, 5, 4, and 3 nucleotides formed filter-bound complexes to similar degrees, whereas a dinucleotide failed to enter into complexes that were retained by the filter. *Modified from Leder P, Nirenberg M. RNA codewords and protein synthesis, III. On the nucleotide sequence of a cysteine and a leucine RNA codeword.* Proc Natl Acad Sci USA *1964;52(6):1521−9.*

The filter-binding procedure that Nirenberg and Leder established, and particularly their demonstration that RNA triplets formed filter-retained specific complexes with a matching ribosome-bound aminoacyl tRNA, was a breakthrough development. It is mostly this methodology that enabled Nirenberg and his associates (and in parallel Khorana and his group) to decipher within about a year the complete genetic code. Leder and Nirenberg made the first step in this direction almost immediately after they established the filter-binding assay. In a paper in the Aug. 1964 issue of the *Proceedings of the National Academy of Sciences,* they showed that the triplet GUU, but not its sequence isomers UGU or UUG, promoted binding of ribosome-associated ^{14}C-valyl-tRNA to cellulose nitrate filters.[46] The specificity of this binding was verified when ribosome-associated ^{14}C-phenylalanyl-tRNA (which formed robust complexes with poly(U)) or ^{14}C-leucyl tRNA failed to form filter-retained complexes with any one of the three tested triplets. Also, the filters did not retain any of the tRNA-linked amino acids when triplets were replaced by the dinucleotides GU, UG, or UU (Fig. 10.17A). GUU was identified, therefore, as a codon for valine.

(A)

TABLE 1
CODEWORD SPECIFICITY

Expt.	Addition, mμmoles base residues	C¹⁴-Amino-Acyl-sRNA Bound to Ribosomes, μμmoles		
		C¹⁴-Val-sRNA	C¹⁴-Phe-sRNA	C¹⁴-Leu-sRNA
1	None	0.38	0.22	0.37
	4.7 Poly U	0.23	4.73	0.22
	4.7 Poly UG	2.65	1.93	0.24
2	None	0.40	0.22	0.62
	4.7 GpUpU	1.11	0.27	0.56
	4.7 UpGpU	0.40	0.25	0.55
	4.7 UpUpG	0.37	0.25	0.44
3	None	0.18	0.22	0.69
	4.7 GpU	0.20	0.22	0.69
	4.7 UpG	0.19	0.21	0.71
	4.7 UpU	0.18	0.20	0.67

dinucleotides are ineffective

(B)

OLIGONUCLEOTIDE-AMINOACYL-sRNA SPECIFICITY

	μμMoles C¹⁴-Aminoacyl-sRNA Bound to Ribosomes				
			C¹⁴-Leu-sRNA		
Addition	S³⁵-Cys-sRNA	Unfractionated	II	IB	IA
None	0.29	1.08	0.76	0.40	0.30
UpGpU	1.46	0.96	0.78	0.40	0.28
UpUpG	0.32	1.02	1.74	0.36	0.34
GpUpU	0.34	0.86	0.92	0.34	0.30
pUpUpU	—	0.87	0.80	0.38	0.26
UpUpU	0.32	—	—	—	—
UpG	0.21	0.92	0.92	—	—
GpU	0.34	0.80	0.86	—	—
UpU	0.26	0.88	0.88	—	—

FIGURE 10.17   (A) The triplet GUU encoded valine. Only GUU (red rectangle) but not its sequence isomers UGU or UUG or the dinucleotides GU, UG, or UU promoted the formation of filter-retained, ribosome-bound ¹⁴C-valyl-tRNA (red circle). Control ribosome-bound ¹⁴C-phenylalanyl-tRNA formed filter-retained complex with poly(U) but not with any of the tested tri- or dinucleotides. Also, ribosome-bound ¹⁴C-leucyl tRNA failed to form filter-retained complex with poly(U) or poly(UG) or with any of the triplets or dinucleotides. (B) Degeneracy of the code for leucine. Only one of three different leucine-accepting tRNA species corresponded to the UUG codon. Filter binding was quantified in systems that contained different triplets or dinucleotides and ribosome-bound unfractionated or separated subtypes II, IB, or IA of ¹⁴C-leucyl tRNA. Specific binding of ribosome-bound ³⁵S-cysteinyl tRNA to the filters in the presence of the identified codon UGU served as a positive control. The filters did not bind unfractionated ¹⁴C-leucyl tRNA or its isolated subtypes IB or IA (*green rectangles*). By contrast, type II leucyl tRNA formed with ribosomes and the triplet UUG a specific complex that was retained by the filter (*small red circle*). Panels (A) and (B): *Adapted from Leder P, Nirenberg M. RNA codewords and protein synthesis, III. On the nucleotide sequence of a cysteine and a leucine RNA codeword. Proc Natl Acad Sci USA 1964;52(6):1521−9.*

10.4.3   The Filter-Binding Assay Provided Direct Demonstration of the Degeneracy of the Code

Using the filter-binding methodology, Leder and Nirenberg soon reported that the two sequence isomers, UGU and UUG which did not encrypt valine encoded cysteine and leucine, respectively.[47] Most importantly, they also discovered that leucine was carried by multiple tRNA species. How did they reach this conclusion? In the course of their identification of the codon for leucine, they noticed a paradoxical phenomenon. On the one hand, the synthetic heteropolyribonucleotide poly(UG) directed incorporation of ^{14}C-leucine into protein in a cell-free system.[47] By contrast, however, neither poly (UG) nor any of the isomeric triplets GUU, UGU, or UUG formed filter-retained complexes with ribosome-bound ^{14}C-leucyl tRNA (Fig. 10.17A).

This paradox was resolved when Leder and Nirenberg demonstrated that there were several species of leucyl tRNA and that the codon UUG was recognized by only one of these subtypes.[46] For their experiment, they received resolved species of leucyl tRNA from Bernard Weisblum, who was at the time a postdoctoral fellow in the laboratory of Seymour Benzer and from Günther von Ehernstein of the Johns Hopkins School of Medicine. These two researchers, who took part in the landmark experiment that proved that tRNA determines the identity of amino acid that is incorporated into protein[48] (see Chapter 8: "The Adaptor Hypothesis and the Discovery of Transfer RNA"), used countercurrent distribution to separate different species of leucyl tRNA.[49,50] This technique separated different types of tRNA on the basis of their different proclivities to be dissolved in mixtures of different ratios of immiscible organic and aqueous phases.[51−53] The three leucyl tRNA fractions marked IA, IB, and II that Weisblum and von Ehrenstein separated comprised 29%, 55%, and 15%, respectively, of the total leucine tRNA acceptor capacity.[47] In their experiment Leder and Nirenberg compared retention by filters of complexes of different base triplets or dinucleotides with ribosome-bound unfractionated or resolved subfractions IA, IB, and II of ^{14}C-leucyl tRNA. Results showed that the unfractionated ^{14}C-leucyl tRNA or its subtypes IA and IB did not bind to the filters with any of the tested triplets or nucleotide pairs. By contrast, however, ribosome-bound fraction II ^{14}C-leucyl tRNA formed complexes exclusively with the triplet UUG. Thus the triplet UUG was identified as a codon that was recognized by the minor type II of leucyl tRNA (Fig. 10.17B). Dilution of this minor UUG-recognizing type of leucyl tRNA (15% of the total) by the predominant other subspecies of leucyl tRNA that did not interact with UUG explained, therefore, the lack of complex formation between UUG and the unfractionated leucyl tRNA. Most importantly, this experiment demonstrated that multiple species of leucine-accepting tRNA existed, each recognizing a different codon for the same amino acid. This observation provided, therefore, direct evidence for degeneracy of the code.[47]

Just a few months after Leder and Nirenberg published their evidence for the degeneracy of the code,[47] their findings were substantiated and expanded by Weisblum, Gonano, von Ehernstein, and Benzer.[49] These four investigators separated by the countercurrent distribution technique five different leucine-accepting tRNA species. Two of these leucyl tRNA species, subtypes

I and IIB, similarly transferred leucine into protein under the direction of a poly(UC) template. To examine whether these two leucine-accepting tRNA species recognized different codons in natural mRNA, subtypes I and IIB of the tRNA were charged with ^{3}H- and with ^{14}C-leucine, respectively. The double-labeled leucyl tRNA molecules were added to an extract of rabbit reticulocytes that carried out cell-free synthesis of hemoglobin. The radiolabeled hemoglobin that was synthesized was isolated, α globin chains were separated and digested with trypsin, and ^{3}H- and ^{14}C-labels of leucine were detected in tryptic peptides of the protein. ^{3}H-leucine that was originally carried by leucyl tRNA type I was spotted in multiple peptides. By contrast, ^{14}C-leucine which was carried by type IIB tRNA was exclusively restricted to only one single α globin peptide.[49] The transfer of leucine from the two tRNA species into different positions along the chain of α globin was in line with the idea that each type of leucyl tRNA corresponded to a different leucine coding triplet in the α globin mRNA. This experiment directly validated, therefore, the general principle of the degeneracy of the code.[49]

10.4.4   1964−66: The Nirenberg Group Completed the Deciphering of the Code

Starting in late 1964, the Nirenberg laboratory launched a systematic quest to complete the full bilingual dictionary of codon sequences and of their corresponding amino acids. Their general methodology involved the use of *E. coli* extracts to charge tRNA with 19 unlabeled and a single radiolabeled amino acid. Twenty different mixtures, each containing a different radioactive amino acid, were interacted with ribosomes and a tested RNA triplet. The filter retention assay was then applied to determine which tRNA-linked-radiolabeled amino acid formed a filter-bound ternary complex with the ribosomes and the tested triplet. Applying this approach, Nirenberg, his post-doctoral fellows, and technicians built up a catalogue of degenerate RNA codons and their corresponding amino acids. The experimental results were neatly entered into Nirenberg's logbook (Fig. 10.18A, see also Section 10.5) and were published piecemeal in a series of papers that appeared between late 1964 and early 1966.[45−47,54−58] Tabulation of partial results in one representative paper is shown in Fig. 10.18B.

In a Jun. 1966 Cold Spring Harbor Symposium that was dedicated to the genetic code Nirenberg presented a nearly complete table of the code.[59] As described later, Khorana and his associates also imparted in the same symposium a virtually identical code that they deciphered independently by testing in the filter-binding assay chemically synthesized RNA triplets.[60] Thus the two laboratories independently verified the code, which Nirenberg and Khorana presented in its final definitive form in their respective 1968 Nobel lectures (Fig. 10.18C).

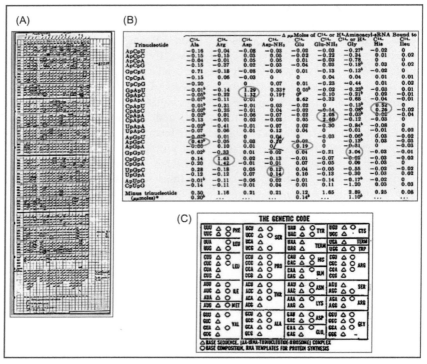

FIGURE 10.18 (A) Facsimile of a page from Nirenberg's laboratory logbook with tabulation of results of the retention on filters of complexes of ribosome-bound aminoacyl tRNA and tested RNA triplets. (B) Part of a table charting results of binding of different ribosome-bound tRNA-linked-radiolabeled amino acids to various RNA triplets. Instances of significant binding to the filters that signified recognition of specific triplets by particular aminoacyl tRNA species are *circled in red.* (C) Table of the complete genetic code that Nirenberg presented in his 1968 Nobel Prize lecture.[33] Triplet sequences that were determined by the filter-binding assay were marked by triangles and circles denoted corresponding base compositions of the triplets as determined by translation of synthetic templates in a cell-free protein synthesis system. *Panel (A): From the Marshall W. Nirenberg Collection, National Library of Medicine Profiles in Science, http://profiles.nlm.nih.gov/ps/access/JJBCCS_.jpg. Panel (B): Modified from Nirenberg M, Leder P, Bernfield M, et al. RNA codewords and protein synthesis, VII. On the general nature of the RNA code.* Proc Natl Acad Sci USA *1965;53(5):1161−8.*

In his 1966 Cold Spring Harbor Symposium presentation of the code Nirenberg marked the triplets UAG and UAA as potential termination codons because they failed to promote binding to filters of any ribosome-bound aminoacyl tRNA.[59] Indeed, shortly thereafter, Martin Weigert and Alan Garen[61] and independently Sydney Brenner and associates[62] employed genetic approaches to unequivocally identify the triplets UAG and UAA as termination ("nonsense") codons. Two years later Brenner, Barnett, Katz, and Crick identified the triplet UGA as a third nonsense codon.[63] Also in

1967 the future (2007) Nobel laureate Mario Cappechi[66] (1937−) and then Nirenberg and his associates[64,67] showed that these three "stop codons" terminated translation by binding protein release factors.

Another attribute of the code was uncovered when Brian Clark and Kjeld Marcker of the Cambridge MRC Laboratory of Molecular Biology identified in 1966 the protein synthesis initiation codon. They showed that when the triplet AUG was positioned at the 5′ terminus of bacterial mRNA it initiated protein synthesis by pairing with N-formylmethionyl tRNA.[65,68] However, when this triplet was located internally in the mRNA, it encodes methionine.[69]

10.4.5   The Deciphered Code Raised New Questions

Shortly after its deciphering, some fundamental facets of the code were deliberated.

Crick's Wobble Hypothesis

Taking into account the three termination codons, the number of remaining triplets that could encode amino acid was $64 − 3 = 61$. Naively it could be expected, therefore, that each cell would contain 61 different tRNA species, each having an anticodon complementary to a unique codon in mRNA. Experimental data (compiled in http://gtrnadb.ucsc.edu/) revealed, however, that cells of most of the living species have less than 45 different tRNA types. To explain how fewer than 61 tRNA species recognize the 61 amino acid−encoding triplets, Crick came up in 1966 with his so-called Wobble Hypothesis.[70] The crux of this hypothesis was that whereas the first two bases of a codon formed strict Watson−Crick bonds with the second and third bases in the anticodon, small movement ("wobble") of the first base of the anticodon might have caused the formation of noncanonical hydrogen bonds with the third base of the codon. Crick proposed that when the first base of the anticodon (which pairs the third base of the codon) was either A or C, it also formed Watson−Crick bonding with a complementary base in the codon. However, when U or G occupied the first position of the anticodon, they wobbled such that they formed noncanonical hydrogen bonds with the third base of the codon. This wobble allowed a single anticodon sequence in tRNA to recognize two codons that had two different bases in their third position.[70] Calculation indicated that with the implementation of the wobble principles, the 61 codons could be satisfied by a minimum number of 31 different tRNA species.

In addition to explaining how less than 61 different tRNA types suffice for the translation of 61 codons in mRNA, the wobble hypothesis also illuminated aspects of the recognition of the three-dimensional structure of tRNA by aminoacyl tRNA synthetase and also informed aspects of codon evolution and preference.

Universality of the Code

While he was completing the decoding project, Nirenberg inquired whether all the organisms shared one common code or did different species use different codons for the same amino acids. He and his associates found that the same codons were recognized by tRNA molecules from *E. coli* and from embryos of the amphibian *Xenopus nuriella* and from hamster liver cells.[59,71] Within a short time many other laboratories confirmed that the code was indeed nearly universal. Despite its near universality, some variations in the code were later documented in mitochondria and in certain microorganisms. Thus, for instance, instead of acting as a termination codon, the triplet UGA was found to encode tryptophan in yeast and humans mitochondria. All in all, the near universality of the genetic code opened the still unresolved question of its origin and evolution[72−76] (see also Section 10.8).

10.5 NIRENBERG'S STYLE OF DOING SCIENCE AND HIS PERSONAL DEMEANOR

A treasure trove of primary material on Nirenberg's life and work, papers, letters, and images is posted online in a dedicated site of the National Library of Medicine "Profiles in Science" project: http://profiles.nlm.nih. gov/ps/retrieve/Collection/CID/JJ. Here we touch on only few aspects of Nirenberg's scientific and personal qualities.

In the course of the charmed years 1961−1966, Nirenberg's code-deciphering endeavor expanded progressively. In the first year and a half of the project Nirenberg worked alone to single-handedly perfect the conditions for cell-free protein synthesis. At that early stage his objective was not yet to decrypt the code but to synthesize penicillinase in vitro. The turning point was the arrival of Heinrich Matthaei and the introduction of two "technical" modifications that changed the course of the work. The first change was that instead of monitoring cell-free protein synthesis by gauging increased activity of in vitro−synthesized penicillinase, it was measured by the incorporation of a radiolabeled amino acid into acid-insoluble material. The second modification was the addition of exogenous RNA to the cell-free system. Such RNA was first an mRNA-containing preparation of *E. coli* rRNA and later synthetic poly(U).[10,15] Experimentation with the modified systems led in turn to the breakthrough discovery that poly(U) encoded polyphenylalanine and thus that UUU was the codon for phenylalanine. This discovery put Nirenberg's work under the limelight of the scientific community and of the general public, and more importantly, it made him focus on the mission of breaking the genetic code. Recognizing the importance of the project, DeWitt Stetten (1909−1990) who was the Associate Director for Research at the NIH NIAMD, assigned to Nirenberg a technician—Linda Greenhouse—and new postdoctoral fellows joined in shortly thereafter. When the project expanded and the available laboratory space and personnel became restrictive, Nirenberg moved to another Institute within the NIH (NIAMDD) that offered more space, postdoctoral fellows, and technicians. Nirenberg described this expansion phase in his interview with Ruth Harris: (http://history.nih.gov/archives/downloads/Nirenberg%20oral%20history%20Chap%204%20-%20Race%20to%20decode.pdf).

As mentioned earlier, Nirenberg had the advantage of working in a supportive environment and several of his peers contributed to the progress of the deciphering project. Among those were Gordon Tomkins, Leon Heppel, Maxine Singer, and Robert Martin. In addition, because of the obvious importance of the work that was going on in his laboratory, Nirenberg attracted superb young researchers who came to the NIH in the mid-1960s to be trained in biomedical research. As history proved, many of the postdoctoral fellows that took part in the decoding project went on to make their personal mark in various areas of biomedicine. Among those were Philip (Phil) Leder (Box 10.5), Thomas (Tom) Caskey, Samuel Barondes, Sidney (Sid) Pestka, Merton (Mert) Bernfield, Edward (Ed) Scolnick, Oliver W. (Bill) Jones, and others (Nirenberg listed all of the 27 postdoctoral fellows in a 2004 scientific memoir[14]). Nirenberg also had the dedicated help of able technicians; first the aforementioned Linda Greenhouse and at later times Norma Zabriskie Heaton and Teresa Caryk.

The exceptionally exciting times in the Nirenberg laboratory during the deciphering of the code, the physical and personal makeup of the laboratory, and a portrayal of Nirenberg himself were vividly depicted by Norma Zabriskie Heaton's in her tribute to Nirenberg at a 2010 NIH memorial service in his honor (cited online at http://caroltorgan.com/genetic-code-decipherer-lab-life/). Some elements that were brought up in her following homage are illustrated in Fig. 10.19.

I would like to take you back to those early days and share with you what it was like working in the lab when the genetic code was deciphered. The lab work was intensive, exciting, and downright fun. But it could also be frustrating, stressful and tedious. How many people have the opportunity to work on a project that is so significant and fulfilling? The challenge sometimes was to see beyond the repetition to the goal. Sometimes I thought that if I had to do one more binding assay protocol I'd scream. But the repetition was just Marshall's obsession with accuracy and reproducibility.

Marshall was modest, soft-spoken and kind. He was a true gentleman and it was his nature to be generous with praise. He was very creative and had so many ideas—he liked to try them out with fast little experiments just to see where they might lead—he called them "quickies." Marshall was very focused and demanding. He had a meticulous and painstaking approach to data analysis and when you discussed your results with him, he expected you to know your experiments inside and out, backward and forward. No detail was

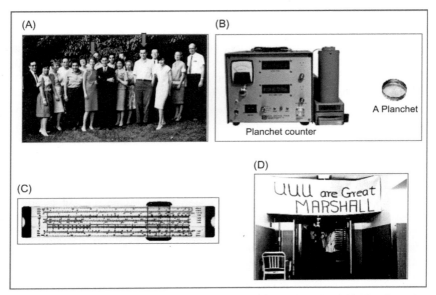

FIGURE 10.19 (A) Group portrait of the Nirenberg laboratory in early 1964. The picture has many of the persons (and some of their spouses) that Norma Heaton mentioned in her tribute. Red arrows mark Norma Heaton and Marshall Nirenberg. The photo is reproduced from Ref. 14. (B) Planchet radioactivity counter. Dried filters were placed in aluminum planchets such as the one shown at right out of scale. The filter-containing planchet was placed in a sliding tray at the bottom of the brown gas-flow cylinder of the counter, and the tray was pushed in to start the counting of radioactivity. Cumulative counts and counts per minutes appeared in the two red windows of the counter and the numbers were copied by hand into a notebook (no printer yet!). (C) A sliding rule such as those that were routinely used to calculate results (no electronic calculators yet!). (D) A party at the Nirenberg laboratory on the day that he was proclaimed laureate of the 1968 Nobel Prize. The banner aptly said: "UUU are Great MARSHALL." *Panel (A): From Nirenberg M. Historical review: deciphering the genetic code—a personal account.* Trends Biochem Sci *2004;29(1):46–54. Panel (D): From the Marshall W. Nirenberg Collection, National Library of Medicine Profiles in Science, http://profiles.nlm.nih.gov/ps/access/JJBCCQ_.jpg.*

overlooked or insignificant to him, and you had better know which solutions you used, who made them, when they were made, how they were made and the lot number of every reagent. Marshall didn't keep the same hours as the rest of the lab. He might arrive anywhere from mid-morning to early afternoon and he would always work very late into the night. "How goes it?" was a typical greeting. He wrote out the protocols to follow and left them on the lab bench. His handwriting could be challenging to decipher, but, like learning a foreign language, it became easier with time. He told me it was because he was a natural lefty, forced to learn to write with his right hand.

    The labs were all crowded—we didn't work side by side—it was elbow to elbow. It was so busy that Theresa labeled our TCA reagent in Ukrainian so that it didn't wander off. Marshall had a tiny office in the back part of our single module lab just wide enough for his desk, two chairs and a file cabinet, and you usually had to clear off a chair in order to sit down.

Marshall and Phil Leder devised what became known as the binding assay. This was the assay that we went on to use to decipher the 64 triplets of the genetic code. Marshall was always pleased when we could end the week with a successful experiment, so when I ran the first binding assay just before the Thanksgiving holiday in 1963, it was a high note. I will never forget hearing the shouts, especially from Ed Scolnick and Mert Bernfield, when the latest codon result would come off the Nuclear Chicago planchet counter tucked into a tiny alcove in the hallway. Marshall was always just as delighted but his reaction was more restrained—he might pump his fist and say "wonderful".

Some of you will recall that back in the olden days, there were no calculators so we used slide rules to get our results. This was before the days of personal computers and spreadsheets so tabulating a summary of our results as each triplet codon was deciphered was challenging. I taped enough data paper together to create very large charts and drew the columns and rows with a ruler. Then, I painstakingly entered all the data from our experiments by hand. The resulting charts would become the "Rosetta Stone of the Genetic Code" and some of them are now in the Smithsonian and the Library of Medicine.

Manuscripts were written and rewritten, typed and retyped . . . . . . often up until the last moment before the deadline. This was before the days of e-mail and fax machines, so when Marshall and the secretary finished his paper for the Lasker Foundation award at 3 a.m. the morning of its deadline, Marshall left me a plane ticket and I flew to New York and hand delivered his paper to the Lasker Foundation headquarters. The very next morning was Wednesday October 16th, 1968 and it was announced that Marshall had won the Nobel Prize.

10.6   THE INDEPENDENT DECRYPTION OF THE CODE BY THE MASTER BIOORGANIC CHEMIST HAR GOBIND KHORANA

In parallel to Nirenberg, the University of Wisconsin bioorganic chemist Har Gobind Khorana also attacked the problem of the genetic code (Box 10.6).

BOX 10.6 Har Gobind Khorana

Khorana was born in 1922 to Hindu parents in a Punjab village in a region of present day Pakistan. He was homeschooled by his father, the village "patwari"—an equivalent of taxation official. In the respective years 1943 and 1945 he got his BSc and MSc degrees from Punjab University in Lahore. Moving to England, he earned in 1948 a PhD degree from the University of Liverpool. He then did research in several institutions starting at the Swiss Federal Institute of

(Continued)

BOX 10.6 (Continued)

Technology in Zürich (1948–49) and continuing in Cambridge (1950–51) where he started his studies of nucleic acids under the 1957 Nobel Prize laureate Lord Alexander Todd (1907–97). In 1952 he was recruited by the University of British Columbia in Canada and in 1960 he moved to the University of Wisconsin, Madison. There he worked with a large group of researchers in the Institute for Enzyme Research to decipher the code. After completing the decoding work in 1965 and having been awarded the 1968 Nobel Prize in Physiology and Medicine, he embarked on the stepwise synthesis of a gene for yeast alanine tRNA. This work, which was completed in 1972, was followed by total synthesis of a functioning gene for the precursor to *E. coli* tyrosine suppressor tRNA. In 1970 he was made the Sloan Professor of Biology and Chemistry at the MIT where he studied bacterial and mammalian rhodopsin, a subject that he pursued until his retirement in 2007 (Fig. 10.20).

Har Gobind Khorana
(1922–2011)

FIGURE 10.20   Har Gobind Khorana.

Khorana drew inspiration to pursue the question of the code from the Nirenberg and Matthaei 1961 poly(U) experiment.[15] However, whereas Nirenberg took a purely biochemical approach to the synthesis of first polyribonucleotides and later of RNA triplets, Khorana created these molecules by combining organic synthesis with enzyme-catalyzed synthesis. The two laboratories similarly determined the coding specificities of polyribonucleotides and of RNA triplets by, respectively, determining their coding specificities in cell-free peptide synthesis system and by the formation of specific filter-retained complexes of ribosome-bound aminoacyl tRNA and RNA triplets.[45]

The intense parallel efforts of Nirenberg and Khorana culminated in 1965–66 with the announcements by the two laboratories of their successful deciphering of the complete code. Satisfactorily, although Nirenberg and

Khorana used different methods to prepare their RNA triplets, the final codes that they exposed were identical. This last section of the chapter is dedicated to the work and findings of Khorana and his group.

10.6.1 Khorana's Stepwise Synthesis of Coding Polyribonucleotides

Khorana combined organochemical and enzymatic methods to synthesize RNA molecules of defined nucleotide sequences and then used these RNA chains as templates for cell-free protein synthesis. At the time that the work started, the chemistry of direct synthesis of RNA was still undeveloped (see Section 10.6.3). Efforts in the Khorana laboratory between the late 1950s and the first third of the 1960s were, therefore, dedicated in large part to the development of organochemical methods for the synthesis of short chains of oligodeoxyribonucleotides with predetermined nucleotide sequences. These chemically synthesized short DNA chains either served as direct templates for the synthesis of RNA by RNA polymerase or were first extended by DNA polymerase into long polydeoxyribonucleotides that were then transcribed by RNA polymerase into polyribonucleotide chains (Fig. 10.21A). Khorana and his large group of postdoctoral fellows dedicated several years to the development of methods for the chemical synthesis of DNA oligomers and to their enzyme-catalyzed polymerization and transcription into polyribonucleotides. They documented the stepwise evolution of these procedures in scores of papers that were mostly published in the *Journal of the American Chemical Society*. Also, Khorana outlined in several review articles the principles of his strategies for the synthesis of polyribonucleotides of predetermined base sequences.[79–81]

Chemical Synthesis of Short DNA Chains

Chemical synthesis of DNA oligomers proceeded along the following steps: (1) Protecting groups were placed at unwanted potential reactive groups of deoxynucleotides such as the primary or secondary hydroxyl groups in the deoxyribose ring, amino groups in the purine or pyrimidine rings, and phosphoryl in the phosphomonoester group (for selected papers on various protection strategies that the Khorana group took, see Refs. 82–87). (2) The protected deoxyribonucleotides were used as substrates in condensation reactions that ultimately yielded short DNA chains with predetermined deoxyribonucleotide sequences. In general, condensation of nucleotides occurred through the activation of a phosphoryl group in one protected mononucleotide such that it attacked a 3'-hydroxyl group in another protected nucleotide, forming a phosphodiester-linked dinucleotide. Similar condensation procedures were employed to also extend the dinucleotides by one or two additional nucleotides. The resultant core units that consisted of predetermined

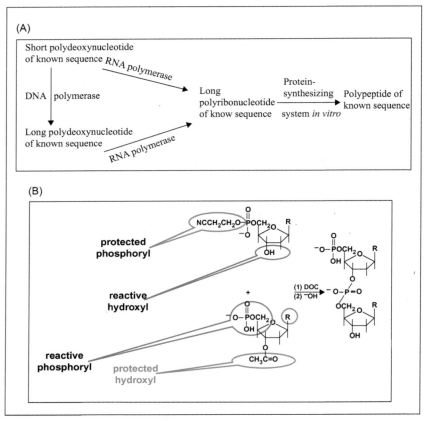

(A)

Short polydeoxynucleotide
of known sequence *RNA polymerase*

DNA | polymerase   →   Long
polyribonucleotide
of know sequence

Protein-
synthesizing   Polypeptide of
———————→   known sequence
system *in vitro*

Long polydeoxynucleotide
of known sequence   *RNA polymerase*

(B)

protected
phosphoryl

reactive
hydroxyl

reactive
phosphoryl   protected
hydroxyl

FIGURE 10.21   (A) Strategy of synthesis of polyribonucleotides of predetermined nucleotide sequences and their translation into protein. Oligodeoxyribonucleotides that were chemically synthesized were either transcribed directly by RNA polymerase into complementary RNA chains or extended by DNA polymerase into longer polydeoxyribonucleotides that were then transcribed by RNA polymerase into polyribonucleotide chains. The RNA transcripts of the short or extended oligodeoxyribonucleotides served as templates for cell-free peptide synthesis. Scheme reproduced from Ref. 77. (B) Organochemical synthesis of AT, GT or CT dinucleotides. Reactive 5' phosphoryl group in a protected R' mononucleotide such as *N*-benzoyladenine, *N*-acetylguanine, or *N*-anisoylcytosine, attacked in the presence of the condensing agent dicyclohexylcarbodiimide (DCC) the 3' hydroxyl group in thymidine (marked "R") forming a phosphodiester bond (the reactive groups are *circled in red*). Both the attacked and attacking mononucleotides in this example had protecting groups (*circled in green*) on the hydroxyl (3'-*O*-acetyl) or phosphoryl groups (5'-β-cyanoethyl) of their respective deoxyribose rings. *Panel (A): From Khorana HG. Nucleic acid synthesis in the study of the genetic code. Nobel Nobel lecture December 12, 1968. Amsterdam: Elsevier Publishing Company; 1972. Panel (B): Modified from Ohtsuka E, Moon MW, Khorana HG. Studies on polynucleotides. XLIII. The synthesis of deoxyribopolynucleotides containing repeating dinucleotide sequences.* J Am Chem Soc 1965;**87** *(13):2956−70.*

two, three, or four nucleotides were then ligated into longer chains of repeating units of di-, tri-, or tetranucleotides. Combinations of nucleotides in the synthesized DNA were chosen on the basis of the predicted structure of its RNA transcripts such that combinations that were predicted to form internal Watson–Crick hydrogen bonds were excluded.[81] General principles of the chemical synthesis of DNA oligomers[81] and technical details of the synthesis of specific repeating sequences[78,88−92] were documented in a series of papers that the Khorana laboratory published in the first half of the 1960s. (3) In the last step the polymerization products were purified and their base sequences were verified.

Fig. 10.21B exemplifies the chemical synthesis of a short DNA chain with a preset dinucleotide repeat. The first step of this process was condensation of two protected mononucleotides into a dinucleotide of predetermined arrangement. In the shown case, thymidine was reacted in the presence of the condensing agent dicyclohexylcarbodiimide (DCC) with each of the remaining three nucleotides to form AT, GT, or CT dinucleotides.[78] Khorana explored two approaches for the creation of extended chains of alternating nucleotides. One strategy was a one-by-one condensation of additional nucleotides into a chain of increasing length. The other tactic was the joining together of short blocks of di-, tri-, or tetranucleotides. Thus, for instance, incubating each dinucleotide with DCC for 7 days generated longer chains of dinucleotide repeats. Found to be more efficient,[93,94] the linking together of blocks of two, three, or four nucleotides became the strategy of choice. After removal of the protecting groups with ammonia, oligomeric products of different lengths were separated by column chromatography and their compositions and nucleotide sequences were verified by enzymatic degradation.[78] Modifications of this general procedure allowed synthesis of seven sets of DNA chains of repeating trinucleotides[90−92,95,96] and two sets of tetranucleotides[94,97] (Fig. 10.22A).

Enzyme-Catalyzed Synthesis of DNA and RNA Chains of Predetermined Sequences

Transcription by RNA polymerase of the chemically synthesized short DNA chains was employed to produce RNA chains of predetermined repeating nucleotide sequences. This was done in one of two ways. The synthetic DNA oligomers either served as direct templates for RNA polymerase or they were first replicated and extended by DNA polymerase to generate long DNA chains that were then transcribed by RNA polymerase. Experience in the Khorana laboratory proved that the use of DNA polymerase-generated long DNA molecules as templates for RNA polymerase was more economical and that their transcription yielded higher amounts of RNA.[98] The replication and extension by DNA polymerase I from *E. coli* of short DNA chains of different defined nucleotides sequences were described in several papers.[99,100] Complementary short DNA chains of dissimilar or equal lengths

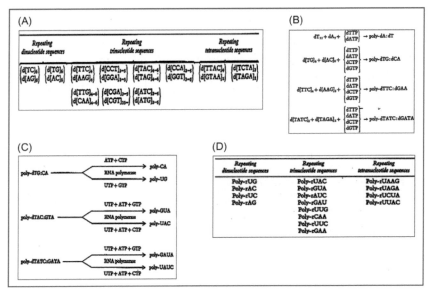

FIGURE 10.22   (A) Chemically synthesized short chains of repeating two, three, and four deoxyribonucleotides. (B) DNA polymerase-catalyzed synthesis of repeating nucleotide DNA sequences. Complementary chemically synthesized short DNA chains of equal or different lengths were annealed and extended and amplified by *Escherichia coli* DNA polymerase I. (C) Synthesis by RNA polymerase of RNA chains with repeating nucleotide sequences. Duplex DNA molecules that were prepared as in (B) served as templates for RNA polymerase. The identity of the transcribed DNA strand was determined by choice of complementing ribonucleotide triphosphate substrates. (D) RNA polymerase-generated RNA chains of repeating nucleotide sequences. Panels (A)–(D) are from Ref. 77. *Panels (A)–(D): From Khorana HG.* Nucleic acid synthesis in the study of the genetic code. Nobel Nobel lecture December 12, 1968. *Amsterdam: Elsevier Publishing Company; 1972.*

were annealed and then replicated through priming or slippage of the replicated strands to yield longer duplex DNA molecules (Fig. 10.22B). The generated long DNA chains were next used as templates for RNA polymerase. Furnishing the RNA polymerase with chosen complementary ribonucleotide triphosphates forced the enzyme to transcribe only one of the two complementary DNA strands[101–104] (Fig. 10.22C). Using this practice, the Khorana group produced RNA polymers that consisted of predetermined repeats of two, three, or four nucleotides (Fig. 10.22D).

10.6.2   The Coding Problem Seized: Determination of the Amino Acid Composition of Peptides Translated In Vitro From Synthetic Polyribonucleotide Templates

The laboratories of Khorana, Nirenberg, and Ochoa initially attacked the coding problem by taking a seemingly similar approach of determining the amino acid–coding specificities of synthetic polyribonucleotide templates. However, whereas the Nirenberg and Ochoa groups generated such templates with PNP, Khorana employed organochemical methods to synthesize RNA molecules with predetermined nucleotide sequences. Specifically, Khorana synthesized RNA that was comprised of defined di-, tri-, or tetranucleotide repeats that were then translated in vitro to produce respective co-peptides that were composed of two, three, or four repeating amino acids. Thus, although Nirenberg, Ochoa, and Khorana employed a similar general strategy of using polyribonucleotide templates in cell-free protein synthesis system, Khorana had the advantage of using chemically synthesized RNA templates whose nucleotide sequences were precisely predetermined. By contrast, the sequences of the RNA templates that Nirenberg and Ochoa synthesized with RNA PNP could be only statistically estimated.

Initially Khorana and his associates conducted cell-free incorporation of labeled amino acids into protein under the direction of polyribonucleotide templates that were comprised of dinucleotide repeats. They found that because of the two alternative reading frames of such RNA sequences, their peptide products were comprised of pairs of alternating amino acids. For instance, the synthetic template poly(UC) directed the incorporation of both ^{14}C-serine and ^{14}C-leucine and both amino acids had to be present in the reaction mixture in order that synthesis would take place (Fig. 10.23A).

This result indicated that the two alternative reading frames of poly(UC), that is, CUC and UCU directed synthesis of a leucine–serine co-peptide. Similar synthesis of co-peptides of pairs of other amino acids was documented for poly(AG), poly(UG), and poly(AC) (Fig. 10.23B). Last, RNA sequences of three nucleotide repeats had three alternative reading frames. For example, the template poly(UUC) in which three triplet codons could be read; UUC, CUU, and UCU directed synthesis of a co-peptide of three amino acids: phenylalanine, serine, and leucine. Other trinucleotide repeats encoded co-peptides of triads of other amino acids (Fig. 10.23C).

Several facets of these experiments should be noted:

Limits to Codon Assignment: Similar to Nirenberg's experiments of the same type (Section 10.3.6), the analysis of polypeptides that were translated in vitro from synthetic polyribonucleotide templates enabled Khorana to determine the base compositions of codons but not their exact nucleotide sequences. For instance, one each of the two triplets, CUC and UCU, encoded serine and leucine. However, it could not be determined which the codon for serine was and which triplet encoded leucine (Fig. 10.23B). Yet, in some cases codon assignments could be logically deduced by identifying a common amino acid in peptides that were translated from templates that had two or three different repeating sequences.

Direct Identification of Chain Terminating Codons: An important aspect of Khorana's results was his success in the direct identification of chain terminating (nonsense) triplets. At the time that he was conducting experiments

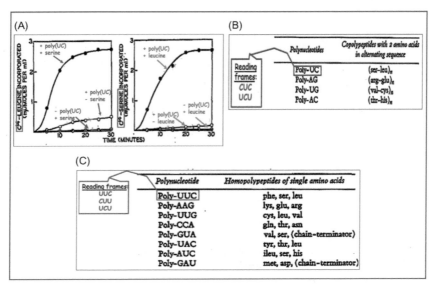

FIGURE 10.23   (A) Poly(UC) template directed incorporation in vitro of both leucine and ser-ine. The cell-free protein synthesis was furnished with either ^{14}C-leucine or ^{14}C-serine. Polypeptide synthesis depended on the presence of the template and of both amino acids (^{14}C-leucine + unlabeled serine or ^{14}C-serine + unlabeled leucine). Thus, poly(UC) appeared to encode a leucine−serine co-peptide. (B) The coding specificities of four dinucleotide repeats in RNA. Experiments such as in (A) were conducted for the shown four dinucleotide repeats—all of which encoded co-polypeptides of two alternating amino acids. The observed result was explained by the two alternative reading frames of every dinucleotide repeat such as the in the shown example of the CUC and UCU triplets in poly(UC). (C) The three alternative reading frames of trinucleotide repeats directed the synthesis of co-peptides of three different amino acids or of two amino acids and a termination signal. *Panels (A)−(C): Modified from Khorana HG. Synthetic nucleic acids and the genetic code. JAMA 1968;206(9):1978−82.*

on the in vitro translation of synthetic polyribonucleotides, genetic analyses by Weigart and Garen[61] and by Brenner and associates[62] of *"ochre"* and *"amber"* mutations in *E. coli* identified the two respective triplets UAA and UAG as polypeptide-terminating signals. To demonstrate directly that these trinucleotides acted as stop codons, Khorana translated in vitro synthetic RNA templates that were comprised of tetranucleotide repeats. One repeat template, poly(UUAC) contained a UAA trinucleotide and another template, poly(UAGA) contained a UAG triplet. UAA and UAG that appeared in every fourth position of the reading frame of the respective RNA templates were predicted to terminate translation. Products of the translation of these templates would be expected to be peptides of only two or three amino acids. Indeed, consistent with this prediction, cell-free translation of the UAA- and UAG-containing respective templates poly(GUAA) and poly(GAUA) failed to yield acid-insoluble peptides.[106]

Identification of the Initiation Codon: As already noted in the context of Nirenberg's work, translation in vitro of synthetic polyribonucleotides was feasible only because the reaction mixtures contained higher than physiological concentrations of magnesium ions. However, in 1964 and 1965 it was discovered that protein synthesis under physiological conditions of lower magnesium concentration was initiated by the binding of *N*-formylmethionyl tRNA to the 5' terminal codon in mRNA.[65,68,69] Khorana and associates showed that polypeptide synthesis in cell-free system that contained low (4 mM) concentration of magnesium required the presence of an initiator codon in the RNA template and of *N*-formylmethionyl tRNA.[107] By assessing the capacity of several synthetic polyribonucleotides to direct cell-free polypeptide synthesis at low magnesium concentration and by monitoring their ability to dictate incorporation of *N*-formyl methionine at the N-terminus of the synthesized polypeptides, the Khorana group identified AUG, GUG, and GUA as chain initiation signals.[107] Although their results were experimentally valid, it was established later that only AUG acted as the physiological initiation codon.

10.6.3 Khorana's Crowning Achievement: Decryption of the Highly Degenerate Code and Assignment of Codons for All the 20 Amino Acids

In the earlier stages of his work, Khorana circumvented technical difficulties that prescribed direct chemical synthesis of RNA by first preparing polydeoxyribonucleotides and then using RNA polymerase to generate RNA chains. The main obstacle to organochemical synthesis of RNA was the daunting problem that the 2'-hydroxyl group in the ribose ring created. In time, however, the Khorana group developed methods that enabled chemical synthesis of short chains of RNA. Most importantly, they succeeded in chemically synthesizing all the 64 RNA triplets. In the ultimate step Khorana applied the Nirenberg and Lederfilter-binding assay to determine the amino acid—coding specificities of the synthetic triplets.

Starting in the early 1960 the Khorana laboratory explored several alternative approaches to the chemical synthesis of oligoribonucleotides.[108–116] By late 1965 Khorana settled on the most satisfactory method for stepwise chemical synthesis of all the 64 possible RNA triplets. Essential features of the chosen approach are charted in Fig. 10.24A.

In the first step, a diribonucleotide was synthesized by condensing in the presence of DCC a protected ribonucleoside 3'-phosphate with a protected ribonucleoside that carried a free 5'-hydroxyl group. In the next step a third ribonucleoside 3'-phosphate was condensed with the dinucleotide. Ultimately the protecting groups were removed by ammonia and the trinucleotides were resolved using paper chromatography and electrophoresis.[117]

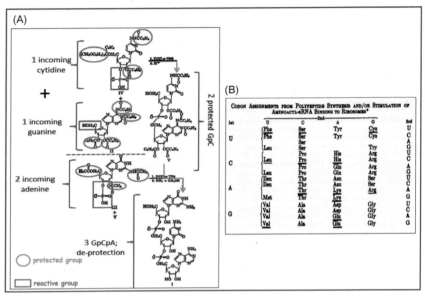

FIGURE 10.24 (A) Stepwise synthesis of aguanylyl $(3' \to 5')$ cytidyl $(3' \to 5')$ adenosine. The condensation steps are numbered in their order of execution and protected groups are *circled in green* and reactive groups are bordered by *red rectangles*. Scheme adapted from Ref. 117. (B) Khorana's initial tabulation of the genetic code. Codon assignments that were determined by the filter-binding assay are not underlined. The assignments of CUC for leucine and AGA for arginine *(thin underlines)* were made on the basis of cell-free translation of synthetic templates. The assignments marked by thick lines were derived from results of both in vitro polypeptide synthesis and filter binding. The Table is reproduced from Ref. 118. *Panel (A): Modified from Lohrmann R, Soll D, Hayatsu H, Ohtsuka E, Khorana HG. Studies on polynucleotides. LI. Syntheses of the 64 possible ribotrinucleotides derived from the four major ribomononucleotides. J Am Chem Soc 1966;88 (4):819–29. Panel (B): From Söll D, Ohtsuka E, Jones DS, et al. Studies on polynucleotides, XLIX. Stimulation of the binding of aminoacyl-sRNA's to ribosomes by ribotrinucleotides and a survey of codon assignments for 20 amino acids. Proc Natl Acad Sci USA 1965;54(5):1378–85.*

Using their 64 synthetic trinucleotides, the Khorana group went on to assign codons for each amino acid mostly by employing the filter-binding assay of Nirenberg and Leder,[45] and in some cases by determining the amino acid compositions of polypeptides that were translated in a cell-free system from synthetic polyribonucleotide templates. For the binding assay, formation of filter-retained ternary complex with a specific ribosome-bound aminoacyl tRNA of each RNA triplet was determined for each RNA triplet.

Results of the filter-binding assays and some data that were obtained by cell-free peptide synthesis enabled Khorana and his associates to list in a single 1965 paper the codons for all the 20 amino acids.[118] The highly degenerate code that they acquired was largely equal to the code that Nirenberg decrypted (compare Fig. 10.24B with Fig. 10.18C, respectively). After he completed the deciphering the code in *E. coli*, Khorana predicted that it would prove to be the same in other organisms. This prognostication, which had also been made by Nirenberg (Section 10.4.5), proved in time to be fundamentally correct.

10.7   AFTER BREAKING THE CODE, KHORANA UNDERTOOK NEW DIFFICULT CHALLENGES

In 1968 Khorana, who was then 46 years old, shared a Nobel Prize in Physiology and Medicine with Nirenberg and Holley. This award marked the end of a period of around-the-clock double-shift efforts by his large team at the University of Wisconsin to decipher the code. Almost immediately after finishing the code work, Khorana launched a new ambitious project to chemically synthesize a complete gene for transfer RNA. This work, which had started in the Institute for Enzyme Research in the University of Wisconsin, was completed after Khorana moved in 1970 to the MIT as its Alfred P. Sloan Professor of Biology and Chemistry. The monumental task of total chemical synthesis of the double-stranded gene for yeast alanine tRNA was completed in 1972. The Dec. 28, 1972, issue of the *Journal of Molecular Biology* was entirely occupied by 15 sequential papers that described individual steps of the synthesis. The first of these articles outlined the strategy that Khorana had taken for the synthesis of the 77 nucleotide—long tRNAgene.[119] In the first stage 15 oligodeoxynucleotide segments of 5—20 nucleotides each were synthesized. In the next phase these segments were ligated into three sections of the duplex DNA. The ultimate step was then the joining together of the three sections into a complete 77 unit-long duplex DNA.[119] Although the synthetic alanine tRNA gene possessed the correct nucleotide sequence, it lacked appropriate signals for the initiation and termination of transcription. In the next few years the Khorana group undertook the total chemical synthesis of the 126 nucleotide—long DNA for the precursor to *E. coli* tyrosine suppressor tRNA. This work was finished in 1979 and the intricate steps of the synthesis were detailed in 17 back-to-back papers in the *Journal of Biological Chemistry*. In the last of this series of papers, Khorana and associates reported the successful in vitro transcription of the synthetic gene and the processing of the primary transcript into mature tRNA.[120] The prodigious achievement of chemically synthesizing a complete functioning gene won popular acclaim. With his picture on the cover of the Sep. 16, 1976, issue of *Time* magazine, Khorana and his work received a glowing description:

> *Like the sages of his native India, Organic Chemist and Nobel Laureate Har Gobind Khorana is an extremely patient man. Nine years ago, he began working on the chemical synthesis of a single gene—the basic unit of heredity. By 1970 he had constructed a yeast-cell gene identical to the original—except for*

one thing: it lacked the vital "start" and "stop" signals to make it function in a living cell. Last week members of Khorana's team at the Massachusetts Institute of Technology disclosed that his goal had finally been achieved [...].

Khorana's commitment to new challenges was demonstrated in the last phase of his scientific career. At that time he ended his lifetime work on nucleic acids to explore signaling pathways of vision in vertebrates. Initially he studied light-sensitive bacteria, focusing on the purple membranes of *Halobacterium halobium*. He and his many associates characterized rhodopsin from this bacterium and clarified the mechanism of the pumping of protons through the membrane by the light-activated bacteriorhodopsin. These investigations led in turn to studies on the structure and function of mammalian rhodopsin and on mutations in this protein that were associated with the disorder *retinitis pigmentosa*.

Khorana's exceptional enthusiasm for science, his pioneering spirit, drive, modesty, and remarkable career were vividly portrayed after his 2011 death by colleagues and past members of his group.[121–124]

10.8 POSTSCRIPT: THE UNSOLVED QUESTIONS OF THE ORIGIN OF THE CODE AND OF THE EVOLUTION OF THE TRANSLATION MACHINERY

The efforts to first define the code (see Chapter 10: "Defining the Genetic Code") and then to decrypt it (this chapter) bore far-reaching consequences. The tabulation of all the code words and their corresponding encoded amino acids was one of the greatest achievements of the "golden age" of molecular biology. In 1966, soon after the code became fully deciphered, the 1958 Nobel laureate George Beadle (1903–89) and his writer wife Muriel Beadle-McClure (1915–94) conveyed in their book *The Language of Life* the sense of wonder that the discovered code had raised[125]:

The DNA code has revealed our possession of a language much older than hieroglyphics, a language as old as life itself, a language that is the most living language of all—even if its letters are invisible and its words are buried deep in the cells of our bodies.

Space would not permit a full discussion of the immense impact that the breaking of the code had on the development and achievements of modern biology, biotechnology, and medicine. These accomplishments reach far beyond the confines of the laboratory and have direct bearing on many aspects of our lives. To remain, however, within the realm of basic science, it must be pointed out that almost immediately after the challenge of breaking the code was met, an even more complex problem of the origin of the nearly universal code was raised. Broadly, knowledge of the extant code and of the translation machinery does not provide direct clues about the historical processes that led to the selection of a virtually universal code and to its evolution before, or side-by-side with the translation apparatus. Treatment of this unresolved enigma is much beyond the scope of this chapter. The interested reader may wish to consult selected early[72,73,126-131] and more current[74-76,132-135] speculations, theories, and experiments on the origin of the code and of the translation apparatus.

REFERENCES

1. Siekevitz P. Uptake of radioactive alanine in vitro into the proteins of rat liver fractions. *J Biol Chem* 1952;**195**(2):549-65.

2. Nismann B, Bergmann FH, Berg P. Observations on amino acid-dependent exchanges of inorganic pyrophosphate and ATP. *Biochim Biophys Acta* 1957;**26**(3):639-40.

3. Spiegelman S. Nucleic acids and the synthesis of proteins. In: McElroy WD, Glass B, editors. *A symposium on the chemical basis of heredity*. Baltimore, MD: The Johns Hopkins Press; 1957. p. 232-67.

4. Brown GL, Brown AV. Fractionation of deoxyribonucleic acids and reproduction of T2 bacteriophage. *Symp Soc Exp Biol* 1958;**12**:6-30.

5. Simkin JL. Protein biosynthesis. *Annu Rev Biochem* 1959;**28**:145-70.

6. Lamborg MR, Zamecnik PC. Amino acid incorporation into protein by extracts of *E. coli*. *Biochim Biophys Acta* 1960;**42**:206-11.

7. Tissieres A, Schlessinger D, Gros F. Amino acid incorporation into proteins by *Escherichia coli* ribosomes. *Proc Natl Acad Sci USA* 1960;**46**(11):1450-63.

8. Portugal FH. *The least likely man: Marshall Nirenberg and the discovery of the genetic code*. Cambridge, MA: MIT Press; 2015.

9. Grolle J. Des ganzen wirklichkeit. *Der Spiegel* 2012;**1**:128-30.

10. Matthaei H, Nirenberg MW. The dependence of cell-free protein synthesis in *E. coli* upon RNA prepared from ribosomes. *Biochem Biophys Res Commun* 1961;**4**(6):404-8.

11. Matthaei JH, Nirenberg MW. Characteristics and stabilization of DNAase-sensitive protein synthesis in *E. coli* extracts. *Proc Natl Acad Sci USA* 1961;**47**(10):1580-8.

12. Jacob F, Monod J. Genetic regulatory mechanisms in the synthesis of proteins. *J Mol Biol* 1961;**3**(3):318-56.

13. Kameyama T, David Novelli G. The cell-free synthesis of β-galactosidase by *Escherichia coli*. *Biochem Biophys Res Commun* 1960;**2**(6):393-6.

14. Nirenberg M. Historical review: deciphering the genetic code—a personal account. *Trends Biochem Sci* 2004;**29**(1):46-54.

15. Nirenberg MW, Matthaei JH. The dependence of cell-free protein synthesis in *E. coli* upon naturally occurring or synthetic polyribonucleotides. *Proc Natl Acad Sci USA* 1961;**47**(10):1588-602.

16. Judson HF. *The eighth day of creation*. Expanded ed. New York, NY: Cold Spring Harbor Laboratory Press; 1996.

17. Tsugita A, Fraenkel-Conrat H, Nirenberg MW, Matthaei JH. Demonstration of the messenger role of viral RNA. *Proc Natl Acad Sci USA* 1962;**48**(5):846-53.

18. Nirenberg MW, Matthaei JH. The dependence of cell-free protein synthesis in *E. coli* upon naturally occurring or synthetic template RNA. In: Engelhardt VA, editor. *Biological structure and function at the molecular level*. Moscow: MacMillan Co; 1961. p. 184-9.

19. Kay LE. *Who wrote the book of life? A history of the genetic code.* Stanford, CA: Stanford University Press; 2000.

20. Jenkins T, Arthur G. Steinberg, 1912–2006. *Am J Hum Genet* 2007;**80**(6):100913.

21. Tiselius A. *Discours solonnel. Les prix nobel en 1961.* Stockholm: Norstedt & Söner; 1962. p. 14–15.

22. Nirenberg MW, Matthaei JH, Jones OW. An intermediate in the biosynthesis of polyphenylalanine directed by synthetic template RNA. *Proc Natl Acad Sci USA* 1962;**48**(1):104–9.

23. Roberts RB. Alternative codes and templates. *Proc Natl Acad Sci USA* 1962;**48** (5):897–900.

24. Maxwell ES. Stimulation of amino acid incorporation into protein by natural and synthetic polyribonucleotides in a mammalian cell-free system. *Proc Natl Acad Sci USA* 1962;**48** (8):1639–43.

25. Lengyel P. Memories of a senior scientist: on passing the fiftieth anniversary of the beginning of deciphering the genetic code. *Annu Rev Microbiol* 2012;**66**:27–38.

26. Brenner S. RNA, ribosomes, and protein synthesis. *Cold Spring Harb Symp Quant Biol* 1961;**26**:101–10.

27. Grunberg-Manago M, Ochoa S. Enzymatic synthesis and breakdown of polynucleotides; polynycleotide phosphorylase. *J Am Chem Soc* 1955;**77**(11):3165–6.

28. Grunberg-Manago M, Oritz PJ, Ochoa S. Enzymatic synthesis of nucleic acid like polynucleotides. *Science* 1955;**122**(3176):907–10.

29. Grunberg-Manago M. Severo Ochoa: 24 September 1905–1 November 1993. *Biographical Memoirs of Fellows of the Royal Society* 1997;**43**:350–65.

30. Gardner RS, Wahba AJ, Basilio C, Miller RS, Lengyel P, Speyer JF. Synthetic polynucleotides and the amino acid code. VII. *Proc Natl Acad Sci USA* 1962;**48**(12):2087–94.

31. Martin RG, Matthaei JH, Jones OW, Nirenberg MW. Ribonucleotide composition of the genetic code. *Biochem Biophys Res Commun* 1962;**6**(6):410–14.

32. Matthaei JH, Jones OW, Martin RG, Nirenberg MW. Characteristics and composition of RNA coding units. *Proc Natl Acad Sci USA* 1962;**48**(4):666–77.

33. Nirenberg M. *The genetic code. Nobel lecture December 12, 1968.* Amsterdam: Elsevier Publishing Company; 1972.

34. Lengyel P, Speyer JF, Ochoa S. Synthetic polynucleotides and the amino acid code. *Proc Natl Acad Sci USA* 1961;**47**(12):1936–42.

35. Speyer JF, Lengyel P, Basilio C, Ochoa S. Synthetic polynucleotides and the amino acid code. II. *Proc Natl Acad Sci USA* 1962;**48**(1):63–8.

36. Jones Jr OW, Nirenberg MW. Qualitative survey of RNA codewords. *Proc Natl Acad Sci USA* 1962;**48**(12):2115–23.

37. Lengyel P, Speyer JF, Basilio C, Ochoa S. Synthetic polynucleotides and the amino acid code. III. *Proc Natl Acad Sci USA* 1962;**48**(2):282–4.

38. Speyer JF, Lengyel P, Basilio C, Ochoa S. Synthetic polynucleotides and the amino acid code. IV. *Proc Natl Acad Sci USA* 1962;**48**(3):441–8.

39. Tsugita A, Fraenkel-Conrat H. The composition of proteins of chemically evoked mutants of TMV RNA. *J Mol Biol* 1962;**4**(2):73–82.

40. Wittmann HG. Ansätze zur entschlüsselung des genetischen codes. *Naturwissenschafen* 1961;**48**(24):729–34.

41. Crick FH. *The recent excitement in the coding problem. Prog Nucleic Acids Res.* New York, NY: Academic Press; 1963. p. 163–217.

42. Leder P, Singer MF, Brimacombe RL. Synthesis of trinucleoside diphosphates with polynucleotide phosphorylase. *Biochemistry* 1965;**4**(8):1561–7.

43. Heppel LA, Whitfeld PR, Markham R. Nucleotide exchange reactions catalysed by ribonuclease and spleen phosphodiesterase. II. Synthesis of polynucleotides. *Biochem J* 1955;**60**(1):8−15.

44. Kaji A, Kaji H. Specific interaction of soluble RNA with polyribonucleic acid induced polysomes. *Biochem Biophys Res Commun* 1963;**13**(3):186−92.

45. Nirenberg M, Leder P. RNA codewords and protein synthesis. The effect of trinucleotides upon the binding of sRNA to ribosomes. *Science* 1964;**145**(3639):1399−407.

46. Leder P, Nirenberg M. RNA codewords and protein synthesis. II. Nucleotide sequence of a valine RNA codeword. *Proc Natl Acad Sci USA* 1964;**52**(2):420−7.

47. Leder P, Nirenberg M. RNA codewords and protein synthesis, III. On the nucleotide sequence of a cysteine and a leucine RNA codeword. *Proc Natl Acad Sci USA* 1964;**52**(6):1521−9.

48. Chapeville F, Lipmann F, Von Ehrenstein G, Weisblum B, Ray Jr WJ, Benzer S. On the role of soluble ribonucleic acid in coding for amino acids. *Proc Natl Acad Sci USA* 1962;**48**(6):1086−92.

49. Weisblum B, Gonano F, Von EG, Benzer S. A demonstration of coding degeneracy for leucine in the synthesis of protein. *Proc Natl Acad Sci USA* 1965;**53**(2):328−34.

50. Weisblum B. Back to Camelot: defining the specific role of tRNA in protein synthesis. *Trends Biochem Sci* 1999;**24**(6):247−50.

51. Zachau HG, Tada M, Lawson WB, Schweiger M. Fraktionierung der iöslichen ribonucleinsäure. *Biochim Biophys Acta* 1961;**53**(1):221−3.

52. Doctor BP, Apgar J, Holley RW. Fractionation of yeast amino acid-acceptor ribonucleic acids by counter-current distribution. *J Biol Chem* 1961;**236**(4):1117−20.

53. Apgar J, Holley RW, Merrill SH. Countercurrent distribution of yeast "soluble" ribonucleic acids in a modification of the Kirby system. *Biochim Biophys Acta* 1961;**53**(1):220−1.

54. Bernfield MR, Nirenberg MW. RNA codewords and protein synthesis. The nucleotide sequences of multiple codewords for phenylalanine, serine, leucine, and proline. *Science* 1965;**147**(3657):479−84.

55. Trupin JS, Rottman FM, Brimacombe RL, Leder P, Bernfield MR, Nirenberg MW. RNA codewords and protein synthesis, VI. On the nucleotide sequences of degenerate codeword sets for isoleucine, tyrosine, asparagine, and lysine. *Proc Natl Acad Sci USA* 1965;**53**(4):807−11.

56. Nirenberg M, Leder P, Bernfield M, et al. RNA codewords and protein synthesis, VII. On the general nature of the RNA code. *Proc Natl Acad Sci USA* 1965;**53**(5):1161−8.

57. Brimacombe R, Trupin J, Nirenberg M, Leder P, Bernfield M, Jaouni T. RNA codewords and protein synthesis, VIII. Nucleotide sequences of synonym codons for arginine, valine, cysteine, and alanine. *Proc Natl Acad Sci USA* 1965;**54**(3):954−60.

58. Kellogg DA, Doctor BP, Loebel JE, Nirenberg MW. RNA codons and protein synthesis. IX. Synonym codon recognition by multiple species of valine-, alanine-, and methionine-sRNA. *Proc Natl Acad Sci USA* 1966;**55**(4):912−19.

59. Nirenberg M, Caskey T, Marshall R, et al. The RNA code and protein synthesis. *Cold Spring Harb Symp Quant Biol* 1966;**31**:11−24.

60. Söll D, Cherayil J, Jones DS, et al. sRNA specificity for codon recognition as studied by the ribosomal binding technique. *Cold Spring Harb Symp Quant Biol* 1966;**31**:51−61.

61. Weigert MG, Garen A. Base composition of nonsense codons in *E. coli*. Evidence from amino-acid substitutions at a tryptophan site in alkaline phosphatase. *Nature* 1965;**206**(4988):992−4.

62. Brenner S, Stretton AO, Kaplan S. Genetic code: the "nonsense" triplets for chain termination and their suppression. *Nature* 1965;**206**(4988):994−8.

63. Brenner S, Barnett L, Katz ER, Crick FH. UGA: a third nonsense triplet in the genetic code. *Nature* 1967;**213**(5075):449−50.

64. Caskey CT, Tompkins R, Scolnick E, Caryk T, Nirenberg M. Sequential translation of tri-nucleotide codons for the initiation and termination of protein synthesis. *Science* 1968;**162** (3849):135−8.

65. Marcker K, Sanger F. *N*-Formyl-methionyl-S-RNA. *J Mol Biol* 1964;**8**(6):835−40.

66. Capecchi MR. Polypeptide chain termination in vitro: isolation of a release factor. *Proc Natl Acad Sci USA* 1967;**58**(3):1144−51.

67. Scolnick E, Tompkins R, Caskey T, Nirenberg M. Release factors differing in specificity for terminator codons. *Proc Natl Acad Sci USA* 1968;**61**(2):768−74.

68. Marcker K. The formation of *N*-formyl-methionyl-sRNA. *J Mol Biol* 1965;**14**(1):63−70.

69. Clark BF, Marcker KA. The role of *N*-formyl-methionyl-sRNA in protein biosynthesis. *J Mol Biol* 1966;**17**(2):394−406.

70. Crick FH. Codon−anticodon pairing: the wobble hypothesis. *J Mol Biol* 1966;**19** (2):548−55.

71. Marshall RE, Caskey CT, Nirenberg M. Fine structure of RNA codewords recognized by bacterial, amphibian, and mammalian transfer RNA. *Science* 1967;**155**(3764):820−6.

72. Woese CR, Dugre DH, Saxinger WC, Dugre SA. The molecular basis for the genetic code. *Proc Natl Acad Sci USA* 1966;**55**(4):966−74.

73. Crick FH. The origin of the genetic code. *J Mol Biol* 1968;**38**(3):367−80.

74. Wong JT. Coevolution theory of the genetic code at age thirty. *BioEssays* 2005;**27** (4):416−25.

75. Yarus M, Caporaso JG, Knight R. Origins of the genetic code: the escaped triplet theory. *Annu Rev Biochem* 2005;**74**:179−98.

76. Koonin EV, Novozhilov AS. Origin and evolution of the genetic code: the universal enigma. *IUBMB Life* 2009;**61**(2):99−111.

77. Khorana HG. *Nucleic acid synthesis in the study of the genetic code. Nobel Nobel lecture December 12, 1968.* Amsterdam: Elsevier Publishing Company; 1972.

78. Ohtsuka E, Moon MW, Khorana HG. Studies on polynucleotides. XLIII. The synthesis of deoxyribopolynucleotides containing repeating dinucleotide sequences. *J Am Chem Soc* 1965;**87**(13):2956−70.

79. Khorana HG. Chemical and enzymatic synthesis of polynucleotides. In: Chargaff E, Davidson JN, editors. *The nucleic acids.* New York, NY: Academic Press; 1960. p. 105−46.

80. Khorana HG. *Some recent developments in the chemistry of phosphate esters of biological interest.* New York, NY: John Wiley & Sons; 1961.

81. Khorana HG, Jacob TM, Moon MW, Narang SA, Ohtsuka E. Studies on polynucleotides XLII. The synthesis of deoxyribopolynucleotides containing repeating nucleotide sequences. Introduction and general considerations. *J Am Chem Soc* 1965;**87**(13):2954−6.

82. Gilham PT, Khorana HG. Studies on polynucleotides. I. A new and general method for the chemical synthesis of the C5″-C3″ internucleotidic linkage. Syntheses of deoxyribo-dinucleotides. *J Am Chem Soc* 1958;**80**(23):6212−22.

83. Gilham PT, Khorana HG. Studies on polynucleotides. V.1 Stepwise synthesis of oligonu-cleotides. syntheses of thymidylyl-(5′ → 3′)-thymidylyl-(5′ → 3′)-thymidine and deoxycyti-dylyl-(5′ → 3′)-deoxyadenylyl-(5′ → 3′)-thymidine2. *J Am Chem Soc* 1959;**81**(17):4647−50.

84. Weimann G, Khorana HG. Studies on polynucleotides. XIII. Stepwise synthesis of deoxyribo-oligonucleotides. An alternative general approach and the synthesis of thymidine di-, tri- and tetranucleotides bearing 3′-phosphomonoester end groups. *J Am Chem Soc* 1962;**84**(3):419−30.

85. Schaller H, Khorana HG. Studies on polynucleotides. XXV. The stepwise synthesis of specific deoxyribopolynucleotides (5). Further studies on the synthesis of internucleotide bond by the carbodiimide method. The synthesis of suitably protected dinucleotides as intermediates in the synthesis of higher oligonucleotides. *J Am Chem Soc* 1963;**85**(23):3828−35.

86. Schaller H, Khorana HG. Studies on polynucleotides. XXVII. The stepwise synthesis of specific deoxyribopolynucleotides (7). The synthesis of polynucleotides containing deoxycytidine and deoxyguanosine in specific sequences and of homologous deoxycytidine polynucleotides terminating in thymidine. *J Am Chem Soc* 1963;**85**(23):3841−51.

87. Schaller H, Weimann G, Lerch B, Khorana HG. Studies on polynucleotides. XXIV. The stepwise synthesis of specific deoxyribopolynucleotides (4). Protected derivatives of deoxyribonucleosides and new syntheses of eoxyribonucleoside-3' phosphates. *J Am Chem Soc* 1963;**85**(23):3821−7.

88. Weimann G, Schaller H, Khorana HG. Studies on olynucleotides. XXVI. The stepwise synthesis of specific deoxyribopolynucleotides (6). The synthesis of thymidylyl-(3′ → 5′)-deoxyadenylyl-(3′ → 5′)-thymidylyl-(3′ → 5′)-thymidylyl-(3′ → 5′)-thymidine and of polynucleotides containing thymidine and deoxyadenosine in alternating sequence. *J Am Chem Soc* 1963;**85**(23):3835−41.

89. Jacob TM, Khorana HG. Studies on polynucleotides. XXXVII. The synthesis of specific deoxyribopolynucleotides. Further examination of the approach involving stepwise synthesis. *J Am Chem Soc* 1965;**87**(2):368−74.

90. Jacob TM, Khorana HG. Studies on polynucleotides. XLIV. The synthesis of dodecanucleotides containing the repeating trinucleotide sequence thymidylyl-(3′ → 5′)-thymidylyl-(3′ → 5′)-deoxycytidine. *J Am Chem Soc* 1965;**87**(13):2971−81.

91. Narang SA, Jacob TM, Khorana HG. Studies on Polynucleotides. XLVI. The synthesis of hexanucleotides containing the repeating trinucleotide sequences deoxycytidylyl-(3′ → 5′)-deoxyadenylyl-(3′ → 5′)-deoxyadenosine and deoxyguanylyl-(3′ → 5′)-deoxyadenylyl-(3′ → 5′)-deoxyadenosine. *J Am Chem Soc* 1965;**87**(13):2988−95.

92. Narang SA, Khorana HG. Studies on polynucleotides. XLV. The synthesis of dodecanucleotides containing the repeating trinucleotide sequence thymidylyl-(3′ → 5′)-thymidylyl-(3′ → 5′)-deoxyinosine. *J Am Chem Soc* 1965;**87**(13):2981−8.

93. Khorana HG. Synthesis in the study of nucleic acids. The Fourth Jubilee Lecture. *Biochem J* 1968;**109**(5):709−25.

94. Kössel H, Moon MW, Khorana HG. Studies on polynucleotides. LX. The use of preformed dinucleotide blocks in stepwise synthesis of deoxyribopolynucleotides. *J Am Chem Soc* 1967;**89**(9):2148−54.

95. Narang SA, Jacob TM, Khorana HG. Studies on polynucleotides. LXII. Deoxyribopolynucleotides containing repeating trinucleotide sequences (4). Preparation of suitably Protected deoxyribotrinucleotides. *J Am Chem Soc* 1967;**89**(9):2158−66.

96. Narang SA, Jacob TM, Khorana HG. Studies on polynucleotides. LXIII. Deoxyribopolynucleotides containing repeating trinucleotide sequences (5). The polymerization of protected deoxyribotrinucleotides. *J Am Chem Soc* 1967;**89**(9):2167−77.

97. Ohtsuka E, Khorana HG. Studies on polynucleotides. LXVI. The synthesis of deoxyribopolynucleotides containing repeating tetranucleotide sequences (3). A further study of the synthetic approach involving condensation of preformed oligonucleotide blocks. *J Am Chem Soc* 1967;**89**(9):2195−202.

98. Khorana HG, Buchi H, Jacob TM, Kossel H, Narang SA, Ohtsuka E. Studies on polynucleotides. LXI. Polynucleotide synthesis in relation to the genetic code. General introduction. *J Am Chem Soc* 1967;**89**(9):2154−8.

99. Falaschi A, Adler J, Khorana HG. Chemically synthesized deoxypolynucleotides as templates for ribonucleic acid polymerase. *J Biol Chem* 1963;**238**(9):3080−5.

100. Byrd C, Ohtsuka E, Moon MW, Khorana HG. Synthetic deoxyribo-oligonucleotides as templates for the DNA polymerase of *Escherichia coli*: new DNA-like 1-polymers containing repeating nucleotide sequences. *Proc Natl Acad Sci USA* 1965;**53**(1):79−86.

101. Nishimura S, Jacob TM, Khorana HG. Synthetic deoxyribopolynucleotides as templates for ribonucleic acid polymerase: the formation and characterization of a ribopolynucleotide with a repeating trinucleotide sequence. *Proc Natl Acad Sci USA* 1964;**52**(6):1494−501.

102. Mehrotra BD, Khorana HG. Studies on polynucleotides. XI. Synthetic deoxyribopolynucleotides as templates for ribonucleic acid polymerase: the influence of temperature on template function. *J Biol Chem* 1965;**240**(4):1750−3.

103. Nishimura S, Jones DS, Khorana HG. Studies on polynucleotides. XLVIII. The in vitro synthesis of a co-polypeptide containing two amino acids in alternating sequence dependent upon a DNA-like polymer containing two nucleotides in alternating sequence. *J Mol Biol* 1965;**13**(1):302−24.

104. Nishimura S, Jones DS, Ohtsuka E, Hayatsu H, Jacob TM, Khorana HG. Studies on polynucleotides: XLVII. The in vitro synthesis of homopeptides as directed by a ribopolynucleotide containing a repeating trinucleotide sequence. New codon sequences for lysine, glutamic acid and arginine. *J Mol Biol* 1965;**13**(1):283−301.

105. Khorana HG. Synthetic nucleic acids and the genetic code. *Jama* 1968;**206**(9):1978−82.

106. Kossel H, Morgan AR, Khorana HG. Studies on polynucleotides. LXXIII. Synthesis in vitro of polypeptides containing repeating tetrapeptide sequences dependent upon DNA-like polymers containing repeating tetranucleotide sequences: direction of reading of messenger RNA. *J Mol Biol* 1967;**26**(3):449−75.

107. Ghosh HP, Söll D, Khorana HG. Studies on polynucleotides: LXVII. Initiation of protein synthesis in vitro as studied by using ribopolynucleotides with repeating nucleotide sequences as messengers. *J Mol Biol* 1967;**25**(2):275−98.

108. Smith M, Rammler DH, Goldberg IH, Khorana HG. Studies on polynucleotides. XIV Specific synthesis of the C3′-C5′ interribonucleotide linkage. Syntheses of uridylyl-(3′5′)-uridine and uridylyl-(3′5′)-adenosine. *J Am Chem Soc* 1962;**84**(3):430−40.

109. Rammler DH, Khorana HG. Studies on Polynucleotides. XVI. Specific synthesis of the C3′-C5′ interribonucleotidic linkage. Examination of routes involving protected ribonucleosides and ribonucleoside-3′ phosphates. Syntheses of uridylyl-(3′ → 5′)-adenosine, uridylyl-(3′ → 5′)-cytidine, adenylyl-(3′ → 5′)-adenosine and related compounds. *J Am Chem Soc* 1962;**84**(16):3112−22.

110. Rammler DH, Khorana HG. Studies on polynucleotides. XX. Amino acid acceptor ribonucleic acids. The synthesis and properties of 2′ (or 3′-*O*-(DL-phenylalanyl)-adenosine, 2′ (or 3′)-*O*-(DL-phenylalanyl)-uridine and related compounds. *J Am Chem Soc* 1963;**85**(13):1997−2002.

111. Lapidot Y, Khorana HG. Studies on polynucleotides. XXVIII. The specific synthesis of C3′-C5′-linked ribooligonucleotides. The stepwise synthesis of uridylyl-(3′ → 5′)-adenylyl-(3′ → 5′)-uridylyl-(3′ → 5′)-uridine. *J Am Chem Soc* 1963;**85**(23):3852−7.

112. Lapidot Y, Khorana HG. Studies on polynucleotides. XXIX. The specific synthesis of C3′-C5′-linked ribooligonucleotides. Homologous adenine oligonucleotides. *J Am Chem Soc* 1963;**85**(23):3857−62.

113. Coutsogeorgopoulos C, Khorana HG. Studies on polynucleotides. XXXI. The specific synthesis of C3'-C5'-linked ribopolynucleotides. 2. A further study of the synthesis of uridine polynucleotides. *J Am Chem Soc* 1964;**86**(14):2926−32.

114. Lohrmann R, Khorana HG. Studies on Polynucleotides. XXXIV. The specific synthesis of C3'-C5'-linked ribooligonucleotides. New protected derivatives of ribonucleosides and ribonucleoside 3'-phosphates. Further syntheses of diribonucleoside phosphates. *J Am Chem Soc* 1964;**86**(19):4188−94.

115. Söll D, Khorana HG. Studies on polynucleotides. XXXV. The specific synthesis of C3'-C5'-linked ribooligonucleotides. VIII. The synthesis of ribodinucleotides bearing 3'-phosphomonoester groups. *J Am Chem Soc* 1965;**87**(2):350−9.

116. Söll D, Khorana HG. Studies on polynucleotides. XXXVI. The specific synthesis of C3'-C5'-linked ribooligonucleotides. IX. The synthesis of ribodinucleotides bearing 3'-phosphomonoester groups. *J Am Chem Soc* 1965;**87**(2):360−7.

117. Lohrmann R, Soll D, Hayatsu H, Ohtsuka E, Khorana HG. Studies on polynucleotides. LI. Syntheses of the 64 possible ribotrinucleotides derived from the four major ribomononucleotides. *J Am Chem Soc* 1966;**88**(4):819−29.

118. Söll D, Ohtsuka E, Jones DS, et al. Studies on polynucleotides, XLIX. Stimulation of the binding of aminoacyl-sRNA's to ribosomes by ribotrinucleotides and a survey of codon assignments for 20 amino acids. *Proc Natl Acad Sci USA* 1965;**54**(5):1378−85.

119. Khorana HG, Agarwal KL, Büchi H, et al. CIII. Total synthesis of the structural gene for an alanine transfer ribonucleic acid from yeast. *J Mol Biol* 1972;**72**(2):209−17.

120. Sekiya T, Contreras R, Takeya T, Khorana HG. Total synthesis of a tyrosine suppressor transfer RNA gene. XVII. Transcription, in vitro, of the synthetic gene and processing of the primary transcript to transfer RNA. *J Biol Chem* 1979;**254**(13):5802−16.

121. Ansari AZ, Rosner MR, Adler J. Har Gobind Khorana 1922−2011. *Cell* 2011;**147** (7):1433−5.

122. Caruthers M, Wells R. Retrospective. Har Gobind Khorana (1922−2011). *Science* 2011;**334**(6062):1511.

123. RajBhandary UL. Har Gobind Khorana (1922−2011). *Nature* 2011;**480**(7377):322.

124. Sakmar TP. Har Gobind Khorana (1922−2011): pioneering spirit. *PLoS Biol* 2012;**10**(2): e1001273.

125. Beadle G, Beadle M. *The language of life: an introduction to the science of genetics.* London: Victor Gollancz; 1966.

126. Woese C. On the evolution of the genetic code. *Proc Natl Acad Sci USA* 1965;**54** (6):1546−52.

127. Wong J. A co-evolution theory of the genetic code. *Proc Natl Acad Sci USA* 1975;**72** (5):1909−12.

128. Crick F, Brenner S, Klug A, Pieczenik G. A speculation on the origin of protein synthesis. *Orig Life* 1976;**7**(4):389−97.

129. Szathmary E. Coding coenzyme handles: a hypothesis for the origin of the genetic code. *Proc Natl Acad Sci USA* 1993;**90**(21):9916−20.

130. Szathmary E. The origin of the genetic code: amino acids as cofactors in an RNA world. *Trends Genet* 1999;**15**(6):223−9.

131. Yarus M. Amino acids as RNA ligands: a direct-RNA-template theory for the code's origin. *J Mol Evol* 1998;**47**(1):109−17.

132. Rodin S, Rodin A. Origin of the genetic code: first aminoacyl-tRNA synthetases could replace isofunctional ribozymes when only the second base of codons was established. *DNA Cell Biol* 2006;**25**(6):365−75.
133. Wolf Y, Koonin E. On the origin of the translation system and the genetic code in the RNA world by means of natural selection, exaptation, and subfunctionalization. *Biol Direct* 2007;**2**(1):14.
134. Rodin S, Rodin A. On the origin of the genetic code: signatures of its primordial complementarity in tRNAs and aminoacyl-tRNA synthetases. *Heredity* 2008;**100**(4):341−55.
135. Rodin A, Szathmary E, Rodin S. On origin of genetic code and tRNA before translation. *Biol Direct* 2011;**6**(1):14.

Chapter 11

The Surprising Discovery of Split Genes and of RNA Splicing

The Discovery of Split Genes and of RNA Splicing in Eukaryotes Defeated the Preconception of the Universality of Gene–mRNA Collinearity

Chapter Outline

M. Fry: Landmark Experiments in Molecular Biology. DOI: http://dx.doi.org/10.1016/B978-0-12-802074-6.00011-4

481

Within a few years after its publication in 1961, the far-reaching predictions of Jacob and Monod's operon model of gene expression[1] were validated in a train of groundbreaking experimental discoveries. In the very same *annus mirabilis* of 1961 DNA-complementing, rapidly turning-over mRNA molecules were detected in phage-infected[2,3] and in uninfected[4,5] bacterial cells. Less than 3 years later it was elegantly demonstrated that the nucleotide sequences of bacteriophage and bacterial genes (and implicitly of their complementary mRNA) were collinear with the amino acid sequences of their respective product proteins.[6-8] The string of landmark discoveries that were made between the late 1950s and mid-1960s also included the discovery of the role of transfer RNA (tRNA) as an adaptor in the translation of nucleotide sequences of mRNA into the amino acid sequences of proteins (see Chapter 8: "The Adaptor Hypothesis and the Discovery of Transfer RNA") and the deciphering of the genetic code (see Chapter 7: "Defining the Genetic Code"). There are hardly similar examples in the history of biology for such a rapid succession of theoretical and experimental triumphs that re-formed much of the thinking about the molecular basis of inheritance and of gene expression. Yet, one less constructive aspect of these revolutionary developments was the period's emergent belief that all the living organisms universally shared the same molecular mechanisms of gene expression as those that were uncovered in prokaryotes. The edict 'Anything found to be true of *Escherichia coli* must also be true of elephants' that the charismatic and tremendously influential Jacques Monod coined already in 1954 appeared to become an article of faith in the molecular biology community. A case in point was the paradigmatic rapidly turning-over collinear mRNA of bacteria and their viruses. Being of the mind that all mRNA molecules were unstable, researchers disregarded for a long time Hämmerling's reports of stable informational "substances" (mRNA in today's idiom) that dictated protein synthesis in the cytoplasm of *Acetabularia*[9,10] (see Chapter 9: "The Discovery and Rediscovery of Prokaryotic Messenger RNA"). Another case of conceptual rigidity was the conviction, based on the prokaryotic model, that genes, mRNA, and their product proteins were all collinear. This chapter traces the long and winding experimental road that led in the end to the realization that unlike prokaryotic mRNA, molecules of mRNA in eukaryotes were not the collinear transcripts of genes but rather products of splicing of much larger primary transcripts.

Much of the historical background to the discovery of mRNA in eukaryotic cells and of their split genes and RNA splicing were covered in a 2014 Cold Spring Harbor meeting titled *mRNA: From discovery to synthesis and regulation in bacteria and eukaryotes*. Material that was presented in this meeting is available online at http://library.cshl.edu/Meetings/mRNA/.

11.1 FIRST CLUE: MOST OF THE NUCLEAR RNA BROKE DOWN IN THE NUCLEUS WITHOUT EVER REACHING THE CYTOPLASM

The Oxford University cell biologist Henry Harris (Box 11.1) was studying in the late 1950s the intracellular fate of radioactively labeled proteins.

BOX 11.1 Henry Harris

After studying medicine at the Royal Prince Alfred Hospital in Sydney, Henry Harris, an Australian-British cell biologist, moved to the Sir William Dunn School of Pathology at Oxford University at which he received a DPhil degree in 1954. Remaining in Oxford for the length of his career—he became in 1979 its Regius Professor of Medicine. In the course of his studies on cancer cells he discovered giant nuclear RNA molecules that were largely broken down in the nucleus without ever reaching the cytoplasm. In later pioneering experiments on cell fusion he discovered that fusion of cancer with normal cells yielded nonmalignant hybrids. This observation was an early indication for the existence of genes that suppress malignancy (suppressor genes). He vividly described his adventures in science in a book *The Balance of Improbabilities. A Scientific Life*[11] (Fig. 11.1).

Henry Harris
(1925–)

FIGURE 11.1   Henry Harris.

Contrary to the incorrect 1955 assertion of Hogness, Cohn, and Monod that proteins are stable,[12] and in agreement with an earlier conclusion of Rudolph Schoenheimer[13] (1898–1941), Harris' results indicated that proteins were turning over in several cell types.[14] At about the same time Angus Graham and Louis (Lou) Siminovitch of the University of Toronto contended that in analogy with the claimed stability of proteins, RNA too was stable in dividing animal cells.[15,16] Because his own experimental results revealed that rather than being stable as claimed, proteins were degraded, Harris took on to also assess the parallel claim that cellular RNA was stable.[17] In early experiments he tracked in nondividing macrophages the incorporation of ^{14}C adenine into DNA and RNA and followed the fate of the radioactively labeled nucleic acids. Results showed that the DNA was neither synthesized nor broken down in these resting cells, whereas a fraction of the RNA was rapidly synthesized but was also rapidly broken down.[18] To monitor the fate of RNA in the cell nucleus and in the cytoplasm, Harris next applied autoradiography to view radioactively labeled RNA in nondividing fibroblasts and macrophages. The nondividing cells that were fed with ^{3}H adenosine did not replicate their DNA and the labeled adenosine was incorporated into RNA only. In the actual experiment cells were either exposed to the radioactive precursor for just 30 minutes ("pulse") or were incubated for additional 8 hours after the pulse in a medium that contained an excess of nonradioactive adenosine ("chase"). Both pulse-labeled and chased cells were fixed and then autoradiographed.[19]

As shown in Fig. 11.2A, most of the pulse-labeled RNA was localized in the nuclei and after 8 hours of chase, the radioactive RNA disappeared from the nuclei without being translocated to the cytoplasm.

Quantification of the nuclear and cytoplasmic radioactive RNA during the chase period revealed rapid degradation of the nuclear RNA and much lower turnover of its cytoplasmic counterpart (Fig. 11.2B). Notably the disappearance of the labeled RNA from the nuclei was not accompanied by its reciprocal enrichment in the cytoplasm. This lack of significant re-localization suggested that there was no blanket translocation of RNA from nucleus to cytoplasm in these nondividing cells. The question of transport of RNA from nucleus to cytoplasm was intently addressed in subsequent studies of the Harris laboratory. Analysis of labeled RNA in dividing HeLa cells revealed that less than 10% of the radioactivity in the nuclear RNA ended up in the cytoplasm. Importantly too, the base compositions of the nuclear and cytoplasmic RNA were found to be significantly different.[20,21] In another experiment the antibiotic actinomycin D blocked de novo RNA synthesis immediately after the RNA was pulse labeled (see Fig. 11.7). Results showed that most of the radiolabeled RNA was broken down in the nucleus and that the nucleotide composition of its minor portion that reached the cytoplasm was different from the base composition of the nuclear RNA.[22] All in all, these findings stood in discord with a view of simple translocation of primary mRNA transcripts from nucleus to cytoplasm. Importantly, they also raised the

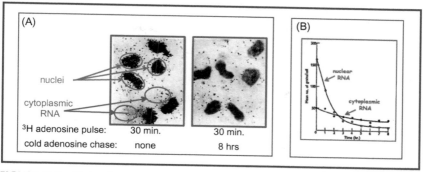

FIGURE 11.2   Nuclear but not cytoplasmic RNA of macrophages is rapidly broken down. (A) Autoradiography of ^{3}H-adenosine pulse-labeled and of chased macrophages. After 30 minutes of pulse labeling, the nuclei were filled with radioactive RNA. However, following an 8-hour chase with an excess of unlabeled adenosine the nuclei were virtually devoid of radioactive RNA. By contrast, the number of radioactive grains in the cytoplasm after the pulse declined only minimally at the end of the chase period.(B) Kinetics of the turnover of nuclear and cytoplasmic radioactive RNA. Macrophages that were pulse labeled with ^{3}H-adenosine and then subjected to chase were autoradiographed at different times during the chase period. Numbers of grains of radioactive RNA were counted in the cell nuclei and cytoplasm and plotted against hours of chase. *Panels (A) and (B): Modified from Harris H. Turnover of nuclear and cytoplasmic ribonucleic acid in two types of animal cell, with some further observations on the nucleolus.* Biochem J *1959;73(2):362—9.*

question of the significance of the rapid degradation of most of the nuclear RNA. Because these results were in stark contrast with the then consensually accepted model of prokaryotic mRNA, they were hard to swallow. Indeed, in a retrospective short chronicle of the early 1960s controversies on animal cell RNA, Harris described his disputes with scientists who adhered to the prokaryotic mRNA dogma and thus found faults with his experimental results or with their interpretation.[17] In a 1964 symposium at the Institute of Microbiology of Rutgers University, Harris stood his grounds and ventured to speculate on the possible significance of the rapid breakdown of RNA in the nucleus.[23] With our present day knowledge of split genes in eukaryotes and the excision of noncoding sequences from primary nuclear transcripts, Harris' conjectures in 1964 appear now farsighted despite their inaccurate details (Ref. 23, p. 469):

> *Only a small proportion of the RNA made in the nucleus of higher-plant and animal cells serves as a template for the synthesis of protein. This RNA is characterized by its ability to assume a form which protects it from intranuclear degradation. Most of the nuclear RNA, however, is made of parts of the DNA which do not contain information for the synthesis of specific proteins. This RNA does not assume the configuration necessary for protection from degradation and is eliminated within the cell nucleus. It plays no role in the synthesis of cell protein, but serves as a background on which mutation and selection may operate to produce new templates for protein synthesis.*

11.2   A SECOND FINDING: GIANT MOLECULES OF UNSTABLE RNA WERE DETECTED IN THE NUCLEUS

The laboratory of James Darnell (Box 11.2) was a source of many of the discoveries that were made from the early 1960s to the mid-1970s on the structure and metabolism of nuclear and cytoplasmic RNA of animal cells and their viruses.

During an early sojourn in the NIH laboratory of the cell biologist Harry Eagle (1905−92), Darnell acquired experience in growing animal cells in culture and propagating, quantifying, and analyzing RNA of viruses. His subsequent year in 1959 at the Pasteur Institute coincided with the conceptualization by Monod and Jacob of the idea of mRNA.[25] Thus, when Darnell established his own independent laboratory at the MIT, he undertook to examine RNA of animal cells. With his first postdoctoral fellow Klaus Scherrer (later at the Pasteur Institute), they exposed HeLa cells to ^{14}C-uridine for different lengths of time and then extracted the total cell RNA with hot phenol that contained the detergent sodium dodecyl sulfate (SDS).[26] Next, the extracted radiolabeled RNA and unlabeled ribosomal RNA (rRNA)

BOX 11.2 James Darnell

Darnell earned an MD degree in 1955 from Washington University School of Medicine. At an early phase of his research career, he studied poliovirus at the NIH laboratory of Harry Eagle. Following a short stay with François Jacob at the Pasteur Institute in Paris, he served at different times on the faculties of the MIT, the Albert Einstein College of Medicine, and Columbia University and since 1974 until his retirement he was a Professor in Rockefeller University. For many years he was a leading student of heterogeneous nuclear RNA (hnRNA) and of mRNA of animal cells and their viruses. Yet, despite the recognition that hnRNA was precursor to mRNA, his studies did not lead to comprehension that the large-sized hnRNA was spliced to produce shorter mRNA molecules. His later investigations on the regulation of gene expression in animal cells culminated in the discovery of the STAT and the Jak-STAT pathway of transcription control. He engagingly described his career in a recent review[24] (Fig. 11.3).

James Darnell
(1930−)

FIGURE 11.3   James Darnell.

and tRNA were resolved by sucrose gradient centrifugation (Fig. 9.17), and the distribution of the unlabeled and radioactive RNA among the gradient fractions was monitored. This analysis revealed that after exposure to ^{14}C-uridine for 5 or 30 minutes, the population of labeled RNA molecules migrated in the gradient as a heterogeneous peak of approximately 45S. This mass of the RNA was later shown to correspond to chains of 12,000–14,000 nucleotides in length; sizes that were much greater than the 2000 and 5000 nucleotide-long 18S and 28S rRNA subunits, respectively (Fig. 11.4A). Labeling of the RNA for the longer period of 60 minutes exposed in addition to the 45S species, a slower migrating peak of ∼35S that was still larger than mature rRNA. This second peak was shown later to be a large-size precursor to shorter mature rRNA subunits. The processing of these precursor

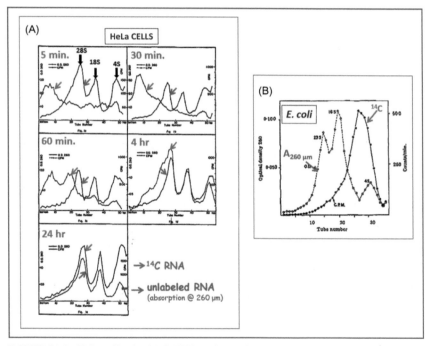

FIGURE 11.4 HeLa cell pulse-labeled RNA migrated in sucrose gradient as a population of giant molecules. (A) RNA that was extracted by phenol/SDS from HeLa cells after their exposure to ^{14}C-uridine for the indicated lengths of time was resolved by sucrose gradient centrifugation. Unlabeled 28S and 18S chains of rRNA and 4S tRNA were detected in fractions of the gradient by their adsorption at 260 mμ. The labeled RNA was detected by the counting of radioactivity in the fractions. (B) Size distribution in sucrose gradient of pulse-labeled *Escherichia coli* RNA. *Panel (A): Modified from Scherrer K, Darnell JE. Sedimentation characteristics of rapidly labelled RNA from HeLa cells.* Biochem Biophys Res Commun 1962;7(6):486–90. *Panel (B): Adapted from Gros F, Hiatt H, Gilbert W, Kurland CG, Risebrough RW, Watson JD. Unstable ribonucleic acid revealed by pulse labelling of Escherichia coli.* Nature 1961;**190**(4776):581–5.

molecules into the mature rRNA chains is beyond the scope of this chapter and the interested reader may consult some excellent review articles on the subject.[27,28] Radioactive RNA that was isolated from cells after their exposure to ^{14}C-uridine for even longer periods ($\geq$4 hours) comigrated mainly or exclusively with the 18S and 28S rRNA chains[26] (Fig. 11.4A).

The novelty of detecting giant RNA molecules in the pulse-labeled HeLa cells was underscored by a comparison with the size distribution of pulse-labeled RNA in a bacterial cell. As was shown by Gros et al.,[4] pulse-labeled RNA of *Escherichia coli* cells migrated in sucrose gradient as a wide peak of $\sim$12S, a significantly lower size than that of both the 23S and 16S chains of bacterial rRNA (Fig. 11.4B).

Unsurprisingly the conspicuous discord between the apparent sizes of short-lived RNA in prokaryotic and in animal cells was disconcerting. In a retrospective account Scherrer reminisced about Watson's reaction to his and Darnell's observations[29]:

> At that time, only a few investigators were interested in the molecular biology of animal cells; the 'serious' research was in E. coli and bacteriophages. James Watson visited MIT frequently, and would discuss our strange results. One day, he told me "To work with animal cells, you got to be a hero or a fool!" In spite of Watson's ambiguous comment, I continued to extract RNA from HeLa cells to be analyzed on sucrose gradients.

Watson's equivocal comment notwithstanding, just a few weeks after the publication of Darnell and Scherrer's results, their findings were largely corroborated by Howard Hiatt (Box 9.6) who was a short time before member of the Watson team that identified mRNA in *E. coli*.[4] Rather than using mammalian cells in culture, Hiatt followed the fate of labeled RNA in cells of resting and regenerating rat liver. He labeled RNA by injecting rats with[14]C-orotic acid which was converted in vivo into uracil that was incorporated into RNA. Rats were sacrificed at different times after the introduction of the radioactive precursor, their liver cells were isolated, cell nuclei and cytoplasm were separated, and their RNA was extracted. Sucrose gradient resolution of the RNA revealed major differences between the populations of labeled RNA molecules in the nucleus and in the cytoplasm. After short-term labeling, the nucleus contained giant RNA molecules whose heterogeneous size exceeded the sizes of the rRNA chains (Fig. 11.5). The giant RNA species that was observed in the cell nucleus was aptly named heterogeneous nuclear RNA (hnRNA). The population of hnRNA molecules was found to diminish progressively with longer periods of labeling so-much-so that the hnRNA peak became undetectable after 3—17 hours of exposure of the cells to ^{14}C-orotic acid, and radioactivity was found only in the peaks of rRNA and tRNA (Fig. 11.5; left-hand panel). Parallel examination of the radioactively labeled RNA in the cytoplasm revealed a distinctly different picture. There, the earliest appearing label was in the 4S peak of

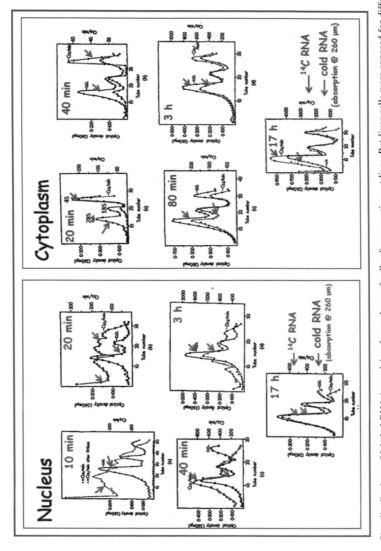

FIGURE 11.5 Size distributions of labeled RNA in nuclei and cytoplasm of cells of regenerating rat liver. Rat liver cells were exposed for different lengths of time to ^{14}C-orotic acid, and RNA that was extracted from their nuclear and cytoplasmic fractions was resolved by sucrose gradient centrifugation. Fractions of the gradients were monitored for absorption at 260 mμ of rRNA and tRNA and for radioactivity of the radiolabeled RNA. *Modified from Hiatt HH. A rapidly labeled RNA in rat liver nuclei. J Mol Biol 1962;5(2):217–29.*

Base composition of RNA of liver cytoplasm and nuclei					
Fraction	Interval between ³²PO₄ and killing	Moles/100 moles of total nucleotides§			
		AMP	GMP	CMP	UMP
Cytoplasm	40 min	17·0	30·7	32·8	19·5
	4 hr	17·5	32·4	30·7	19·4
Nuclei (SDS-extracted)	40 min	14·3	34·7	26·6	24·4
	4 hr	16·3	30·9	28·1	24·7

FIGURE 11.6 Base compositions of radioactively labeled nuclear and cytoplasmic RNA after exposure of liver cells to ^{14}C-orotic acid for short (40 minutes) and long (4 hours) periods of time. Following short-term labeling, the relative proportions of the four bases in RNA from the cytoplasm (*red ellipses*) were significantly different from their proportions in the nuclear RNA (*blue ellipses*). By contrast, after long-term labeling the base compositions of cytoplasmic (*red rectangles*) and nuclear (*blue rectangles*) RNA became more similar. *Modified from Hiatt HH. A rapidly labeled RNA in rat liver nuclei. J Mol Biol 1962;5(2):217–29.*

tRNA and at longer times radioactivity accumulated under the peaks of 18S and 28S rRNA. Most importantly, however, there was no trace in the cytoplasm of the giant RNA that was clearly sighted in the nucleus (Fig. 11.5; right-hand panel). Even more significantly, analysis revealed distinctly different base compositions of pulse (40 minutes)-labeled nuclear and cytoplasmic RNA fractions. The base compositions of nuclear and cytoplasmic RNA fractions became more similar after a longer labeling period of 4 hours (Fig. 11.6).

Hence, it appeared that in addition to having much larger sizes than cytoplasmic rRNA, the pulse-labeled hnRNA was distinguished from rRNA by its overall base composition. Conversely, in addition to the overlapping sizes of longer-term labeled nuclear RNA and cytoplasmic rRNA (Fig. 11.5; left-handed panel), their base compositions were also similar.

With the benefit of hindsight, today's observer recognizes that being unprocessed precursors to mRNA, the base sequences of the short-lived giant hnRNA molecules were different from those of the cytoplasmic mRNA. However, at the time of his study Hiatt assumed that both the nucleus and cytoplasm were sites of protein synthesis and that each of these cellular compartments contained different classes of mRNA molecules that served as templates for different types of proteins. Specifically, he proposed that the small-sized stable mRNA that supposedly encoded exported proteins were translated in the cytoplasm, whereas the nucleus was the site of translation of changing sets of renewable heterogeneous mRNA molecules[30]:

The large quantity of a relatively small variety of proteins which are exported by the liver may be synthesized on relatively stable templates in the cytoplasm. In the nucleus, on the other hand, a renewable messenger would afford the cell a capacity to alter rapidly its function, as might be necessary for example, during liver regeneration following partial hepatectomy. Here a labile messenger would serve the same function which current concepts ascribe to it in the bacterium.

11.3 FIRST HINTS THAT THE LABILE GIANT NUCLEAR RNA MAY BE RELATED TO MESSENGER RNA

In a 1963 follow-up study Scherrer, Latham, and Darnell traced the fate of pulse-labeled nuclear RNA after the exposure of HeLa cells to the transcription inhibitor actinomycin D (Fig. 11.7A).[31] Through its association with transcription initiation complex binding sites in DNA, actinomycin D prevents elongation of RNA chains by RNA polymerase. The introduction of actinomycin D enabled, therefore, monitoring of the fate of preexisting RNA molecules without the complication of an ongoing accumulation of newly made RNA molecules.

Scherrer, Latham, and Darnell exposed HeLa cells to ^{14}C-uridine for 30 minutes and then blocked the synthesis of new RNA chains by adding actinomycin D to the medium. Quantification of acid-precipitable RNA at different times during the incorporation of ^{14}C-uridine and after the introduction of actinomycin D revealed that about one-third of the labeled RNA was rapidly degraded after the addition of the transcription-blocking antibiotic, whereas the rest of the RNA remained stable (Fig. 11.7B). No similar loss of radioactive RNA was observed when actinomycin was added to cells that were exposed to ^{14}C-uridine for 18 hours. Approximate calculation indicated that the unstable component of the short term—labeled RNA amounted to about 0.5% of the total RNA content of the cells.[31]

Resolution of pulse-labeled RNA by sucrose gradient sedimentation indicated that in addition to the flow of radioactivity from the 45S and 35S peaks

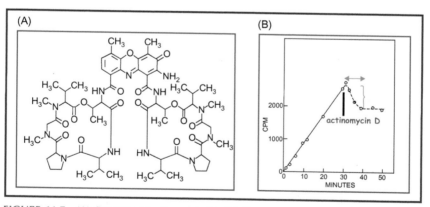

FIGURE 11.7 (A) Structure of actinomycin D. (B) Short-term (30 minutes) incorporation of ^{14}C-uridine into RNA was followed by the inhibition of synthesis of new RNA chains by actinomycin D. Acid-precipitable ^{14}C-labeled RNA was quantified at different intervals before and after the introduction of actinomycin D. Marked in *red* is a ~25—30% fraction of the prelabeled RNA that was degraded within 10 minutes after the addition of actinomycin D. *Adapted from Scherrer K, Latham H, Darnell JE. Demonstration of an unstable RNA and of a precursor to ribosomal RNA in HeLa cells. Proc Natl Acad Sci USA 1963;49(2):240–8.*

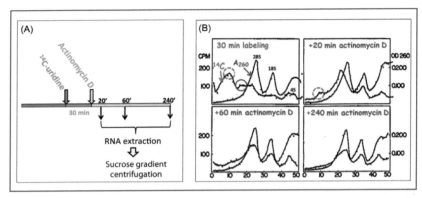

FIGURE 11.8   Fates of short and long term—labeled HeLa cell RNA in the presence of actinomycin D. (A) Design of the experiment. HeLa cells incorporated ^{14}C-uridine into RNA for 30
minutes until actinomycin D was added to block the synthesis of new RNA molecules. Total cell
RNA that was extracted from the cells at the end of the pulse period and at different times after
the introduction of actinomycin D was resolved by sucrose gradient sedimentation. (B) Sucrose
gradient profiles of the RNA before the addition of actinomycin D and at different times after its
introduction. Unlabeled RNA was traced in collected fractions of the gradient by the determination of adsorption at 260 mμ and pulse-labeled RNA was detected by counting radioactivity in
the fractions. Note that the 45S hnRNA peak (*red circle*) and the 35S peak (*blue circle*) disappeared already 20 minutes after the introduction of actinomycin D. *Adapted from Scherrer K,
Latham H, Darnell JE. Demonstration of an unstable RNA and of a precursor to ribosomal RNA
in HeLa cells.* Proc Natl Acad Sci USA *1963;**49**(2):240—8.*

into the stable 28S and 18S rRNA chains, portions of the 45S and the 35S
RNA were degraded within 20 minutes after the addition of actinomycin D.
By contrast, the 28S and 18S rRNA and 4S tRNA species remained
stable for the duration of the experiment (Fig. 11.8).

Most suggestively, Scherrer, Latham, and Darnell followed the procedure
of Hall and Spiegelman[3] to show that ^{32}P pulse-labeled RNA from HeLa
cells formed RNase-resistant hybrids with homologous DNA. The hybridized
RNA was found to have a significantly lower GC ratio than the total RNA.
It should be remarked here that in 1963 the existence of mRNA in animal
cells was not yet a proven fact. Yet, Scherrer, Latham, and Darnell argued
that because of the short lifetime of a fraction of the pulse-labeled RNA and
its complementarity to DNA, it was possible that chains of mRNA were
present in the population of the 45S and 35S giant molecules. Based on their
gathered data, they offered the following interim conclusions[31]:

> [...] (1) Ribosomal RNA in the HeLa cell is formed as part of very large RNA
> molecules which are then converted into smaller pieces. (2) There is in these
> cells a fraction that is rapidly synthesized and degraded, but the majority of
> the RNA is metabolically stable. (3) Of the total RNA a small fraction, which
> preferentially hybridizes with DNA, has a base ration nearer to that of DNA

than the majority of the RNA. This fraction possible corresponds to the DNA-like RNA described by others in animal cells [here the authors referred to previously published 1962 supporting reports[32,33]]

11.4 RELATIONSHIP OF POLYSOME-ASSOCIATED mRNA TO hnRNA

Evidence that the cytoplasm of animal cells contained mRNA molecules was gathered in the first half of the 1960s. The Darnell group identified in a series of papers a small fraction (~5%) of the cytoplasmic RNA whose properties were compatible with mRNA.[34-38] Their studies were initiated with the identification of poly-ribosomal structures ("polysomes") as the platform for protein synthesis. Darnell's early work with the then graduate student Jonathan (Jon) Warner (later at the Albert Einstein College of Medicine) indicated that poliovirus RNA directed in vitro synthesis of protein in structures that were larger than single ribosomes.[39] This observation was clarified when Warner joined Paul Knopf in the MIT laboratory of Alex Rich to demonstrate that protein synthesis in reticulocytes occurred on polysomes in which several individual ribosomes assembled along a single translated mRNA molecule into structures that were named polyribosomes (polysomes).[40] With globin being the overwhelmingly major protein that reticulocytes synthesize, their polysomes were shown to consist of five ribosomes—a number that corresponded to the length of globin mRNA.[40] The longer chains of polio virus mRNA were was shown shortly thereafter to be translated on polysomes that had a higher number of ribosomes.[34]

11.4.1 Cytoplasmic mRNA Detected

Joining forces with Sheldon Penman (1930–), an MIT high-energy physicist-turned molecular biologist, and with two postdoctoral fellows, Scherrer and Yechiel Becker, Darnell investigated the cytoplasmic RNA that directed protein synthesis in uninfected and in polio virus–infected HeLa cells. As illustrated in Fig. 11.9A, nascent proteins were found to be associated with the polysomal structures. Importantly, pulse-labeled RNA that was extracted from the polysomes migrated in a sucrose gradient as a wide peak of 8-20S that was distinct from the 28S and 18S rRNA species (Fig. 11.9B).

This rapidly labeled mammalian RNA was roughly similar in size to *E. coli* and phage pulse-labeled mRNA[2,4] but it was much shorter than the 45S rapidly labeled hnRNA (Figs. 11.4, 11.5 and 11.8). Importantly the base composition of the polysome-associated cytoplasmic pulse-labeled RNA was similar to that of the DNA but it differed from the composition of rRNA.[34] Based on its DNA-like base composition and its prokaryotic mRNA-like size, the briefly labeled polysome-bound RNA was presumed to represent

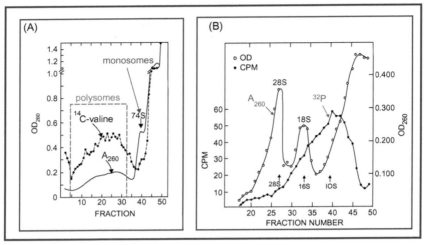

FIGURE 11.9   HeLa cells synthesize proteins in vivo and the polysomes-associated pulse-labeled RNA is distinct from rRNA. (A) Newly made proteins are associated with polysomes. HeLa cells were exposed for 1 minute to ^{14}C-valine, incorporation was stopped, and the separate cytoplasmic fraction was resolved by sucrose gradient centrifugation. Results indicated that the nascent-labeled protein was associated with the polysomal fraction, whereas no newly made protein was associated with individual ribosomes ("monosomes"). (B) HeLa cell RNA was pulse labeled for 40 minutes by ^{32}P, polysomes were resolved by sucrose gradient sedimentation as in (A), and their extracted RNA was centrifuged in a sucrose gradient. Detection of the adsorption at 260 mμ of the unlabeled RNA and of the radioactivity of the pulse-labeled RNA showed that the polysomes-associated labeled DNA had lower size than the 28S and 18S rRNA subunits. Sucrose gradient profiles in (A) and (B) were adapted from Ref. 34. *Panels (A) and (B): Modified from Penman S, Scherrer K, Becker Y, Darnell JE. Polyribosomes in normal and poliovirus-infected HeLa cells and their relationship to messenger-RNA.* Proc Natl Acad Sci USA *1963;49(5):654–62.*

mammalian cell mRNA. In subsequent works Darnell and associates substantiated the assumption that mammalian cell ribosomes associated with cytoplasmic mRNA to form polysomes that were engaged in the synthesis of proteins.[34-38] Notably, these results also established the distinct difference between the sizes and fates of the cytoplasmic mRNA and of the rapidly made giant hnRNA molecules that were quickly degraded in the nucleus without ever reaching the cytoplasm.[36,38]

11.4.2   Emerging Hints of Precursor–Product Relationship Between hnRNA and mRNA

By separating nuclei from nucleoli that contained the bulk of rRNA precursor molecules, it became possible to obtain hnRNA that was free of rRNA. The base composition of such hnRNA was found to be similar to the compositions of both the genomic DNA and of cytoplasmic mRNA.[38] The Caltech group of the 1975 Nobel laureate Renato Dulbecco (1914–2012) detected polyoma virus-specific RNA in induced tumor cells.[41] Branching from this finding, Darnell and his postdoctoral fellow Uno Lindberg showed that Simian Virus 40 (SV40) produced in the nucleus virus-specific RNA that was larger than SV40-specific polysomal RNA.[42] Although this finding was restricted to virus-specific RNA, it suggested a possible precursor–product relationship between the hnRNA and cytoplasmic mRNA. All in all, the accumulated information invited at the end of the 1960s a definitive answer to the question of the relationship between hnRNA in the nucleus and mRNA in the cytoplasm.

11.5   CONSERVED TAGS AT THE 3'- AND 5'-ENDS OF hnRNA AND OF mRNA SUBSTANTIATED THEIR PRECURSOR– PRODUCT RELATIONSHIP

Several principal points about the metabolism of RNA in animal cells became evident by the end of the 1960s. (1) The cell nucleus contained short-living giant RNA transcripts. (2) Rapidly made RNA of much smaller size was associated with the protein-synthesizing polysomes in the cytoplasm. (3) The association of the rapidly made cytoplasmic RNA with polysomes, its size and DNA-like base composition were characteristic of mRNA. (4) DNA-like base composition of the rapidly labeled hnRNA indicated that it contained mRNA sequences. Despite these indications of kinship between hnRNA and the cytoplasmic mRNA, their exact relationship had yet to be determined. A nearly definitive evidence that hnRNA was precursor to mRNA was obtained in a series of discoveries in the early to mid-1970s.

11.5.1   Both hnRNA and mRNA in Animal Cells Were Found to Have Tracts of Polyadenylic Acid at Their 3'-Ends

A Poly(A) Polymerase Was Discovered

The University of Pittsburgh biochemist Mary Edmonds (Box 11.3) identified and isolated in 1959–60 an enzyme from thymus that utilized ATP to generate chains of polyadenylic acid [poly(A)] with no need for a template.[43]

The ~100-fold partially purified poly(A) polymerase utilized ATP to synthesize poly(A) in vitro in a Mg^{2+}-dependent reaction: $nATP \rightarrow poly(A) + nPP_i$. The enzyme acted in a substrate-specific manner such that only ATP but not CTP, UTP, or GTP were utilized for the polymerization of corresponding ribonucleotide monophosphates (rNMPs).[43]

RNA-Linked Poly(A) Chains Were Detected

Almost a decade after she isolated and characterized poly(A) polymerase from thymus tissue, Edmonds and Grace Caramela characterized RNA-linked poly (A) chains that they isolated from Ehrlich ascites cells.[44]Fig. 11.11A is a schematic illustration of the experimental procedure that the two investigators

employed to isolate in vivo-synthesized poly(A)-tagged RNA molecules. After growing the cells in the presence of a radioactive ^{32}P isotope, the total cellular RNA was extracted and passed through a poly(T)-cellulose column. The cellulose-born poly(T) bound RNA molecules that had poly(A) tracts. Unbound RNA molecules were washed off the column by salt solution and molecules that remained in association with the poly(T) were released by passing through the column a buffer with no salt (Fig. 11.11A). To determine RNA base compositions, the separately collected poly(T) unbound and bound RNA fractions were hydrolyzed by alkali and rNMP products of hydrolysis were resolved by ion exchange column chromatography and quantified.

As is evident from the tabulated results in Fig. 11.11B, the unbound RNA fraction was comprised of roughly equal proportions of all the four rNMPs. By contrast, the bound RNA that represented only ~1% of the total cell RNA was highly enriched in AMP which comprised ~75% of the total rNMP content of this fraction (Fig. 11.11B). Thus the poly(dT) column appeared to have trapped RNA molecules that had poly(A) tracts. Edmonds and Caramela also discovered that the majority of these poly(A)-containing RNA molecules was located in the cell nucleus. Also the poly(A)-tagged RNA fraction was more resistant to nucleolytic digestion by T1 RNase than the bulk RNA. Sucrose gradient sedimentation of the nuclease-resistant, poly(A)-containing RNA revealed its size to be in the 8-10S range.[44]

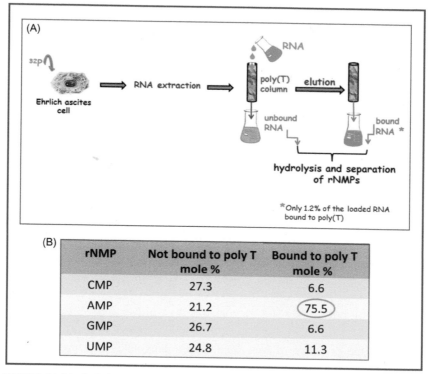

FIGURE 11.11   Isolation and characterization of a minor poly(A)-rich RNA fraction from Ehrlich ascites cells. (A) Scheme of the design of the Edmonds and Caramela[44] experiment. (B) Relative content of the four rNMPs in hydrolyzates of unbound and poly (T)-bound fractions of the cell RNA. *Results derived from Edmonds M, Caramela MG. The isolation and characterization of adenosine monophosphate-rich polynucleotides synthesized by Ehrlich ascites cells. J Biol Chem 1969;244(5):1314–24.*

Both hnRNA and mRNA Were Found to Have Poly(A) Tracts

Less than 2 years after Edmonds isolated poly(A)-tagged RNA, her laboratory[45] and the groups of Darnell at Columbia University[46] and of George Brawerman (1927–) at Tufts University[47] published three back-to-back papers in the *Proceedings of the National Academy*—all showing that both eukaryotic mRNA and hnRNA carried poly(A) tails at their 3′-ends. Similar to Edmonds' previous report,[44] Darnell and associates found that poly(A)-tagged RNA resisted digestion by RNase in solutions of high ionic strength.[46] Taking advantage of this relative resistance to hydrolysis by RNase, Darnell's team showed that both nuclear and cytoplasmic RNA had poly(A) chains. In this experiment (Fig. 11.12A) total HeLa cell RNA was pulse labeled with ^{14}C-uridine while parallel short-term exposure of the cells to ^{3}H-adenosine favorably labeled RNA molecules that had poly(A) tracts.

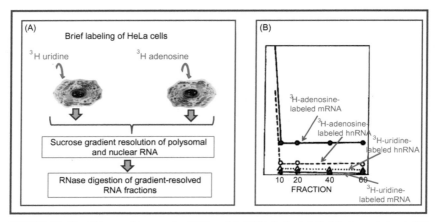

FIGURE 11.12 RNase resistance of mRNA and hnRNA indicated that both carried poly(A) tails. (A) Schematic depiction of the experiment.[46](B) Sucrose gradient−resolved fractions of ³H-adenosine-labeled polysomal mRNA and nuclear hnRNA were more resistant to RNase hydrolysis than ³H-uridine-labeled bulk cytoplasmic and nuclear RNA. *Adapted from Darnell JE, Wall R, Tushinski RJ. An adenylic acid-rich sequence in messenger RNA of HeLa cells and its possible relationship to reiterated sites in DNA. Proc Natl Acad Sci USA 1971;68 (6):1321−5.*

Incorporation of the radioactively labeled precursors into RNA was conducted in the presence of low dose of actinomycin D that preferentially inhibited the synthesis of rRNA. After termination of the radioactive labeling, the cells were lysed, their nuclei and cytoplasm were separated, and the nuclear sap and polysomes were resolved by sucrose gradient sedimentation. Fractions of gradients with resolved nuclear sap or polysomes that contained total ³H-uridine-labeled RNA or poly(A)-enriched ³H-adenosine-labeled RNA were subjected to RNase digestion at high ionic strength. As illustrated in Fig. 11.12B, RNase completely hydrolyzed the total RNA from both polysomes and nuclei. By contrast, all the gradient-resolved fractions of the poly(A)-enriched polysomal RNA had a substantial RNase-resistant component. Similarly, fractions of the ³H-adenosine-labeled nuclear RNA also contained a smaller but significant share of RNase-resistant molecules. This result was consistent with the idea that both the polysome-associated cytoplasmic mRNA and the nuclear hnRNA had molecules that carried poly(A) tracts.

hnRNA and mRNA Were Found to Have 150−200 Bases-Long Poly(A) Tails at Their 3′-Ends

The main properties of the poly(A) tracts of hnRNA and mRNA were defined between 1971 and 1973. The Brawerman group reported that the isolated RNase-resistant remainder of mRNA migrated in sucrose gradients as a 4S species that corresponded to polynucleotide chains of 50–150 bases.[47] More accurate measurements that were made shortly thereafter showed the poly(A) tract to be comprised of 150–250 nucleotides.[48] It was also revealed that the addition of a poly(A) tail to RNA molecules was an actinomycin-resistant posttranscriptional event.[48] In addition, analysis showed that the poly(A) chains were covalently linked to the 3′-ends of the RNA molecules and that the terminal poly(A) tails of hnRNA and mRNA were of similar size.[49]

The Presence of Poly(A) Tails in Both hnRNA and mRNA Was in Line With Their Precursor–Product Relationship

Two lines of evidence were consistent with respective precursor–product relationship between the poly(A)-bearing hnRNA and the similarly end-tagged mRNA molecules. First, the inhibition of the addition of poly(A) tails to hnRNA by the adenosine analog cordycepin (3′-deoxyadenosine) also depressed the appearance of mRNA in the cytoplasm.[48] Second, monitoring of the kinetics of poly(A) addition revealed that poly(A) chains appeared first in hnRNA and only later in mRNA. This nucleus-to-cytoplasm migration of poly(A)-tagged RNA molecules also gave credence to the notion that hnRNA was the precursor of mRNA.[50] All in all, considering the data that became available in 1971, Mary Edmonds and Darnell separately raised ideas on the nature of hnRNA and mRNA and on their relationship. Edmonds speculated that[45]:

> Previously it has been established that the HnRNA turns over very rapidly within HeLa nuclei; less than 10% of this RNA is likely to be transported to the cytoplasm, if any is [...]. We observe approximately a 10-fold difference in poly(A) content between HnRNA and mRNA [...]. We therefore raise the possibility that each molecule of HnRNA transcribed contains within it one sequence of potential mRNA adjacent to a small poly(A) sequence. The HnRNA molecule would be rapidly degraded after or during transcription, and only a small portion of it including mRNA and poly(A) would be conserved and exported to the cytoplasm.

Based on their observation that poly(A) was present in both hnRNA and mRNA, Darnell and his associates made the following suggestions[48]:

> Most HnRNA is not a precursor of mRNA but is synthesized and degraded within the cell nucleus [...]. Any scheme for the use of a portion of the HnRNA to form a portion of the mRNA must distinguish between HnRNA destined for mRNA and the bulk of HnRNA. The attachment of poly(A) could be part of such a selective process. [...] The following scheme of mRNA biosynthesis in HeLa cells seems justified by these experiments. After transcription, HnRNA molecules are selected by some unspecified mechanism for conversion to mRNA. A poly(A) segment is attached, probably at a terminus of the HnRNA molecule. One or more nucleases might first recognize and cleave off the mRNA plus poly(A) unit and

then destroy the remainder of the HnRNA, or might simply destroy all the HnRNA except for this unit. This posttranscriptional modification is necessary for mRNA molecules to reach the cytoplasm.

Essentially, therefore, Darnell assumed that the conversion of hnRNA to mRNA involved cleavage and degradation of the 5′ portion of hnRNA and conservation of its poly(A) tagged 3′-end as mature mRNA. This model was put under question when the subsequently discovered 5′-caps of hnRNA were found to also be conserved in mRNA (see Section 11.5.2).

Although this chapter dealt with the poly(A) tracts in hnRNA and mRNA mainly as markers of the 3′-ends of these two classes of eukaryotic RNA, this section cannot be concluded without a mention of later discoveries on the metabolism and cellular functions of these 3′-tails. It was found that the poly(A) tail binds a specific poly(A)-binding protein which promotes the export of mRNA from the nucleus to the cytoplasm stimulates translation and inhibits its degradation.[51] This poly(A)-binding protein also associates with and recruits other proteins that affect translation.[52] One such protein is initiation factor-4G, which in turn enlists the 40S ribosomal subunit.[53]

11.5.2   The Plot Thickened: Both hnRNA and mRNA in Animal Cells Were Found to Have a Cap Structure at Their 5′-Ends

A succession of discoveries that were made in the mid-1970s in several laboratories revealed that additionally to the poly(A) tails at the 3′-termini of mRNA and hnRNA of animal cells and their viruses, these two categories of RNA had also a unique structure, that was designated "cap," at their 5′-ends.

Many researchers contributed to the discovery and the characterization of cap structures in RNA from diverse cells and viruses (the history of the discovery of cap structures was comprehensively reviewed in Ref. 54). Two pioneering students of RNA caps were Aaron Shatkin and Robert (Bob) Perry (Box 11.4).

Discovery of Cap Structures in Viral RNA

Interestingly, a unique blocked and methylated 5′-terminal structure was first detected small nuclear RNA from hepatoma cells and not in mRNA or hnRNA.[55] This structure consisted of trimethyl guanine that was linked by a pyrophosphate bridge to a methylated adenosine residue at the 5′-end of the RNA. However, since small nuclear RNA had no known function at the time, the significance of this early observation did not gain proper recognition. At about the same time, Perry and his longtime associate Dawn Kelly found that mRNA in mouse L-cells was methylated and that hnRNA also had low level of methylation.[56] Coincidentally, Shatkin together with a group from the National Institute of Genetics in Mishima, Japan showed that the 5′-ends of the double-stranded RNA of reoviruses were blocked by

BOX 11.4 Aaron Shatkin and Robert Perry

Aaron Shatkin obtained in 1956 a bachelor's degree from Bowdoin College and won in 1961 a PhD degree under the guidance of Edward Tatum at the Rockefeller University. His lifelong interest in animal viruses began with his studies at the NIH on the biogenesis of mRNA of vaccinia virus. There he also started to study double-stranded RNA reoviruses, a subject that became a major focus of his investigations. After moving to the Roche Institute of Molecular Biology, he found with Amiya Banerjee that the reovirus genome consisted of 10 separate double-stranded RNA "chromosomes." These genomic RNA fragments were copied by the viral RNA polymerase to produce transcripts that functioned as both mRNAs and as templates for synthesis of progeny fragments of the genome. This work led to his discovery with Yasuhiro Furuichi of the 5'-terminal m⁷GpppN cap structure in the viral mRNA. Subsequent studies in his laboratory established the importance of the cap for mRNA stability, uncovered the RNA ligation pathway, and described splicing of tRNA in animal cells.

A native of Chicago, Robert Perry earned in 1951 a bachelor's degree in mathematics from Northwestern University and in 1956 a PhD in biophysics from the University of Chicago. In 1960 he joined the Fox Chase Institute of Cancer Research in Philadelphia at which he remained until his retirement in 2004. Perry's lifelong research interest was the metabolism of RNA in animal cells. In 1962 he discovered that rRNA was made in the nucleolus. Shortly thereafter he, side-by-side with Darnell and Sheldon Penman, demonstrated that mature rRNA was the product of the processing of larger pre-rRNA precursor molecules. Following his subsequent discovery of methylated bases and sugar residues in mRNA, he, Fritz Rottman, and Shatkin identified the methylated cap structure of mRNA. In later work Perry and associates followed the fate of transcripts of immunoglobulin genes and studied ribosome biogenesis (Fig. 11.13).

Aaron J. Shatkin
(1934–2012)

Robert P. Perry
(1931–2013)

FIGURE 11.13   Aaron J. Shatkin and Robert P. Perry.

2'-*O*-methyl guanosine.[57] Similarly a methylguanosine residue was also identified by others in the 5'-end of the RNA genome of silkworm cytoplasmic polyhedrosis virus (CPV).[58] Yasuhiro Furuichi of the Mishima Institute of Genetics complemented the identification of methylated ribonucleotides in the 5'-termini of viral RNA by detecting methyltransferase activity in purified particles of CPV.[59] This enzyme, which was promptly detected in various other viruses, transferred methyl group from S-adenosylmethionine to the 5'-end of the viral RNA.

The Structure of the Cap Was Defined

By 1975 several groups identified the site of methylation in RNA and elucidated the structure of 5'-caps in mRNA of different viruses.[60-62] Common to all the caps was a 7-methylguanosine residue that was linked via a 5'-5' triphosphate bridge to the first nucleotide in the RNA chain (Fig. 11.14).

A specific enzyme, guanylyl transferase, was found to add the blocking guanosine residue to 5'-ends of 20−100 nucleotides long chains of nascent pre-mRNA during their transcription. Because the 5'-termini of the RNA molecules ended in a phosphate group, the bond that linked GTP to the RNA molecule was an unusual 5'-5' triphosphate. Additionally to the blocking 5' methylated guanosine residue, the first few different ribonucleotides of the mRNA itself had also methyl groups at variable locations.

Caps Were Also Identified in mRNA and hnRNA of Animal Cell

Within few months after their initial report on the structure of viral RNA cap, Shatkin and associates collaborated with the Darnell laboratory to show

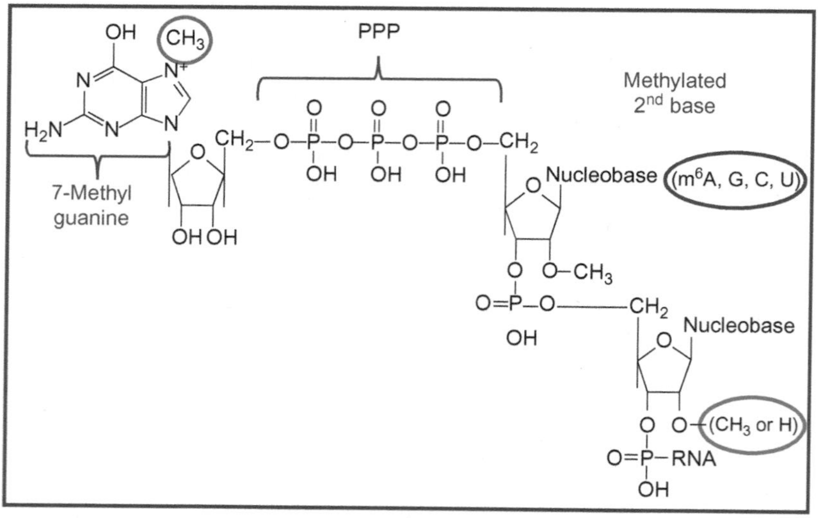

FIGURE 11.14   Structure of the RNA 5'-cap. In all the cap types the 5'-terminal base is 7-methyl (*red circle*) guanine linked by a triphosphate bridge to the first ribonucleotide in the RNA. The first nucleotide in the RNA chain (A, G, C, or U) is also methylated at position 6 (*blue ellipse*) and the ribose residue of the next ribonucleotide may or may not be methylated (*green ellipse*). The next two ribonucleotides may also be variably methylated.

that HeLa cell mRNA also carried a cap at its 5′-end.[63] Caps were concurrently identified in mRNA from mouse L-cells by Perry and associates,[64] in Novikoff hepatoma mRNA by the Michigan State University laboratory of Fritz Rottman (1937–2013),[65] and in HeLa cell mRNA by Bernard Moss (1937–) and his associates at the NIH.[66,67]

Almost immediately after the identification of caps in mRNA, studies of L and HeLa cells in the laboratories of Perry[68] and of Shatkin and Darnell,[69] respectively, revealed that hnRNA molecules were also capped. Some principal attributes of mRNA and hnRNA caps were brought to light in short order. Ribonucleotides at the 5′-ends of these types of RNA were found to be methylated to different extents and at different positions. Caps have been detected in both hnRNA molecules that had or were devoid of poly(A) tails. The presence of caps in hnRNA molecules that still had not acquired a 3′ tail of poly(A) inferred that capping preceded polyadenylation. Significantly, the assumed kinship between hnRNA and mRNA was reinforced by finding that the majority of the caps in hnRNA of L-cell were conserved in their mRNA.[70] Finally, Shatkin, Darnell, and their associates found that hnRNA contained internally positioned methylated nucleotides in excess of caps and that the internally methylated residues of hnRNA were absent from molecules of mRNA.[69] Considering the different ratios of caps to internally positioned methylated nucleotides in hnRNA and in mRNA, these authors raised the following suggestion[69]:

The present findings that hnRNA contains caps and internal methylations in excess of caps are compatible with the proposal that mRNA in cultured cells derives in part from higher molecular weight nuclear RNA.

Though insightful, this proposal did not take the extra step of suggesting that hnRNA may be spliced to yield mRNA (see Sections 4.3 and 11.6).

Functions of Caps Were Identified at a Later Time

Much had been learned since the discovery of caps about their metabolism and biological roles. Although the focus of this chapter is on caps and poly(A) tails as mere markers of the respective 5′- and 3′-ends of both hnRNA and mRNA, subsequently discovered functions of the caps deserve mention. Caps were found to combine with two nuclear proteins, CBP 20 and CBP 80 (CBP stands for Cap-Binding Protein) to form a Cap Binding Complex (CBC).[71,72] The CBC protein affects splicing of the pre-mRNA and contributes to the transport of pre-mRNA from the nucleus to the cytoplasm. Once the capped mRNA reaches the cytoplasm, its translation requires binding of the initiation factor eIF4A to the cap structure.[73] Notably, evidence showed that capping, polyadenylation, and splicing of hnRNA are interdepended.[74]

11.5.3   A Puzzle Surfaced: Both hnRNA and mRNA Had Caps and Poly(A) Tracts at Their Respective 5'- and 3'-Ends and yet mRNA Was Several Folds Shorter Than hnRNA

The discovered 5'-caps and 3'-poly(A) tails in hnRNA and in mRNA and the identification of internally methylated bases in hnRNA but not in mRNA raised a baffling conundrum: How could internally methylated giant molecules of hnRNA be transmuted into shorter chains of mRNA that had the same terminal tags but no internal methylated bases? This puzzle was aptly described by Darnell in a 1999 retrospective account[75]:

> At the end of these experiments (published in February 1976) [Darnell referred here to Ref. 69] we had clear knowledge that poly(A)-containing mRNA had one cap and one poly(A) per 1500 bases, whereas the polyadenylated fraction of hnRNA had an average of one cap and one poly(A) per 5000 bases. And in fact the largest hnRNA had one cap and one poly(A) per 10,000 bases.

Additionally to the described paradox of the different sizes of hnRNA and mRNA despite their conserved 5'- and 3'-terminal caps and poly(A) tails, there was the question of the internal methylated bases in hnRNA that were absent from mRNA.[69] These components of the riddle of the conversion of hnRNA into mRNA are illustrated in Fig. 11.15.

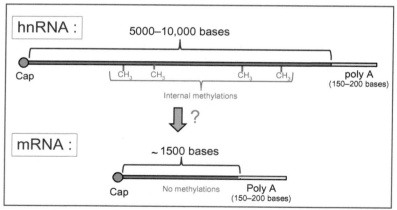

FIGURE 11.15   Relationship between hnRNA and mRNA as it was known in the mid-1970s (diagram not in scale). Each molecule of hnRNA and of mRNA had at its 5'- and 3'-ends, respectively, a single cap and one poly(A) tail. In addition, some internal bases in hnRNA were methylated, whereas mRNA was devoid of methylated internal bases. Finally the hnRNA molecules were approximately three- to sevenfold longer than those of mRNA. Since it was already understood at the time that hnRNA was a precursor to mRNA, the question was how could 5'- and 3'-terminally tagged and internally methylated large molecules of hnRNA converted into similarly tagged and unmethylated shorter molecules of cytoplasmic mRNA (*arrow with a question mark*).

With the benefit of hindsight the solution to this problem appears to a present day reader to be self-evident; that is, that internal sections of hnRNA are removed and the remaining RNA segments are ligated to yield shorter, unmethylated mRNA molecules. Yet, although such mechanism could have been inferred from the available experimental evidence, it was not postulated at the time even as a theoretical possibility. Section 11.7 discusses possible reasons for the failure to predict from the existing empirical data a splicing mechanism. Historically, though, with no theoretical model in existence, splicing was discovered by experiment alone.

11.6 SPLIT GENES DISCOVERED: ADENOVIRUS mRNA WAS FOUND TO BE NONCOLLINEAR WITH THE TRANSCRIBED GENE

Two laboratories provided in 1977 independent experimental solution to the riddle of the conversion of hnRNA into mRNA. The future (1993) Nobel laureates Philip Sharp at the MIT and Richard Roberts in the Cold Spring Harbor Laboratory (Box 11.5) independently made the surprising discovery that unlike the collinear prokaryotic mRNA, in eukaryotes collinear primary hnRNA transcripts were spliced to yield noncollinear mRNA molecules.

Historically, however, both the Sharp and Roberts teams made their groundbreaking discovery not in animal cells but in the simpler experimental system of human adenovirus—a double-stranded DNA virus that causes upper respiratory infections. The advantage of the virus system over cells

---

BOX 11.5 Philip Sharp and Richard Roberts

Born and raised in a rural farming community in Kentucky, Phillip Sharp attended a small liberal arts college (Union) in his native state and did a PhD thesis on the physical chemistry of DNA at the University of Illinois. During his postdoctoral work in the Caltech laboratory of Norman Davidson, he used the technique of electron microscopy of DNA heteroduplexes to identify deletions in DNA. A variant of this experimental procedure was later instrumental in his discovery of split genes and of RNA splicing. During a second postdoctoral period at Cold Spring Harbor, he studied adenovirus gene expression and constructed with others a restriction map of the viral genome. On the invitation of Salvador Luria he joined in 1974 the MIT Department of Biology. There he continued his studies on adenovirus gene expression, focusing on the quantification of early and late viral mRNA molecules. These studies were capped by his 1977 discovery with Susan Berget and Clair Moore of split viral genes. For this achievement he shared with Richard Roberts the 1993 Nobel Prize in Physiology and Medicine.

(Continued)

BOX 11.5 (Continued)

Raised and educated in Bath, UK, Roberts did his PhD thesis in Sheffield University and did postdoctoral research at Harvard. Working initially on tRNA sequencing, he changed course to study and utilize restriction enzymes. After his recruitment by Watson to the Cold Spring Harbor Laboratory, he constructed with others a restriction map of adenovirus and looked for initiation and termination signals in the viral mRNA. Shortly after the discovery of capping, he made the surprising finding that all the late adenovirus mRNA species started with the same 5′ sequence, although their main bodies were mapped far away from this 5′-terminal sequence. He and his associates then demonstrated by electron microscopy the split nature of adenovirus genes. For his independent discovery of RNA splicing, he shared with Philip Sharp the 1993 Nobel Prize in Physiology and Medicine (Fig. 11.16).

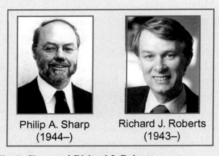

Philip A. Sharp (1944–)   Richard J. Roberts (1943–)

FIGURE 11.16   Philip A. Sharp and Richard J. Roberts.

was that although the fraction of transcripts of any specific cellular gene is small and thus hard to detect, viral gene transcripts comprise 30−50% of the nuclear pulse-labeled RNA in infected cells.

After the discovery of the noncollinearity of adenovirus mRNA, methods were developed to follow the fate of transcripts of specific cellular genes. Use of these methods revealed that cellular mRNA molecules in eukaryotes were also noncollinear with the transcribed genes.

11.6.1   Structure of Adenovirus and the Transcription of Its DNA

Adenoviruses are nonenveloped viruses that have a nonsegmented double-stranded DNA genome. To attach to surface receptors on the host cell, the virus uses knob domains of fibers that are secured to bases of penton protein on its capsid. The most abundant viral capsid protein is the homo-trimeric hexon[76] (Fig. 11.17A). Adenoviruses penetrate the host cell through endosomes that acidify, disband, and release virions into the cytoplasm. With the

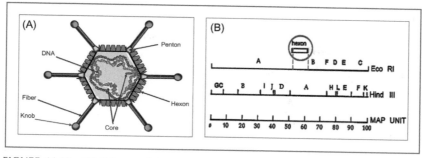

FIGURE 11.17 (A) Scheme of the structure of adenovirus. (B) Restriction maps of human adenovirus-2. The vertical lines mark cleavage sites by either *Eco*RI or *Hind*III restriction endonucleases. Capital letters designate the product restriction fragments. Position of the hexon gene is marked by a *red circle. Adapted from Berget SM, Moore C, Sharp PA. Spliced segments at the 5' terminus of adenovirus 2 late mRNA.* Proc Natl Acad Sci USA *1977;74(8):3171—5.*

assistance of microtubules, the viruses are then transported through the nuclear pore complex into the nucleus.[77] The virus particles disassemble in the nucleus and release their DNA which then associates with histone molecules.

Transcription of adenovirus genes in the nucleus is followed by migration of mature viral mRNA molecules to the cytoplasm, their association there with polysomes and their translation.[79] Even before split genes were discovered, some properties of the adenovirus RNA appeared already to be inconsistent with the prokaryotic-based dogma of collinear mRNA. First, after identifying 3'-poly(A) tagged adenovirus-2 mRNA molecules,[80] Darnell and associates showed in 1973 that the size of the viral RNA in nuclei of infected cells was much larger than that of the cytoplasmic mRNA.[81] Moreover, in "hybridization competition" experiments they demonstrated that excessive amounts of the cytoplasmic viral mRNA excluded only part of the nuclear RNA from its hybrids with the viral genome. This suggested that the primary nuclear transcripts included some viral sequences that were not represented in the mature mRNA molecules.[81]

11.6.2 Restriction Fragments of Adenovirus DNA Were Essential Tools in the Discovery of Split Genes and RNA Splicing

Restriction fragments of the adenovirus genome and their ordered arrangement in maps were crucial elements of the experimental system that was used in the discovery of the noncollinear nature of the viral mRNA, of split genes, and of RNA splicing. A large Cold Spring Harbor Laboratory team that included Roberts and Sharp and by the group of Lennart Philipson at Uppsala University (1929—2011) provided in a string of technical papers detailed restriction maps of the human adenovirus-2 genome. In these studies the viral DNA was digested by specific restriction nucleases to produce a series of fragments that were then hybridized to specific early or late viral mRNA species. The fragments were then assembled in order into a map of the full genome of the virus and each fragment was functionally identified by the mRNA species that it encoded. Fig. 11.17B shows map of ordered fragments that were generated by the restriction nucleases *Eco*RI and *Hin*dIII. Complete genomic maps of adenovirus were presented in the 1974 Cold Spring Harbor Symposium on tumor viruses.[82−84] A technique that was developed at about the same time enabled the separation of the two strands of adenovirus DNA fragments such that the transcribed strand could be identified by hybridization.[85]

The use of restriction fragments and the identification of transcribed strands unveiled attributes of the nuclear and cytoplasmic adenovirus RNA that did not conform to the simple picture of a collinear mRNA. First, Steven Bachenheimer (later at the University of North Carolina) and Darnell reported in late 1975 that adenovirus-2 produced primary nuclear transcripts of 25−30,000 base pairs that spanned all or most of its genome.[86] Damaging in vivo the transcribing viral genome by UV light resulted in preferential loss of distal segments of the RNA transcripts. By exploiting this phenomenon Darnell and associates showed that late mRNA molecules were derived from long nuclear transcripts.[87,88] Hybridization of the large nuclear transcripts with ordered restriction fragments of the viral DNA yielded the surprising result that most of these RNA molecules started from a common point in the genome. Finally, processing of the giant nuclear transcripts within a few minutes was found to generate much smaller molecules of mRNA.[86] All in all, these data were in line with the idea that long molecules of nuclear adenovirus RNA served as precursors to the shorter mRNA molecules. It also became evident that portions of the nuclear primary transcripts were lost during their conversion into the shorter mRNA chains.

11.6.3 Electron Microscopy of Nucleic Acids Was Another Crucial Tool in the Discovery of the Noncollinearity of Adenovirus mRNA

Side-by-side with the employment of restriction fragments of adenoviruses DNA, an equally essential tool in the discovery of split genes in eukaryotes and of the noncollinearity of their mRNA was electron microscopy of DNA−RNA heteroduplexes. This technique enabled direct sighting of sections of intragenic DNA sequences that were absent from mature mRNA molecules. To better appreciate the contribution of electron microscopy of DNA−RNA hybrids to the discovery of the noncollinear eukaryotic mRNA, a description of the evolution and principles of this technique must first be introduced.

Electron Microscopy of DNA–DNA Heteroduplexes

Electron microscopy of nucleic acids was applied already in the late 1960s to map deletion mutations in DNA. When the Stanford University molecular biologist Ronald (Ron) Davis (1941–) was still a postdoctoral fellow in the Caltech laboratory of Norman Davidson (1916–2002), he developed a technique to visualize deletions in DNA by electron microscopy and to measure their size with considerable accuracy.[89] Fig. 11.18A illustrates the principle of this technique.

Wild-type DNA was mixed together with DNA from a deletion mutant. Denaturation of the two types of DNA and their subsequent reannealing resulted in a mixture that contained homoduplexes of wild type or mutant DNA strands and also heteroduplexes between one strand of wild type and one strand of mutant DNA. Since the wild-type strand could not find a complementary sequence at the site of the deletion in the mutant strand, it formed a single-stranded loop at that position. The loop was then visualized by electron microscopy and its contours were measured to yield rather accurate estimate of the size of the deletion.[89] When Philip Sharp joined the Davidson laboratory, he used this technique to visualize and measure deletions in DNA of the phage φX174.[90] Electron micrographs from that study of heteroduplexes of wild-type DNA and mutant DNA are shown in Fig. 11.18B.

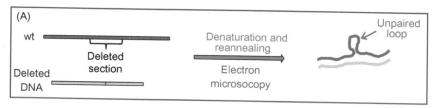

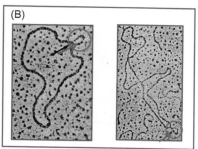

FIGURE 11.18 Mapping of deletions in DNA by electron microscopy of heteroduplexes between wild-type DNA and DNA of a deletion mutant. (A) Scheme of the procedure. (B) Electron micrographs of two heteroduplexes. *Red circles* mark loops of single-stranded wild-type DNA that have no complementary sequences at the region of the deletion in the mutant DNA. *Adapted from Davis RW, Davidson N. Electron-microscopic visualization of deletion mutations.* Proc Natl Acad Sci USA *1968;60(1):243–50.*

Electron Microscopy of DNA–RNA Heteroduplexes

Visualization of DNA–DNA heteroduplexes evolved within several years into the imaging of DNA–RNA heteroduplexes to identify DNA regions that were represented in or missing from their RNA transcripts. This technique, termed R-loop mapping, was developed in the Stanford University laboratory of David Hogness (1925–) by Raymond White (1943–)[91] and in collaboration with Ronald Davies.[92] The principle of this method is illustrated in Fig. 11.19A.

RNA was annealed to DNA in the presence of formamide. By forming hydrogen bonds with DNA strands this agent competes with Watson–Crick base pairs and disrupts duplex DNA. Under borderline conditions of defined formamide concentration and temperature, DNA double strands were found to be less stable than RNA–DNA hybrids. Thus, in a mixture of DNA and complementary RNA, the entrance of a homologous RNA strand into an RNA–DNA heteroduplex displaced the noncomplementary DNA strand from the DNA duplex. As a result, a circlet of the dislocated DNA strand, designated R-loop, was formed at the site of the RNA–DNA heteroduplex. Electron microscopy was then employed to visualize and to map the segment

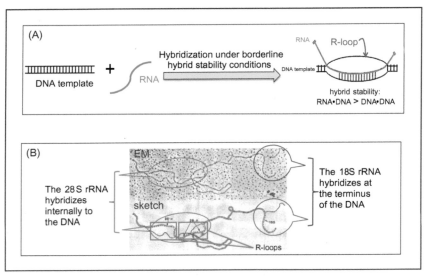

FIGURE 11.19   (A) Scheme of the R-loop technique for the mapping of transcribed tracts in DNA. See text for details. (B) Electron micrograph (upper panel) and tracing (lower panel) of *Drosophila* rRNA–DNA heteroduplex. The 18S rRNA displaced a continuous DNA stretch of the encoding genes, whereas the 28S rRNA formed two R-loops by displacing two DNA tracts (*green rectangles*). These two blocks of rRNA-encoding DNA were interrupted by the marked stretch of nontranscribed DNA sequence. *Modified from White RL, Hogness DS. R loop mapping of the 18S and 28S sequences in the long and short repeating units of* Drosophila melanogaster *rDNA. Cell 1977;10(2):177–92.*

in the DNA that was homologous to the hybridized RNA. White and Hogness originally applied their innovative technique to map the rRNA genes in *Drosophila*. In this pioneering work they demonstrated that the encoding gene for the 28S rRNA was interrupted in its midst by a 5-kilobase (kb) insertion of a nontranscribed DNA tract[91] (Fig. 11.19B).

11.6.4   Discovery of Split Adenovirus Genes and of Their Noncollinear mRNA

The Work of Philip Sharp

Philip Sharp had left the Cold Spring Harbor Laboratory in 1974 to join the MIT Department of Biology. Sharp later testified that by sharing the same floor with the laboratories of David Baltimore, Robert Weinberg, Nancy Hopkins, and David Housman, he found himself in an ideal scientific environment. Initial efforts in his MIT laboratory were dedicated to the mapping of regions in adeno- virus DNA that encode specific mRNA molecules at different stages of the lytic cycle. To identify which DNA fragment encoded each of the mRNA species, viral mRNA that was isolated at different times after infection was hybridized to restriction fragments of adenovirus DNA. It was this line of investigation that led to the eventual landmark discovery of the segmented structure of a viral gene and of the processing of its primary transcript into a noncollinear mRNA.

The historical discovery of split gene was made in the course of a study of the abundant late mRNA that encodes hexon—the major capsid protein of adenovirus-2 (Fig. 11.17A). Previous to this study the location of the hexon gene in the viral genome (Fig. 11.17B) was determined in 1975 by a Cold Spring Harbor team.[93] Susan Berget (later at the Baylor College of Medicine), Claire Moore, and Philip Sharp reported the landmark results of their classical experiment in the Aug. 1977 issue of the *Proceedings of the National Academy*. In this work they purified hexon mRNA, annealed it to homologous double-stranded *Hind*III restriction fragment of the viral DNA and viewed the formed heteroduplexes by electron microscopy (Fig. 11.20).[78] Electron micrographs of the hybrids revealed R-loops of displaced DNA strand at the site of pairing between the mRNA and a com- plementary DNA sequence[78] (Fig. 11.19). Consistently, however, small single-stranded tails were observed at both the 5′ and 3′ ends of the R-loops. The 3′ single-stranded tails could be explained by the presence of a poly(A) tract in the mRNA that had no complementary sequence in the DNA, whereas the existence of 5′ tails was not readily explicable.

One possibility was that this leader tail represented displacement of an RNA tract from the heteroduplex by a 5′ DNA sequence that re-formed a DNA–DNA duplex. This option was proven incorrect by showing that essentially identical loops and tails were formed when the mRNA was annealed to a single-stranded *Hind*III fragment of the viral DNA. Being

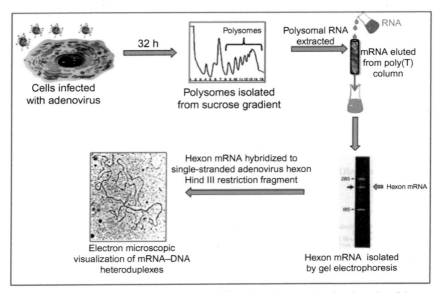

FIGURE 11.20   Scheme of the experiment that revealed that cytoplasmic adenovirus-2 hexon mRNA was not collinear with its encoding gene. Late viral RNA was isolated 32 hours after infection from resolved polysomes and poly(A)-tagged mRNA molecules were purified by affinity chromatography on poly(T) column. The major late viral mRNA species, hexon mRNA, was resolved by gel electrophoresis, its band was visualized, excised from the gel, and purified. The hexon mRNA was then hybridized to homologous restriction fragments of the viral DNA in the presence of formamide and under borderline conditions of duplex stability. Due to the higher stability of RNA–DNA heteroduplexes over DNA duplex, hexon mRNA replaced a DNA strand in hybrid with a complementary region in the DNA. As a result, an R-loop of the displaced DNA strand could be visualized by electron microscopy. *Image of the gel and the electron micrograph were derived from Berget SM, Moore C, Sharp PA. Spliced segments at the 5′ terminus of adenovirus 2 late mRNA.* Proc Natl Acad Sci USA *1977;74(8):3171–5.*

devoid of a complementary second strand, such a fragment could not reform DNA duplex and displace the RNA strand from the heteroduplex.[78] A series of additional control experiments excluded other, less likely, interpretations for the presence of the single-stranded RNA tails. The only remaining plausible explanation for the 5′ single-stranded RNA tail was that the DNA did not have a sequence that was complementary to a 5′ tract in the mRNA. If this was the case, then this 5′-end tract and the body of the hexon mRNA could not be transcribed from a contiguous stretch of DNA. That this was indeed the case was demonstrated by hybridizing hexon mRNA to an *Eco*RI DNA fragment (Fig. 11.17B). Unlike the *Hind*III fragment, the *Eco*RI segment included sequences that were much upstream to the body of the hexon gene. Electron micrographs of the hybrid (Fig. 11.21A) indicated that the 5′ region of the hexon mRNA formed a heteroduplex with three short DNA sequences that were situated upstream to the body of the hexon gene.

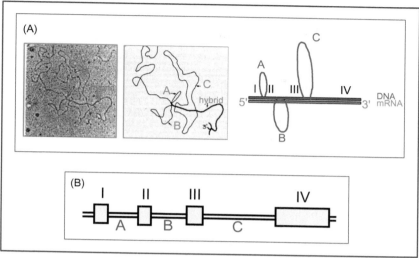

FIGURE 11.21 (A) Electron micrograph and tracing of heteroduplex between adenovirus-2 hexon mRNA and *Eco*RI fragment of the viral DNA that included the body of the hexon gene and upstream 5′ sequences. Visible loops of displaced single-stranded DNA are marked **A**, **B**, and **C**. The right-hand sketch is a schematic depiction of the heteroduplex regions **I**, **II**, **III**, and **IV** and of the loops of intervening DNA sequences that had no homologous tracts in the mRNA. (B) Scheme of the hexon gene. Coding sequences **II**, **III**, and **IV** that remain in the mature mRNA (rectangles) were later named *exons*. The exons are separated from one another by noncoding intervening sequences: **A**, **B**, and **C**. These segments, later named *introns*, are excised from the primary hnRNA transcript and are thus absent from the mRNA. *Panel (A): Adapted from Berget SM, Moore C, Sharp PA. Spliced segments at the 5′ terminus of adenovirus 2 late mRNA. Proc Natl Acad Sci USA 1977;74(8):3171–5 and Sharp PA. Split genes and RNA splicing. Nobel lecture. Cell 1994;77(6):805–15. Panel (B): Adapted from Sharp PA. Split genes and RNA splicing. Nobel lecture. Cell 1994;77(6):805–15.*

As a result, three loops of displaced single-stranded DNA, marked **A**, **B**, and **C** in Fig. 11.21A, were visible in the electron microscope. It was suggested, therefore, that the 5′ sequence of the mRNA and its body were transcribed from four nonadjoining DNA sequences. Identification of a similar 5′ single-stranded tail in an R-loop of mRNA of a different late viral protein (the so-called 100 K protein) suggested that a common 5′ sequence was probably attached to the bodies of more than one late mRNA species.[78] This possibility had been proven to be correct in the parallel work of Richard Roberts and associates (*vide infra*). The overall implication was that the viral genes were split into coding tracts that were later named *exons*, and into intervening noncoding sequences that were named *introns* (Fig. 11.21B). The full gene, including both exons and introns, was transcribed into a large hnRNA primary transcript that was processed by splicing whereby introns were excised from the contiguous hnRNA transcript and the severed exons were ligated into much shorter noncollinear mRNA molecule.[78]

Less than a year after the description of the fragmented nature of adeno-virus late gene, Sharp and Arnold Berk isolated mRNA transcripts of seven early genes of this virus and showed by hybridization that they too were pro-ducts of splicing of larger transcripts of split genes. It was also reported in this work that some of the mRNA molecules were products of alternative splicing of single genes.[95]

The Work of Richard Roberts

In a remarkable instance of happenstance, right at the time that the Sharp team discovered split genes of adenovirus and proposed that their transcripts were processed into noncollinear mRNA, the Cold Spring Harbor group of Richard Roberts independently obtained similar results. Based on their experimental observations, Roberts and associates also reached the conclusion that adenovi-rus genes were fragmented and that that mRNA molecules were noncollinear products of the splicing of larger primary transcripts. Initial experiments of Roberts with Richard Gelinas (later at the Institute of systems Biology, Seattle) appeared in the Jul. 1977 issue of the journal *Cell*. In it they documen-ted the outcome of the T1 RNase digestion of a mixture of at least 12 distinct species of adenovirus-2 late mRNA. Product oligonucleotides that represented the 5′-termini of the mRNA molecules were selectively retained and then eluted from a column of dihydroxyboryl-cellulose. Sequencing of these 5′ oli-gomers yielded the unexpected result that the different individual mRNA spe-cies shared a single common capped sequence, albeit with different states of methylation, $^{m7}GpppACU(C_4U_3)G$. The same 5′ oligomeric sequence was also identified in individual late mRNA species that were isolated from their hybrids with different regions of the viral genome.[96] The full significance of this observation became clearer when the Roberts team carried out electron microscopic R-loop mapping of hybrids between the adenovirus DNA and the viral mRNA that was isolated from polysomes of infected cells. Results of this experiment were reported in a second paper that was published in *Cell* in Sep. of 1977. The title of this work reflected the authors' astonishment at their own findings: "An Amazing Sequence Arrangement at the 5′-ends of Adenovirus 2 Messenger RNA."[97] The general strategy that they employed in this study was similar to the one that Sharp and his associates used. Polysomal RNA that was isolated at late stages of adenovirus-2 infection was annealed to restriction fragments of the viral DNA. The arrangement of R-loops as visualized by electron microscopy of the RNA−DNA heteroduplexes revealed that the 5′-ends of several late viral mRNA molecules, including hexon mRNA, were transcribed from three separate short DNA sequences that were situated upstream to the main body of the gene.[97] In discussing their findings Roberts and associates implied that the juxtaposition in mRNA of sequences that were mapped away from one another in the encoding gene was an inherent feature of the biosynthesis of the viral mRNA.

Very soon after the parallel discoveries of Sharp and of Roberts, many investigators reported that cellular genes of eukaryotes, and not just of their viruses, were also split and that the splicing of their primary hnRNA transcripts produced mRNA molecules that were not collinear with their encoding genes. The splicing of a large primary transcript to produce much shorter mRNA solved the mystery of the different sizes of hnRNA and mRNA and explained how the 5′ cap structure and the 3′ poly(A) tail of hnRNA were conserved in mRNA despite its greatly reduced length. Excision of inner portions of the hnRNA also explained why its internally methylated residues were absent from the mature mRNA. The far-reaching mechanistic and evolutionary implications of the discovery of the split structure of eukaryotic genes are beyond the scope of this chapter. Discussion is here limited to the historic shift from the prokaryotic paradigm of collinearity between gene, mRNA, and protein product to the more complex noncollinearity between eukaryotic genes and their mRNA.

11.7 POSTSCRIPT: WHY DID THEORY FAIL TO PREDICT SPLICING?

Empirical data that was gathered since the mid-1960s established by the mid-1970s many of the molecular details of the metabolism of RNA in eukaryotic cells. The main elements that were then recognized at that time were illustrated in Fig. 11.15. It was well understood by then that whereas the nucleus contained large-sized molecules of unstable primary hnRNA transcripts, the cytoplasmic polysomes were associated with much shorter chains of mRNA. The discovery of poly(A) tracts at the 3′ termini of both hnRNA and mRNA and of cap structures at the 5′-ends of both RNA types shaped an understanding that terminal tags that were attached to both ends of hnRNA were conserved in mRNA. Hybridization competition experiments revealed that mRNA sequences were present in hnRNA, whereas some sequences of hnRNA were absent from homologous mRNA. Finally, although hnRNA molecules were internally methylated, mRNA had no such methylated residues. Much of these data convinced scholars that the rapidly turning-over nuclear hnRNA molecules served as precursors to the cytoplasmic mRNA. Yet, despite the multiple clues, it was not speculated that internal tracts might be excised from hnRNA and its coding sequences be rejoined. Such mechanism could provide an explanation to the production of shorter 5′ capped and 3′ poly(A)-tailed mature mRNA molecules. Yet, no such theoretical model was offered and it was, therefore, experiment alone that ultimately revealed that this indeed was the case. Why then did theory fail to foresee splicing despite the substantial empirical indications for such mechanism?

Integration of the available experimental data into a theory of split genes and of splicing undoubtedly needed a leap of daring inventiveness that was not trivial (except for those who enjoy the wisdom of hindsight). Yet, it can also be argued that the development of such an idea was hindered by the entrenched concept of collinearity of mRNA molecules with their transcribed genes. That this was the case was best articulated by James Darnell, the discoverer of many of the attributes of eukaryotic hnRNA and mRNA and of their precursor–product relationship. In a retrospective 1999 address he provided a frank explanation for the failure of his leading laboratory to realize that splicing was a mechanism that could explain how hnRNA was converted into mRNA (the pertinent assertion is underlined)[75]:

> *Despite biochemical evidence from HeLa cell nuclear RNA (>10 kb capped and polyadenylated molecules) and knowing of the existence of adenovirus major transcript, it had remained too great a leap of imagination for our group to suggest that the two mRNA signposts, the 5' cap and the 3' poly(A), were brought together in the smaller mRNA by splicing, leaving out intervening molecules. Rather, the discussion in our papers concentrated on how mRNA might be processed from each end of the large capped, poly(A)-containing nuclear molecule. We were still in the grips of an earlier era of molecular biology in which important watchword was collinearity, the central tenet of which was that an mRNA transcript would match the stretch of DNA from which it was copied.*

Indeed, as already discussed in Section 9.14, the elegantly demonstrated collinearity of genes and mRNA in phage and bacteria metamorphosed into a universal rule for all organisms. This generalization was befitting of the zeitgeist that was best reflected in Jacque Monod's famous authoritative sweeping statement: *"Anything found to be true of* E. coli *must also be true of elephants."* It was, therefore, intellectually difficult to shake off a preconception of shared basic molecular workings among all living organisms. In this case of a seeking for similarities between prokaryotes and eukaryotes and overlooking their differences, one is reminded of the words of the father of empiricism, Francis Bacon (1561–1626) [*book I, (1620)*section 55, 323]:

> *The greatest and, perhaps, radical distinction between different men's dispositions for philosophy and the sciences is this, that some are more vigorous and active in observing differences of things others in observing their resemblances [...] each of them readily falls into excess by catching either the nice distinctions or shadows of resemblance.*

More than three centuries later, Erwin Chargaff wrote about the benefits and risks of generalizations in science[98]:

> *Generalizations in science are both necessary and hazardous; they carry a semblance of finality which conceals their essentially provisional character: they drive forward, as they retard; they add, but they also take away.*

These words appear to be applicable to the ultimately disproved generalization of a universal collinearity of gene, mRNA, and protein.

REFERENCES

1. Jacob F, Monod J. Genetic regulatory mechanisms in the synthesis of proteins. *J Mol Biol* 1961;**3**(3):318–56.
2. Brenner S, Jacob F, Meselson M. An unstable intermediate carrying information from genes to ribosomes for protein synthesis. *Nature* 1961;**190**(4776):576–81.
3. Hall BD, Spiegelman S. Sequence complementarity of T2-DNA and T2-specific RNA. *Proc Natl Acad Sci USA* 1961;**47**(2):137–63.
4. Gros F, Hiatt H, Gilbert W, Kurland CG, Risebrough RW, Watson JD. Unstable ribonucleic acid revealed by pulse labelling of *Escherichia coli*. *Nature* 1961;**190** (4776):581–5.
5. Hayashi M, Spiegelman S. The selective synthesis of informational RNA in bacteria. *Proc Natl Acad Sci USA* 1961;**47**(10):1564–80.
6. Yanofsky C, Carlton BC, Guest JR, Helinski DR, Henning U. On the collinearity of gene structure and protein structure. *Proc Natl Acad Sci USA* 1964;**51**(2):266–72.
7. Yanofsky C, Drapeau GR, Guest JR, Carlton BC. The complete amino acid sequence of the tryptophan synthase A protein (alpha subunit) and its colinear relationship with the genetic map of the A gene. *Proc Natl Acad Sci USA* 1967;**57**(2):296–8.
8. Sarabhai AS, Stretton AO, Brenner S, Bolle A. Co-linearity of the gene with the polypeptide chain. *Nature* 1964;**201**(4914):13–17.
9. Hämmerling J. Über formbildende substanzen bei *Acetabularia mediterranea*, ihre räumliche und zeitliche verteilung und ihre herkunft. *Wilhelm Roux' Arch Entwicklungsmech Organ* 1934;**131**:1–81.
10. Hämmerling J. Nucleo-cytoplasmic relationships in the development of *Acetabularia*. *Intl Rev Cytol* 1953;**2**:475–98.
11. Harris H. *The Balance of improbabilities. A scientific life*. NY: Oxford University Press; 1987.
12. Hogness DS, Cohn M, Monod J. Studies on the induced synthesis of beta-galactosidase in *Escherichia coli*: the kinetics and mechanism of sulfur incorporation. *Biochim Biophys Acta* 1955;**16**(1):99–116.
13. Schoenheimer R. *The dynamic state of body constituents*. Cambridge, MA: Harvard University Press; 1942.
14. Harris H, Watts JW. Turnover of protein in a non-multiplying animal cell. *Nature* 1958;**181**(4623):1582–4.
15. Graham AF, Siminovitch L. Significance of ribonucleic acid and deoxyribonucleic acid turnover studies. *J Histochem Cytochem* 1956;**4**(6):508–15.
16. Graham AF, Siminovitch L. Conservation of RNA and DNA phosphorus in strain L (Earle) mouse cells. *Biochim Biophys Acta* 1957;**26**(2):427–8.
17. Harris H. An RNA heresy in the fifties. *Trends Biochem Sci* 1994;**19**(7):303–5.
18. Watts JW, Harris H. Turnover of nucleic acids in a non-multiplying animal cell. *Biochem J* 1959;**72**(1):147–53.
19. Harris H. Turnover of nuclear and cytoplasmic ribonucleic acid in two types of animal cell, with some further observations on the nucleolus. *Biochem J* 1959;**73**(2):362–9.

20. Harris H, Watts JW. The relationship between nuclear and cytoplasmic ribonucleic acid. *Proc R Soc Lond Biol Sci B* 1962;**156**(962):109−21.

21. Harris H, Fisher HW, Rodgers A, Spencer T, Watts JW. An examination of the ribonucleic acids in the HeLa cell with special reference to current theory about the transfer of information from nucleus to cytoplasm. *Proc R Soc Lond Biol Sci B* 1963;**157**(967):177−98.

22. Harris H. Rapidly labelled ribonucleic acid in the cell nucleus. *Nature* 1963;**198** (4876):184−5.

23. Harris H. The short-lived RNA in the cell nucleus and its possible role in evolution. In: Bryson V, Vogel G, editors. *Evolving genes and proteins*. New York, NY: Academic Press; 1965. p. 629.

24. Darnell Jr JE. Joys and surprises of a career studying eukaryotic gene expression. *J Biol Chem* 2013;**288**(18):12957−66.

25. Darnell JE. Special Achievement in Medical Science Award. The surprises of mammalian molecular cell biology. *Nat Med* 2002;**8**(10):1068−71.

26. Scherrer K, Darnell JE. Sedimentation characteristics of rapidly labelled RNA from HeLa cells. *Biochem Biophys Res Commun* 1962;**7**(6):486−90.

27. Nazar R. Ribosomal RNA processing and ribosome biogenesis in eukaryotes. *IUBMB Life* 2004;**56**(8):457−65.

28. McCann K, Baserga S. Making ribosomes: pre-rRNA transcription and processing. In: Sesma A, von der Haar T, editors. *Fungal RNA biology*. Switzerland: Springer International Publishing; 2014. p. 217−32.

29. Scherrer K. Historical review: the discovery of 'giant' RNA and RNA processing: 40 years of enigma. *Trends Biochem Sci* 2003;**28**(10):566−71.

30. Hiatt HH. A rapidly labeled RNA in rat liver nuclei. *J Mol Biol* 1962;**5**(2):217−29.

31. Scherrer K, Latham H, Darnell JE. Demonstration of an unstable RNA and of a precursor to ribosomal RNA in HeLa cells. *Proc Natl Acad Sci USA* 1963;**49**(2):240−8.

32. Sibatani A, De Kloet SR, Allfrey VG, Mirsky AE. Isolation of a nuclear RNA fraction resembling DNA in its base composition. *Proc Natl Acad Sci USA* 1962;**48**(3):471−7.

33. Georgiev GP, Mantieva VL. The isolation of DNA-like RNA and ribosomal RNA from the nucleolo-chromosomal apparatus of mammalian cells. *Biochim Biophys Acta* 1962;**61** (8):153−4.

34. Penman S, Scherrer K, Becker Y, Darnell JE. Polyribosomes in normal and poliovirus-infected HeLa cells and their relationship to messenger-RNA. *Proc Natl Acad Sci USA* 1963;**49**(5):654−62.

35. Girard M, Latham H, Penman S, Darnell JE. Entrance of newly formed messenger RNA and ribosomes into HeLa cell cytoplasm. *J Mol Biol* 1965;**11**(2):187−201.

36. Warner JR, Soeiro R, Birnboim HC, Girard M, Darnell JE. Rapidly labeled HeLa cell nuclear RNA. I. Identification by zone sedimentation of a heterogeneous fraction separate from ribosomal precursor RNA. *J Mol Biol* 1966;**19**(2):349−61.

37. Latham H, Darnell JE. Distribution of mRNA in the cytoplasmic polyribosomes of the HeLa cell. *J Mol Biol* 1965;**14**(1):1−12.

38. Soeiro R, Birnboim HC, Darnell JE. Rapidly labeled HeLa cell nuclear RNA. II. Base composition and cellular localization of a heterogeneous RNA fraction. *J Mol Biol* 1966;**19** (2):362−72.

39. Warner J, Madden MJ, Darnell JE. The interaction of poliovirus RNA with *Escherichia coli* ribosomes. *Virology* 1963;**19**(3):393−9.

40. Warner JR, Knopf PM, Rich A. A multiple ribosomal structure in protein synthesis. *Proc Natl Acad Sci USA* 1963;**49**(1):122−9.

41. Benjamin TL. Virus-specific RNA in cells productively infected or transformed by polyoma virus. *J Mol Biol* 1966;**16**(2):359−73.

42. Lindberg U, Darnell JE. SV40-specific RNA in the nucleus and polyribosomes of transformed cells. *Proc Natl Acad Sci USA* 1970;**65**(4):1089−96.

43. Edmonds M, Abrams R. Polynucleotide biosynthesis: formation of a sequence of adenylate units from adenosine triphosphate by an enzyme from thymus nuclei. *J Biol Chem* 1960;**235**(4):1142−9.

44. Edmonds M, Caramela MG. The isolation and characterization of adenosine monophosphate-rich polynucleotides synthesized by Ehrlich ascites cells. *J Biol Chem* 1969;**244**(5):1314−24.

45. Edmonds M, Vaughan Jr MH, Nakazato H. Polyadenylic acid sequences in the heterogeneous nuclear RNA and rapidly-labeled polyribosomal RNA of HeLa cells: possible evidence for a precursor relationship. *Proc Natl Acad Sci USA* 1971;**68**(6):1336−40.

46. Darnell JE, Wall R, Tushinski RJ. An adenylic acid-rich sequence in messenger RNA of HeLa cells and its possible relationship to reiterated sites in DNA. *Proc Natl Acad Sci USA* 1971;**68**(6):1321−5.

47. Lee SY, Mendecki J, Brawerman G. A polynucleotide segment rich in adenylic acid in the rapidly-labeled polyribosomal RNA component of mouse sarcoma 180 ascites cells. *Proc Natl Acad Sci USA* 1971;**68**(6):1331−5.

48. Darnell JE, Philipson L, Wall R, Adesnik M. Polyadenylic acid sequences: role in conversion of nuclear RNA into messenger RNA. *Science* 1971;**174**(4008):507−10.

49. Molloy GR, Darnell JE. Characterization of the poly(adenylic acid) regions and the adjacent nucleotides in heterogeneous nuclear ribonucleic acid and messenger ribonucleic acid from HeLa cells. *Biochemistry* 1973;**12**(12):2324−30.

50. Jelinek W, Adesnik M, Salditt M, et al. Further evidence on the nuclear origin and transfer to the cytoplasm of polyadenylic acid sequences in mammalian cell RNA. *J Mol Biol* 1973;**75**(3):515−32.

51. Coller JM, Gray NK, Wickens MP. mRNA stabilization by poly(A) binding protein is independent of poly(A) and requires translation. *Genes Dev* 1998;**12**(20):3226−35.

52. Vinciguerra P, Stutz F. mRNA export: an assembly line from genes to nuclear pores. *Curr Opin Cell Biol* 2004;**16**(3):285−92.

53. Gray NK, Coller JM, Dickson KS, Wickens M. Multiple portions of poly(A)-binding protein stimulate translation in vivo. *EMBO J* 2000;**19**(17):4723−33.

54. Banerjee AK. 5′-Terminal cap structure in eukaryotic messenger ribonucleic acids. *Microbiol Rev* 1980;**44**(2):175−205.

55. Reddy R, Ro-Choi TS, Henning D, Busch H. Primary sequence of U-1 nuclear ribonucleic acid of Novikoff hepatoma ascites cells. *J Biol Chem* 1974;**249**(20):6486−94.

56. Perry RP, Kelley DE. Existence of methylated messenger RNA in mouse L cells. *Cell* 1974;**1**(1):37−42.

57. Miura K, Watanabe K, Sugiura M, Shatkin AJ. The 5′-terminal nucleotide sequences of the double-stranded RNA of human reovirus. *Proc Natl Acad Sci USA* 1974;**71**(10):3979−83.

58. Miura K, Watanabe K, Sugiura M. 5′-Terminal nucleotide sequences of the double-stranded RNA of silkworm cytoplasmic polyhedrosis virus. *J Mol Biol* 1974;**86**(1):31−48.

59. Furuichi Y. Methylation-coupled transcription by virus-associated transcriptase of cytoplasmic polyhedrosis virus containing double-stranded RNA. *Nucleic Acids Res* 1974;**1**(6):802−9.

60. Furuichi Y, Morgan M, Muthukrishnan S, Shatkin AJ. Reovirus messenger RNA contains a methylated, blocked 5'-terminal structure: m-7G(5')ppp(5')G-MpCp. *Proc Natl Acad Sci USA* 1975;**72**(1):362–6.

61. Furuichi Y, Miura K. A blocked structure at the 5' terminus of mRNA from cytoplasmic polyhedrosis virus. *Nature* 1975;**253**(5490):374–5.

62. Wei CM, Moss B. Methylated nucleotides block 5'-terminus of vaccinia virus messenger RNA. *Proc Natl Acad Sci USA* 1975;**72**(1):318–22.

63. Furuichi Y, Morgan M, Shatkin AJ, Jelinek W, Salditt-Georgieff M, Darnell JE. Methylated, blocked 5 termini in HeLa cell mRNA. *Proc Natl Acad Sci USA* 1975;**72** (5):1904–8.

64. Perry RP, Kelley DE, Friderici K, Rottman F. The methylated constituents of L cell messenger RNA: evidence for an unusual cluster at the 5' terminus. *Cell* 1975;**4**(4):387–94.

65. Desrosiers RC, Friderici KH, Rottman FM. Characterization of Novikoff hepatoma mRNA methylation and heterogeneity in the methylated 5' terminus. *Biochemistry* 1975;**14** (20):4367–74.

66. Wei CM, Gershowitz A, Moss B. Methylated nucleotides block 5' terminus of HeLa cell messenger RNA. *Cell* 1975;**4**(4):379–86.

67. Wei CM, Gershowitz A, Moss B. 5'-Terminal and internal methylated nucleotide sequences in HeLa cell mRNA. *Biochemistry* 1976;**15**(2):397–401.

68. Perry RP, Kelley DE. Methylated constituents of heterogeneous nuclear RNA: presence in blocked 5' terminal structures. *Cell* 1975;**6**(1):13–19.

69. Salditt-Georgieff M, Jelinek W, Darnell JE, Furuichi Y, Morgan M, Shatkin A. Methyl labeling of HeLa cell hnRNA: a comparison with mRNA. *Cell* 1976;**7**(2):227–37.

70. Perry RP, Kelley DE. Kinetics of formation of 5' terminal caps in mRNA. *Cell* 1976;**8** (3):433–42.

71. Gonatopoulos-Pournatzis T, Cowling VH. Cap-binding complex (CBC). *Biochem J* 2014; **457**(2):231–42.

72. Cowling VH. Regulation of mRNA cap methylation. *Biochem J* 2010;**425**(Pt 2):295–302.

73. Sonenberg N. eIF4E, the mRNA cap-binding protein: from basic discovery to translational research. *Biochem Cell Biol* 2008;**86**(2):178–83.

74. Proudfoot NJ, Furger A, Dye MJ. Integrating mRNA processing with transcription. *Cell* 2002;**108**(4):501–12.

75. Darnell Jr JE, Wilson EB. Lecture, 1998. Eukaryotic RNAs: once more from the beginning. *Mol Biol Cell* 1999;**10**(6):1685–92.

76. Athappily FK, Murali R, Rux JJ, Cai Z, Burnett RM. The refined crystal structure of hexon, the major coat protein of adenovirus type 2, at 2·9 Å resolution. *J Mol Biol* 1994;**242**(4):430–55.

77. Dales S. Early events in cell-animal virus interactions. *Bact Rev* 1973;**37**(2):103–35.

78. Berget SM, Moore C, Sharp PA. Spliced segments at the 5' terminus of adenovirus 2 late mRNA. *Proc Natl Acad Sci USA* 1977;**74**(8):3171–5.

79. Green M. Oncogenic viruses. *Annu Rev Biochem* 1970;**39**:701–56.

80. Philipson L, Wall R, Glickman G, Darnell JE. Addition of polyadenylate sequences to virus-specific RNA during adenovirus replication. *Proc Natl Acad Sci USA* 1971;**68** (11):2806–9.

81. Wall R, Philipson L, Darnell JE. Processing of adenovirus specific nuclear RNA during virus replication. *Virology* 1972;**50**(1):27–34.

82. Mulder C, Arrand JR, Delius H, et al. Cleavage maps of DNA from adenovirus types 2 and 5 by restriction endonucleases *Eco*RI and *Hpa*I. *Cold Spring Harb Symp Quant Biol* 1974;**39**:397−400.

83. Sharp PA, Gallimore PH, Flint SJ. Mapping of Adenovirus 2 RNA sequences in lytically infected cells and transformed cell lines. *Cold Spring Harb Symp Quant Biol* 1974; **39**:457−74.

84. Philipson L, Pettersson U, Lindberg U, Tibbetts C, Vennström B, Persson T. RNA synthesis and processing in adenovirus-infected cells. *Cold Spring Harb Symp Quant Biol* 1974; **39**:447−56.

85. Tibbetts C, Pettersson U. Complementary strand-specific sequences from unique fragments of adenovirus type 2 DNA for hybridization-mapping experiments. *J Mol Biol* 1974;**88** (4):767−84.

86. Bachenheimer S, Darnell JE. Adenovirus-2 mRNA is transcribed as part of a high-molecular-weight precursor RNA. *Proc Natl Acad Sci USA* 1975;**72**(11):4445−9.

87. Goldberg S, Weber J, Darnell Jr JE. The definition of a large viral transcription unit late in Ad2 infection of HeLa cells: mapping by effects of ultraviolet irradiation. *Cell* 1977;**10** (4):617−21.

88. Goldberg S, Nevins J, Darnell JE. Evidence from UV transcription mapping that late adenovirus type 2 mRNA is derived from a large precursor molecule. *J Virol* 1978;**25**(3): 806−10.

89. Davis RW, Davidson N. Electron-microscopic visualization of deletion mutations. *Proc Natl Acad Sci USA* 1968;**60**(1):243−50.

90. Kim J, Sharp PA, Davidson N. Electron microscope studies of heteroduplex DNA from a deletion mutant of bacteriophage phiX-174. *Proc Natl Acad Sci USA* 1972;**69**(7):1948−52.

91. White RL, Hogness DS. R loop mapping of the 18S and 28S sequences in the long and short repeating units of *Drosophila melanogaster* rDNA. *Cell* 1977;**10**(2):177−92.

92. Thomas M, White RL, Davis RW. Hybridization of RNA to double-stranded DNA: formation of R-loops. *Proc Natl Acad Sci USA* 1976;**73**(7):2294−8.

93. Lewis JB, Atkins JF, Anderson CW, Baum PR, Gesteland RF. Mapping of late adenovirus genes by cell-free translation of RNA selected by hybridization to specific DNA fragments. *Proc Natl Acad Sci USA* 1975;**72**(4):1344−8.

94. Sharp PA. Split genes and RNA splicing. Nobel lecture. *Cell* 1994;**77**(6):805−15.

95. Berk AJ, Sharp PA. Structure of the adenovirus 2 early mRNAs. *Cell* 1978;**14** (3):695−711.

96. Gelinas RE, Roberts RJ. One predominant 5′-undecanucleotide in adenovirus 2 late messenger RNAs. *Cell* 1977;**11**(3):533−44.

97. Chow LT, Gelinas RE, Broker TR, Roberts RJ. An amazing sequence arrangement at the 5′ ends of adenovirus 2 messenger RNA. *Cell* 1977;**12**(1):1−8.

98. Chargaff E. Chemical specificity of nucleic acids and mechanism of their enzymatic degradation. *Experientia* 1950;**6**(6):201−9.

Chapter 12

Postscript: On the Place of Theory and Experiment in Molecular Biology

Chapter Outline

Theory and experiment had manifestly different relative contributions to discoveries that were described in this volume. Detailed examination of the questions of theory and experimentation in biology is much beyond the scope of this short epilogue. The interested reader may consult other sources for wide-ranging philosophical implications of this subject.[1−6] Discussion here is restricted to reexamination of the roles of theory and experiment in three representative discoveries that were covered in previous chapters. Before getting down to the examination of specific cases, the meaning of "theory" in this inquiry should be clarified. Of the different definitions of "theory," it is used here in its narrowest prescriptive connotation: a conjecture about what should be. Thus rather than a broad speculation that ventures to explain and find order in nature, the inference of "theory" in this discussion is closer to "hypothesis": that is, an experimentally testable conjecture.

12.1 THREE CASES ILLUSTRATE THAT THE POWER OF THEORY DIMINISHES WITH INCREASED COMPLEXITY OF AN INVESTIGATED PROBLEM

The three examined cases are: the identification of the semiconservative mode of DNA replication (see Chapter 6: "Meselson and Stahl Proved That DNA Is

M. Fry: Landmark Experiments in Molecular Biology. DOI: http://dx.doi.org/10.1016/B978-0-12-802074-6.00012-6

Replicated in a Semiconservative Fashion"), the definition and deciphering of the genetic code (see Chapter 7: "Defining the Genetic Code" and Chapter 10: "The Deciphering of the Genetic Code," respectively), and the discovery of messenger RNA (mRNA) in prokaryotes (see Chapter 9: "The Discovery and Rediscovery of Prokaryotic Messenger RNA"). The historical record shows that theory had an essential part in the elucidation of the semiconservative mode of DNA replication. By contrast, theory played a limited role in the definition of the genetic code and had no part in its deciphering. Finally, despite suggestive empirical evidence, theory failed until late to predict the existence of mRNA. Only after the mass of accumulated experimental data became overwhelming did the hypothesis of mRNA crystallize and its detection by targeted experiments was made possible. The discrepant relative contributions of theory and experiment in these three cases were due to multifaceted scientific, sociological, and personal factors. Focusing on just the scientific aspect of the story, it becomes evident that the power of theory in these and other cases was inversely proportional to the complexity of the question at hand. Thus as the complexity of problems increased, theory became less potent and the part of experiment turned out to be dominant or absolute.

12.2   SEMICONSERVATIVE DNA REPLICATION

The identification of the semiconservative mechanism of DNA replication is a rare case in biological research in which theory heralded and prescribed experiments to test it. The search for the mode of DNA replication had two distinct phases. In the first stage three alternative theoretical models of DNA replication were formulated, and in a second phase these models were subjected to critical experimental testing. The first period (1953—55) produced the three competing abstract models of semiconservative,[7] dispersive,[8] and conservative replication of DNA.[9,10] In the next stage (1955—58) model-guided experiments culminated in empirical validation of the semiconservative model of DNA replication and in virtual refutation of the two competing models. The effective contribution of theory in this case can be attributed to the relative straightforwardness of the question at hand. Once Watson and Crick showed that DNA is a double helix[11] and have pointed to the centrality of strand complementarity in DNA replication and inheritance,[7] the mode of replication of DNA became a question of mechanics. The mechanical nature of the problem is underscored by the fact that practically nothing was known at the time about the enzymology of DNA replication. Indeed, this information proved to be unnecessary for either the construction of the alternative theoretical models of the mode of replication or for the experimental demonstration of its semiconservative mechanism. The competing models of fully conservative, semiconservative, or patch-wise dispersive replication were the only imaginable plausible alternatives. These models set up a formal basis for the design of critical experiments to identify the actual mode of DNA duplication. Although early experiments to identify the correct model were thwarted by technical complications,[12–14] the employment of equilibrium centrifugation enabled Matthew Meselson and Franklin Stahl to convincingly refute the dispersive and conservative models and validate the semiconservative mode of replication.[15] In a recent philosophical analysis, Weber argued that the Meselson and Stahl experiment represents an experimental version of inference to the best explanation. His contention was that as such it comes close to the status of a "crucial experiment" that decisively chooses among a group of alternative hypotheses.[6]

In sum, the success of theory in guiding definitive experiment to determine the mode of DNA replication can be attributed to the relative simplicity of the problem. Effective models could be constructed because the mechanics of replication involved very few and known variables. Thus the basic tenets of all the three models were that DNA was a double-strand helix and that the products of replication were faithful copies of the two complementary strands. The three outwardly equally credible models of the mechanics of replication were all based on these two elements. Importantly, because the expected outcomes of these models of replication were different, they could be put to test by experiment. In sum, the relative simplicity of the problem of the mechanics of the mode of DNA replication and the testability of the offered abstract models account in this case for the successful coupling of predictive theories with their experimental testing.

12.3   DEFINITION AND DECIPHERING OF THE GENETIC CODE

The definition of features of the genetic code posed a greater challenge than the mechanical problem of the mode of DNA replication. Yet, as happened with the question of replication, efforts to define the genetic code have also begun by a phase of theorization. Adopting information theory concepts, the question of the genetic code was treated as a number theory problem.[16] Physicists and mathematicians took this approach to generate abstract combinatorial models that did not readily succumb to critical experimental testing. Models of such kind were proposed for instance by George Gamow in his Jul. 8, 1953, letter to Watson and Crick (facsimile 1 in Ref. 17), and in his May 27, 1954, letter to Crick (http://profiles.nlm.nih.gov/ps/retrieve/ResourceMetadata/SCBBWS#transcript). Generally, these abstract models paid no attention to the reality of the then largely unknown inherent biochemical intricacies of the code and of its translation. The only early attempt to treat the code as a biochemical entity was Gamow's proposed key-and-lock fit between the code-bearing nucleic acids and amino acids.[18–20] Crick handily rejected this idea for its lack of stereochemical grounding (Crick's unpublished 1955 communication to the RNA Tie Club: http://profiles.nlm.nih.gov/SC/B/B/G/F/_/scbbgf.pdf). However, he too was unable to foretell at that early stage the complex biochemical reality of the code and its translation. Thus, although the code problem appeared initially to be amenable to theoretical modeling, it was quickly realized that theory alone could not grapple with its complexity. This understanding was presciently articulated by Brenner already in 1957 (Wellcome Library Archives; Closed stores Arch. &MSSPP/CRI/H2/24: Box 79; Ref. no. ebaf293b-7f02-46c8-a16f-c45333219e28):

> ... As far as we can see the coding problem is now an experimental problem and the role of theory will be to organize and make sense of the facts as they come out of the laboratory.

Indeed, rather than by theoretical leaps, the principal features of the code were gradually exposed by painstaking experimental work. Its nonoverlapping nature was revealed by the cataloging of dipeptide combinations in experimentally determined sequences of diverse proteins[21] and by empirical demonstration that point mutations in TMV RNA caused substitutions of only single amino acids.[22] As to the number of nucleotides in a code word, theoretical arguments favored triplet codons. However, theory alone could not answer how 64 different trinucleotides encoded just 20 amino acids. It remained for experimentalists to ultimately prove that the triplet-based code was degenerate and commaless.[23] Notably though, theory was not completely absent from the pursuit of the code. One facet of the problem was how to wed the different chemistry of the nucleic acid template with that of the translated protein. Here history provided an exceptional example of prophetic theoretical prediction in the form of Crick's adaptor hypothesis.[24] Yet, parallel to and without awareness of this abstract hypothesis,[25] Paul Zamecnik and Mahlon Hoagland conducted biochemical experiments that culminated in the identification of the actual adaptor molecule transfer RNA (tRNA) and of the enzymes that charge it with amino acids.[26,27] Thus, Crick's brilliant theoretical leap did not guide in actual fact the experimental discovery of tRNA.

The principally experiment-based definition of the code as a string of non-overlapping and commaless degenerate nucleotide triplets as well as the identification of tRNA and its activating enzymes, all provided a basis for the subsequent purely experimental second phase of the deciphering of the code.[28,29] The limited role of theory in defining the genetic code can be retrospectively explained by the complexity of the problem. The intricate biochemistry of the code and its translation involved multiple unknown variables that theory alone was largely unable to predict and that only experiment could uncover.

12.4   IDENTIFICATION AND DETECTION OF PROKARYOTIC mRNA

Unlike the problems of the mechanism of DNA replication and of the genetic code, theory had a negligible role in the discovery of mRNA. This case offers an example of the difficulty that theory faces in predicting the existence of a yet unknown biological pathway or component. A series of findings between the late 1930s and the early 1950s revealed that RNA was the likely direct template for protein synthesis. Initial studies found positive correlation between levels of RNA in cells and the extent of cellular protein synthesis.[30−34] In parallel, the bulk of the cellular RNA was found to be contained within ribosomes that were also identified as the site of protein synthesis.[35−38] In addition, (false) analogy was drawn between ribosomes and ribosome-size viruses whose proteins are encoded by their own RNA genomes. These observations led to formulation of the "one gene−one ribosome−one protein" idea which claimed that the RNA component of each ribosome was a transcript of a specific gene that served in turn as the direct template for the synthesis of an encoded particular protein.[24] With no challenging alternative theories, this model was the universally accepted paradigm throughout most of the 1950s. The "one gene−one ribosome−one protein" hypothesis became so entrenched that it resisted several contradicting experimental findings (see Chapter 9: "The Discovery and Rediscovery of Prokaryotic Messenger RNA"). Indeed, this hypothesis persisted as an unquestioned article of faith even after Ken Volkin and Lazarus Astrachan detected a minor fraction of unstable RNA in phage-infected bacteria.[39−42] A first theoretical challenge to the "one gene−one ribosome−one protein" dogma grew out of the indirect PaJaMo experiment of Arthur Pardee, Françoise Jacob, and Jacque Monod. Results of this experiment raised the speculation that a labile template molecule might serve as an intermediary ("messenger") that carries genetic information from the encoding DNA to the protein synthesis machinery.[43,44] However, this conjecture, which did not specify the chemical nature of the unstable messenger, was only one of several alternative interpretations of results of the PaJaMo experiment. Finally, however, the accumulated weight of the interpreted PaJaMo result, the direct observations of Volkin and Astrachan and other lines of evidence precipitated a revelatory moment at which the mRNA hypothesis was crystallized. Guided by this conjecture, direct experiments were designed and executed to detect a minor fraction of nonribosomal short-lived RNA molecules in phage-infected bacteria.[45] At the same time, independent experiments that were not guided by a clearly formulated mRNA hypothesis discovered similarly unstable RNA molecules in uninfected bacteria.[46] Subsequent demonstration of complementarity between these labile RNA molecules and DNA indicated that they represented mRNA transcripts of the DNA.[47,48] Because the complex details of RNA and protein syntheses were largely unknown in the 1950s, it is retrospectively understandable why it was difficult and even impossible to develop a predictive theory of mRNA. Thus unawareness of the existence of essential variables defied the framing of productive theoretical models of transmission of genetic information from DNA to protein. This case offers, therefore, an excellent example of the defeat of theory in the face of a complex biological problem and of its successful solution by experiments.

12.5   THE COMPLEXITY OF BIOLOGY CONSTRAINS THE POWER OF THEORY

Although productive conjectures played an essential part in the elucidation of the semiconservative mode of DNA replication, their contribution to the definition of the genetic code was limited and they had no significant role in the discovery of mRNA. It appears, therefore, that theoretical considerations became less effective as the complexity of a system and its number of unknowns grew. Moreover, theoretical models of more complex systems often threw investigators off the right track. As detailed in Chapter 9, "The Discovery and Rediscovery of Prokaryotic Messenger RNA," "one gene−one ribosome−one protein" was more of an article of faith then a testable and tested working hypothesis. It thus misled researchers through most of the 1950s, blinding them to opposing experimental results and hindering formulation of possible alternative models of information transmission from DNA to protein. The case of the split nature of eukaryotic genes and of RNA splicing offers an even more conspicuous example of a misleading percept holding back the resolution of a complex system. Experiments in the mid-1960s demonstrated that the amino acid sequences of phage[49] and bacterial[50,51] proteins were colinear with the base sequences of their encoding genes. On the basis of this finding, it became paradigmatically believed that colinearity between the sequences of gene, mRNA, and protein was universally maintained in all the living organisms[52] (see Chapter 9: "The Discovery and Rediscovery of Prokaryotic Messenger RNA" and Chapter 11: "The Surprising Discovery of Split Genes and of RNA Splicing"). Although it had no experimental support, this belief was not put to test by targeted experiments. Further, it became so engrained that the accumulated weight of discrepant experimental evidence failed to topple it. Data gathered in the late 1960s and in the early 1970s showed that the respective 5′ and 3′ terminal cap and poly(A) of heterogeneous nuclear RNA (hnRNA) were preserved in the much shorter cytoplasmic mRNA molecules, whereas inner methylated bases of the hnRNA were absent from the mRNA (Fig. 11.15). Although these results appear in hindsight to have been highly suggestive of the possibility of splicing, they had never resulted in a theoretical proposal that the processing of primary RNA transcripts in nuclei of eukaryotic cells involved removal of internal segments from the hnRNA. Eventually, it was experiment alone, with no guiding theory that uncovered the fragmented structure of eukaryotic genes and of the production of spliced mRNA molecules by excision of introns from the hnRNA transcripts.[53−55] Why were the split structure of eukaryotic genes and RNA splicing so slow to be discovered and why had not theory predicted them? One reason was that as in other complex systems, theory was defeated by the intricacies of the fragmented structure of genes and of the processing of RNA. Another impeding factor was the rooted faith that gene, mRNA, and protein were colinear in every living organism. This preconception seems to have blinded researchers to the significance of experimental evidence that indicated non-colinearity in eukaryotes.[52]

Biology, even at its molecular level, is too complex in most cases to allow effective theorization. It is enlightening to read in this context what Leslie Orgel and Francis Crick, the great theoretician of molecular biology, wrote about the relative places of predictive theory and experiment in biology.[56] In 1968 Crick[57] and separately Orgel[58] raised speculations on the origin and evolution of the genetic code. Twenty-five years later, in 1993, they revisited their conjectures and examined them in light of experimental discoveries that have been made in the elapsed time. Referring to the discovery by Thomas Cech and Sidney Altman of ribozyme as remnants of the RNA world in present-day organisms, Orgel and Crick forthrightly declared that their 1968 theoretical speculations failed to foresee future directions and were thus incapable of dictating which experiments should be done[56]:

> *How clearly did we anticipate the exciting experimental discoveries of the last decade? We must confess that we did not anticipate them at all! We discussed a number of hypothetical schemes for the origins of our genetic system and touched on each of the major features of the RNA world hypothesis. However, we did not ourselves search for, nor did we encourage others to search for, relics of the RNA world in contemporary organisms. [...]*

Orgel and Crick then went on to assess the place of theory and experiment in biology:

> *How much influence did our speculations have on the subsequent development of the subject? Very little. We doubt that Cech and Altman were aware of our papers when they made their important discoveries. The lesson is clear: speculation is fun, but even correct hypotheses without experimental follow-up are unlikely to have much effect on the development of biology.*

REFERENCES

1. Schaffner KF. *Discovery and explanation in biology and medicine.* Chicago: University of Chicago Press; 1993.
2. Schaffner KF. Interactions among theory, experiment, and technology in molecular biology. *PSA: Proc Bienn Meeting Philos Sci Assoc* 1994;**2**:192−205.
3. Schaffner KF. Theory structure and knowledge representation in molecular biology. In: Sarkar S, editor. *The philosophy and history of molecular biology: new perspectives.* Dordrecht, Boston, London: Kluwer; 1996. pp. 27−65.
4. Weber M. *Philosophy of experimental biology.* Cambridge: Cambridge University Press; 2005.
5. Weber M. Experimentation. In: Sarkar S, Plutynski A, editors. *A companion to the philosophy of biology.* Hoboken, NJ: Wiley-Blackwell; 2008. pp. 472−88.

6. Weber M. The crux of crucial experiments: Duhem's problems and inference to the best explanation. *Br J Philos Sci* 2009;**60**(1):19−49.

7. Watson JD, Crick FH. Genetical implications of the structure of deoxyribonucleic acid. *Nature* 1953;**171**(4361):964−7.

8. Delbruck M. On the replication of desoxyribonucleic acid (DNA). *Proc Natl Acad Sci USA* 1954;**40**(9):783−8.

9. Bloch DP. A possible mechanism for the replication of the helical structure of desoxyribonucleic acid. *Proc Natl Acad Sci USA* 1955;**41**(12):1058−64.

10. Butler JA. The action of ionizing radiations on biological materials; facts and theories. *Radiation Res* 1956;**4**(1):20−32.

11. Watson JD, Crick FH. Molecular structure of nucleic acids: a structure for deoxyribose nucleic acid. *Nature* 1953;**171**(4356):737−8.

12. Stent GS, Jerne NK. The distribution of parental phosphorus atoms among bacteriophage progeny. *Proc Natl Acad Sci USA* 1955;**41**(10):704−9.

13. Fuerst CR, Stent GS. Inactivation of bacteria by decay of incorporated radioactive phosphorus. *J Gen Physiol* 1956;**40**(1):73−90.

14. Levinthal C. The mechanism of DNA replication and genetic recombination in phage. *Proc Natl Acad Sci USA* 1956;**42**(7):394−404.

15. Meselson M, Stahl FW. The replication of DNA in *Escherichia coli*. *Proc Natl Acad Sci USA* 1958;**44**(7):671−82.

16. Kay LE. *Who wrote the book of life? A history of the genetic code*. Stanford, CA: Stanford University Press; 2000.

17. Watson JD. *Genes, girls, and Gamow: After the Double Helix*. New York, NY: Vintage Books; 2001.

18. Gamow G. Possible relation between deoxyribonucleic acid and protein structures. *Nature* 1954;**173**(4398):318.

19. Gamow G. Possible mathematical relation between deoxyribonucleic acid and proteins. *Biol Medd Kgl Dan Vidensk Selsk* 1954;**22**(3):1−11.

20. Gamow G, Rich A, Ycas M. The problem of information transfer from the nucleic acids to proteins. *Adv Biol Med Phys* 1956;**4**:23−68.

21. Brenner S. On the impossibility of all overlapping triplet codes in information transfer from nucleic acid to proteins. *Proc Natl Acad Sci USA* 1957;**43**(8):687−94.

22. Tsugita A, Fraenkel-Conrat H. The amino acid composition and C-terminal sequence of a chemically evoked mutant of TMV. *Proc Natl Acad Sci USA* 1960;**46**(5):636−42.

23. Crick FH, Barnett L, Brenner S, Watts-Tobin RJ. General nature of the genetic code for proteins. *Nature* 1961;**192**(4809):1227−32.

24. Crick FH. On protein synthesis. *Symp Soc Exp Biol* 1958;**12**:138−63.

25. Hoagland M. Enter transfer RNA. *Nature* 2004;**431**(7006):249.

26. Hoagland MB, Zamecnik PC, Stephenson ML. Intermediate reactions in protein biosynthesis. *Biochim Biophys Acta* 1957;**24**(1):215−16.

27. Hoagland MB, Stephenson ML, Scott JF, Hecht LI, Zamecnik PC. A soluble ribonucleic acid intermediate in protein synthesis. *J Biol Chem* 1958;**231**(1):241−57.

28. Nirenberg M. *The genetic code. Nobel lecture, December 12, 1968*. Amsterdam: Elsevier Publishing Company; 1972.

29. Khorana HG, editor. *Nucleic acid synthesis in the study of the genetic code. Nobel lecture, December 12, 1968*. Amsterdam: Elsevier Publishing Company; 1972.

30. Caspersson T, Schultz J. Nucleic acid metabolism of the chromosomes in relation to gene reproduction. *Nature* 1938;**142**(3589):294−5.

31. Caspersson T, Schultz J. Pentose nucleotides in the cytoplasm of growing tissues. *Nature* 1939;**143**(3623):602−3.

32. Caspersson T. The relations between nucleic acid and protein synthesis. *Symp Soc Exp Biol* 1947;**1**:127−51.

33. Brachet J. The metabolism of nucleic acids during embryonic development. *Cold Spring Harbor Symp Quant Biol* 1947;**12**:18−27.

34. Brachet J. The localization and the role of ribonucleic acid in the cell. *Ann NY Acad Sci* 1950;**50**(8):861−9.

35. Hultin T. Incorporation in vivo of ^{15}N-labeled glycine into liver fractions of newly hatched chicks. *Exptl Cell Res* 1950;**1**(3):376−81.

36. Keller EG. Turnover of proteins of cell fractions of adult rat liver in vivo. *Fed Proc* 1951;**10**:206.

37. Siekevitz P, Zamecnik PC. In vitro incorporation of 1-C14-DL-alanine into proteins of rat liver granular fractions. *Fed Proc* 1951;**10**:246.

38. Siekevitz P. Uptake of radioactive alanine in vitro into the proteins of rat liver fractions. *J Biol Chem* 1952;**195**(2):549−65.

39. Volkin E, Astrachan L. Phosphorus incorporation in *Escherichia coli* ribo-nucleic acid after infection with bacteriophage T2. *Virology* 1956;**2**(2):149−61.

40. Volkin E, Astrachan L. Intracellular distribution of labeled ribonucleic acid after phage infection of *Escherichia coli*. *Virology* 1956;**2**(4):433−7.

41. Volkin E, Astrachan L. RNA metabolism in T2-infected *Escherichia coli*. In: McElroy WD, Glass B, editors. *A symposium on the chemical basis of heredity*. Baltimore, MD: The Johns Hopkins Press; 1957. pp. 686−95.

42. Volkin E, Astrachan L, Countryman JL. Metabolism of RNA phosphorus in *Escherichia coli* infected with bacteriophage T7. *Virology* 1958;**6**(2):545−55.

43. Pardee AB, Jacob F, Monod J. The genetic control and cytoplasmic expression of "Inducibility" in the synthesis of β-galactosidase by *E. coli*. *J Mol Biol* 1959;**1**(2):165−78.

44. Riley M, Pardee AB, Jacob F, Monod J. On the expression of a structural gene. *J Mol Biol* 1960;**2**(4):216−25.

45. Brenner S, Jacob F, Meselson M. An unstable intermediate carrying information from genes to ribosomes for protein synthesis. *Nature* 1961;**190**(4776):576−81.

46. Gros F, Hiatt H, Gilbert W, Kurland CG, Risebrough RW, Watson JD. Unstable ribonucleic acid revealed by pulse labelling of *Escherichia coli*. *Nature* 1961;**190**(4776):581−5.

47. Hall BD, Spiegelman S. Sequence complementarity of T2-DNA and T2-specific RNA. *Proc Natl Acad Sci USA* 1961;**47**(2):137−63.

48. Hayashi M, Spiegelman S. The selective synthesis of informational RNA in bacteria. *Proc Natl Acad Sci USA* 1961;**47**(10):1564−80.

49. Sarabhai AS, Stretton AO, Brenner S, Bolle A. Co-linearity of the gene with the polypeptide chain. *Nature* 1964;**201**(4914):13−17.

50. Yanofsky C, Carlton BC, Guest JR, Helinski DR, Henning U. On the colinearity of gene structure and protein structure. *Proc Natl Acad Sci USA* 1964;**51**(2):266−72.

51. Yanofsky C, Drapeau GR, Guest JR, Carlton BC. The complete amino acid sequence of the tryptophan synthase A protein (alpha subunit) and its colinear relationship with the genetic map of the A gene. *Proc Natl Acad Sci USA* 1967;**57**(2):296−8.

52. Darnell Jr JE, Wilson EB. Lecture, 1998. Eukaryotic RNAs: once more from the beginning. *Mol Biol Cell* 1999;**10**(6):1685−92.

53. Berget SM, Moore C, Sharp PA. Spliced segments at the 5′ terminus of adenovirus 2 late mRNA. *Proc Natl Acad Sci USA* 1977;**74**(8):3171−5.

54. Gelinas RE, Roberts RJ. One predominant 5'-undecanucleotide in adenovirus 2 late messenger RNAs. *Cell* 1977;**11**(3):533—44.

55. Chow LT, Gelinas RE, Broker TR, Roberts RJ. An amazing sequence arrangement at the 5' ends of adenovirus 2 messenger RNA. *Cell* 1977;**12**(1):1—8.

56. Orgel LE, Crick FH. Anticipating an RNA world. Some past speculations on the origin of life: where are they today? *FASEB J* 1993;**7**(1):238—9.

57. Crick FH. The origin of the genetic code. *J Mol Biol* 1968;**38**(3):367—79.

58. Orgel LE. Evolution of the genetic apparatus. *J Mol Biol* 1968;**38**(3):381—93.

Index

Note: Page numbers followed by "*b*," "*f*," and "*t*" refer to boxes, figures, and tables, respectively.